the category of the technical is what is right, not what is beautiful, and the category of the beautiful is what is aesthetic, not what is right. the category of information is the true not the beautiful. and the category of use is the useful not the technical.

certainly the product we are looking for is one that functions technically and is formally attractive, durable in use and intelligible in function, meaning and origin.

otl aicher

Aicher, Otl: the world as design.
Berlin: Ernst & Sohn, 1994, p. 66

StadiumATLAS

Technical Recommendations for Grandstands in Modern Stadia

Stefan Nixdorf

Ernst & Sohn
A Wiley Company

Foreword

For some time now, I have been aware of the project of a *StadiumATLAS* by Dr (eng.) Stefan Nixdorf. In the immediate follow-up of the Football World Cup 2006 in Germany it became the subject of discussions and an intense exchange of experiences with the author. Therefore, I am pleased to see the book completed now.

Without a doubt, the book will be able to fill the gaps that arise inevitably in the context of the competition guidelines issued by FIFA, UEFA, DFB and DFL. Those guidelines are technical recommendations by nature and tend to constitute a considerable challenge to planners; often enough there are no model solutions. This applies in particular to the calculation of sightlines for grandstand (structures), to the increasing requirements on the part of the media with their high-quality equipment, to the demands on spectator service and catering for VIP and executive areas, to safety zones and many other issues.

The *StadiumATLAS* will represent a true help for all parties involved in the construction and operation of a grandstand, and it will facilitate the finding and effective creation of relevant solutions.

The comprehensive scope of the book is complemented by a discussion of the history of stadium construction, by indications given on the first steps of planning, by advice on the issue of grass or artificial turf, by the chapter on 'Light and architecture' and by other topics from multifunctionality to adaptable building components.

Personally, I am convinced that this reference book will be widely and internationally distributed, and that it will take its place in many a specialist library.

Horst R. Schmidt
Chairman / Managing Director
Deutscher Fußball-Bund e.V. (DFB)
Frankfurt am Main, October 2007

Preface

This *StadiumATLAS* aims to be a planning guide for the construction of spectator stands in modern sports and event complexes.

A methodological comparison of the venues of the FIFA World Cup 2006™ in Germany is complemented by a catalogue of technical recommendations and requirements for the new erection or the modernization of multifunctional sports arenas on the basis of current European building regulations.

Apart from all essential aspects of planning, the main focus lies on developing future concepts for the construction of grandstands.

Requirements for the building type 'spectator facility' have changed significantly during the past decades. Improving convenience for spectators and a better commercial exploitation have become guiding principles for the design of new sports complexes.

The disappearance of athletics tracks has moved the stands closer to the pitch, which does create the desired effect of a more dense atmosphere among the auditorium, but at the same time changes viewing conditions. Parameters of planning such as eye-point or sightline elevation have become decisive quality standards in the assessment of grandstands.

In the light of the growing complexity of mutlipurpose utilization, the influences affecting the geometry of the stadium structure become ever more perplexing.

Decisions and guidelines from different spheres of responsibility, in particular regarding safety-relevant issues such as the set-up of the 'first row', have a strong but often unnoticed effect on the elevational geometry of a sightline profile.

In this handbook, the principles of building regulations and the guidelines of the major sports associations are analyzed and interrelated in order to clarify relationships and enable critical conclusions.

The *StadiumATLAS* aims to illustrate the constructional and geometrical effects of certain specifications, and to facilitate decision-making for planners and clients, regarding important parameters of stadium design.

Constellations of operators, users and owners may vary from object to object. Thus, dealing with individual requirements in the run-up of each project seems particularly important. On top of that, a modern sports and event venue has to accommodate various essential facilities. Suitability for large events (such as FIFA-World Cup or UEFA EURO Final Round, the athletics competitions of IAAF, IOC or other regionally important sports events) is regulated by various requirement catalogues.

In stadium planning, many necessary measures may be realized on a temporary basis; a stadium's capability to carry out diverse types of events without unreasonable efforts of conversion must be guaranteed. At the same time, in its function as an assembly space, a stadium must yield adequate revenue.

The architecture of stadia, arenas or theatres is subject to much more than merely satisfying utilization or business aspects: it is primarily a question of 'emotion', as proven by the positive experiences of the FIFA World Cup 2006™.

In 1908, German architect Hermann Muthesius writes that a structure evolves into a work of art only when it goes beyond meeting pure necessity.[1] And according to Roman master builder Vitruvius architecture is based on three principles: firmitas (stability, quality, reliability) utilitas (utility, usefulness, orientation) venustas (grace, beauty, genius, effect).[2] 'Architecture is the harmony of all parts which is achieved by making it impossible to take anything away, to add or alter without destroying the whole', writes Leon Battista Alberti in 1452.[3]

The *StadiumATLAS* aims to be a planning guide which introduces and combines the various levels, areas and effects decision-making. It merely provides the structural 'receptacle' for the architectural 'genius' of future generations of stadium structures. There is no intention to belatedly analyze and assess the twelve World Cup arenas. All plans for these complexes have been issued for construction, approved and tested.[4] They provide orientation for designs and final planning of future stadium complexes and, as built objects, possess an absolute reference character. Therefore, all theoretical considerations and scientific discussions are derived from and lead back into practice.

Stefan Nixdorf
Aachen, 2007

The word architecture is a combination of the Greek words *arché* (= beginning, source, base) and *techné* (= art, trade, technology, tectonics).

Content

Foreword 4
Preface 5

Chapter 01
A history of stadium construction
A brief historical survey of building typologies and definitions of terms 10
A historical definition of terms 10
Circus and amphitheatre 12
Marcus Pollio Vitruvius 13
The theatre of Alberti 14
From the baroque to the loggia theatre 16
Divergences between geometric viewing angles 17
Historical classifications 18
The Olympic idea 20
Generations of stadia 21
The origins of football 22

Chapter 02
Preliminary planning
A general overview of the measures to be taken in the run-up of constructing a sports and event venue 24
Event and motivation 24
Preliminary planning 25
Combined GRW-procedure 26
Noise control and traffic 27
Accessibility and stationary traffic 28

Chapter 03
Organization of use
Determination of user groups and safety aspects related to a sports complex 32
Distribution of functions 32
Safety in sports stadia 33
Zoning and sectors 34
Sector division 35
Entrance areas 36
Admission control system 36

Chapter 04
Auditorium
A definition of all planning parameters relevant for a grandstand structure 38
Spectator comfort 38
Distribution of functions at World Cup Stadium Cologne 39
Definition of seating accommodation 40
Measurements of seats 40
Barriers and sightline 41
Barrier types 42
Seating types 43
Safety barriers for seats 44
Standing accommodation 45
Radial gangways in standing areas 46
Crush barriers 47
Places for wheelchair users 48
Location and number of places 49
Space requirements for wheelchair places 51

Chapter 05
VIP areas & corporate hospitality
Different concepts of hospitality and the special function of executive lounges and boxes 52
VIP lounges 52
Executive capacities 56
VIP boxes 57
World Cup Hospitality Concept 2006 58
Concepts for private hospitality boxes 59
Definition of the term 'box tier' 62
Lounge concepts 64
Surface area demands for lounges 67

Chapter 06
Stadium catering
Concepts of culinary service for spectators 68
Stadium catering 68
Kiosk concepts 71
First-aid points 71
Production in the central kitchen 72
General WC areas 74

Chapter 07
Media facilities
Working areas and requirements of the media and press 76
Media 76
Working areas for the media 77
Commentators' positions 78
Photographers 79
Written press (print media) 80
TV production 80
Mixed Zone 82
Press centre in the stadium 83
FIFA – TV compound 84
Temporary media facilities 85
Camera positions 88
Space requirements of camera positions 90
Super slow motion 93
High Definition TV (16:9) 93

Chapter 08
Players' area and changing rooms
Facilities for players, referees and team officials 94
Players' area 94
Players' bench and tunnel 96

Chapter 09
Stadium administration

Essential functional rooms such as stadium control room and other ancillary rooms   98
Stadium control room   98
Other ancillary rooms   99

Chapter 10
Planning principles

A listing of essential regulations and recommendations for the construction of spectator facilities   100
Regulations and Recommendations   100
Mandatory construction specifications   101
Definition of EN/DIN   103
The German MVStättV 2005 (Model regulation for places of assembly)   104
Dimensioning of escape routes   105
Egress times   107
Holding capacity   108

Chapter 11
Pre-dimensioning and circulation

Shedding light on the relationship between circulation systems and holding capacity   110
External circulation systems   110
Internal circulation systems   112
Radial gangway types   113
Reduction factor   115
Block definition for seating accommodation   116
Circulation areas   117
Capacity pre-dimensioning   118

Chapter 12
The Modulor 'EN'

The definition of a typical stadium spectator as an anthropometric minimum standard and contractually binding reference size for the planning and construction of spectator stands   122
The Modulor 'EN' (European Norm)   122
Architecture and harmony   123
The 'Golden Section'   124
Le Corbusier's Modulor   126
'Homo bene figuratus'   127
Anthropometrics   128

Chapter 13
Physiology of viewing

An evaluation of visual acuity, spatial perception and visual angle zones based on an examination of the human eye   130
Physiology of the eye   130
Physiology of visual acuity (resolution)   131
Physiology of spatial viewing   133
Human field of perception   134
Visual perception   134
LIVE-effect   135
Viewing distance   136
Visual angle zones   138
Optimum viewing circle   140

Chapter 14
Layout types and undulation

Defining geometrical families and explaining the geometrical context of ground plans and of inclination profiles for grandstands   142
Layout types   142
Theatre undulation   146

Chapter 15
Principles of sightline calculation

A methodical approach to determine optimized sightline profiles for places of congregation   150
Guidelines for sightline profiles   150
General terminology   151
Calculation of single riser height   152
Sightline elevation ['C' value]   153
The term 'sightline'   154
Stand elevation   155
Eye-point height in relation to seating/standing place A_S 156
Seating row depth and width   158
Sightline elevation 'C'   159
Distance of the 'first row' D 161

Chapter 16
Inclination and 'polygonal transformation'

The optimization of viewing conditions by means of maximum slope and 'polygonal transformation' of an eye-point curve   162
General introduction   162
Radial gangway versus stairs   162
Request for deviation   164
Maximum gradient of stands   165
Safety compensation (20 cm)   166
Limitation of rows (maximum inclination)   167
'Polygonal transformation'   168
Continuous difference in rise   169
Parameter studies   171

Chapter 17
Parameter studies
Geometrical relations between the sightline parameters eye-point height, tread width, 'C' value and distance of the 'first row' 172
Calculation of a grandstand profile 172
Parameter: height and distance 173
Adjustment of 'C' value 174
Sightline summary 176
Five-step method 177

Chapter 18
Playing fields
A summary of the major sports types and relevant dimensions as a means to determine the minimum distance of a grandstand 178
Facilities for athletics competitions 178
Orientation 179
Athletics arena 181
Form type 01 (IAAF standard 400-m running track) 182
Form type 02 (IAAF Double Bend Track 40/70) 182
Form type 03 (IAAF Double Bend Track 60) 183
Form type 04 (IAAF Double Bend Track 74/53) 183
Football playing field 184
Baseball playing field 185
American football playing field 186
Rugby playing field 187
Cricket playing field 188
Australian Football playing field 189
Hockey playing field 190
Playing fields for indoor sports 191
Minimum angle of view at the advertising hoarding 194
System variants for perimeter board advertising 195

Chapter 19
Securing of the playing field
The role of the 'first row' and securing of the stadium interior toward the grandstand 196
Significance of the 'first row' 196
Securing measures for the playing field 196
Principles of pitch securing 197
Option 1: Elevation of the 'first row' 200
Option 2: Installing a fence system 200
Option 3: Moats 201
Option 4: Presence of security staff 201
Mobile security measures 202
Service ring with downward slope 204
Rescue or relief gates 205
Securing against streakers vs. securing against panic 206

Chapter 20
Adaptable stadia
Structural conversion of modern sports and event facilities in the light of multifunctional utilization 208
The adaptable coliseum 208
Multifunctional stadium interior 210
Multipurpose hall 212
Movable grass pitch 214
Adaptable stadia 216
Retractable lower tier 218
Convertible roofs 220
Retractable roof membranes 222
Adapting capacity 224
Supplementary utilizations 225

Chapter 21
Light and architecture
Light orchestration and media-compatible floodlighting for event venues 226
Allianz Arena, Munich 226
Comparison PTFE/PVC 227
Light and structure 228
Floodlight planning 229
'Ring of Fire' – Berlin 231
The development of floodlight technology 232
Illuminance E 233
Sports association guidelines 233
Safety facilities 235

Chapter 22
Natural versus artificial turf
Shedding light on a question of faith and the current technological status of developments for the playing field 236
Natural versus artificial turf 236
The artificial turf fibre 238
'FIFA recommended 2 Star' 239
Natural turf 240
'GreenGoal' – FIFA WM 2006™ 241
Turf exchange 242
Factors supporting turf growth 243

Chapter 23
The planning of significant roof structures
An overview of structural systems for fast orientation 244
Structural systematics 245
Hybrid structural systems 245
Families of structural systems 246
Linear systems 248
'Simulated' spatial systems 250
'True' spatial systems 251
FIFA requirements for the roof 253
FIFA scoreboard 254
Visual communication system 254

Chapter 24
Grandstand profiles of built examples
A comparison of the stadia for
FIFA World Cup 2006™ in Germany   256
World Cup Stadium Berlin   258
Olympic Stadium Berlin –
Two-tier stadium (athletics)   262
World Cup Stadium Dortmund   268
Signal Iduna Park, Dortmund –
Two-tier stadium (football)   270
World Cup Stadium Frankfurt   274
Commerzbank Arena, Frankfurt –
Two-tier stadium (football)   278
World Cup Stadium Gelsenkirchen   282
Veltins Arena, Gelsenkirchen –
Two-tier arena (multipurpose hall)   286
World Cup Stadium Hamburg   290
AOL Arena, Hamburg –
Three-tier stadium (football)   292
World Cup Stadium Hanover   296
AWD Arena, Hanover –
Two-tier stadium (football)   298
World Cup Stadium Kaiserslautern   302
Fritz Walter Stadium, Kaiserslautern –
One-tier stadium (football)   304
World Cup Stadium Cologne   308
RheinEnergieStadion, Cologne –
Two-tier stadium (football)   310
World Cup Stadium Leipzig   314
Zentralstadion, Leipzig –
Two-tier stadium (football)   316
World Cup Stadium Munich   320
Allianz Arena, Munich –
Three-tier stadium (football)   322
World Cup Stadium Nuremberg   326
World Cup Stadium Nuremberg –
Two-tier stadium (athletics)   328
World Cup Stadium Stuttgart   332
Gottlieb Daimler Stadium, Stuttgart –
Two-tier stadium (athletics)   334

Chapter 25
Final evaluation and compendium
An assessment summary of the twelve German
World Cup stadia and selected stadia in
Korea / Japan, Portugal, Austria / Switzerland,
South Africa   338
Stadium evaluation   338
FIFA WorldCup 2002™   346
UEFA EURO 2004™   348
FIFA WorldCup 2006™   351
UEFA EURO 2008™   352
FIFA WorldCup 2010™   355

Appendix
Epilogue   357
Notes   358
Bibliography   362
Sources   364
Acknowledgement   366
Architects of the Football World Cup stadia,
Germany 2006   367
List of advertisers   367

Chapter 01

001 Antique stadium at Olympia, Greece
002 Theatre at Epidauros, Greece

A history of stadium construction
A brief historical survey of building typologies and definitions of terms

A historical definition of terms

Indispensable for a better understanding of the modern and contemporary stadium is a brief backward glance at historic prototypes.

The stadium

The word 'stadium' is derived from the Greek word *stadion*, which referred first of all to an ancient unit of length ranging, depending upon local conditions, between 177 and 192 meters. The term was also extended to refer to the settings for athletic competitions as a whole, that is to say, the track together with its surrounding spectator tiers. The Greeks, then, already used the term *stadion* to refer to all athletics tracks taking the forms of elongated rectangles with semicircular termini, and later to the settings for other types of sporting events. (1 unit of length = 1 *stadion*)

In accordance with IAAF (International Association of Athletics Federations) standards, every stadium intended for international competitions should have a 400-m track surrounding a grass playing field. Alongside the track and in the remaining curved sectors are found areas for high jump and throwing. The term 'stadium' today refers to any facility used for athletic competitions, quite independently of the type of sport involved. The conversion of many track-and-field sports facilities into 'pure' soccer stadia has severely constricted the availability of multi-use facilities. The circulating stands allow the public to observe the competition, whether from seats or from standing positions.

Today, many stadia are multifunctional, which is to say, they can also be used for concerts and other large-scale events such as religious conferences or company gatherings. As a rule, the playing field is surrounded by an ascending and generally open roof structure. In many regulations, an open playing surface is designated as a 'stadium'.[1]

The arena

An arena (the name is derived from the Latin word for 'sand') was originally a 'place of combat, strewn with sand, and set in an amphitheatre, circus or stadium'.[2]

It serves as a venue for athletic competitions and cultural events. Spurred on by the possibility of naming venues after sponsors who pay for the privilege, the term is used today not only for the space of activity itself, but also for the entire surrounding architectural structure. Influenced by the advertising industry, the term has been taken over from American English, and arenas are today often named after their principal commercial sponsors.

Increasingly, this convention is associated with the refinancing of construction measures of venues for sports and other events. Examples are the AOL Arena in Hamburg, formerly known as the Volksparkstadion, and the AufSchalke Arena, the successor to the neighbouring Parkstadion Gelsenkirchen, which was renamed 'Veltins Arena' in mid-2005 as a venue for sporting events, concerts and other entertainment events. For the FIFA World Cup 2006™, all names were made uniform and consistently referred to the respective city of the event, for example 'World Cup Stadium Cologne'.

Definitions of uses

In the city-states of early Greek antiquity, stadia 'served at first as places for cultic martial combat, and later as venues for the athletic contests which served the patriotic interests and above all the self-validation of the ruling oligarchy'.[3] Seating capacity often corresponded to the entire population of a given city. Taking place at four-year intervals between 776 BC and 393 AD in the ancient stadium in Olympia on the Peloponnesian Peninsula – and so to speak the sole 'authentic' Olympic stadium – were the 'Games of Olympia.'

The stadium offered viewing places for approximately 30,000 spectators. It was, however, not possible to be seated. In order to improve the views of spectators, earth walls were built up around the track. Small stone tribunes or platforms were installed only for the use of judges on the southern slope, and one for the highest priestess of the Temple of Hera on an altar on the northern slope.

The Stadium of Delphi, dating from the fifth century BC, is one of the best-preserved examples from Greek antiquity. The facility measures approximately 177 m in length and exploits the sloping topography for the earth wall of the tribune. Here too, for use by notables in attendance stepped structures were installed in the form of stone seat rows. These offered unobstructed views of the U-shaped running track.

'Arising from the aristocratic cult of death in the Roman stadia were the gladiator games. These mass festivals were tools of the centrally organized ruling structure of the 'Imperium Romanum,' and must have served to distract and control the populations, concentrated in metropolises, by means of 'panem et circenses'.[4]

Constructed already during the early period of the Roman empire, around 500 BC, was the 'Circus Maximus', the largest circus in ancient Rome, generally a setting for chariot races and less frequently for gladiatorial combats or animal hunts. To begin with spectators sat on the slopes of both sides of the valley and later on simple wooden tribunes; here also, stone seats were installed only for senators. The facility was enlarged on numerous occasions up to the 4th century AD and in its final form measured 3 ½ stadia (635 m) in length and 110 m in width.

The elongated track itself had a length of 590 m and formed an oblong oval which was divided at the centre by a wall, called the 'spina'. The seating capacity of the Circus Maximus is estimated to have been more than 300,000. Although stadia were also set up during the Roman era, less importance was attributed in Italy to Greek-style athletic competitions. Attracting greater attention, in fact, was the theatre, a building type which the Romans also adopted from the Greeks.

Two of the best-known examples of Greek theatres are the Dionysus Theatre in Athens, located on the southern slope of the Acropolis (and inaugurated in 534 BC) with its 13,000 to 17,000 seats; and the theatre in Epidauros, dating from 300 BC, with its 14,000 seats whose great acoustics make it a model for contemporary theatre buildings and concert halls.

In essence, the Roman theatre took over all of the important elements of the Greek theatre and further developed this Hellenistic prototype. While in Greece it was customary to integrate the theatre space organically into a natural setting (i.e. embedding the spectator tribunes in the slope of a hill), the Roman theatre inaugurated the self-sufficient theatre building, now closed off from its surroundings.

Here the architectural structure stands in sharp contrast to the natural surroundings and Roman technology freed it from its dependency on the structure of the site. It seems likely that this building type had its origin in the amphitheatres of the Campania region, which was also the location of the first documented gladiatorial games in 64 BC. The so-called 'Double Theatre' of Curio, constructed in Rome in the year 50 BC, consists of two semicircular auditoria which can be shifted together to form a closed 'cavea.'

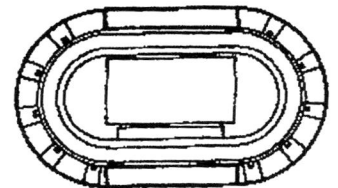
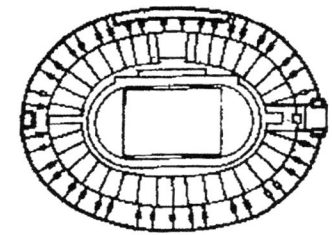

003 **Ground plan scheme drawings**
   a. Panathenaic Stadium, Athens 1896
   b. White City Stadium, London 1908
   c. Olympic Stadium, Berlin 1936

004 **Antique Stadium at Delphi, Greece**

01 A history of stadium construction

Circus and amphitheatre

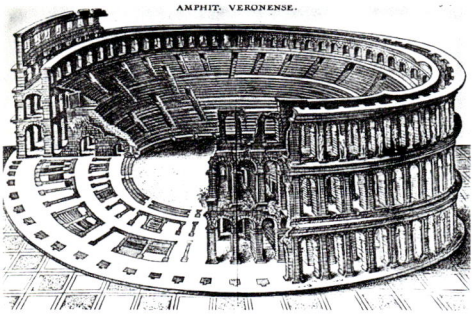

005 **Circus Maximus, Rome**
(picture of model)
006 **Amphitheatre of Verona**
Christophorus Coriolanus, 1672

In essence, an amphitheatre (from the Greek *amphi* = double) is formed by the 'joining together' of two theatres and the omission of the stage section. The semicircle of the Roman theatre is thereby closed to form an oval composed of *arena* and *cavea* (Latin for the auditorium of a theatre). The resulting ellipsoid ground plan allows spectators to be seated closer to the events.

According to Leon Battista Alberti, both forms can be traced back to the classical theatre. In his view, a circus is not much different from a theatre 'whose wings have been extended in length and set parallel on relationship to one another' and set above a small columned hall. An amphitheatre, on the other hand, consists of two theatres 'whose rows of steps have been connected at the ends in order to form a circulating passage'.[5] Lacking a stage podium, the amphitheatre surrounds an empty central field.

In many cases, this took the form of an elevated hollow floor whose underground passageways facilitated the maintenance of the activity surface set above them. The structure of the amphitheatre was similar to that of the terrace and its system of accesses corresponded exactly to the model of the classical theatre building.

The differences lay in the use of the interior surfaces of the circus, all the way from theatre performances to animal hunts and gladiatorial combat. In the rounded geometries involved, Alberti perceives the advantage that 'the tormented beast was unable to find any corner into which it might be able to retreat'.

The total dimensions of length to width of a 'cavea' is given by Alberti, resulting from the experiences of his predecessors, as 7/8 by 3/3. The circus, he writes, is divided and halved at the centre by a wall. Its surface should be at least 60 ells wide.

According to Vitruvius, an ell is almost six hands in breadth, while a foot corresponds to four hands. Based, then, on the Ionian foot (= 29.6 cm) for the measurement of lengths, one hand equals 7.4 cm x 6 = 44.4 cm. An ell, then, measures approximately 26.5 m, so that the length of the central area of the circus is seven times that of its width, or 187 m (hence, the unit of measurement known as the 'stadion'.)

The first row of seats in the circus is raised at least 6 feet in height (= 1.77 m, or the height of one Roman man). The construction of the tribunes, corresponding to 1/5 – 1/6 of the width of the central field, then, is circa 5.3 m = 18 feet divided by 2.5 feet per width of a step = approximately 7 rows of seats.

Still, at this point in his discussion of the circus, Alberti deviates from his description of an amphitheatre with a rise of at least 7 feet. Arriving at a mean value out of 6 and 7 feet, the result is 1.92 m.

Note: Today's units of measurement, for example the English foot = 30.48 cm, lies 3 % above that of the Roman system of measurement from the time of the Emperor Augustus (an Ionic foot = 29.6 cm). The architect Le Corbusier adapted his system of measurement for his 'Modulor 2' to the English foot (cf. the section on Modulor 'EN' in chapter 12).
When we now compare Alberti's specifications for the elevation of the first row, it becomes evident that 1.92 m augmented by approximately 3 % = 2.0 m, which corresponds exactly to the height that is today (i.e. as of 2004) regarded as the DFB (German Football Federation) standard for technical security in the construction of modern sports stadia, provided the first row is to be secured by means of a rise.[6]

Marcus Pollio Vitruvius

In his *Ten Books on Architecture*, Vitruvius formulates the basic conditions for theatre construction. As a Roman master builder he was quite familiar with Greek models, however, the chapter devoted to specifying the proportions recommended for the steps of the stands is oriented less to the provision of the necessary viewing conditions, but instead focuses on the propagation of sound. 'Voice is a flowing breath of air, perceptible to the hearing by contact. (…) It moves in an endless number of circular rounds, like the innumerably increasing circular waves which appear when a stone is thrown into smooth water (…) but the voice not only proceeds horizontally, but also ascends vertically by regular stages.'[7]

Apparently, the Roman and Greek master builders attributed greater importance to the 'comprehensibility of the spoken word' than to the provision of adequate views. Vitruvius as well was more concerned with shaping a theatre that would be devoted to spoken performances. In the theatres of antiquity, large choruses and dance ensembles dominated the stage. The individual actor, who emerged from the mass of the chorus, usually wore a high cothurnus (Greek = *kothornos*), a thick soled, high boot which in Roman times resembled a stilt. In order to amplify their voices, the actors wore masks into which mouthpieces had been built. Hence, the use of facial expressions and of rapid movements was essentially ruled out.

Vitruvius discusses the theatre (meaning the place where spectators sat) and its 'healthfulness'. He describes the form of the theatre as being comprised of one part for spectators, who are arranged in a circular configuration, and one for the stage, which it set rectangular to the other. He discusses climatic conditions, saying that air and wind movements should not pass freely through the circular auditorium, and that the theatre (the stands) should never be positioned in 'unwholesome quarters'. In order to avoid excessive heat in their theatres, the Romans stretched sailcloth above them, in keeping with Campanian customs, while Pompey even arranged to have water flowing along the aisles and between the seats. In Greece, many stadia were set on topographically advantageous slopes, or even mountains, since this not only reduced construction costs, but also provided security against the collapse of buildings. In Rome, theatres were made of wood, at least to begin with. As an example, Vitruvius cites the temporary tribunes of the splendid theatre of M. Scaurus, built in 695 AD with a holding capacity of 80,000.

Vitruvius expresses himself as follows on the arrangement of the rows of seats and aisles:

'The curved cross-aisles should be constructed in proportionate relation, it is thought, to the height of the theatre, but not higher than the footway of the passage is broad. … In short, it should be so contrived that a line drawn from the lowest to the highest seat will touch the top edges and angles of all the seats. Thus the voice will meet with no obstruction. The different entrances ought to be numerous and spacious, the upper not connected with the lower, but built in a continuous straight line from all parts of the house, without turnings, so that the people may not be crowded together when let out from shows, but may have separate exits from all parts without obstructions.'[8]

Newton in his Vitruvius has been interpreted to mean that in the called for proportions for the steps of a tier should be 'twice as broad as high'.[9] A straight line should run along the front edges of the steps, so that all of the steps have the same degree of rise. The drawing from Newton's Vitruvius (see previous page) clarifies the handling of the horizontal distribution aisle, which becomes necessary when the spectators of a tier must be distributed from the entrance to the gangway. Conspicuous is the fact that stadium visitors were expected to sit or stand on gently elevated stepped stone benches lacking individual seats. Apparently, the front edges of the stepped rows run along a line, just as the Roman Vitruvius had demanded.

The result of this is an ascending linear gradient which, in the view of the ancients, has highly positive acoustic effects but which is disadvantageous vis-à-vis the creation of adequate sightlines. That is, unless the initial value of the sightline elevation '*C*' does not fall below the prescribed minimal value in the course of the ascent of the tiers. This linear configuration of the seating is no longer current today but, in principle, can still be found in the valid regulations published by German Institute for Standards concerning the construction of spectator facilities.[10]

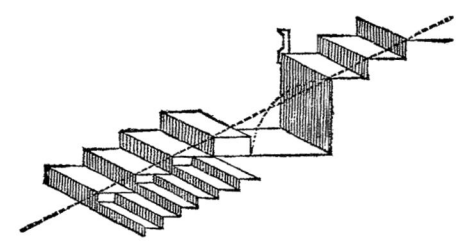

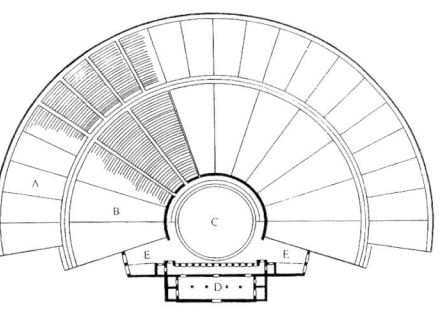

007 **Ratio of the step rises**
  Seats and steps in a theatre, according to Vitruvius (drawn by Newton)

008 **Consummated geometry, Theatre at Epidauros, Greece**
  A the upper, narrower cavea
  B 12 segments of the lower cavea
  C the circular orchestra
  D 'proscenium' stage house
  E lateral termini of the forestage ('proscenium')

01 A history of stadium construction

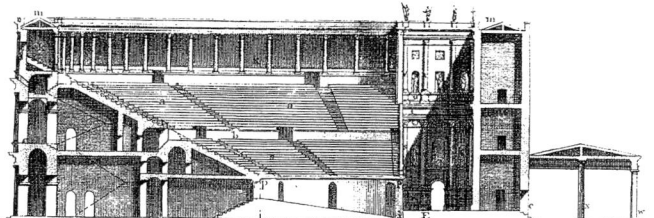

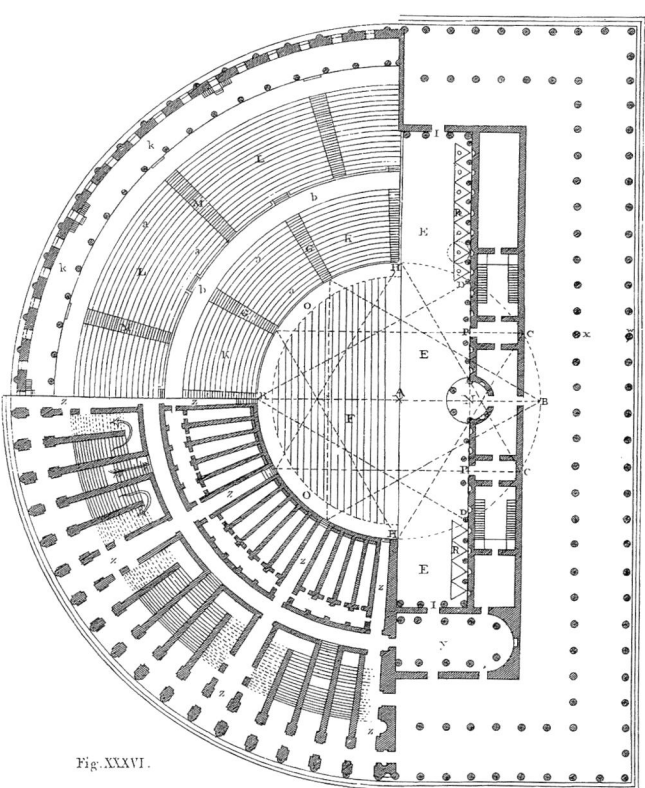

Fig. XXXVI.

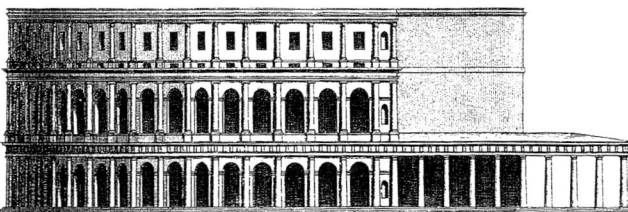

The theatre of Alberti

Discussed now as our second historical representative of stadium building after the Roman master builder Vitruvius is the Italian humanist and architect Leon Battista Alberti. A writer, mathematician and architectural theorist of the Renaissance, he lived between 1404 and 1472. His conception of geometrical perspective influenced occidental painting well into the 19th century. He brought together his investigations of Roman buildings in a 10-volume theoretical treatment of architecture.

De re aedificatoria libri decem, which dates from 1452, is also known in English as the *Ten Books on Architecture*. It is quite evident that the model for this treatise was Vitruvius' body of writings. Even today, Alberti's discussions of an *ideal theatre* and the amphitheatre type derived from it may be regarded as models for contemporary event arenas. His descriptions of the ornamentation of public secular buildings and those concerning the ideal theatre, which he regarded as being in every sense perfected and time-tested, can be translated as follows into the language of modern sport and events facilities. Consistent with the Roman unit of measurement (according to Augustus), the unit of length employed is an ionic foot = 16 x 1.85 cm (a finger width) = 29.6 cm.

According to Alberti's ideal descriptions, the form of the ground plan of the (stone) theatre corresponds to a semicircle. The depth of the tiers corresponds to 2/3 of the diameter of the central surface, or 'orchestra'.[11]

a) The rise of the first row is, according to Alberti, 1/9 of the diameter of the central surface, while for smaller theatres it is at least 7 feet in height, or circa 2.07 m.

b) The depth of the risers is 2.5 feet = ca. 74 cm, and they are 1.5 feet in height = 44.4 cm (this corresponds, again, to the length of a Roman ell).

For the rise-to-run ratio, this results in a proportion of 1:1.65 and a tier inclination of 31°. This corresponds approximately to the average slope of the upper tiers of a comparable modern stadium.

c) The number and width of the gangways must be adapted to the capacity of the respective theatre. Alberti specifies the number of main entrances as seven, of which one serves as the principal access to the central surface. Set among these are to be an adequate number of additional entrances, depending on the size of the theatre.

d) In larger theatres, the terraces are to be subdivided into three tiers. This is affected through so-called 'connected concourses'. This refers to horizontal gangways that have the width of two risers (= 1.48 m), and are carved into the terraces, rather than set on top of them.

e) The access staircases located below the tribunes should have two qualities promoting efficient circulation: they should facilitate quick and easy access to the upper tiers, while simultaneously allowing a more leisurely ascent.

f) The semicircular surface at the centre should be reserved for honoured guests and should be outfitted with comfortable and 'lavishly decorated' seats, insofar as this is appropriate to the building type in question.

In order to ensure that honorary guests are able to see the action well, the stage surface used by actors and chorus should be no higher than 5 feet (= 1.48 m). If places of honour are not required for the central area, then the somewhat larger acting surface can be built at an elevation of up to six ells (6 x 44.4 cm = ca. 2.67 m).

The crowning 'portico', the so-called colonnade, which is set at the upper end of the tribune, with its orientation toward the interior, not only serves as a covered walkway, but also assumes (in Alberti's opinion) an acoustic function, improving the comprehensibility of spoken language by throwing back the sound. The voices are allowed to sound fuller by means of acoustic vessels set into niches of the semicircular rear wall (called 'ring wall') behind the *summa gradatio*, the uppermost loge. An additional possibility is seen by Vitruvius in the arrangement of *cella echea (vasa)* beneath the steps of the tribune, small hollow spaces within which the acoustic vessels can be oriented toward openings in the front face of the steppings.

As a cloth sail suspended above the spectators, the tent roof, or so-called *velum*, casts a shadow and is spanned above the tribunes from ship's masts. Constructed underneath this colonnade for static purposes are additional columns and arcaded passageways which serve as exterior façades. This to some extent two-storied *portico* functions as a lobby, a circulation route and as a place to linger. It serves not only the training and recreation of the *ancients*, as Albert writes, but also to shelter visitors from rain.

Note: A portico is a single- or multi-naved porch that is open on one side and supported by columns. Also referred to as porticos are colonnades found in upper storeys that are opened toward the exterior.

For in fact, such an amphitheatre is constructed in order to allow the people to be impressed with themselves, to see themselves at their best. … To satisfy this general requirement is the task of the architect. He creates such a crater by means of art, and as simply as possible, in order to allow the people themselves to act as its ornamentation. When the people are assembled there, they look at one another in wonder, for they are otherwise accustomed to seeing one another only in passing, to encountering one another in throngs without order or discipline. But here, the mass of people sees itself, this many headed, multiple, fluctuating beast, with its errant movements, unified into noble body, fashioned into a unity, bound into a mass and fixed into a single shape, one animated by a single spirit.

The simplicity of the oval form is perceptible in the most pleasing way o every eye, and every spectator serves as a measure of the vastness of the whole. And when the building is seen in an empty state, there is no longer any indication of its scale, and one hardly knows any longer whether it is large or small.

J. W. Goethe [12]

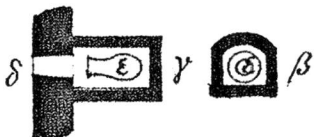

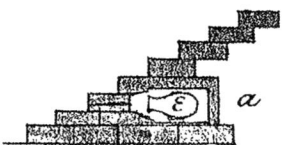

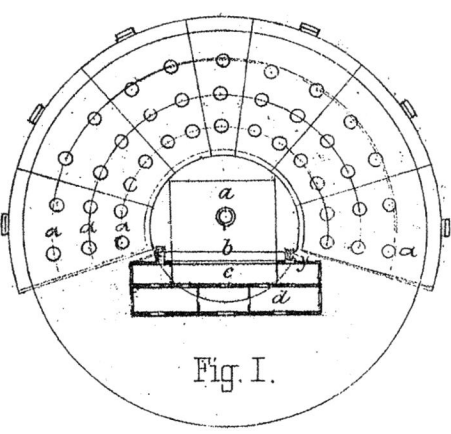

page 14:
009 **Ideal theatre according to L. B. Alberti (drawing by Newton) from Vitruvius – *De architectura libra decem* (A. Rode)**

010 **Drawing of the Coliseum, 1725
Seating and standing stairs
'Cella echea (vasa)'**
Acoustic vessels located below the tribune steps (section and plan)

From the baroque to the loggia theatre

011 **Study model of the Theatre at Epidauros, 4th – 3rd centuries BC**
Model: Roland Reimann, Darmstadt, Lehrstuhl für Raumkunst und Lichtgestaltung, Technical University Munich

012 **Historical photo of the Arena of Verona**

In the theatre halls of princely palaces, with their spatial unification of auditorium and stage, it was a question of 'seeing and being seen.'

Unlike today, in the Baroque era, spectators visited theatres not primarily in order to view a play, nor to take in an opera production, corresponding to their particular interests and tastes, but instead entered the princely theatre as invited guests in attendance at a festive event.

As Hans Gussman writes in *Theategebäude*: 'In contrast to the Greek theatre, with its seating arrangement covering three quarters of a circle, the courtly society of the baroque era in its loggias enjoyed no consistent geometrical point of view, since the stage was been shifted from the spatial focus to the edge of the auditorium.' [13]

The amphitheatre-style arrangement of the seats, which lay in a ring-shaped configuration around the performance area, resulted in good views of the entire surface of the stage. 'It has been confirmed that the acoustics in such theatres were quite good, and that the unity and consistency of the arrangement of the spectator seats – which also facilitated mutual views between spectators – supplied all of the preconditions for a communal experience.' [14]

The architectonic shape of this theatre is a stage formed as a circle (orchestra), around which the spectator seats were arranged in a concentric manner. For all spectators, the perceptibility of the point of focus is reduced here to the simplest formula. By contrast, the Baroque space – ostensibly the ideal form for self-presentation in a courtly society – is the very antithesis of the 'democratic' Greek theatre. 'The spectators in the tiers enjoy no explicit viewing relationships to the play. They celebrate themselves via the theatre.' [15]

The French Revolution was decisive for the evolution of the theatre. The individual – with its relationships to bourgeois society, and with its personal problems – was shifted now to the centre of the spectator's interest. Play and spectator entered into a direct contentual relationship involving acting, seeing and hearing. With the partial (and later total) darkening of the auditorium, the representative festival of courtly society was transformed gradually into a theatre in the modern sense, and in the sense of the Greek word, which meant 'to watch attentively, to regard with astonishment.'

Theatre: Fr. *théâtre*, Lat. *theatreum*, Gr. *théatron* = viewing place, from *theasthai* = to view (from http://de.wikipedia.org)

The subsequent change in theatre construction is demonstrated clearly by P. O. Gellinek with reference to various parameters. The theatrical performance itself becomes increasingly important, and the gaze of the spectator becomes increasingly concentrated on the stage. In order to improve sight axes, the floor acquires a slope, as does the ceiling of the auditorium, in this case for reasons related to acoustics. Social encounters, accordingly, are displaced now to corridors and foyers.

Despite this development, and despite the introduction of partially amphitheatre-style stalls, modern theatre plans right up to the mid-20th century continued to retain the older basic shape of a 'concentric' auditorium. Only through German composer Wilhelm Richard Wagner (1813 – 1883) was a transformation subsequently affected in the field of opera.

Text, action and music now joined to form a 'higher unity'. The spatial configuration is altered, for with the increasing significance of the action taking place on the stage, spectators must be shifted closer to the proceedings, and good views become a primary concern. According to P. O. Gellinek, Wagner demanded for the first time 'that each spectator must be able to see the entire stage.' [16]

Gussman replied to the question of 'stall, tiers, or boxes' by having the decision depend upon the desired relationship between artist and spectator. 'Considered spatially, the closer the connection between spectator and stage, the more gently it must be formed.' [17]

W. Kallmorgen gives the dimensions of the ground plan of a theatre lobby as being 1.5 to 2 m^2 per person. [18] Today as well, this value corresponds to the mean value of the 12 built examples of the FIFA World Cup 2006™ in Germany.

Divergences between geometric viewing angles

In stadia or theatres containing tribunes, or raised viewing platforms, the plan of the seating or standing tiers strongly conditions the sightlines available to spectators. In principle, the 'true sightline' toward the point of focus (e.g. a certain point along the central axis of the stage) is distinguished from the 'geometric sight reference point', which is determined by the orientation of a seat fixed to the stand.[19] The degree of the possible deviation of a viewer's sight direction from the seating axis is termed the 'sightline difference', which in turn depends on the radius of curvature of the seating rows.

The principle of 'aligned' single seats is used in the construction of terraced seating accommodations in both theatres and stadia. The principal direction of sight that is determined by the ergonomics of the formed seat is referred to as the 'orthogonal sight reference point'.

The latter is located axially in relation to the alignment of the seating shell, which normally runs perpendicular to the rear of a row of fixed seats. Ideally, this principal sightline is identical with this axis. The sightline difference resulting from the divergence of the two should not exceed a certain degree, for comfortable protracted viewing is possible only within a given margin.

The angle of divergence between the sightline of a seating position properly aligned toward the centre of the pitch and the seat axis should not exceed 30°.

The quality of seating is seriously diminished if the viewer is compelled by alignment of the ergonomic seating shell to sit in a twisted position in order to view events on the stage. Ideally, viewers should be able to follow events by moving their eyes, but without moving their heads. An angle of 30° corresponds to the demands on visual capacity specified by driver's licence regulations for daytime central visual acuity. By combining eye (2 x 30°) and head movements (2 x 30°), the functional field of vision may be widened to 120°.

In 1950, based on medical studies of polychromatic vision for a 'no-eye-movement', Harrold Burris-Meyer arrived at a viewing angle of 40° angle for theatre viewing.

In general usage, 'convergence' denotes the combination of views and aims; it involves common striving and mutual accommodation. In order to determine the sightline divergence between the true sightline and the geometrical sight reference point of an aligned seat, an observation point must be defined:

Ideally, the 'convergence point' corresponds to the central point of the playing field (i.e. the kick-off point in football).

The divergence corresponds to the horizontal angle between the neutral position of the head and the concentric gaze of the spectator toward the centre of the field.

In cases where a longitudinal athletics track is to be incorporated into a stadium, the above principle leads to the adoption of an approximately oval form, or ideally, a purely circular one. In that case, the front edges of the step will run not parallel to the playing field, but instead perpendicular to the line of convergence.

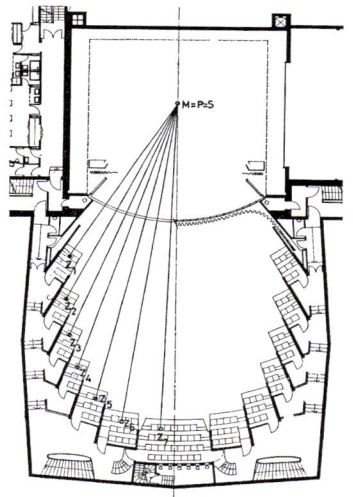

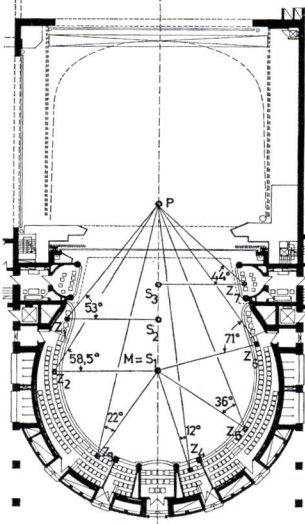

013 **Riphan Opera, auditorium, Cologne**
014 **Bolshoi Theatre, auditorium, Moscow**
(currently)

plans:
015 **Sight reference point *P***
Municipal Theatre, Lünen
016 **Sightline divergence**
National Theatre, Munich

01 A history of stadium construction

Historical classifications

017 **Final design for the sports complex on the Oberwiesenfeld, Munich**
Behnisch and Partner (1972)
018 **Olympic Stadium, Munich (1972)**

During the millennia following the genesis of the arena, the task of constructing places of assembly and the motivations for doing so have continued to evolve. In 1974, in his dissertation *Regie und Selbsterfahrung der Massen* [The Management and the Self-Experience of the Masses], Franz Joachim Verspohl attempted to contextualise the architectural type of the stadium art-historically: 'Although the structures belonging to this genre are among the most elaborate constructive achievements of the 20th century, the significance of individual stadium buildings in terms of the history of style has never been assessed independently by art historians. This systematic investigation of the prehistory of this building type, one reaching back into the Renaissance, compensates for the absence of attempts to mediate between and to distinguish between the tasks performed by the modern large-scale sports facility and the amphitheatre of antiquity.' [20]

With the Olympic Games that took place in Munich in 1972, the early 1970s seemed an appropriate time for art-historical investigations, socio-political discussion and historical classifications of this building type. The motto adopted at the time, 'joyful games', received a singular response in the design produced by the architectural firm Behnisch and Partner of Stuttgart. In collaboration with the engineer Frei Otto they produced a stadium architecture which – with its freely formed transparent cable net roof suspended above a topographically varied landscape surface – reflected the ideal of 'sport as a youthful game with a cheerful mass choreography' and echoed the motto of the Olympic Games of 1972 in a striking manner.

'This was a built social vision that was justifiably declared a landmark', as V. Marg puts it. [21]

Again and again, the image of the human being, its role in society, and the modes of intercourse between and among individuals are expressed in various architectural idioms. Hardly any other architectural form mirrors the publicly significant face of a society more perfectly than its spaces of congregation. Such facilities become contemporary witnesses of the attitudes and forms of consciousness of human beings toward one another.

Emerging under French absolutism was the idea that surviving historical tradition and architectural remains of the stadia of antiquity were to be conserved and reactivated.

At the start of the French Revolution, the architect Etienne-Louis Boullée presented his design for a spheroid coliseum capable of seating 300,000 citizens. The plans for this imposing venue were produced during the loyalist period of the Ancien Régime, and were intended to satisfy both moral and political ends. In a treatise Boullée writes: 'Under the eyes of all, the spirit of the citizens rises and is purified … truly, what could be more gripping than to see this marvellous arena filled with shining youths, who attempt to distinguish themselves with all forms of physical exercise, displaying all of their bodily skills in races, for example, and manifesting through military marches the will to defend the fatherland.' [22]

The same point of view is also summarized by Boissy D'Anglas in his essay 'Essai sur les fates nationales et quelques la sur les artes', says Verspohl, who also cites Maximilien de Robespierre (died 1794), the French politician and one of the most influential men of the French Revolution. 'There exists, however, one type of institution which must be valued as an essential component of the education of the people. I mean the national festival. And when the people gather together, this makes them better; for when gathered together, people want to please one another, and this they can only do by means of things which they themselves value.' [23]

The most important changes introduced by Boullée in his plan affected the form of the ground plan. He adopted not the oval of the Roman Coliseum, but instead the circle. According to Verspohl, this modification is not an instance of false imitation, but was instead predetermined by an ideological predisposition toward the circle as a form.

Citing Alberti and invoking statements dating from the Renaissance, Verspohl writes that the circle was regarded as the purest symbol of nature. Although duocentric ground plans (i.e. ellipsoid and oval) were feasible during this period, architectural theory was dominated by a preference for monocentric circular forms. This probably resulted from a centrist, hierarchical image of both world and society, while elongated forms might instead be associated with the duocentric conception of the world and of society. [24]

The choice on the part of Boullée for circular plans, and of the sphere as an ideal form, is based on a reception of conservative Renaissance theory. Rising in 1790 on the Champ de Mars in Paris was the largest stadium of modern times. Erected for the Festival of the Federation was an earth-floored stadium with an enormous wooden portal. Here, 50,000 representatives of the newly-formed 'departements' swore an oath to the new constitution. This staging of the political self-understanding of an entire nation, this 'image of fraternity' of a 'grand nation', was to receive a new and imposing form of expression by means of architecture as a common space for more than 500,000 citizens.

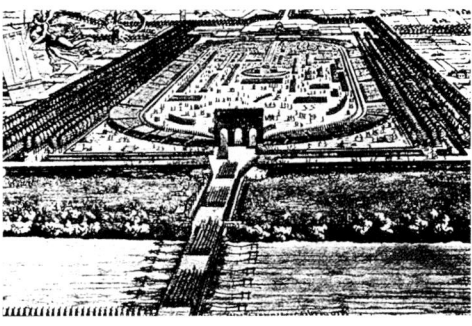

019 **Festival of the Federation, Champ de Mars, Paris (1790)**

Design of a coliseum, E. L. Boullée
020 **External view**
021 **Drawing, section (300,000 spectators)**

01 A history of stadium construction

022 **Olympic Stadium, Athens 1896**
023 **White City Stadium, London 1908**
024 **German Stadium, Berlin 1916**

025 **Olympic rings**

The Olympic idea

The 'Olympiades', as forerunners of the modern Olympic Games, were organized in Greece between 1859 in 1888; their initial impulse provided by Prince Otto of Wittelsbach. Taking the Munich Oktoberfest as model, which at the time combined the exhibition of agricultural products and the staging of athletic competitions, they were oriented toward the athletic games of antiquity. As in Bavaria, which had become a kingdom in 1806, the young Greek kingdom sought to strengthen national identity by means of a central festival.

During the 19th century, individual consciousness of national history became ever stronger, and Neoclassicism elevated the imitation of classical antiquity to a programme. One of the spiritual fathers of the movement was the German Johann Joachim Winckelmann (1717–1768), the founder of scientific archaeology. During excavations carried out in Athens, German archaeologists discovered the former hippodrome of Herodes Atticus. They rebuilt it as a foundation of Greek merchant Averoff for the opening events of the Panhellenic Games of 1896, which were attended by 40,000 visitors. Pierre de Coubertin was one of the most important initiators of a nationalistic sports movements under the aegis of the motto: 'The cult of honour and selflessness, the work of perfectibility and social satisfaction.'

In the ensuing decades, the Olympic ideal of *citius, altius, fortius* (Latin for 'faster, higher, stronger/further') stimulated a genuine boom in the field of stadium construction, for many of Europe's larger cities applied to host the games, each desiring an opportunity to present its nation. The young German empire also pursued the idea of hosting a peaceful athletics competition between the nations in order to develop its identity and applied to host the Olympic Games in Berlin in 1916. When approval was received, work began on a 'German Stadium' in the middle of Grunewald's horse racetrack. World War I meant the cancellation of the Berlin games, but a renewed application by the Weimar Republic to host the Olympic Games in 1936 met with success.

Following the seizure of power in Germany by the National Socialists in 1993, Walter and Werner March, sons of the deceased architect Otto March, directed the re-planning, which involved renaming the 'German Stadium' into 'Reichssportfeld'. In contrast to the Athenian Hippodrome, Otto March's design, with its 30,000 spectator seats, is designed to accommodate Olympic disciplines. There were spacious stadia for all types of sports and athletic competitions, a cycling track, and the large playing surface with gymnastics and playing fields at the centre. Even a swimming stadium, with its 100-m lane on the back straight of the north side of the stadium, lies across from the imperial loge on the honorary tier of the south side. In contrast to most contemporary sports stadia, ball sports received no particular emphasis. For urban planning reasons, and for the sake of developing the total ensemble, the Berlin Stadium deviates from the classical north-south orientation of the long axis.

'An internationally admired and exemplary complex … was created on this site …, a complex that was arranged in a classical monumental manner forming a unique urban ensemble.' (V. Marg, 2004).[25]

The 1936 games took place publicly, under the eyes of the world. The most modern communication technologies, such as radio and television transmissions, were introduced for the first time. Leni Riefenstahl's film tribute to the beauty of

Generations of stadia

the human form gave expression to the 'Aryan cult of the body'.

The entire facility of the Reichssportfeld served the political goals of popular self-awareness, with everything staged and organized by the National Socialist Minister of Propaganda. Belonging to this staging was the basin containing the Olympic fire burning 'on a Delphic tripod above the marathon gate, the bell tower in the bearing of the stadium axis with this inscription on its bell 'I call upon the youth of the world', placed above the Langemarckhalle in remembrance of the thousands of young killed in action during World War I in Flanders.

The lights shining into the night-time sky during the closing ceremonies on the neighbouring Field of Mars contributed to the 'orchestration of the masses for athletic training in preparation of the heroic death. This staged self-awareness of the masses made was overwhelming and grave. Three years later, Hitler began the Second World War II that would devastate Europe.'[26]

No flame was ever set at the Olympic Games of antiquity, although it is likely that this was the custom at Athenian sporting events. In 1928, an Olympic flame was lit for the first time at the Summer Games in Amsterdam. In accordance with an idea proposed by Carl Diem, the first torch race was organized at the Berlin Olympic Summer Games in 1936. From this time onward, the torch race and the lighting of the Olympic flame have played a vital part at every opening of Olympic summer games.

'The sports stadium as we know it today developed in industrial locations in England and on the European continent', writes Barry Lowe, head of the London planning team of HOKSVE. COM.[27] He provides a chronological sketch describing the phases of typological development. Translated to the prevalent circumstances in Germany and in free association to Rod Sheard, the author has undertaken the following classifications.

The first generation

Towards the end of the 19th century, the competition rules for certain types of sports were codified for the first time, making possible fair contests between different teams and allowing the pertinent requirements to be enforced at many locations and in many different kinds of buildings. The laying out of the rail system in the wake of advancing industrialization made travel far easier, both for athletes and for their supporters, thereby heightening the public's interest in sports. In 1896, the modern Olympic Games were called to life. For stadia such as London's White City Stadium, constructed in 1908, or for the predecessor design for 'Deutsches Stadion Berlin,' dating from 1916, the concept of 'multi-functionality' is no mere nomenclature, but instead a constructive organizational principle. The open-air stadium is capable of accommodating many different types of sports. Wooden cycling racetracks were integrated within the large ring formed by racetracks, and a competition-size swimming pool was incorporated into one of the long stands. Laid out in Germany in the 1920s were numerous sports parks which incorporated a number of special stadia in a landscaped park to create a larger ensemble. Such facilities gave expression to the rising social role of sports activities.

Stadium generations according to Rod Sheard, senior principal of HOK sport:[28]

First Generation Stadia
The history of the modern stadium dates back to the codification of sports in the second half of the 19th century. The First Generation of stadia placed the emphasis on accommodating large numbers of spectators, with minimal concern for the quality of the facilities or the comfort of those spectators.

Second Generation Stadia: The Influence of Television
Television, which had been developed in the 1930s, began broadcasting sports events. The Second Generation of stadia was the response, placing greater emphasis on the comfort of spectators and improving support facilities in the venue. However, these stadia were still largely concrete bowls and a great many of the worlds sporting venues remain as Second Generation stadia.

Third Generation Stadia: The Family Stadia
The Third Generation of stadia emerged in the early 1990s, developing more user-friendly facilities to lure the entire family. Sport was the focus, but not the only attraction, and the principle source of revenue for the sporting clubs changed, shifting from turnstile receipts towards merchandising and television. ... [Sports Grounds], which set a new standard for the quality of spectator facilities, with an abundance of bars, food outlets and supporters' shops.

Fourth Generation Stadia: Corporate Sponsorship and the Media
It became clear that stadia could make money if the design, funding and management were integrated. Stadia should not be regarded as a drain on a city's finances. A new era was emerging, of which the new Telstra Dome in Melbourne is a classic example. This is a truly Fourth Generation stadium, with an opening roof, moving seating tiers and a below-pitch car park. This is a blueprint for the city of the future.

Fifth Generation Stadia: Urban Regeneration
Stadia have come of age. They have grown into buildings that can be used as catalysts for the planned and strategic growth of 21st-century cities. Stadia have become powerful symbols of our culture, our aspirations and, sometimes, of our failures. We need to learn how to use them wisely.

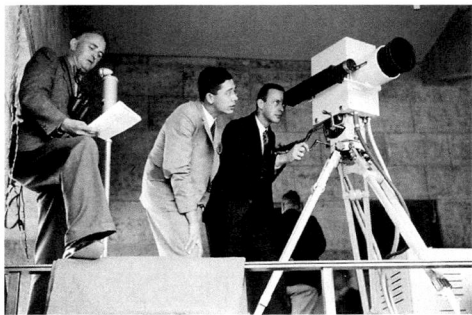

026 **East-west axis,**
Berlin Olympic Stadium 1936

027 **Radio/television transmission,**
Berlin on the final day of the Olympic Games in 1936

01 A history of stadium construction

028 Müngersdorf Sports Park, Cologne 1927
029 New build, Müngersdorfer Stadion 1975
030 Reichssportfeld, aerial photograph 1936

The second generation

After World War II many stadia were remodelled or constructed anew. Their constructive alterations were often based on the principle of the hillside stadium or one with earth foundations whose lower tiers were set either directly on the slope or on top of a substructure. The larger sporting events of the Olympic Games of 1972 and the subsequent FIFA World Cup Football Championship of 1974 stimulated another building boom for sports facilities in Germany. The discipline of football, as the basis for planning modern sports facilities, achieved its final breakthrough with the so-called 'miracle of Bern,' when the German national team won the Football World Cup in Switzerland on July 4, 1954.

At that time, nonetheless, athletics and football enjoyed equal ranking as planning factors and were, as a rule, integrated on equal terms in stadium construction with the intention of creating inclusive facilities capable of accommodating all athletic disciplines. Exceptions were the Westfalenstadion in Dortmund, and the 'Betzenberg' in Kaiserslautern, both of which might be termed prototypes of the 'pure' football stadium.

Here, predominantly radial, reinforced concrete prefab structures project above the upper tiers (ex.: Müngersdorfer Stadion in Cologne, 1975). Such partially or totally covered structures responded to the growing demands for comfort on the part of spectators. Floodlight systems soon became the standard and made sport events compatible for television broadcasting. As a rule, spectators were provided with only basic utilities, with bathroom facilities and a number of kiosks.

The third generation

By virtue of the profound transformations entailed by their multifunctional structures of use and increasingly heightened demands, the contemporary stadium or arena can be referred to as the 'next generation'.

'The latest modernizations and new builds of football stadia for the World Cup 2006 pursue a marketing regime for professional football that is typical of the time, one that brings about an altered self-awareness on the part of the masses.

For in place of popular stadia for all types of sports, we now see the maximised marketing of privatised entertainment, mostly involving exclusively football arenas which are subdivided for use by diverse consumer groups:

VIP groups, businesspeople, the press, average citizens and fans respectively are separated into different consumer groups. The stands too are clearly distinguished according to class, and the same is true of lounges, boxes, elevators and approaches. The self-experience of the masses now takes the form of a frenzy of collective emotion, paid for individually and segregated by class, an orgiastic high-volume roar that is echoed by the closed roofs of the tiers. All of this is reminiscent of the wise old Bohemian proverb 'When the flags flutter, reason is in the trumpets'.'[29]

The origins of football

In 1583, Philip Stubbes wrote the following about football in his *Anatomie of Abuses:* 'A devilish pastime which fosters the growth of envy, resentment and wickedness, and which even leads, on occasion, to fights, murder, manslaughter and great losses of blood.'[30]

Around 500 years later, on the occasion of a football tournament held on St. Peter's Square in Rome, Pope Benedict XVI stated that 'Soccer means the emergence from the enslaved seriousness of everyday life into the free seriousness of something that need not exist at all, and is therefore so beautiful.'[31]

Ever since Monsieur Coubertin propagated the Olympic idea, the emphasis has continued to shift gradually from athletics to the more popular football. A striking instance of this is the absence of athletics tracks from most of the stadia erected in Germany for the FIFA soccer World Cup of 2006. A brief retrospective should help to illuminate the extensive history of this sport.

The following information has been drawn from a series published in 2005 by the *Süddeutsche Zeitung:*

China The ancient football game known as 'Tsu Chu' developed in China around 2500 BC In this game, a silken net was suspended between bamboo poles. The players attempt to catch the ball in the net. Freely translated, 'Tsu' means 'to kick a ball with the feet,' and 'Chu' means 'a stuffed ball made of animal hide'.

Egypt An Egyptian tomb drawing dating from ca. 2000 BC shows a man juggling a ball.

Japan Played as early as 600 BC – 380 AD was 'Kemari', which bears a remarkable resemblance to contemporary forms of football. Political as well as religious motivations are said to have played a role in the game, according to the Samurai warrior's code of honour. The objective of the game is to pass the ball between two poles.

Greek antiquity/the Roman Empire: In the early 6th century BC, the Greeks played a mixture of football and handball, a game they called *Episkyros*. The Romans competed in a similar sport, which they called *Harpastum*, one disseminated throughout the entire Roman Empire and still played, it is said, in France as late as the fifth century AD.

Central America Among the pre-Columbian civilizations of the 16th century, football games were played by the Aztecs, Mayas and Zapotecs. On playing fields located near temples, religious leaders gave instructions to players, and in some cases, if the game was lost, participants were even sacrificed on altars.

The Italian Renaissance A mixture of rugby and football was popular in Florence during the 15th and 16th centuries. The game *Giuco dal Calcio fiorent* was played primarily by aristocrats. (Originally, a severed head served as a playing ball.)

Medieval England In the so-called 'motherland of football', it was the Normans who inaugurated this athletic competition in 1066. The sport served as a test of strength, as well as a form of comparison between rival towns. In 1314, King Edward banned the game, which he regarded as excessively brutal. The popularity of the sport, nonetheless, remained undiminished. Recognized in the early 16th century, football was even recommended after church services on Sundays.

The American colonies In New England, the Pilgrims encountered a type of ball game being played by the original inhabitants. The Native Americans called their game *Pasuckquakkohowog*, and it was played on the hard sand of the beach after the ebbing of the tide. The conditions resemble today's rules: two teams compete against one another, weapons are forbidden, and bets were taken on the outcome. A type of football was played in Virginia around 1600, quite soon after the construction of the earliest English settlements on American soil. Here too, the game was banned and hence played only sporadically until 1827, when Harvard students made the sport socially acceptable once more.

The modern game According to E. Augustin, the representatives of 14 London clubs and schools met on October 26, 1863 in the Freemasons Tavern on Great Queen Street near Lincoln Inn Field, in order to found the Football Association. They created the first comprehensive regulatory apparatus. In 1871, the Football Association Challenge became the first organized competition in the world.

031 **Greek wall relief, circa 2000 BC**
Episkyros

Chapter 02

032 **Inner-city location**
Santiago Bernabéu / Real Madrid CF
(80,350 seats) Spain 1947

033 **Suburban location**
Stadion Delle Alpi – Torino Calcio
(71,000 seats) Italy 1990

Preliminary planning
A general overview of the measures to be taken in the run-up of constructing a sports and event venue

Event and motivation

Large sports events are the main driving force behind the building boom in sports stadia, from Olympic Games and World Championships in athletics to FIFA Football World Cups and international UEFA tournaments. The importance of such events and their role for the prestige and image of a country is huge, as are the generated energies which flow into the construction of sports stadia and arenas. The global broadcasting of such competitions fosters the widely held notion of a positive revenue situation. 'Around 90 % of consulted experts anticipate positive economic impulses, around 40 % expect an improvement of the venue location', according to WirtschaftsWoche.[1]

Time schedule and application

The build-up tends to be considerable, frequently starting several years in advance. An initial written application by DFB (Deutscher Fußball Bund) to host the Football World Cup 2006 was submitted to FIFA already in 1993. More than 20 delegations from the larger cities took part in preparatory meetings in 1997 / 98.

As a consequence of the high demand for tickets for World Cup 1998 in France, FIFA had raised the minimum capacity for its tournaments from 30,000 to 40,000 spectators.

Sixteen cities applied to DFB as hosts for the World Cup 2006. FIFA reached a decision on 6 July 2000 in a 12:11 vote. A maximum of twelve venues was permitted and confirmed officially on 14 April 2002 by the LOC.

Initially, eight athletics stadia, seven football stadia and one combination stadium (Bremen with its telescopic lower tier) had applied. Competition ensured that 'athletics tracks were finally sacrificed for the sake of football suitability.'[2]

Following the submission of extensive documentation, and after meeting the requirements of FIFA Technical Recommendations (1995 / 2002), suitability for the World Cup was verified during several venue tours by FIFA / LOC, after which local authorities received confirmation of construction. 'In the framework of modernization or new construction of stadia and arenas, construction work should be completed two years prior to the tournament'.[3]

The example of the FIFA World Cup in Germany 2006 shows that the host of this tournament, held every four years, is in fact determined six years in advance; so preliminary considerations, regarding suitability and competence of venues have to be undertaken even earlier.

Since Müngersdorfer Stadion had not been completed in time for World Cup 1974, this time the authorities of Cologne committed themselves early to build a new stadium – even before official confirmation of Germany as host country. In December 2000, a combined competition was initiated and decided upon in two successive three-month phases, with the first cut of the spade taking place in December 2001. After 4 half-year construction phases, RheinEnergie Stadion was inaugurated in March 2003, hosted the ConFedCup in 2005 and, as one of twelve venues, the FIFA World Cup in 2006.

Preliminary planning

At the beginning of planning a new complex, the type of business operation must be determined; allocating the following utilizations is reliant on the legal status of the client: public operation, e.g. municipal or national, or private operation such as clubs, companies and persons, or combined types. In a second step, the economic potential of the stadium enterprise is analysed in a cost/business plan. The business costs may include revenue through spectators or other visitors, through clubs, advertising (including television), concessions or leasing. Expenses may list staff costs, operating/material costs (energy), cleaning appliances, machines and maintenance; the same for follow-up costs, capital costs (depreciation, taxes, reference value of capital costs), advertising and costs for renovation/new construction.[4]

If there is no external investor to balance arising costs, the planned utilization profile alone needs to ensure the running operation and the maintenance of facilities.

1) Determination of utilization and essential structural requirements:
– Special-purpose stadium: priority on football (or similar), with/without open-air concerts or conferences
– Multi-purpose stadium: athletics and football utilization, with/without open-air concerts and conferences
– Multifunctional arena: football-compatible venue with broad range of events and, if needed, closing roof

2) Determination of type of construction:
– New build: construction at a new location
– Complete conversion: construction at the present location during the season or with temporary relocation
– Modernisation: redevelopment of existing complex and, if need be, extension through complementary uses or adjacent ensemble buildings

3) Decision on utilization profile:
– Mono-functional sports stadium: basic utilization without further uses
– Multifunctional sports and event venue: several types of utilization in one stadium with complementary functions
– Multifunctional sports and leisure park: one or several special stadia or multifunctional arenas with adjacent architectural ensemble (as in Olympic Games Sydney 2000, Athens 2004)

The decision on the type of new build or rebuild depends on the demand for renovation. Since individual utilization requirements vary, the lifespan of a stadium cannot be determined precisely. The example of Germany (Olympics/World Cup in 1972/74 and World Cup Germany in 2006) serves to show that, generally, after around 30 years venues are being refurbished; in this case 3 new builds, 5 complete makeovers/conversions and 4 modernizations/extensions.

The cost of adapting to new operation requirements or technical standards, either as rationalization measure or to implement ecological aspects, is always subject to future profitability of investment.

Security and comfort are the first and foremost objectives of modern stadium planning. Convenience is an expression of the increased expectations and demands of visitors and spectators on a modern event venue. Good viewing conditions and a dense stadium atmosphere are the most desired features of a future stadium structure. The proximity of the stand to the action on the pitch helps to enhance the live experience of a stadium, in spite of increasing media presence and coverage.

Preparation and decisive planning factors:[5]
– Leisure time available and mobility trends in the general public
– Change in demographics and participation
– Change in attraction of sports and non-sports leisure time activities
– Change in spectator participation and their demands on comfort and facilities
– Decrease of public financing and increased privatization of event operation
– Improvement of revenue situation due to new event and marketing strategies
– Change of utilization profile of a venue
– Growing influence of media due to commercial possibilities through global broadcasting/distribution

034 **Olympic Stadium Berlin (during construction)**
Renovation and preservation

Analysis of the 'genius loci'

a) Traffic/transport links and address
b) Topography of the selected location
c) Climate and weather conditions
d) Consideration of historic developments and existing structures
e) Integration into a 'sports park' or urban context

Combined GRW-procedure

From the German 'Principles and Guidelines for Competitions in the Fields of Space Planning, Urban Planning and Architecture'
– GRW 1995 –[6]
Amended version from 22 Dec 2003

Competition types

Idea competitions
Realization competitions

Open competitions
Two-phase competitions
Multi-level competitions

Restricted competitions
Partly open competitions (qualified)
Invitation competitions
Cooperative procedures

Simplified procedures

'The combined competition for the RheinEnergieStadion in Cologne' by Dr Eschenbruch, Kapellmann und Partner, in an interview with Dr Hepermann, STRUKTUR GmbH [from 5 Jan 2005, here summarized and quotes in excerpts].

Preliminary remark
RheinEnergieStadion Cologne was realized on the basis of a combined competition according to GRW 95 for contracting of planning and construction services. According to Dr Eschenbruch, this model does encounter reservations regarding contract award regulations. (Still, this process was completed successfully for stadium construction in Cologne.)

STRUKTUR GmbH (in cooperation with the practice FSW, Dortmund and with employees of the city of Cologne as well as the Kölner Sportstätten GmbH) entirely prepared and executed the combined competition. The main priority was placed on the process-related features and producing the documents for the invitation to tender, i.e. in particular the specification including the performance programme, as foundation for offers by general contractors. Dr Hepermann was also responsible for project management during the procedure.

Combined process
Dr Hepermann: 'A decisive feature of a combined competition is that each architect taking part in the competition needs to prepare his draft for the two competition phases in close cooperation with one or several general contractors.

In the second phase of competition, the architect needs to ensure that at least one general contractor's binding offer is submitted for realization. This means, the architect may or must select a general contractor of his own choice. As a result of the procedure the client is not only provided with the drafts of the architects, but also with the corresponding offers for complete implementation. The competition is entirely anonymous, with notarial assistance in both phases. The names of the architects and general contractors are revealed to the client only after the decision of the competition jury.'

Aside from the combined competition, the two classic alternatives are also examined and assessed in cooperation with the city authorities:
– Standard competition procedure: according to GRW (later possibly with general-contractor tendering)
– Bidding procedure for total contractor: (including draft) with performance programme according to VOB/A.[7]

Since architectural quality is not the highest priority for general contractor (total contractor) offers, the combined competition aims to unite the specific advantages of both classic procedures: a high architectural quality comparable to the standard competition procedure after GRW and the use of know-how of executing firms as well as early cost control and the meeting of deadlines comparable to a qualified total-contractor bidding after VOB/A.

Since the relevant provisions are kept to a minimum in GRW, they contain many potential sources of error on the part of the client as well as room for objections by bidders.

In the run-up of the procedure this becomes an important negative criterion; but in agreement with all parties involved, including legal counsel, the relevant processes are structured and determined in detail. Specifications and numerous special regulations are needed to achieve the desired and intended result.

Normally, the opening date for general contractors' offers, based on VOB/A, takes place only after the decision of the jury. In the interest of a target-oriented decision, the following points should be considered:

Ensuring that costs submitted to the jury by the architect are in accord with the corresponding offer by the general contractor (as yet unknown to the client), e.g. through a binding statement of the architect. General contractors must submit their offers together with the architect's documents for competition phase 2, without being entitled to withdraw up to the opening date, i.e. the end of tender and start of the period during which the tender is binding must be before the opening date.

These and related questions should be resolved with all authorities involved, e.g. the respective contracting body of the city or regional government, architects' association, etc., so that the remaining risk may be deemed acceptable by decision-makers.

In the combined procedure for the World Cup Stadium Cologne, during the first phase altogether 69 architects took part in the competition; in the second phase, after selection by the committee, seven architects with a bidding general contractor each remained.

The whole procedure up to the commissioning of the general contractor took less than 10 months, including all periods of examination and decision-making. Documents were delivered on 1 Dec 2000, after internal preparation (clarification of procedure, provision of documents for first handling phase, etc.) and publications. Documents essentially contained the procedure provisions, the general data and targets in short, as summarized spatial and functional programme.

Noise control and traffic

After a colloquium in January, the first phase was submitted in February; tenders for execution by general contractors were not submitted at this point. After extensive preliminary checking, seven drafts were selected by the jury for the second phase at the beginning of March.

The first phase was internally used for the preparation of the tendering documents for the general contractors' offers, which a notary sent to the participants in mid-March 2001, safeguarding anonymity.

These documents served as a complementary basis for all revisions by the architects and the bidding general contractors in the second phase of competition:

After a separate on-site appointment with each bidder and one possibility for queries which were answered in writing, the architects submitted their revised drafts and the general contractors their binding offers to the notary at the beginning of June.

After preliminary examination of the architects' contributions, the jury convened for phase 2 and determined the winners of the architectural designs: the first prize went to gmp Architects of Gerkan, Marg und Partner, Aachen with general contractor Max Bögl GmbH.

The submission of general contractors' offers took place two days later on 15 June 2001, confirming the cost details which architects had quoted in competition phase 2.

In the follow-up to submission in July 2001, extensive meetings were held with the first and also the fourth prize winner, who was considered for cost reasons. On 11 September 2001, Max Bögl GmbH was commissioned with all implementation works and almost all planning services. As a peculiarity of the combined competition, only the building design up to and including phase 4 HOAI[8] (approval planning) was exempted from the commission of general contractor. In September 2001, the practice of gmp Architects was directly commissioned by the client.

Special features of the draft were as yet unknown when formulating the specifications. Therefore, the invitation for tenders stipulated objectives only, not methods of achieving them; for example, utilization during the entire construction period, i.e. the continuous usability of the stadium for all home matches of 1. FC Köln with a minimum capacity of 30,000.

All relevant provisions, including demarcations of liability and cost limits, were stipulated in the specifications, whereas the implementation of these provisions in planning and execution was determined at a later stage.

The overall planning and construction period starting in September 2001 was marked by target-oriented and constructive cooperation, in order to clarify and swiftly settle the typically arising differences of opinion caused by the conflicting interests of the parties involved.

To realize a time-constricted project, all tasks and responsibilities have to be clearly organized and allocated, and consistently observed.

Final evaluation

Based on the available data today, in particular the final agreement with the general contractor, and according to Dr Hepermann, the standard costs and all specifications of use were observed within the construction period, as well as the requested due dates.

All municipal committees and agencies had a strong part in the successful realization of the project. Architects and building contractors worked to schedule, satisfactorily and reliably.

Depending on the site and the proximity to the city centre), two kinds of stadium typologies emerge: *inner-city* or *suburban* location.

The decision is influenced by the preference of the stadium operator and the municipal concept. Much depends on the status and the general acceptance attached to the stadium, as seen in the discussion of the location for the new build Tivoli Aachen. The project must be compatible with its surroundings, in order to prevent possible legal action taken by local residents against nuisances.

The gathering of large crowds does not only generate commercial opportunities, but also noise which is emitted into the direct vicinity. Therefore, operators of modern sports and event venues need to clarify the noise protection suitability of a potential site.

A special German regulation on noise protection for sports complexes applies to the erection, the composition and the operation of sports venues, inasmuch as they are employed for sports activities.[9] A sports complex includes facilities that are in close proximity and have functionally associated. Arriving and departing traffic as well as ingress and egress times are also included in the period of use of a sports ground. Or as stated in a brief by the German Federal Ministry for the Environment: 'Active and passive sound-proofing alone cannot secure an appropriate level of noise protection. Not least with noise reduction in mind, measures to curtail and shift traffic to more environmentally friendly means of transport will gain importance.'[10]

General requests

The location of a venue (stadium / arena) should have good and travel connections to the inner city and good links to airport and high-speed trains as well as highways and motorways. Proxi-

German 'Rules for the Initiation of Competitions'
– RAW 2004 –[11]

in the fields of space layout,
urban development and engineering

Open competition
Restricted competition
Cooperative competition

Accessibility and stationary traffic

035 **Gottlieb Daimler Stadium, Stuttgart**
Open parking area with around 900 parking spaces
036 **Allianz Arena, Munich**
Main access via the esplanade (multi-storey car park with ca. 10,000 parking spaces)

mity and links to the city centre allow better exploitation of synergetic effects through customer and purchasing potential.

Efficient traffic routes and a reasonable dimensioning of roads, ideally with several main access roads and avoiding residential areas, are essential for the acceptance of a location.

Pedestrian and bike routes have to be provided in sufficient number and width as well as signalling devices and back-up space. Information boards with area maps in strategic positions provide early orientation within the stadium environs, general information, identification of safety zones, and information on rescue accesses and reserved spaces.

Traffic concept

Traffic and transport links to a venue are exceedingly important. Depending on the nature of the event, within a few hours large crowds may gather outside and inside of the arena; after the end of play, though, visitors almost simultaneously push toward the different means of transport, of which the following shall be considered:
- Individual transport: private vehicles, VIPs (VVIPs), bicycles and motorbikes
- Public transport: airplane, long-distance and suburban trains, trams/underground, buses
- Service traffic: athletes and players, media parking areas, shuttles and delivery
- Security personnel (parking and approaches): fire service, paramedics and police

In the construction run-up, a functional transport and traffic concept should be agreed upon with local authorities. Depending on the host, temporary special requirements might apply. UEFA demands an international airport, operational day and night (example: capacity for a final match = circa 80 additional charter flights per day and, if necessary, lifting of night-flying ban).

A dynamic parking space management and parking guidance system are part of modern event operation. FIFA demands the integration of the parking guidance system into the network of the stadium control room, to evaluate the occupancy status of parking spaces, including a registration terminal at the parking areas. A geographical information system (GIS) may also support traffic and parking procedures.

The layout of parking spaces needs to ensure that blocks comprising around 500 cars are clearly identified by signposting and that pedestrian zones or stops for shuttle services are easily accessible. The fitting-out should be low-maintenance, green landscaping with information boards should create a pleasant impression and unrestricted access should be ensured by adequate lighting. Separable areas for private events should be located in close proximity to the stadium build.

Public transport should take up a major part of the traffic volume and the stadium site should have good public transport links. A combination ticket for stadium and public transport tends to promote acceptance, and a cost-free shuttle service to and from Park & Ride places also allows the layout of parking areas which are not within walking distance to the stadium.

Reference values for parking space

A modal split of stadium visitors serves as a reference value for the percentage share of public transport, private or other in stadium traffic (rough preliminary calculation: 40/40/20). Depending on site conditions, municipal construction projects such as stadia must be accompanied by a traffic survey (type and scale of public transport, connecting roads, etc.) In a traffic survey for Müngersdorfer Stadion, Cologne, in 2001, IVV Aachen has made the following distinctions: apart from football and concert events, two special events A/B (mainly group or individual participants) are hosted. As part of the public transport quota, a charter-bus quota is estimated since it is to be considered as stationary traffic.

Football (cap. 45,000): PT 34%, private vehicles 51%, charter bus 8%, other 7%
Concert (cap. 50,000): PT 50%, private vehicles 40%, charter bus 6%, other 4%
Special event A (groups) (cap. 40,000): PT 30%, private vehicles 6%, charter bus 60%, other 4%
Special event B (groups) (cap. 20,000): PT 50%, private vehicles 30%, charter bus 16%, other 4%

Demand for parking spaces is multiplied with an occupancy rate of 2 to 2.5 persons per private vehicle as a recommended parking space figure (buses approx. 50 persons / vehicle).[12] According to the reference and orientation values for stadium-related demand for parking space (Appendix B), the following figures serve to determine the number of necessary parking spaces or garages and are taken as assessment basis by many building regulation authorities.

Assembly places (excluding sports venues):
1 parking space for every 5 seats

Sports and leisure facilities:
– Sports grounds *without* spectator places: one parking space for each 250 to 300 m² of sporting area
– Sports grounds and stadia *with* spectators: one parking space for each 250 to 400 m² sporting area, one parking space for every 10 to 15 visitors

Leisure facilities and sports halls *with* spectators: one parking space for every 50 to 80 m² of surface area, one parking space for every 10 to 15 spectators
– Catering and accommodation facilities
Regional: one parking space for every 4 to 8 seats
Local: one parking space for every 8 to 12 seats
Hotels, guest houses, etc.: one parking space for every 2 to 6 beds
– Example of calculation for parking space figures
50,000 visitors (for 10 to 15 visitors)
→ 3,300 to 5,000 parking spaces
60,000 m² sporting area (for 250 to 400 m²)
→ 150 to 240 parking spaces

Verified parking spaces in the framework of the building permission procedure
→ in total a maximum of 5,240 parking spaces

The reference figures stated by FIFA / UEFA for 50,000 / 60,000 spectators are: 10,000 parking spaces for private vehicles and 500 for buses. UEFA allocates the number of parking spaces for fan buses according to group / quarter / semi-finals and opening match / final at 200 / 250 / 300 / 450. (Numbers may be reduced if there is a sufficient volume of public transport.) The parking spaces should ideally be located on the stadium grounds, brightly lit and clearly signposted. A walking distance of 10 to 15 min is generally viewed as acceptable; at 5 km / h walking speed this results in a distance of 1,000 to 1,500 m. Preferably, parking lots P1, etc. are assigned to a particular spectator sector and colour coded, with separate areas for home / away fans. The location of the parking spaces should be based on accessibility by public transport, and a speedy and organized processing of traffic should be ensured, with direct connection to the motorway.

Space requirements for car parking space
For private vehicles as reference (85 % share), a measurement of 4.70 x 1.75 m may be assumed.[13]
The Regulation on Garages states the following measures:[14]
– minimum length: 5.00 m (incl. projection)
– minimum width:
2.30 m → both sides undeveloped
2.40 m → one side restricted (structurally)
2.50 m → both sides restricted
3.50 m → disabled parking space

037 **RheinEnergieStadion, Cologne**
Main ingress via the axis 'Vorwiesen' (green area in front of stadium)

038 **Zentralstadion, Leipzig**
Open multi-storey car park underneath lower tier stand (ca. 500 parking spaces)

039 **Commerzbank Arena, Frankfurt**
Split-level garage underneath the stand (ca. 1,800 parking spaces)

040 **Road Sign, 315**

page 31:
041 **Allianz Arena, Munich**
Roof landscape, parking and access to 'Esplanade'

Traffic aisle width per parking space (layout)

	2.30 m	2.40 m	2.50 m
for 90° set-up	6.50 m	6.00 m	5.50 m
for 45° set-up	3.50 m	3.25 m	3.00 m

Note: Expert engineers of the German 'Forschungsgesellschaft für Straßen und Verkehrswesen' [Research Society for Roads and Traffic] are currently working on the recommendations for layouts of stationary traffic (2005), which in future will propose a general minimum pitch width for private cars of 2.50 m.[15]

Alternatively, for a better exploitation by systematic feeding of special or only sporadically used parking spaces, W. Steinhart suggests the system 'from back to front', with 5.0-m wide traffic aisles for one-directional traffic and a density of up to 2.30 m per vehicle. The filling of parking areas is supported by mobile barriers and by manual guidance through personnel. A parking space is restricted on only one side by a previously parked vehicle, and in the direction of approach there are only free parking spaces.

Likewise, approval provisions apply for bicycle spaces (e.g. 0.05 spaces per spectator seat and 1.0 spaces for every 250 m² of sporting ground). These shall not be further elaborated here since their accommodation in general does not represent any spatial problems.

Parking spaces for VIPs

should be located in the direct vicinity of the entrances to VIP stands/rooms. A sufficient number of parking spaces should be provided for separately controlled areas, preferably inside the stadium complex. Buses may be withdrawn to a secured location and called back, depending on need. UEFA wishes to match the distance of parking spaces to the celebrity status of the respective group of persons. Gradated after group matches, quarter or semi-finals and opening/final, UEFA requests the following number of parking spaces.
VIP: buses 4 to 8, cars 150 to 250
Commercial partners: buses 4 to 8, cars 70 to 170

Note by DFB/FIFA: For particularly vulnerable persons (VVIPs) sufficient protection against firearms and explosives should be provided in rooms and lounges.

Parking spaces for the media

are to be laid out in addition to the TV compound in the direct vicinity of media areas, segregated from public parking and in the direct external surroundings of the stadium, fenced and monitored. UEFA figures for parking spaces:
Media: 300 to 400 cars

Parking spaces for athletes, participants

Separate parking for athletes and participants in competitions should be planned close to dressing rooms with direct and secured access, avoiding media encounters, separated from public areas and without contact to spectators, otherwise secured by stewards.

Teams, referees and officials park in a non-public and secured area, preferably inside the stadium structure. Team buses may be withdrawn and called back on demand. UEFA makes the following recommendations:
Referees: cars 10
Teams: cars 4, minibuses 2, buses 2
VIP bus: 2 to 4
VIP cars: 30 to 60 (at the stadium)

Parking spaces for service traffic

Operation of a large event venue is ensured by service and support personnel. Sufficient parking space is to be provided for security staff, stewards and suppliers. UEFA:
Employees: cars 20
Signage: trucks 3

Parking spaces for the disabled

At least half of the number of required spectator accommodations for people with disabilities are to be provided as parking spaces.[16] They should be positioned in the vicinity of the stadium, be accessible via separate access routes and be located in an ideal position to the respective entrances. Space requirement is 30 m² per parking space, incl. part of the road lane.

Note: IAKS advises (1993) that except emergency vehicles, all vehicles are to remain outside the main fence.

Chapter 03

Organization of use
Determination of user groups and safety aspects related to a sports complex

Distribution of functions

Modern sports and event complexes represent some of today's most intricate building types, as activities are not reduced to the field or stage, but are also taking place in the ancillary utilization areas. Outside of matches this concerns the stadium administration with its staff and technical personnel responsible for ensuring readiness for events and general maintenance of the sports venue.

During professional matches, the group of active participants is divided into players and team officials as well as accredited media, official partners and suppliers. Referees, coaches, spokespersons for the team and press are part of competition organization; writing press, photographers and radio reporters, TV commentators and technical personnel are part of the media group.

Apart from the responsible event management, operational personnel is made up of stadium control technicians, ticket sales staff and stewards performing ticket checks and body searches; furthermore the personnel for selling merchandise, as well as kiosk, kitchen and catering staff.

To ensure the safe running of an event, stewards and security services must work together closely: including police, fire service, stewards and other security services, first aid and medical rescue services. In part these are organized privately by stadium operators or clubs/associations, and in part by municipal security services responsible for the protection of the general public. All parties should be included in the security planning of a new venue so that relevant structural measures may be incorporated well in advance. Integral solutions and reasonable perspectives are paramount in this respect.

Not only when 'enemy' groups of supporters encounter each other on the stands and in circulation areas do questions on safety arise, but also when it comes to the securing of the general match and event operation. Preventing encounters of specific groups can be vital in this respect.

The new generation of sports complexes is much more complex as compared to the large sports venues for the Olympic Games of 1972 or the Football World Cup of 1974 – the times when stadia of the second generation were built.

Neither their operating nor utilization profiles can be compared to current sports complexes. Therefore, planners of a modern sports venue have a particular responsibility in the coordination of different user groups and their various interests during an event.

Commercial exploitation of large special areas or VIP areas demands segregation from regular spectators: the business area with its prominent guests should only be accessible in a controlled manner. Accredited media representatives should also be able to work undisturbed in their respective areas, and preparation of athletes or players should progress devoid of outside influences. Generally, these areas are all accommodated within one main stand, which makes intersection-free utilization very difficult, more so as work-related possibilities for contact should remain.

Safety in sports stadia

This topic represents one of the most important elements in the planning of sports complexes. Most requirements and regulations are geared to one prerequisite: 'personal security before protection of property'.

For clarification purposes, generally two groups of people and utilizations may be distinguished: the 'active' and the 'inactive'. Part of the first group are those persons indispensable for the preparation, running and maintenance of match operation; part of the second group are all spectators and visitors of a sports or other event, which also includes the special areas inside the sports complex for VIP guests or visitors of other fitness and recreational facilities (restaurants). A general sports ground is an open complex designed for organized competitions as well as unregulated leisure activities. Popular sports facilities are generally intended for groups of active sportsmen and sportswomen, including all school, club, corporate and disabled sports groups as well as all unaffiliated recreational athletes.

The term 'inactive' is applied to all those present without physically taking part in the event; generally relatives, spectators, seniors or disabled groups with other activity profiles. The Stadium ATLAS focuses on modern sports stadia as professional places of assembly with spectator stands for over 5,000 people. In the light of security-related matters, the whole process of an event, from preparation to follow-up, has to be taken into consideration, with attention focusing not only on visitors, but also on employees.

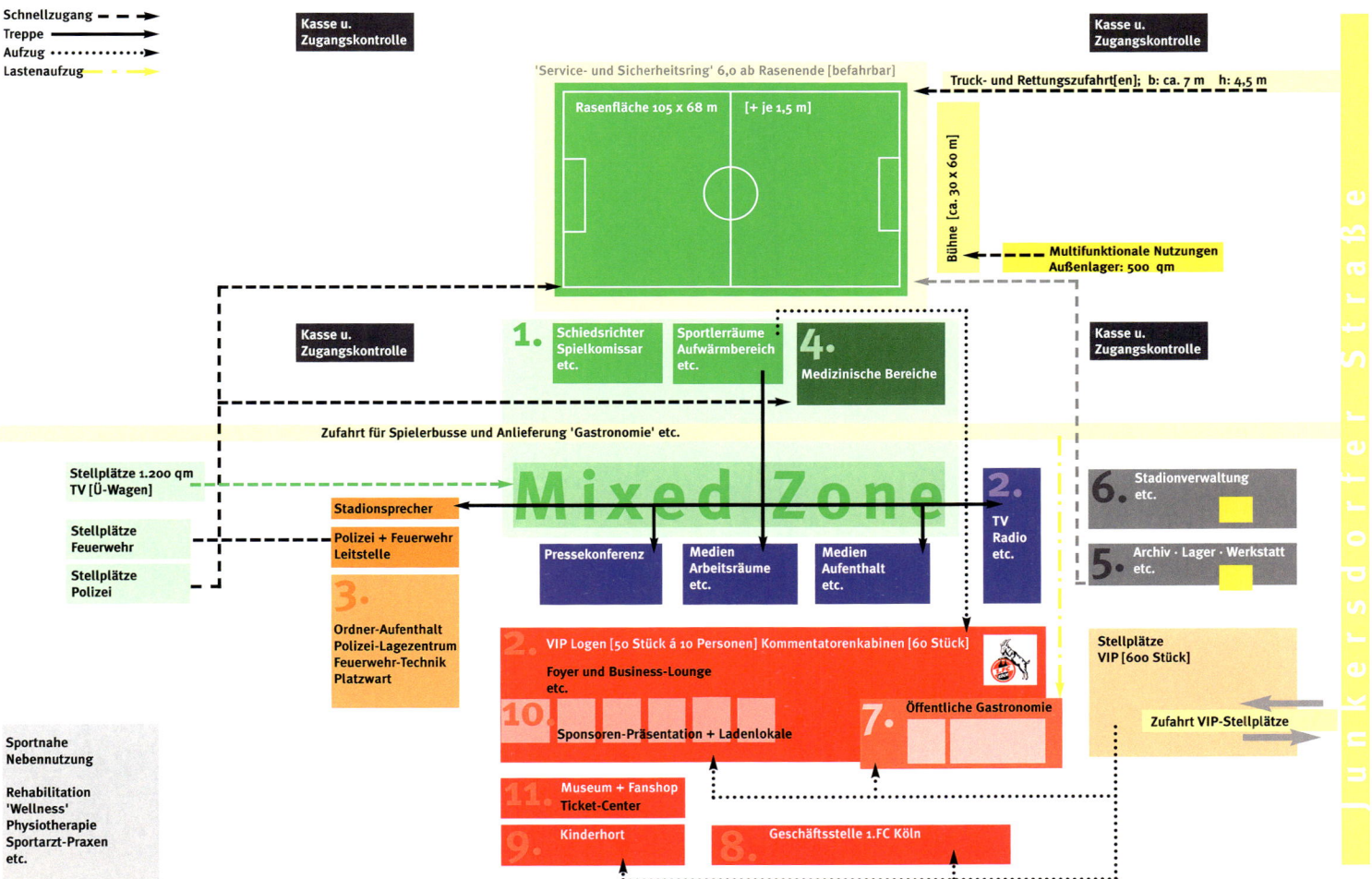

042 **Organigram of functions of a modern FIFA-compatible football stadium;**
Excerpt from the competition initiation for Müngersdorfer Stadion, FSW 2001

Zoning and sectors

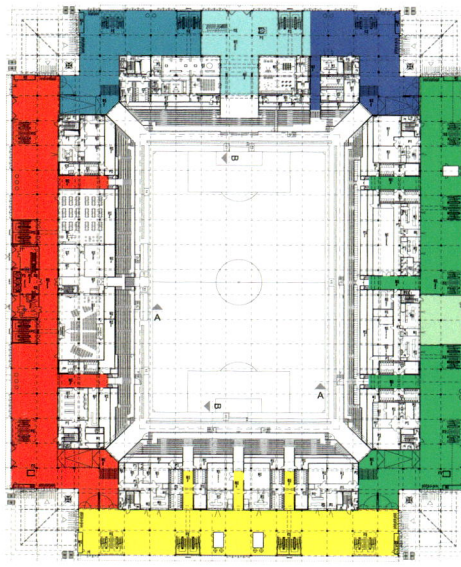

043 **Zoning of a sports field**
Key plan according to DIN 13200-1
044 **Sector divisions**
Drawing of World Cup Stadium Cologne

The entire stadium/arena or the event space is divided into safety rings which are separated into zones and sectors. The term 'zone' denotes a ring-shaped layout of security belts which subdivide the sports grounds even outside the property boundaries. Within these zones certain rights of sovereignty, such as domiciliary rights according to the guidelines of the sports park or stadium rules, and the general rights of admission for spectators apply. In the case of the FIFA World Cup 2006™, special terms of contract apply, and the protective rights of names, product and trademarks of the main sponsors are settled and secured. The following zones are distinguished after their safety level from 1 to 4:

Zone 1: Playing field
Interior area
Central sports and event space as activity area where the event takes place; situated within stadium bowl; serves as place of comparative safety and, in case of disturbances, as relief area. Unroofed in the stadium or permanently covered/convertible in an arena.

Zone 2: Stands
Auditorium
Terraces are sloping structures with standing or seating rows from which spectators view sports or other events. Those areas have separate accesses, are divided into blocks and are controlled.

Zone 3: Stadium
Service area
External access and circulation areas around entrances and exits to various blocks and locations inside the fenced and controlled stadium which is also known as the 'internal safety zone'. It functions as service area with all necessary facilities such as toilets, first aid, catering and souvenir shops (merchandising).

Zone 4: Sports complex
External circulation area
All areas and buildings in the external safety zone including the stadium environs as half-public area; for sports park concepts with supplementary sports and event profiles and the necessary ancillary areas.

Facilities within sectors
Each sector has to accommodate refreshment, toilet and other important facilities and needs to be fitted with dedicated entrances and exits; each of these sectors may be subdivided into smaller areas (cf. chapter 11 on formation of blocks).

Inner and outer safety ring
In a definition by DFB the inner perimeter fence surrounds the central area of the stadium around which the auditorium and the stands are arranged. The outer safety zone is an outer fence enclosing the entire sports ground.[1] It should be difficult to climb over, to penetrate, to crawl under or to remove.

Perimeter fences should be embedded, where possible, in order to keep them from standing out too much; means that facilitate climbing over, like bushes or ticket booths, etc., should also be avoided. All gates (DFB) are to be fitted with an emergency opening mechanism, e.g. for fire fighters. Stadium complexes must have an enclosure at least 2.20 m high to hinder climbing over.[2] Whenever necessary, it should be possible to subdivide the outer safety area into four sectors. According to FIFA, it should be laid out in sufficient distance to the stadium perimeter, with mobile or permanent fencing serving as preliminary barrier for spectator control. Their provision and positioning is determined by the LOC and in coordination with local authorities. (FIFA height minimum is 2.50 m. In future, all fence heights shall be standardized to 2.20 m.)

First check
A sufficient number of entrance gates is to be set up to avoid delays. A first visual ticket check, identity checks and search procedures for spectators of both sexes follow. A storage space for secured items shall be provided, with restricted access.

Second check
located at the stadium entrances, the inner safety ring, with single-passage systems and electronic turnstiles. At this point, a final ticket check follows. In case of conspicuous spectator behaviour, a second identity check is carried out. Cashiers for league matches, etc. are to be included within the inner fencing.

Third/fourth check
This takes place at the stairways to the relevant concourses or at the vomitories of the stand. Stewards may carry out these checks, as part of their task of directing spectators to their seats on the stand.

Sector division

The venue itself is subdivided into four sectors. In principle, these are radial divisions, which may differ according to stadium geometry and grandstand profile. These divisions within the auditorium are mostly executed as mobile or temporary measures, as they only need to be provided and executed in case of need.

Sectors are individually defined by organizers and security staff at every game. A 'forced channelling' into another block (contrary to tickets) may be carried out in case of high-risk matches.[3]

Sector division on the stands

DFB guidelines define the sector subdivision as follows:

Minimum of four separate sectors with dedicated entrances. Partitions of 2.20 m height are to be fitted between standing and seating accommodations, which prevents spectators from changing to other areas (crowd control). In cooperation with the responsible security services, stadium operators/organizers determine the type of barrier and remain accountable. Fencing does not only enforce a negative atmosphere among spectators, but also restricts viewing, an inevitable consequence of the necessary height.

Implementation

Transparent partitions are urgently recommended to counter this effect. Dividing these into two 1.10-m elements would be advisable, in order to allow an on-demand adaptation to the relevant targets of a coordinated safety concept. All employed measures should be dismountable.

Sector division in circulation areas

The sector division is to be continued up to the inner safety zone (stadium fence). To maintain the concept of providing emergency evacuation routes into a neighbouring sector, sufficiently large openings are to be provided as well as wickets to allow accredited persons to change sectors.

Opaque partitions between home and away fans are to be provided for security reasons during regular league operation. This measure has become accepted among security staff as an efficient way to avoid incidents as 'enemy' fan groups generally only approach each other after eye contact. According to police information, the danger for bystanders in this situation lies in being 'overrun'.

'Buffer zones'

In case different spectator groups are put up in one sector, 'buffer zones' or free blocks should be set up between them. After the experience of the UEFA European Championship in Portugal 2004 buffer zones are in principle conceivable. These consist of horizontal climbing-protection devices in the seating area.

Guests and home fans should be placed as far from each other as possible. Partitions should be highly stable, and for guest fans dedicated access areas with marked routes avoiding encounters with other spectators should be provided.[4]

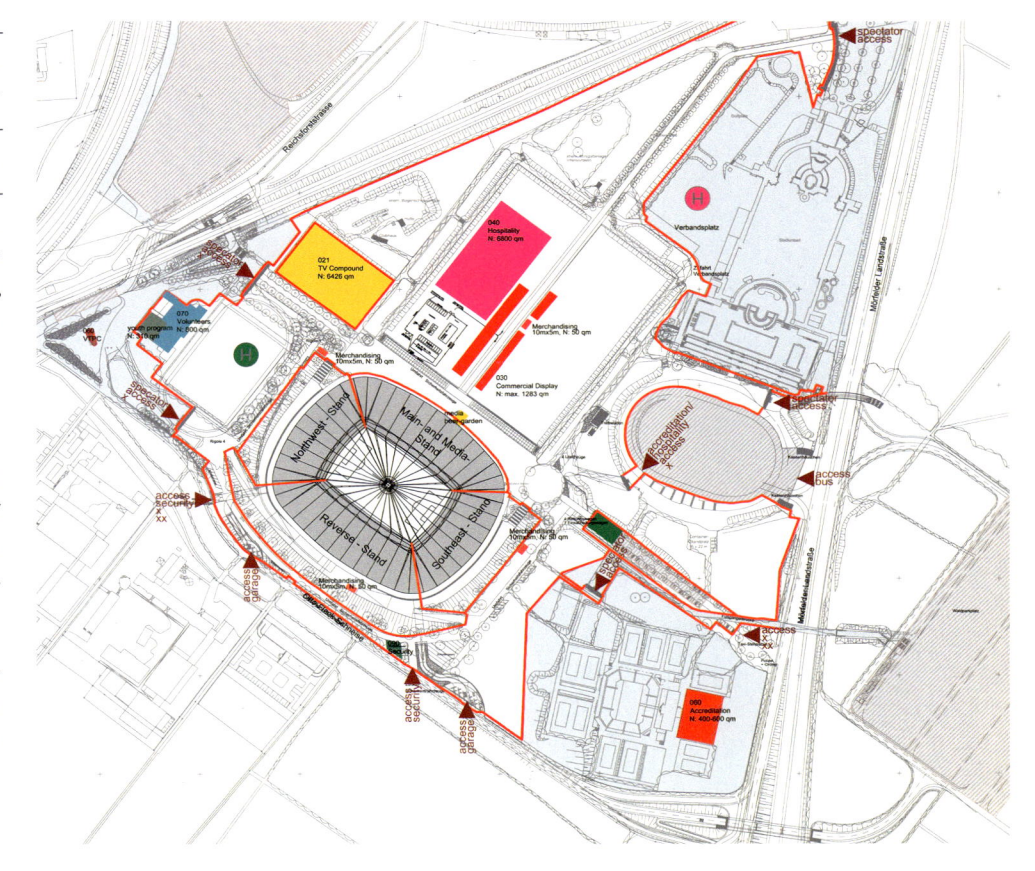

045 **Site plan**
World Cup Stadium Frankfurt with outer and inner safety zone

03 Organization of use

Turnstile systems

Admission capacity per hour (general):
Around 600 to 1,200 persons

According to supplier (KABA Co.)
Turnstile portals: 13 to 16 persons per minute
Turnstiles: 20 to 20 persons per minute

Reference value: overall capacity within one hour, inside the stadium (The facilities should also serve as a second evacuation route in case of emergency.)

Per turnstile system: 660 persons per hour
(according to *Guide to Safety*, 6.7)

046 **Entrance area**
Veltins Arena, Gelsenkirchen, Zone 2
047 **Single-passage and fence system**
Veltins Arena, Gelsenkirchen, Zone 2
048 **Skidata™ turnstile systems**
Allianz Arena, Munich, Zone 2

Entrance areas

FIFA Safety Guidelines (§4) state the requirements for vehicular and pedestrian traffic: to allow it to proceed swiftly and orderly, accesses and exits are to be set up accordingly. Entrances may not be used as exits. In order to avoid congestion through pass-back, 'conspicuous' individuals should be briefly taken to the interior area and then leave under supervision through a separate exit nearby.

The turnstiles and turnstile portals (§9.6) are mechanical single-passage systems that are not permitted as escape routes unless they can be opened easily and fully from the inside. They should be hard-wearing, climate-resistant and vandalism-proof.[5] The equipment comprises anti-pass-back, visual admission by signal management, monitoring by security staff and digital surveillance by the police and emergency centre of operations from where the system may be unlocked in case of emergency.

According to information by Skidata AG, maximum admission is around 800 to 1,000 persons per hour, depending on ticket handling. According to specification, a Skidata turnstile is capable of handling 1,200 persons per hour. Naturally, this is just a theoretical value in reference to actual admission scenarios in a stadium. As empirical value for admission through external units (e.g. six-foot compact gate or gates for the disabled) a number of 500 persons per unit may be assumed.

The relevant official German regulations require railings to ensure single-passage of spectators within the entrance area, including admission checks and searching of persons and their belongings.[6]

Admission control system

Electronic turnstiles with non-touch ticket readers, complemented by barcode readers for day tickets, have become customary and were part of the requirement catalogue of the organizational committee of the FIFA World Cup 2006 in Germany: multifunctional ticket reading units as entry/exit readers for chip and barcode technologies (RFID, Radio Frequency Identification), with contact chip modules or contactless.

As advised by FIFA, the World Cup ticket system should be capable of being implemented afterwards into the electronic ticketing of the stadium. According to the LOC 2006, this requires a consistent information platform for security-relevant data. Individual personal data on admission, re-entry or validity of the ticket shall be verifiable. Information on fan streams in the stadium terraces and their degree of filling up shall facilitate an event-specific response by persons in charge.

The segregation strategy favoured by the sports associations has established pre-selection and separation of spectators as most effective safety strategies. Due to increased use of the internet and telephone orders, stadium tickets are more likely to be distributed online in the future. Ticketing for the FIFA World Cup 2006™ was exclusively handled in this way. There were no sales of tickets at the stadium on the day of the match.

Today, ticket sales for many entertainment and concert events are almost entirely handled in advance, which makes ticket booth sales more and more redundant.

In the wake of progressing methods of telecommunication, in future, digital tickets might be stored on, for example, credit cards or other general storage media.

In any case, a central information point should be provided for any queries on admission control and for the purpose of depositing and picking up tickets. Even when demand or waiting times at ticket booths decrease, people queuing up should not impede the circulation routes and service areas. Availability of tickets should be marked clearly on the outside of the selling points.

The capacity for guest supporters should be 5% (UEFA) or 10% (DFB) of the overall stadium capacity. These figures are relevant for the planning of a stadium to the effect that they determine the size of the terrace sectors to be fenced-in.

Innovative payment systems

'Venue Purse – pay without cash' (skidata™, Salzburg): cashless payment in the stadium is a convenient future trend. Based on a point system, payment of the whole range of goods and services (fan shop, refreshments, etc.) may be processed via a data medium.

Electronic currency has the advantage of prolonging a spectator's sojourn and increasing consumption within the stadium, reducing waiting time at kiosks and gathering of consumption data at the selling points, as basis for future marketing measures (customer relationship programmes, bonus systems). Incorporating non-cash payment into the complex merchandise information system of a stadium operator may optimize the logistics of events.

By introducing a so-called 'Knappenkarte' ['miner card'], in reference to the former term of 'blue-white football miners', the arena Auf-Schalke in Gelsenkirchen and Allianz Arena in Munich have switched to cashless-payment transactions. Via the electronic season ticket, the system registers time and place of admission and provides the option of swift and cashless payment for food and drink.

A 'stadium card' not only serves for admission, but may also function as parking ticket or be valid for public transportation. A one-time stadium guest may or must adapt to cashless payment by buying a day ticket which functions like a cash card that can be charged and recharged at terminals.

Telecommunication

FIFA stipulates in its Technical Recommendations clear requirements for the communication facilities of modern sports stadia: structured and extensible telecommunication lines (optical fibre technology) should be available for EDP. So-called 'highly available operation' (FIFA) should be supported by a server infrastructure, data centre and suitable components fitted at the entrance area (admissions). Cabling should be at a minimum standard of LAN-CAT 5, and the network architecture should be based on LAN (Local Area Network) and W-LAN components (Wide Area Network).

ISDN (Integrated Services Digital Network)

Current communication services (telephone, data, telex) are operated on various special networks. Thanks to digitalization, vatious services can be integrated into one single network. In a first step all narrow-band services (telephone, fax, data, etc.) are united. By introducing optical fibre, all communication services (and moving images) can be transmitted via one connection.

An undisrupted power supply is to be ensured through a suitable UPS-system. The telecommunication equipment employed at the FIFA World Cup 2006™ was supplied centrally by an official World Cup sponsor. Regarding telecommunication infrastructure, UEFA assumes that all visitors of EURO 2008 will be provided with the most up-to-date technology.

Radio communication

DFB accepts an internal telephone network (also mobile) for communication purposes inside the stadium, superseding the former local radio traffic. Duplex systems are still recommended for control point, command posts, police stations, detention rooms, teams and referees as well as the head office.

In an emergency conventional radio systems are obligatory as substitute communication measures. UEFA demands radio traffic for all safety-relevant staff to be guaranteed by the local authorities.

If the stadium structure impedes radio reception, buildings or building parts open to the public, work places and their external areas are to be fitted with aids for orientation.[7]

049 **Sponsors' village (Fan zone)**
World Cup Stadium Cologne, Zone 3
050 **Official FIFA fan shop**
051 **Search procedure**
World Cup Stadium Cologne, Zone 3

Chapter 04

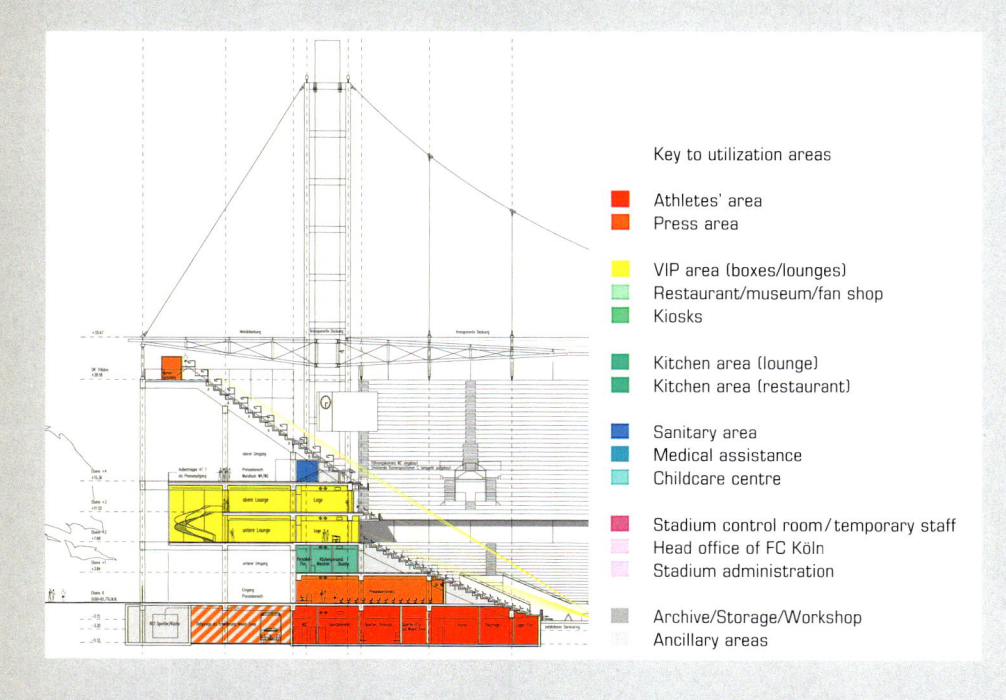

052 System section (main stand east)
World Cup Stadium Cologne

Key to utilization areas
- ■ Athletes' area
- ■ Press area
- ■ VIP area (boxes/lounges)
- ■ Restaurant/museum/fan shop
- ■ Kiosks
- ■ Kitchen area (lounge)
- ■ Kitchen area (restaurant)
- ■ Sanitary area
- ■ Medical assistance
- ■ Childcare centre
- ■ Stadium control room/temporary staff
- ■ Head office of FC Köln
- ■ Stadium administration
- ■ Archive/Storage/Workshop
- ■ Ancillary areas

Auditorium
A definition of all planning parameters relevant for a grandstand structure

Spectator comfort

The auditorium should provide an unrestricted view of the entire activity area.[1] Together with the building layout and the standard of convenience (viewing quality) and of fittings, the quality and type of a stepped stadium structure is largely determined by planning variables such as topography, climate and financing.

Spectator comfort

'Up to the year 2006, spectator demands on convenience within a stadium will continue to rise. As a result, facilities and operations of stadia must meet the highest standards', as stated in FIFA Technical Recommendations 2006.

The stadium concourses should be covered and their dimensions should match the specific capacity. Refreshment stalls should be placed in a clean and attractive environment and be easily accessible and evenly distributed throughout all sectors. Situated close to the standing and seating accommodations, they should invite customers to linger (e.g. with information monitors) and also provide sufficient WC facilities. People queuing up should not be an impediment to other spectators. Supplementary service units like information stands or child-care facilities are also part of modern stadium amenities.

Classic ground-plan layout

The system plan of World Cup Stadium Cologne above illustrates the standard distribution of utilizations in a modern third-generation stadium. Diverging systems or multiple tiers are conceivable. Traditionally, the main areas for active participants, press and executive guests are situated in the western part of the stand and are set-up in vertical succession.

The western location derives historically from the uncovered earth-built tiers of the first generation. Because a low afternoon sun should not impede the viewing of premium seats, a sports complex is traditionally aligned in a north-south direction.

The players' area with its large ancillary utilizations is preferably located at the lower part of the stand.

Due to the provision of 'executive hospitality' areas with box units, which are positioned geometrically sound at the height offset between lower and upper tier, this zone is mostly situated mid-level at the upper end of the lower tier. Depending on the size and number of lounges/boxes, these areas should be centrally accessible. Alternatively, the utilization concept may designate several main areas of use within a typically circular tier of boxes (in Cologne: west – VIP, north – restaurant, east – administration and head office, south – optional area).

The press area is divided into three zones: above/below and outside the stand. These zones should be laid out directly adjacent to one another, with press seats in a centre-line position with a good view and work rooms in the unlit zones underneath the lower tier. Short distances to the mixed zone situated between the players' changing rooms suggest a raised position between executive guests and players.

The general viewing areas run circular throughout all grandstand sectors. Their service units are located at asynchronous intervals to terrace accesses.

page 39:
053 Plan (World Cup Stadium Cologne)
Overview and distribution of utilization
- a. Level –1 Players' area/basement
- b. Level +0 Lower concourse
- c. Level +1 Ancillary service rooms
- d. Level +2 Lounge-level/administration
- e. Level +3 Gallery/upper concourse
- f. Level +4 Upper concourse (west)

Distribution of functions at World Cup Stadium Cologne

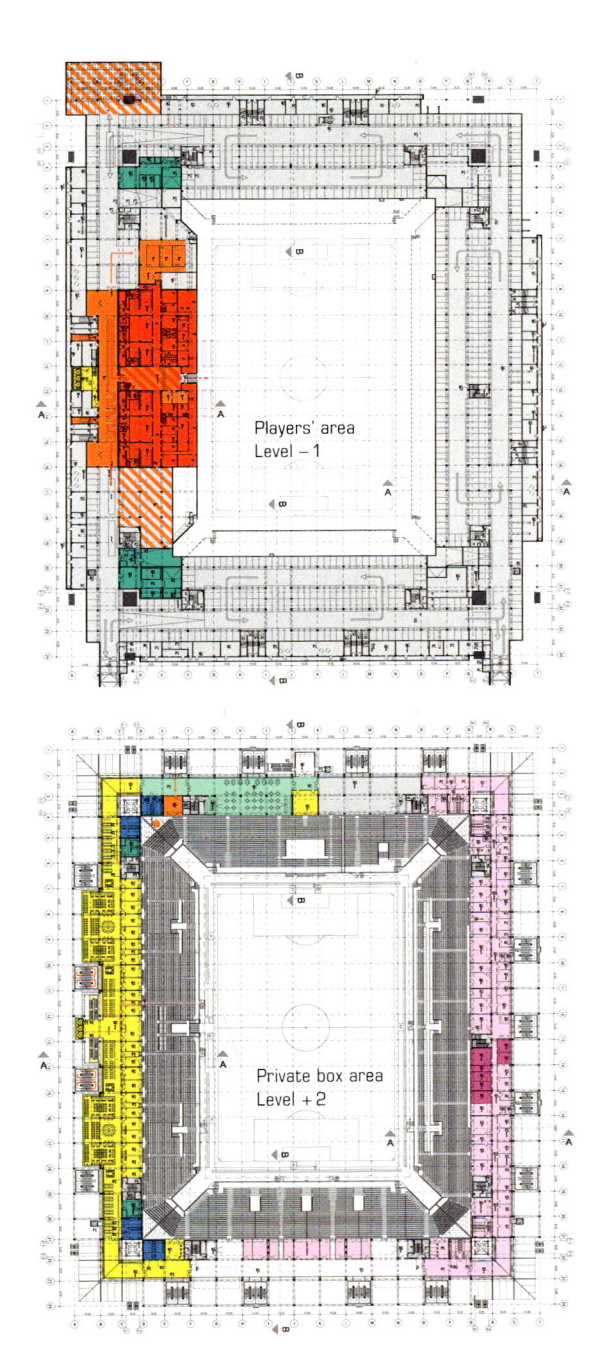

Players' area
Level −1

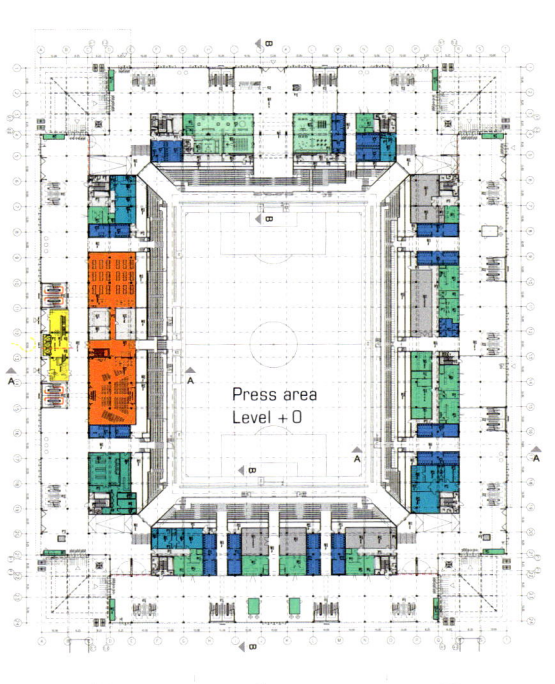

Press area
Level +0

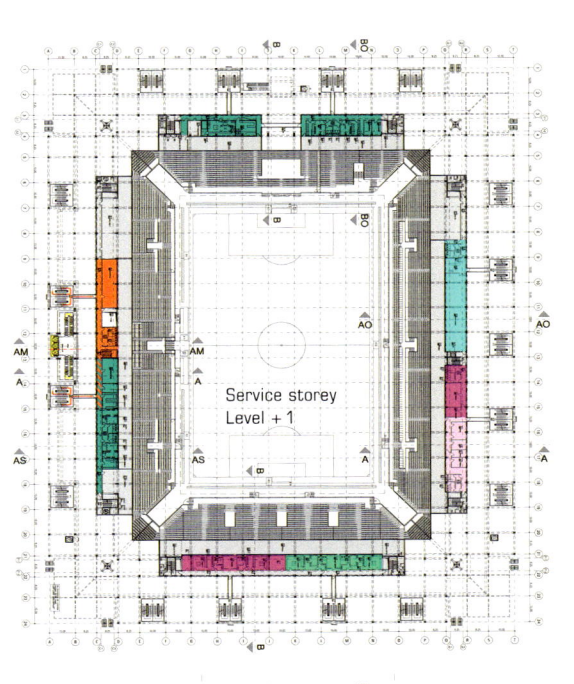

Service storey
Level +1

Private box area
Level +2

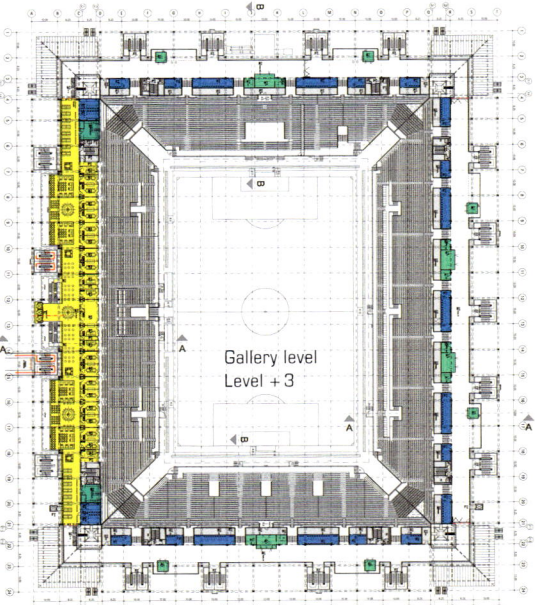

Gallery level
Level +3

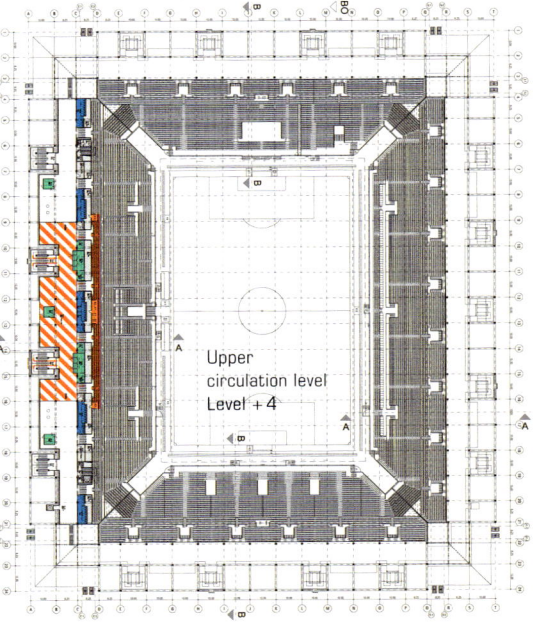

Upper circulation level
Level +4

Definition of seating accommodation

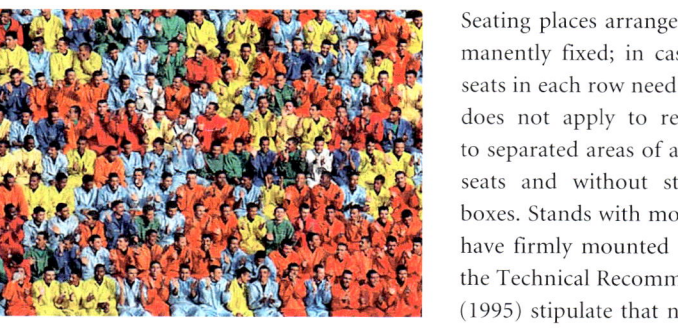

Seating places arranged in rows need to be permanently fixed; in case of a temporary set-up, seats in each row need to be firmly linked.[2] This does not apply to restaurants or canteens or to separated areas of assembly with less than 20 seats and without stairs, such as hospitality boxes. Stands with more than 5,000 places must have firmly mounted single seats. Furthermore, the Technical Recommendations of FIFA/UEFA (1995) stipulate that no standing places are permitted for international games.

Quality of material
In principle, seats need to achieve at least B1 (flame-proof) and for congregation areas with more than 5,000 places, the supporting structure must consist of uninflammable materials.[3]

Construction of seats
Sports associations FIFA, UEFA and DFB point out that seating layouts must comply with legal requirements. Requested are single seat shells which are comfortable and ergonomic, including armrests and tip-up seat bases. Seats should be equipped with backs (minimum height 30 cm); tractor seats are now inadmissible. Breakproof (vandalism-proof), climate resistant and uninflammable materials and permanent fixings are further requirements.

Identification
Seats and rows must be clearly marked, numbered and easy to identify.

Provisional stands
Seating that is intended, due to material, structure and design, only for limited utilization; may in no way be employed for an extended period of time.

Provisional terraces are not admissible for any UEFA competition matches.[4]

Seat benches
Following the evolved standard for seating terraces, FIFA/UEFA no longer permit the use of seat benches. At the same time, these are still admissible in restaurants and canteens as well as spaces with less than 20 seats, such as hospitality boxes.[5]

Allocation
The current German Regulations on Places of Assembly state the following requirements:[6]
– Seats in rows and for standing places: two persons per square metre of surface area
– Standing places in stair rows: two persons per running metre

Measurements of seats

According to DIN 13200-1 (minimum distance from backrest to backrest):[7]
Tread depth: at least 70 cm
Recommended: at least 80 cm

According to MVStättV 2005:[8]
Seat width (grid): at least 50 cm
Clear width: at least 40 cm

Diverging dimensions taken from DIN 13200-1:[9]
Seat width: at least 45 cm
Recommended: at least 50 cm
Seat depth: at least 35 cm
Recommended: at least 40 cm
Clear width: at least 35 cm
Recommended: at least 40 cm
Seat height: maximum 45 cm

Technical Recommendations by FIFA/UEFA:
Seat depth: at least 40 cm
Clear width: at least 40 cm

According to *Green Guide*, seat width is 460 mm + 40 mm armrests = 500 mm to 760 mm (chapter 11.11). Clearway may be reduced in exceptional cases from 400 to 305 mm if less than 7 persons are sitting in one row.

The building design textbook by Neufert (1992) [B030] quotes three seat options:
Work chair: 45.0 to 48.0 cm
Living/dining room chair: 40.0 to 45.0 cm
Small armchair: 37.5 to 40.0 cm

054 **Exemplary spectator stand**

Barriers and sightline

Railing height according to German Standard DIN 18065:[10]
– up to 12 m for residential and other buildings which are not subject to German regulations for workplaces:[11] 90 cm
– up to 12 m for buildings which are subject to the relevant regulations for workplaces: 100 cm (according to industrial safety regulations)
– over 12 m valid for every 110 cm.

Guard rails, crush barriers, fences, gates or glass railings shall be at least 1.10 m high[12], corresponding to the *Guide to Safety*. DIN stipulates the load assumptions for spectator stands at 1.10 m, independent of barrier or fence height.[13]

Stand front edges – Height in front of seating rows

According to the relevant German regulation, measures for front safety barriers are:[14]
– generally 0.90 m
– or 0.80 m for 20-cm width
– or 0.70 m for 50-cm width

Interpolation between these values is permitted, as is a railing height of 0.65 cm if the height difference in front of seating rows does not exceed 1.0 m. In case of a height below 0.90 cm, clearway between railing and front edge should not exceed 53 cm. A safety barrier behind the last seating row should rise at least 1.1 m above the floor level behind it.[15]

Note: Depending on the respective definition of use, the sports complex may be subject to various legal regulations.

Example: A grandstand during an event is subject to different regulations (for assembly places) than after a match (workplace regulation due to cleaning services present).[16] For example, according to the first regulation, the barrier in front of a spectator row may be 90 cm high, but 1.0 m high according to the regulation on workplaces.

This circumstance should be considered during final planning and, if need be, sorted out in advance.

EN/DIN 13200-3 assumes viewing restrictions caused by barrier fencing in front of the terraces.

Scenario 1/2: 90-cm barrier

A standard safety barrier in front of a seating row is normally set up as a solid barrier plus handrail. Height division is subject to planning considerations. Viewing toward the goal line becomes 'critical' in reference to the sightline around box tiers or the front edges of upper tier.

If the overall structure cannot be optimized due to the geometry of a stadium, planners should take care that the space between a mounted handrail (for example a 50-mm round tube or a 50-mm flat steel) and the solid barrier is big enough, and that spectators, despite the necessary height, may at least see above or underneath the handrail. However, the gap should not be too large, in order prevent the passage of a person or child or misuse of the barrier for storage purposes.

The type used in Cologne (20-cm glass railing on 70-cm solid railing $w = 15$ cm) seems to be a good solution. Since only few persons are reliant on a safety barrier, it can be mounted without a handrail and with reduced horizontal load bearing. Alternatively, the entire railing height may be executed as a glass structure (ca. 5 cm), depending on the costs or the decision to use the barrier for advertising purposes.

Scenario 3: 80-cm barrier

The lower railing height results in a widening to $w = 20$ cm. A lower railing also affects the viewing quality around box tiers, etc. (see above). To counter this problem, the barrier has been slanted for the railing model employed in Munich stadium.

Scenario 4: 70-cm barrier

A further widening to 50 cm is rarely employed for even lower barriers, as misuse for storage purposes can hardly be avoided.

Sight restrictions

In the *Guide to Safety at Sports Grounds* (chapter 11.4) sight restrictions are divided into two general categories:

Partial restriction
This applies to seats from which the view may be restricted, for example by a roof support, but not to the extent that spectators have to strain or are encouraged to stand in order to get an improved view.

Serious restriction
This applies to seats from which the view is sufficiently restricted, for example by inadequate sightlines or advertising hoardings, to encourage spectators to stand.

For new plans serious sight restrictions are to be urgently avoided.

Sight obstructions generally affect the security situation of a spectator stand, as dissatisfied visitors standing up or moving out creates a security risk, according to the Scottish Office Sports Council.

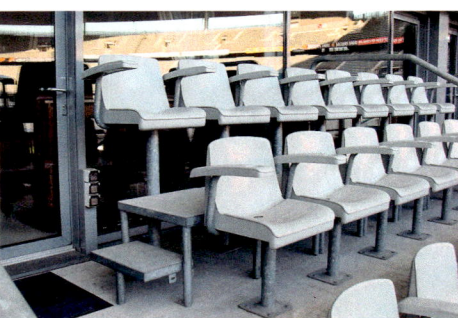

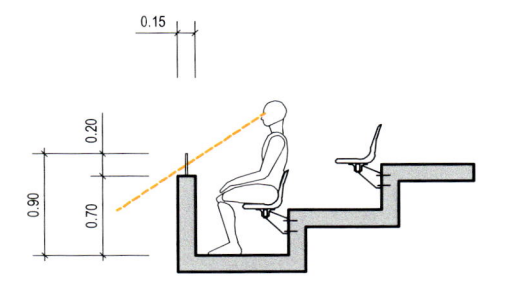

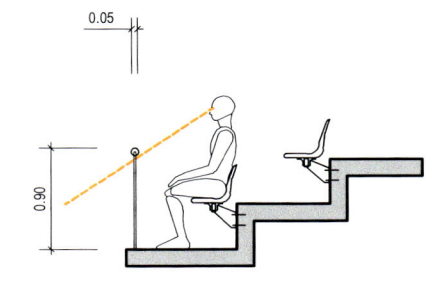

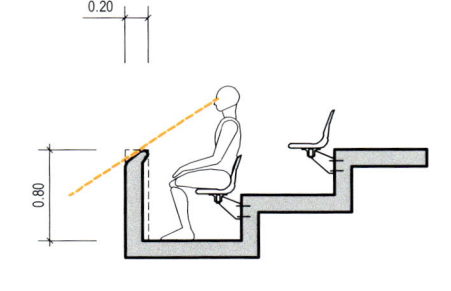

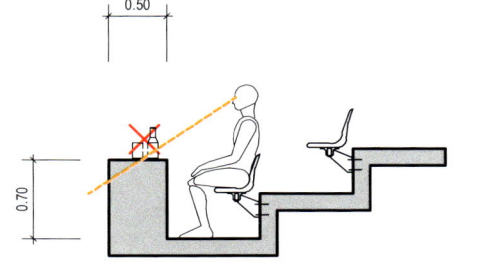

Barrier types

Walkable areas directly bordering subjacent areas are to be protected by barriers, unless they are connected by gangways.[17]

Exceptions to the above mentioned:
– Stage surface toward auditorium
– Stair rows not higher than 50 cm
– Height differences not above 50 cm.

These areas require a reinforcement barrier which projects above the floor of the row behind by at least 65 cm. The relevant EN/DIN leaves the decision on safety barriers to the planners and their evaluation of uses for the respective area.[18] It also limits the differences in height and requires a barrier for heights above 50 cm. At the same time and under certain circumstances, it might also be needed for smaller height differences.

Following the guidelines for standing accommodation, varying from at least 10 cm (MVStättV) to a maximum of 20 cm (DIN 13200-1), barriers will have to be supplied for height differences of over 20 cm. Planning should aim to minimize the risk of people falling, rolling, sliding or skidding through openings in the barrier system.[19]

The upper part of barriers should be so executed as to prevent it from being used for storing items which, if they fall down, might endanger other spectators. If small children are likely to be present, barriers must be designed so as to prevent climbing over and with a distance between barrier elements of not more than 12 cm.

photos:
055 **Glass barrier frame (box tier)**
World Cup Stadium Cologne
056 **Full-glass railing (box area)**
World Cup Stadium Dortmund
057 **Solid barrier (80 / 20 cm)**
World Cup Stadium Gelsenkirchen
058 **Business seats (box area)**
World Cup Stadium, St Denis, Paris

types:
059 **Cologne (h = 90 cm)**
With 20-cm glass barrier (alternative handrail)
060 **Dortmund (h = 90 cm)**
Full glass barrier with handrail (2006)
061 **Munich (h = 80 cm)**
Solid railing with slanted section
062 **Gelsenkirchen (h = 80 cm)**
Solid railing without slanted section

Seating types

Generally, two types of spectator seats may be distinguished: standard seats and business seats. Standard seats are ergonomically formed plastic shells, which are, depending on the supplier's system, fixed to the tread as single seats or in groups of 3 to 10 seats. These are positioned on a mutual traverse to avoid multiple tie points and allow fast dismantling.

The tip-up seat is a widely used variant of the single seat. Approaching and leaving the seat is convenient, as its low height (20–30 cm) in its tipped-up state permits a wider seatway (>40 cm). Tread widths, which can be limited by fittings like safety barriers, could easily provide a minimum clearway of 40 cm. The tip-up mechanism is generally considered higher-maintenance, and investment costs are roughly double.

Business seats belong to the most comfortable seating options in the stadium. They are generally only offered in VIP areas and in box galleries. Armchair-like seats, comparable to cinema seats, are heavily upholstered and tip up due to higher space demands. Armrests provide the comfort of high-priced seating in business areas.

Tread width corresponds to a seating grid of 60 x 90 cm instead of the normal 50 x 80 cm. A comfortable tread depth lies between 90 and 105 cm.

The business or VIP area is one of the main sources of revenue for stadium operators. Therefore, stadia of the third generation focus strongly on the development of private business areas, which can be a strong image factor for the venue. The proliferation of information technology in future will most likely also fit out single seats with more options for digital and media communication.

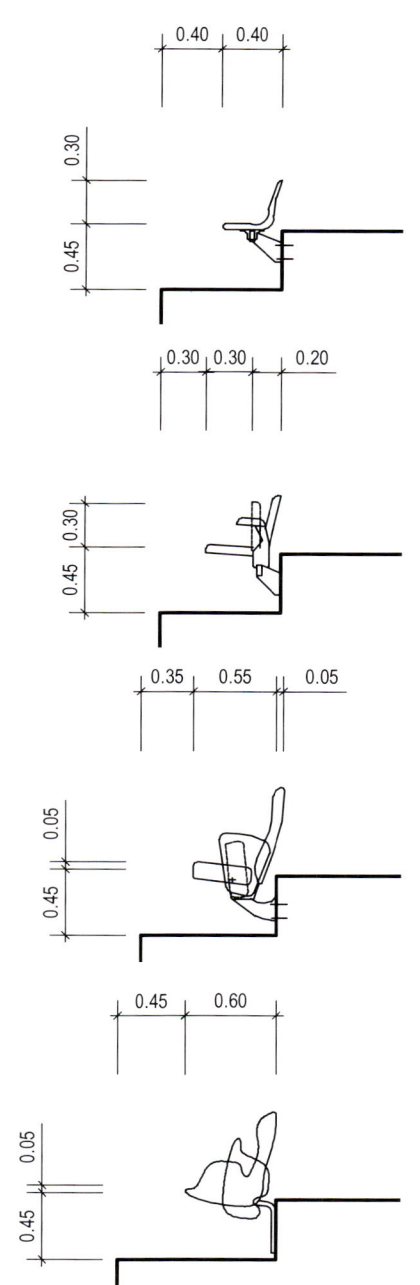

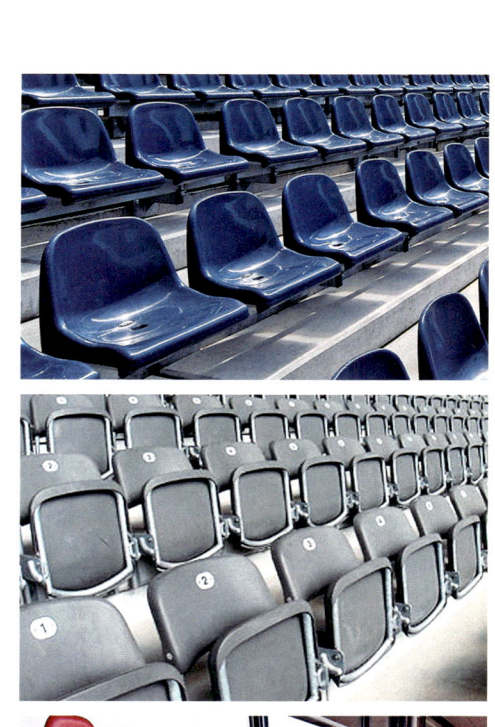

seats:
063 **Single-seat shell**
064 **Single tip-up seat**
065 **Business seat (Cologne)**
066 **Business seat (Munich)**

photos:
067 **Single-seat shell, lower tier**
World Cup Stadium Frankfurt
068 **Tip-up seats, lower tier**
World Cup Stadium Berlin
069 **Box seats, VIP area**
World Cup Stadium Stuttgart
070 **Business seats, VIP area**
World Cup Stadium Munich

Safety barriers for seats

A safety barrier at the step front edges becomes mandatory, if riser height exceeds 50 cm.[21] As a substitute measure for a general safety barrier of 0,90 m or 1,10 m in the seating area, it would be sufficient 'if the backrests of the seats of the seating row in front jut out at least 65 cm from the floor of the stair row behind.'[22]

Construction thickness should be minimized and subtracted from the clear width of a stair row.

According to EN/DIN 13200-1, total tread depth should be at least 80 cm:
40 cm seat + 35 cm minimal clear width = 0,75 m plus ca. 5 cm thickness of safety barrier and fixings = 80 cm.

Note: According to the binding legal regulation, a clear width of at least 0,40 m needs to be maintained.[23]

Note: In the framework of the *Stadium Atlas*, the German regulations on places of assembly (MVStättV 2005) are deemed binding, as they are expected to be accepted as a general standard by all federal states of Germany in the future.

Solution 1
To maintain a clear width of 40 cm, tip-up seats are installed, which in turn results in higher investment costs for the seating area.

Solution 2
The guideline requiring a height of 80 cm + safety barrier is changed in favour of a minimal clearway of just 0,35 m.[24]

Solution 3
The stand structure is to be expanded by the measure of the safety barrier, which would result in a widening of the upper tier by around 6% (5/80 cm) and seems therefore rather uneconomical.

Solution 4
The recommended gross measure of the actual seat depth (40 cm) is given up. The so-called 'gluteal-popliteal space' does not involve further provisions for the front edge of the seat. This measure should be undertaken only in exceptional cases and in agreement with the client. Shorter seat bases need to be tested long-term and testing should in each case be carried out at least for the duration of a regular match.

074 **Minimum clearway**
Sectional drawing of the seating area with safety barrier (compare to World Cup Stadium Cologne)

071 **Protection railing, upper tier**
World Cup Stadium Cologne
072 **Single-seat safety barrier**
World Cup Stadium Kaiserslautern
073 **Elevation of backrest shell (65 cm)**
World Cup Stadium Leipzig

Safety barriers, railings[20]

generally	1.10 m
in front of seating rows	0.90 m
or	0.80 m for a 20-cm width
or	0.70 m for a 50-cm width
or with safety barrier height less than 1.0 m	
	0.65 m (protection)

Standing accommodation

In order to prevent people from moving in and out, standing terraces need to be fitted with appropriate subdivisions. Standing accommodation is to be arranged in blocks of max. 2,500 visitors and subdivided by barriers of at least 2.20 m with dedicated gates.[25]

Block capacities can thus be precisely estimated and controlled. Each block has a dedicated entrance and at least two exits. Flooring should be skid-proof and misuse of any parts of the structure as missiles should be prevented.

According to DFB guidelines for the improvement of security at Bundesliga matches, 'convertible' places should be installed to allow adaptation into seating accommodation.

According to FIFA, standing accommodation should be offered under no circumstances, with international games taking place exclusively in all-seater-stadia.

There are no standing accommodations in the first or premier divisions / leagues in Italy and Britain.[26]

Dimensions for standing accommodation:[27]
Tread depth: at least 35 cm
Recommended: at least 40 cm
Standing place width: at least 50 cm
Spectator ramp, recommended slope: 10% (6°)

Types of standing accommodations
1) Free layout with crush barriers
2) 'Vario-seat' – tip-up seat stirrup ($h = 1,10$ m) in every other row or tip-up seat

Resulting from the above measures is: 50 cm width x 40 cm depth = max. 5 places per m² (excluding circulation areas). This correlates in calculation with the value of two persons per running metre of stair row as stated in the regulation for places of assembly.[28] DIN 13200-1 (7.1) calculates a reduction factor of 6% for the capacity of standing spaces.

The density of seated spectators may not exceed 47 persons per 10 m² (excluding passageways).

The reduction is not further justified, but reappears in the *Guide to Safety*. The Scottish Office apparently assumes a 60-cm grid and relates it to a minimum tread depth of:
35 cm = 0,21 m² → 4,7 places per m² instead of 50-cm grid x 40 cm tread depth = 0,20 m² → 5 places per m² → (−6%).

According to *Green Guide*: gross max. 40 persons / 10 m². Single space at a maximum distance of 12 m from gangway.

Note: In order to fill the space between the first two crush barriers in front of a stage in a practicable manner, in individual cases it has become common approval practice to allow 3 persons per m² within the first two blocks in front of the stage, insofar as the overall capacity of the auditorium does not exceed 2 persons / m² in return. In the wake of each approval plan (request for utilization) consultation with the relevant inspection departments is advisable.[29]

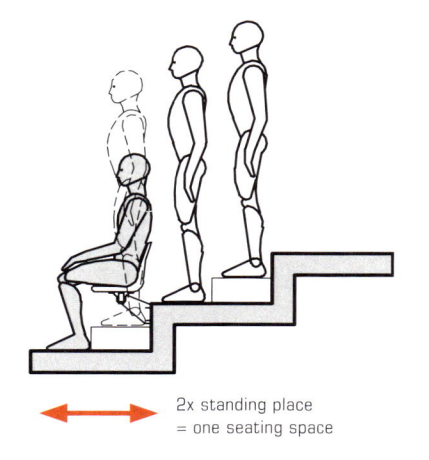

075 **Conversion of standing into seating accommodation**
System sketch

076 **Vario tip-up seat (standing accommodation)**
Müngersdorfer Stadion, Cologne (old)
077 **Vario tip-up seat (standing accommodation)**
World Cup Stadium Hamburg

Layout standing places

We may assume that a staggered place layout (as in theatres) will become the norm for standing accommodations. As there is no single space identification for standing terraces, spectators within their respective block tend to search for an ideal sightline in between the two persons in front.
Within a standing tier, due to the mobility of the masses, particular attention needs to be paid on keeping gangways clear. If need be, stewards should monitor and guarantee this.

Radial gangways in standing areas

A closer consideration of the construction principle of standing accommodation makes sense at this point: EN/DIN 13200-1 allows a maximum slope of 20 cm for standing accommodation, a measure also applicable to normal seating areas under certain conditions.

Standing areas are generally located at the lower tier of a stadium, with home supporters placed at the short sides behind the goal and guest supporters opposite in one of the corner areas. The lower-tier position is not only determined by the proximity to the pitch, but first and foremost by the fact that the lower tier is more gently inclined than the upper tier. Standing terraces are fitted with crush barriers to avoid dangerous movements of surging masses; the lower the tier, the lower the risk of spectators losing their balance.

In EN/DIN 13200-1, standing place depth is recommended at 40 cm (35 cm minimum), i.e. 80 cm standard tread depth divided by two.

Applying this double step concept to the gangway would require a third gangway riser (above 2 x 19 cm = 38 cm) up to the full riser height of 40 cm, for example.

Between gangway and standing area, a step interference arises above a riser height of 38 cm. In this case, the general safety target for evacuation of spectator facilities can obviously not be accomplished, in spite of observing regulations.

Caused by the low height differences at the transition of double standing step to gangway 'triple', standing spectators would be in danger of tripping.

Verification: 40 cm / 3 = 13.5 cm
40 cm / 2 = 20.0 cm

For this reason, as measure for gangways in standing terraces, a double step can be determined as general adaptation for the whole standing accommodation, in agreement with local building permission authorities. In this case, the gangway slope corresponds to the general standing accommodation slope of maximum 2 x 20 cm.

This optimizing measure becomes essential for another reason: a standing/seating row should be accessible from a gangway on the same level, i.e. at least one tread surface should blend into the row.[30] For geometrical reasons this would be impossible in the case of the second standing step (cf. drawing).

Structural advantages of a 'double' step

Standing accommodation is often located next to regular lower-tier seating areas. In contrast to its Roman predecessors no standing places are provided on the upper tier of a modern stadium due to its steeper terrace slope.

The option of converting standing into seating accommodation is required for modern stadia as international games may only take place in all-seater stadia and operators would be hard pressed to relinquish the capacities of standing areas.

Height differences and the tread width of a seat are halved by so-called block steps; the advantage being that areas with standing accommodation may blend smoothly into seating areas.

078 **Standing accommodation World Cup Stadium Nuremberg**
079 **Standing accommodation at AOL Arena, Hamburg**

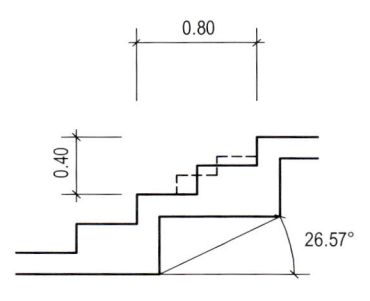

080 **Standing accommodation**
Standing steps (example) 20 on 40 cm
Terrace / gangway
2 risers 20.0 cm x 40.0 cm 'double step'
3 risers 13.3 cm x 26.6 cm 'triple step'

Crush barriers

Evacuation and rescue routes in a standing block are to be kept clear, as are all other gangways. Standing spectator groups are less stable in their position compared to the specified grids for fixed seating places, thus the relevant regulations require:[31]

Height, continuous in front of first row:
$h = 1.10$ m
Distance – successive: after five steps
Distance – sideways: < 5.0 m
Length: at least 3.0 m to 5.50 m
Covering: at least 0.25 m
Position: around front edge of row

Note: The formulation would be clearer with the following requirement: after a maximum of 10 standing rows a crush barrier is fitted, projected after no less than 5 standing rows by the next crush barrier by at least 25 cm.

§ 7 FIFA: Safety Guidelines: Crush barriers must be fitted in stadia with standing accommodation.
§ 9 DFB: Crush barriers in standing areas must be fitted and checked annually.

FIFA Technical Recommendations (2006) require that built-in units such as crush barriers, block fences in standing areas and pitch perimeter fences should be dismountable.[32]

Horizontal loads imposed on barriers

Barriers in areas that are accessible to visitors must withstand the pressure exerted by large groups of people.[33] DIN 13200-3 lists details on the horizontal design loads.[34]

Depending on the slope of the stand, this table assumes a recommended horizontal load of 3.0 to 5.0 kN/m length for barriers which are set up rectangular to the spectator surge.

For international football matches, FIFA/UEFA allow only seating accommodation within the auditorium. The conversion of standing into seating places makes sense for a better exploitation of the stands. At the same time, the conversion efforts should be economically tenable and should be able to be carried out without the use of machinery. Structures withstanding more than 2.0 kN/m are more costly and unwieldy.

Note: DIN proposal 13200-3 (2005) still has draft status. Load assumptions of up to 5 kN/m are stated in table A2. A relevant appeal has been submitted to the standard committee and was being examined at the time of writing.

081 **Conversion of standing into seating accommodation (e. g. for international matches)**
2 x standing to 1 x seating place (in general)
Exception: World Cup Stadium Munich

Attention: Two standing steps share an 80-cm seating step at a grid width of 50 cm. When converting standing into seating places, no further alteration measures are needed. Crush barrier seats or Vario seats simply fold out a seating base. The mechanism is very high-maintenance and must be secured for standing places.

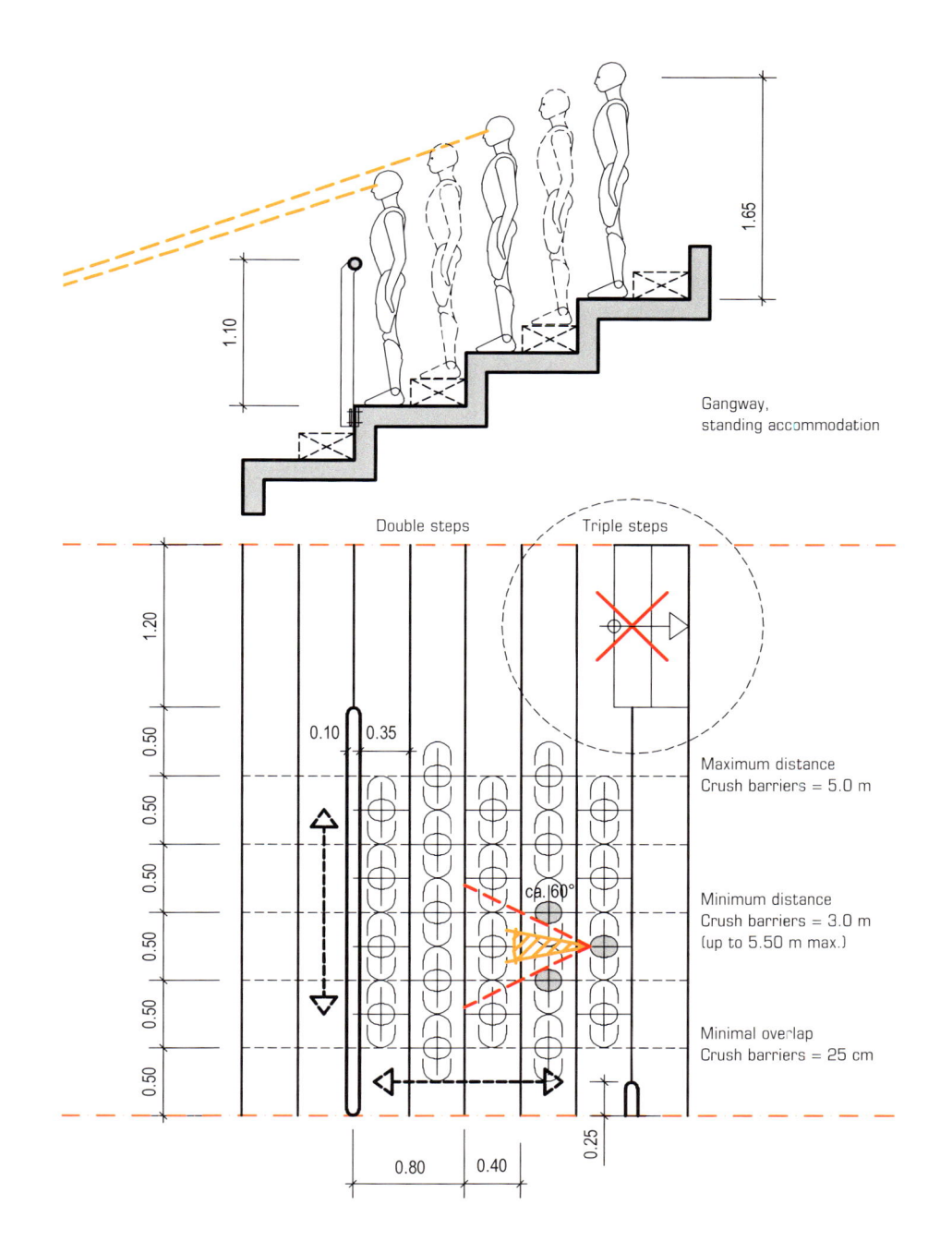

Places for wheelchair users

In January 2006, the Deutsches Institut für Normung [DIN; German Institute for Standards] issued a draft for the new standard DIN 18030 which unites and updates the present regulations on barrier-free construction for residences (DIN 18025) and for the public realm (DIN 18024).

'A barrier-free environment must be designed so as to provide all users, in particular persons with sensory disabilities, with orientation and communication means and to enable a safe and independent use. All information on routing, access (e.g. internal communication systems, elevators) and safety (e.g. emergency or warning systems, escape routes) must be discernible also by persons with sensory disabilities; they should therefore address at least two of the senses of hearing, seeing or tasting.'[35]

The provisions of this standard are designed to ensure the self-determined and independent usability of buildings and other structures. They are valid for new builds and in a more general sense for conversions, modernizations and alterations of use. The term 'barrier-free' is used for buildings which conform to the DIN standard. The respective building regulations of the German federal states still apply and also need to be considered.

Accessibility

Unless for evident and compelling reasons, access to public edifices and complexes as well as main entrances must be barrier-free.[36]

Furthermore, DIN 13200-1 (5.4) demands at least one area to be accessible barrier-free for disabled persons. Such spaces are clearly marked on the ground.

Access routes from parking areas should not intersect main circulation routes and must lead without detours to a dedicated wheelchair access. Routes to wheelchair areas are to be dimensioned broad enough, in order to avoid dangerous jostling around the terraces and to avoid wheelchairs constituting obstacles for other spectators in case of swift emergency evacuation.

Outfit

All places should have a good view of the pitch without obstructions caused by flags or spectators jumping to their feet and should not be set up in the central area (FIFA).

All seats for visually-impaired and mobility-impaired spectators and their helpers are to be weather-protected. Kiosks and a sufficient number of disabled toilets should be located within a short radius (one disabled toilet for ten wheelchair places).

Accompanying persons

One seat is to be planned in the direct vicinity of the wheelchair place, in order to allow helpers to look after this spectator during a match, depending on the severity of the disability.[37] This seat must be provided next to the wheelchair user.[38]

Connecting securely fitted escort seats alternately with one wheelchair place is also recommended.

The alternative position in a previous or following row should guarantee quick access to the disabled person. Fixed tip-up seats directly behind the wheelchair lots would be conceivable; at the same time, the viewing conditions for helpers would be impaired, and standing escorts would be in the field of vision of spectators seated behind them. As stadium seating must be permanently fixed to the structure, 'mobile' seats are not permitted. During an emergency evacuation the relevant persons must be able to use these wheelchair areas as escape routes.

Care for mobility-impaired spectators takes place on site, so planners need to ensure that these spaces are not used as short cuts to the stand by visitors without a mobility handicap. Otherwise, stewards should keep these areas clear. Some clubs support the viewpoint that only central and compact allocation of places for persons with disabilities guarantees satisfactory assistance by voluntary staff.

Technical support

At least two radio transmission systems should be available to visually impaired spectators. Based on previous experiences, a relevant number is to be determined by the clubs and stadium operators. Commentators describe the match via headsets to the above mentioned persons. 'Service counters with closed glazing and duplex systems are to be fitted additionally with an inductive hearing system. To be wheelchair-suitable, the height of the service counter should be 85 cm.'[39]

082 **Stadium interior, Müngersdorfer Stadion, Cologne (old)**
Conventional placement of wheelchair users behind advertising hoarding (nowadays unacceptable by FIFA)

083 **Logo of BBAG**
(BundesBehindertenfanArbeitsGemeinschaft: Working Group of German Disabled Supporters)

Location and number of places

In this regard, experiences with built examples and the wish of architects to plan according to principles of integration sometimes are incompatible.

Wheelchair users should feel as a natural part of the spectator community, but the geometrical conditions of a stepped terrace layout makes it almost impossible to provide equal set-up space for all areas. This applies in particular to upper-tier places. The barrier-free (stepless) accessibility of all essential building parts remains one of the foundations of planning. At the same time, a stadium-wide equal distribution of places for wheelchair users and non-disabled spectators seems impossible.

A even distribution of disabled toilets on all levels is to be considered; to cut distances it would be advisable to provide these close to the areas designated for wheelchair places. Naturally, a segregation of toilet facilities is not intended to exclude wheelchair users but to serve their individual convenience, as toilet facilities are frequented intensely during intervals. It would be generally recommendable, though, to coordinate planning in advance with associations and interest groups.

Minimum figures

At least one percent of places, or a minimum of two places, must be suitable for the disabled.[40] This requirement apparently applies to places of assembly intended for less than 500 persons.

The stadia of World Cup 2006 accommodate on average around 50,000 spectators; the FIFA requirement of at least 50 wheelchair places roughly corresponds to the German regulations on places of assembly.[41] The UEFA Stadia List (item 15) complements the number of spectator places suitable for the disabled to 50 each in at least two

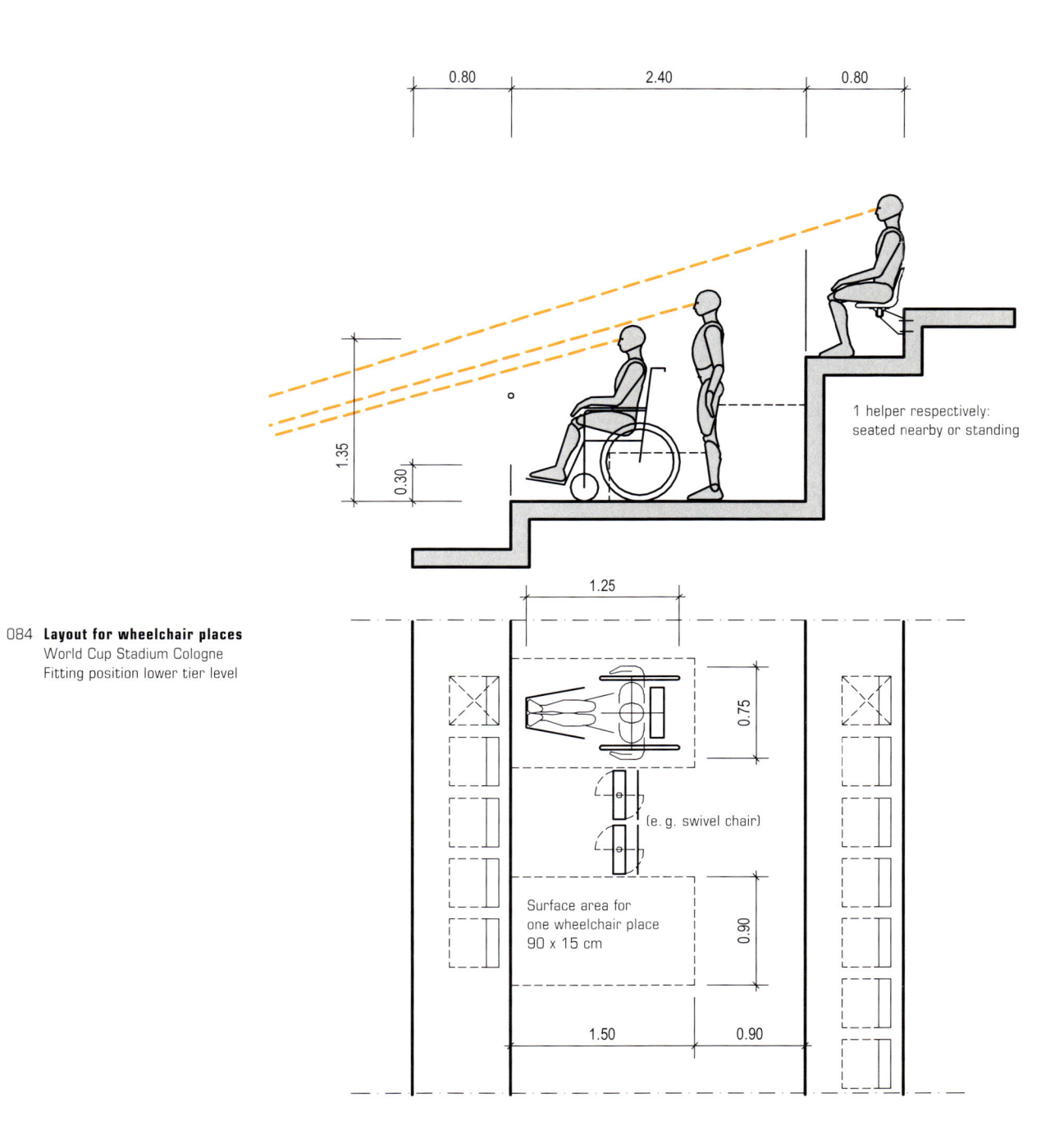

084 **Layout for wheelchair places**
World Cup Stadium Cologne
Fitting position lower tier level

085 World Cup Stadium Munich
Position at the top of lower tier

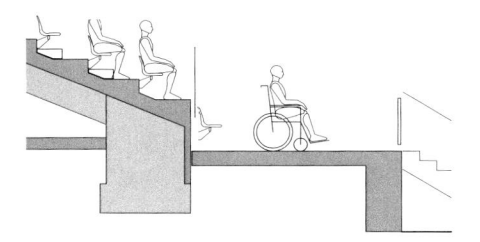

086 WM Stadium Nuremberg
Position at the bottom of lower tier

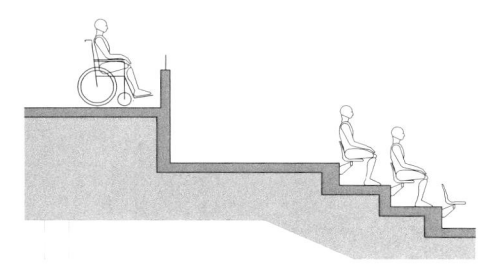

087 World Cup Stadium Frankfurt
Position at the lower tier gallery

sectors. Therefore, the quota of 0.5 % required by DFB should be changed to 1 % (status 2004).

Two different types of placements for wheelchairs are distinguished:
a) at the upper / lower end of a stand
b) on a separately fitted area within the stand.

Regarding item a):

Depending on the height level of the whole stand, wheelchair areas are located at ground level and are accessible from the main circulation area.

In World Cup Stadium Munich, these places are at the upper end of the lower tier (for 200 disabled persons), directly connected to the generally accessible circulation area of the 'grand promenade'. A wheelchair place alternates with a seat for the assistant. The raised seat position of the disabled person and the larger set-up dimensions do not restrict viewing for the person sitting behind.

In World Cup Stadium Nuremberg, wheelchair places for 83 disabled persons are located underneath the main stand (circa 1.40 m at the upper edge of the finished floor level above the terrain upper edge).

The height offset to the stadium visitors sitting behind is big enough to guarantee an unrestricted view over and above the wheelchair and the helper behind.

Regarding item b):

The set-up space is integrated into the sightline profile of the whole grandstand.

In World Cup Stadium Frankfurt, the 120 places for wheelchairs are situated on a separate gallery on the upper concourse, behind a lateral gangway in a raised position. The total height is achieved gradually via walls or ramps.

In World Cup Stadium Cologne, 100 places are located within the lower tier east, accessible at ground level directly from the lower concourse. The area of movement is reserved for wheelchair users and their escorts who are seated in the 'indirect' vicinity.

Note: Sightline geometry must take into account if a wheelchair position with a dimension of 3 x 80 cm = 2.40 m is located in the front of this 'triple step', as the seating base of an electronic wheelchair with 50 to 55 cm is much higher than the standard seating height of 45 cm.

Space requirements for wheelchair places

The currently valid DIN 18024 on barrier-free construction requires the following dimensions for wheelchair places (status 2005):
Set-up width: 0.95 m (FIFA 0.90 m)
Depth: 1.50 m (FIFA 1.40 m)
Clearway: 0.90 m
FIFA rear wall: 0.60 m (minimum)

The size of a standard wheelchair is:
Length: 1.25 m
Width: 0.75 m

The space for movement generally 1.50 m x 1.50 m; they may overlap but not be restricted in their function, provided all other requirements are being met. The measure for clearway, e.g. for doors, is demanded at just 0.90 m.

Ramps (without transverse slope) have a 3–6% longitudinal gradient and are to be interrupted by a 1.50-m deep platform after not more than 6.0 m. Ramps are at least 1.20 m wide. Ancillary routes leading to barrier-free sports/recreational equipment and activity areas must have a clearance of 0.90 m width and 2.30 m height. For a standard tread depth of 80 cm the set-up space is 3 x 0.80 m = 2.40 m. Adding clearway and set-up space amounts to 1.50 m + 0.90 m = 2.40 m.

As circulation surface, the step width sometimes borders on a seating step below. In this case, a wheel stop of 30 cm height is to be fitted and a guard rail must be provided if height difference exceeds 50 cm.

In the course of final planning it should be considered that the above mentioned net width would be insufficient if the passageway is not to be reduced. Due to wheelchair foot rests, an impact protection and front edge nosing must be provided toward the spectator row in front.

Summary

Varying degrees of disability result in individual requirement profiles for each disabled person, also pertaining to the equipment of the wheelchair and affecting the space dimensions of the appliance. Therefore, only a standard measure can be stipulated. (The seating base height of wheelchairs is quoted with 50–55 cm by manufacturers; 55 + 80 cm = 1.35 cm.)

Eye-point height: 1.35 m (new)
Eye-point ground plan: 0.90 m (new), from front-edge position

Emergency evacuation routes

Additional measures recommended by DIN 18030: Installation of fire-proof areas as temporary route stops for the severely mobility impaired; provision of light signal devices for the deaf or hearing impaired; acoustic information for the blind or severely visually impaired.

Wheelchair space

Width 0.90 m (new)
Length 1.50 m (new)
Clearway 0.90 m
Eye-point height 1.35 m (new)

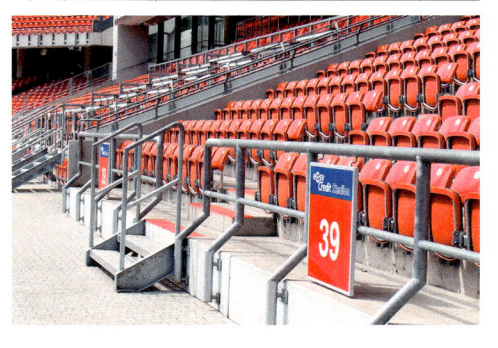

088 **Wheelchair places at the 'grand promenade'**
World Cup Stadium Munich
089 **Wheelchair places at circulation level, lower tier**
World Cup Stadium Berlin
090 **Wheelchair places at lower level, lower tier**
World Cup Stadium Nuremberg

Chapter 05

091 **Entrance VIP area**
Veltins Arena, Gelsenkirchen

092 **Glass bridge leading to executive area**
Gottlieb Daimler Stadium, Stuttgart

VIP areas & corporate hospitality
Different concepts of hospitality and the special function of executive lounges and boxes

VIP lounges

In stadium structures across the centuries there were separate seating areas reserved for dignitaries. Even in the ancient amphitheatres, segregation and a privileged position were imperative. Areas for the 'very important people' in the antique stadium are known today as executive or VIP areas. During the last few decades these types of facilities have evolved greatly and take up many of the construction activities in third-generation stadia. Within the framework of changing utilizations – from mere sports complex to modern multifunctional event venue – private hospitality has become the most profitable source of revenue. Growing commercialization of sports and other events has transformed the previously segregated areas for players and officials into upmarket catering and conference spaces.

Two types of hospitality areas may be distinguished: *lounges* are capable of accommodating a large number of people, whereas *boxes* are normally separate single rooms, generally with direct access to the stadium interior.

The term 'lounge'

Usually, a recreation or waiting room in train stations, airports or hotels is defined by this term. A lounge tends to be spacious and high-quality in design and is mostly fitted with comfortable seating furniture instead of chairs. Frequently, the atmosphere of these premises is characterized by soft background music and dimmed lights. Drinks and occasionally light and simple meals are served. The term 'business lounge' signifies the provision of work facilities, such as internet access for notebooks.[1] Access to a lounge is normally subject to certain conditions: visitors have to be actual guests of the hotel or, in the case of airline lounges, are either business or first-class ticket holders or have a certain customer status.

The term 'business lounge'

This term is predominantly used by business clients who are interested in sports to denote the special areas for honorary guests in sports stadia. These premises mainly serve to strengthen business relations by providing a kind of meeting point for companies to invite, get to know and entertain clients.

The requirement profiles for these areas are largely determined by the procedures during a match day. Customs and features may vary significantly from region to region, and different wishes and demands might be expressed by stadium operators, organizers or clubs.

As they enter the stadium, VIP guests normally pass 3 different stops:

The reception area

Access for honorary guests is mostly separate from public entrances, differing strongly in function and set-up from stadium to stadium. Sometimes separate entrance buildings are provided, for example a light-filled glass hall in the World Cup stadia of Gelsenkirchen or Nuremberg, or a 'Business Centre' (Stuttgart), which already implies the multiple utilization of those areas as conference centres.

In many cases the VIP entrance is also the representative address of the stadium, even outside of matches.

The reception of honorary guests always takes place in a separate zone which is connected to the box area. Depending on how VIPs reach the stadium, they enter through the reception hall or the foyer. From there, delegates should be able to directly and securely access the changing room area. Tunnels (Berlin) or bridge systems (Stuttgart) are provided to keep circulation free of intersections. Guests are greeted and/or accredited at the reception desk and, after checking their tickets, receive a VIP identification, e.g. a colour armband which relates visual information on their respective status.

The lounge

Executive guests then move on to the business lounge where food and drinks are served. Modern lounges generally accommodate several bars and buffets where guests may serve themselves freely. Depending on the catering concept, these areas are equipped with various standing or seating facilities.

If a sports stadium is dominated by a club, great value is attached to visual identification through its symbols and colours. Furnished as comfortable lounges or restaurants, these separate areas can be used and marketed on non-match days, and they should be partitionable so as to be capable of hosting smaller events. In the interest of profitability, private stadium operators are often obliged to enable multiple utilization of these high-quality premises, for example by making them available as conference facilities. Hence, a pleasant appearance and a distinctive appeal are essential for the financial viability of a stadium or arena.

During the planning of technical services great attention should be given to media equipment (screen, light and sound) to meet the demands of various types of events and thus guarantee the adaptability of premises. Access to ancillary areas has to be easy, in order to allow various events taking place at the same time; this also applies to accessibility and fitting-out of toilets, relay kitchens and other facilities.

Commentary on table A
The table relates capacity details by LOC World Cup 2006 (October 2005)
(*) 'Seats' denotes all available seats at FIFA World Cup 2006, calculated from gross capacity minus seats not available at the World Cup (due to viewing restrictions and safety reserves). This capacity is based on current estimates and may change after final definition of requirements.
(**) 'Available tickets' means all tickets minus those which are provided by the LOC to the media, the printing press and the VIP area, or free of charge. Capacity is based on estimates (maximum media demand) and may change after final definition of requirements. 2

093 **VIP driveway through March tunnel (Level – 2)**
Olympic Stadium, Berlin
094 **Business foyer (Level + 0) with two cantilevered stair flights and three glass elevators (grouped)**
RheinEnergieStadion, Cologne
095 **Table A: World Cup spectator capacity**
(column 2/5 according to LOC 2006, Frankfurt)

Table A World Cup Stadium Spectator Capacity	International (seats only)	Standing places	National (total) incl. seats	Seats available (net)	(*/**) Available in percent
01 Berlin	74,200	./.	74,200	63,400	85.4%
02 Dortmund	65,900	27,500	81,250	56,400	85.6%
03 Frankfurt	48,400	8,000	52,100	41,100	84.9%
04 Gelsenkirchen	53,600	16,200	61,500	46,700	87.1%
05 Hamburg	51,300	9,600	56,000	43,300	84.4%
06 Hanover	44,900	8,000	49,500	38,800	86.4%
07 Kaiserslautern	47,700	12,000	48,500	40,000	83.9%
08 Cologne	46,300	9,200	50,700	38,200	82.5%
09 Leipzig	44,300	./.	44,200	37,100	83.7%
10 Munich	65,600	20,000	66,000	55,800	85.1%
11 Nuremberg	43,800	7,800	46,800	39,300	89.7%
12 Stuttgart	53,100	3,500	56,000	45,400	85.5%
Total number	**639,100**	**121,800**	**686,750**	**545,500**	./.
Average number	53,258	12,180	57,229	45,458	85.4%

05 VIP areas & corporate hospitality

Executive tribune

Generally, the number of executive seats corresponds to the catering capacity of the respective lounges. Those comfortably upholstered seats with broad treads are located at the main stand with direct access to the lounge, which often has its own standing accommodation within the auditorium. Depending on the operating concept, tables in the 'business club' are assigned as personalized company tables for a whole season on a ticket quota. The executive stand is generally located at the halfway line on the main stand and in a heightened position to the pitch.

VIP areas need to be segregated from other areas in such a way so as to be accessible only via designated entrances.

According to FIFA, all executive seats must be covered, have an unrestricted view of the pitch and provide high seating comfort: well upholstered, with armrests and adequate room for movement within the seating rows. VIP areas should be extensible on demand.

The gross seating capacity of a stadium is reduced by the amount of seats which are not available for regular ticket sales because they are reserved for media representatives and VIPs, or because they are viewing-restricted or constitute safety reserves.

Contrary to international matches, stadia may offer standing places during the regular season, thereby raising the national total capacity by turning seating into standing places.

The twelve World Cup 2006 stadia have a gross capacity of around 640,000 seats, i.e. approx. 53,000 seats per stadium. Capacity ranges from 74,200 to 43,800 seats. The following figures relate to seats only:

1 x 74,000
2 x 65,000
3 x 50,000 to 55,000
6 x 45,000 to 50,000

Note: The available ticket quota or net capacity is around 15% lower.

096 **Two-storey hospitality box tier**
BayArena, Leverkusen
097 **Doubled hospitality box tier**
Veltins Arena, Gelsenkirchen
098 **Honorary tribune**
Ibrox Park Stadium, Glasgow, GB

More champions to look up to?

Interior ceilings made by OWA

The arena is associated with champions performing in the open air. However, modern stadiums offer much more behind the grandstands. They attract with business, congress and VIP areas, shops, restaurants, offices and leisure centers. Many rooms, many functions, many demands on the building finish, the room ceiling. OWA convinces with professional, integrated ceiling solutions. This is because OWA integrates all types of expertise into the construction of stadiums. Let the champions play when it comes to your stadium project: OWA for ceiling systems.

Odenwald Faserplattenwerk GmbH
Dr.-F.-A.-Freundt-Straße 3 · 63916 Amorbach · Germany · www.owa.de

OWA References

RheinEnergieStadion, Köln
Stade de France, Paris
Palau d'Esports Sant Jordi, Barcelona
Stadium Dnepropetrowsk, Ukraine
VW-Arena, Wolfsburg
New Soccer Stadium, Amsterdam

All you need for ceilings.

Executive capacities

The table on the right summarizes the capacities for executive hospitality at FIFA World Cup 2006 stadia in Germany: at a total of 31,500 places, this amounts to 2,625 seats on average per World Cup stadium.

Individual figures of 620 to 5,600 seats diverge significantly from the average value and thus serve to illustrate the different demand for special seats. The share of executive seats can be roughly calculated with 5% of overall spectator capacity; that of boxes with around 1%.

Seating for honorary guests should be within the 16-m penalty lines to provide these high-priced tickets with adequate viewing comfort.

Planning basis

FIFA Technical Recommendations state the minimum capacities for the VIP area:[3]
Qualifying round: 500 seats (1,000)
Quarter final: 800 seats (1,250)
Semi-final: 800 seats (1,500)
Final: 1,000 seats (2,500)

Based on a stronger interest for matches of the German national team during World Cup 2006, the required number was raised, which reflects the practice of all sports associations. UEFA expects first-class facilities for VIPs and stipulates the following minimum figures for UEFA Euro 2008:
Opening: 500 seats
Group and Quarter finals: 300 seats
Semi-final and final: 700 seats

UEFA Champions League

Adequate room for 200 people is to be reserved in the VIP box as well as an additional external area, the 'Champions Village', with catering facilities of commercial partners (around 4,000 guests for the final of UEFA Champions League alone).

Usually, the capacities of hospitality areas during the regular league are already adapted to the minimum values stated above; the demand of modern stadia at normal operation already lies within the capacities required by FIFA/UEFA.

Exemplary calculation for Cologne Stadium during Bundesliga:
Gross World Cup capacity: 46,500 seats
VIPs = 4.5%: 2,089 seats
Boxes = 1.3%: 596 seats
Gross Bundesliga capacity: 50,700 seats (incl. 9,200 standing places)
Number of boxes = 10%: 50

Note: The number of seats in or in front of each box is on average between 10 and 12.

099 **Table B: World Cup VIP capacity**
(column 3/4 acc. to LOC 2006, Frankfurt)

Table B World Cup Stadium VIP Capacity	Total VIP places	World Cup executive seats	Private boxes	World Cup-related seats, World Cup honorary tribune	World Cup private boxes	Foyer / entrance to business area Size in m²
01 Berlin	5,653	4,758	895	6.4%	1.2%	400
02 Dortmund	1,946	1,784	162	2.7%	0.2%	./.
03 Frankfurt	3,120	2,218	902	4.6%	1.9%	900
04 Gelsenkirchen	3,255	2,438	817	4.5%	1.5%	550
05 Hamburg	2,759	2,179	580	4.2%	1.1%	110
06 Hanover	1,551	1,241	310	2.8%	0.7%	140
07 Kaiserslautern	1,327	993	334	2.1%	0.7%	./.
08 Cologne	2,685	2,089	596	4.5%	1.3%	250
09 Leipzig	2,969	2,660	309	6.0%	0.7%	125
10 Munich	3,449	2,152	1,297	3.3%	2.0%	2 x 270
11 Nuremberg	621	441	180	1.0%	0.4%	190
12 Stuttgart	2,182	1,504	678	2.8%	1.3%	200
Total number	**31,517**	**24,457**	**7,060**	./.	./.	
Average number	2,626	2,038	588	3.7%	1.1%	

VIP boxes

The German word *Loge* is originally French and denotes 'closed room'. It derives from the Latin word *lobia*; the German etymological dictionary further lists as a derivation from *laubia* the term 'lobby', which as foyer (interest group) relates to the honorary area. *Loge* in the German language stands for an enclosed place or segregated auditorium in a theatre.

For the early stadia of modern times, such as the predecessor of the German Stadium in Berlin (1916) or White City Stadium in London (1908), 'multifunction' was not only nomenclature but represented a structural and organizational principle: various sports were being staged within this open-air space. In Berlin, a wooden cycle race track was integrated into the large circle of athletics tracks and one of the long stands accommodated a competition swimming pool. The VIP zone, including the emperor's box, was an integral part of the plan, just as the Panorama Lounge at White City Stadium in London.

Fitting-out stadia of the second generation (in Germany: the Olympic Games 1972 and World Cup 1974) was characterized by basic convenience and service for spectators. The centre of attention was placed on event-related technology (floodlights or scoreboard), whereas roofing was often only provided for VIP or media zones.

Sports grounds today display a significantly higher share of premises for VIPs: the term 'box' [*Loge*] is used mostly for small rentable rooms in VIP areas of sports and event complexes, usually a separate room of around 30 m² size with unrestricted viewing of the stadium bowl.

Fitting-out generally comprises a conference table with sideboard for food and drinks, a bar with cloak room and storage space. Depending on the demand for convenience, small WC cells are installed.

To fully appreciate the acoustic 'live' experience in the auditorium, seats are mostly situated directly in front of the box. Nonetheless, boxes are generally fitted with good soundproofing so that the interior room can be used for business purposes during the event. Depending on the operating concept, individual demands of tenants or clients are accommodated.

Clients have year-round right of use, and service and catering is offered also for conferences and meetings on request. Although special events are often exempted, clients generally have an option on the relevant box tickets. (Example Gelsenkirchen Arena: package price 3,000 Euro; 'Incentive-Lounge' for 20 people, 42 m² with seats and 10 parking spaces.[4]

Sports associations as organizers of large football tournaments or athletics competitions are concerned about the quality of spectator service. Demands on convenience and service have grown considerably in all sports.

100 **Panorama Lounge**
Olympic Stadium, London, 1908
101 **Private box outfit**
Olympic Stadium, Berlin, 2004
102 **VIP meeting point behind box units**
Olympic Stadium, Berlin

World Cup Hospitality Concept 2006

Regarding its utilization profile, a stadium of the third generation is not solely designed for sports or cultural purposes, but rather for maximized revenue.

FIFA exclusively authorized and commissioned one company with the conception, planning, implementation and marketing of the official hospitality programme for FIFA Football World Cup Germany 2006™. In a public invitation to tender in 2003, iSe Hospitality AG won the global exclusive rights to officially distribute hospitality packages, including on-site services for every match of the Football World Cup 2006 in Germany.

The commercial incentive programme *Hospitality* combined premium seats with privileged parking spaces, exceptional catering and personal service as well as small gifts and special entertainment. Packages were divided into four categories:[5]

Sky Box: The incentive package SKYBOX (VIP box) represented the most exclusive offer within the official hospitality programme. This package was designed particularly for business clients wishing to treat their guests in a personal setting with all conveniences.

Elite: The incentive programme ELITE was a single-seat package of the highest level. Only a limited number of premium seats were included in the ELITE programme, which was offered inclusive of all services for most stadia.

Prestige: The PRESTIGE programme as single-seat package provided premium seats, first-class personal assistance and services within or outside the stadium, depending on the respective venue.

Premier: The PREMIER package offer provided single premium seats and first-class services within or outside the stadium.

Note: A discussion of the categories described above makes sense for planning purposes as the relevant quality standards may result in varying demands for protection and segregation. Consequently, routes to the main stand and the VIP areas should be planned so as to avoid intersections between certain spectator groups.

Partitions such as visual boundaries or opaque mobile elements may be temporarily installed. More profound partitioning measures appear less practical because they would have to be adapted to local conditions and the special features of a tournament.

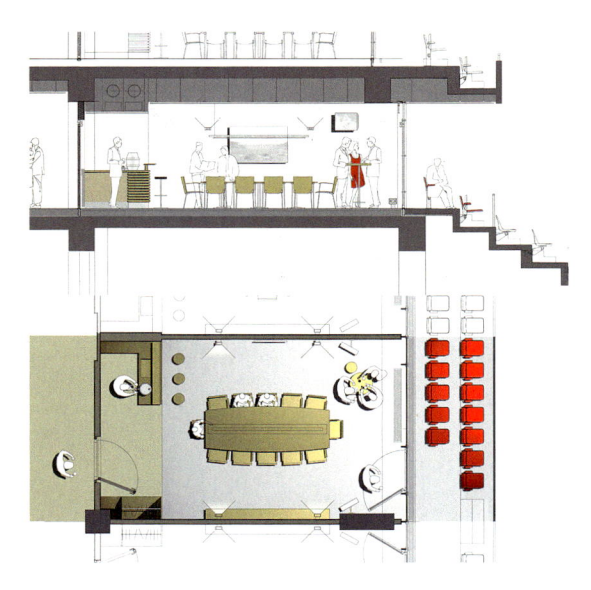

103 **Diagrammatic figure – Proposal for utilization of a VIP box in RheinEnergieStadion, Cologne**
(Sectional drawing / ground plan)

104 **Table C: Box spaces of World Cup stadia**
(Details relate to VIP space with access to stadium bowl and include no special rental space.)

Table C Private boxes	Orientation	Storeys	Location	Standard boxes	Special boxes	Size (m²)	Standard dimensions	Seats outside	Rows
01 Berlin	Circulating	1	L +0	74	13	20	5 x 4 m	10	2
02 Dortmund	East	1	L +4	15	./.	30	4.5 x 6.5 m	12	2
03 Frankfurt	East	2	L +3/+4	74	./.	35	4.5 x 7.5 m	12	2
04 Gelsenkirchen	West	2	L +1/+2	72	36	32	4 x 8 m	10	2
05 Hamburg	East	2	L +2/+3	50	./.	30	4 x 7.5 m	10	2
06 Hanover	East	1	L +2	39	./.	26	4 x 6.5 m	10	2
07 Kaiserslautern	Nord	3–5	L +2	28	./.	17	3 x 5.5 m	9	2
08 Cologne	West	2	L +2/+3	48	2	38	5 x 7.5 m	12	2
09 Leipzig	East	1	L +6	16	./.	35	4 x 8.5 m	18	3
10 Munich	Circulating	1	L +5	100	8	36	4 x 9 (13) m	12	2
11 Nuremberg	West	1	L +2	14	./.	18	3,5 x 5 m	9	2
12 Stuttgart	West	2	L +2/+3	44	4	24	4 x 6 (7) m	15	3
Average number		**2**	**L +2**	**48**	**13**	**28**		**12**	**2**
Minimum		1		14	2	17		9	
Maximum		5		100	36	38		18	

Concepts for private hospitality boxes

The number of boxes at the World Cup 2006 venues varies considerably from 14 up to 100 standard boxes. Demand often depends on the appeal of the playing teams and the operators' marketing skills.

To ensure good viewing, boxes should ideally not be placed behind the goal line; beyond the penalty area lines their marketing potential is reduced.

Assuming an average surface of 30 m² with a standard measure of ca. 4.25 x 7.0 m, only 24 boxes would fit successively in each level. With a preliminary calculation of 10 % of gross capacity, a 50,000-seater stadium can accommodate 50 boxes. This is why most stadia offer two box levels, which also has a significant logistic advantage as the premises may be serviced more easily from the main stand.

In case of heightened demand, other stands or the corner areas need to be connected to the logistics of the main stand. To ensure flexibility of a sports and event venue for future use, many new-build stadia are responding with a continuous 'box tier'. Thus, these high-quality box units can be extended, provided that logistic capacities can be adapted and service routes remain within a certain distance limit.

Allianz Arena in Munich, with a capacity of around 66,000 seats, accommodates over one hundred standard and special boxes – over 1.5 times as many boxes as in the preliminary calculation – which increases the already high marketing potential of this venue. At the same time, the special situation of two home clubs sharing the same stadium results, so to speak, in a doubling of main utilization areas and a much higher number of boxes. Similar to the Olympic Stadium in Berlin, a continuous private box tier is located above the middle tier.

RheinEnergieStadion in Cologne, however, has a two-level box tier in the west stand. Single-level areas north/east/south, which are reserved for special use (e.g. a panorama restaurant), are also linked to the auditorium and provide potential space for further boxes in the future.

Note: Whenever boxes are situated directly at the upper edge of the lower tier and are not separated from the lower tier by a height offset, the special situation of the lower-tier standing accommodations must be taken into account.

In this case, spectators might be standing before the glass front of the hospitality box tier, thus reducing visibility in the special rooms behind.

Fitting out boxes

As a rule, boxes are provided with standard fittings. Clients may, similar to regular rental agreements, realize individual wishes or adaptations (as in Cologne). Another option would be the provision of a high-end structure, finished in the client's corporate design (as in Gelsenkirchen).

Prices for a box in the former arena AufSchalke range between 48,500 to 74,000 € per year.

Spatial relation 'lounge/box'

The location of a box is determined in the planning or operating concept. As a rule, two types may be distinguished:
a) boxes which are directly connected to the rear of VIP-lounge areas
b) a separate VIP-box area which may be connected visually to the executive area via a gallery, but is otherwise operated and accessed completely independently.

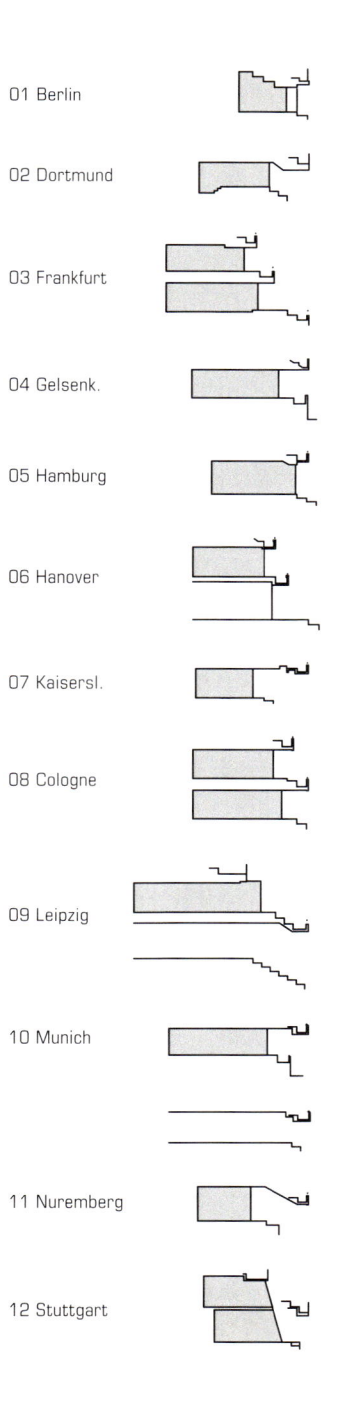

01 Berlin
02 Dortmund
03 Frankfurt
04 Gelsenk.
05 Hamburg
06 Hanover
07 Kaisersl.
08 Cologne
09 Leipzig
10 Munich
11 Nuremberg
12 Stuttgart

105 **Height profile and box tier variants**
World Cup 2006

106 **Prototype hospitality box**
Allianz Arena, Munich

107 **Hospitality box outfit (animation vcube, Aachen)**
RheinEnergieStadion, Cologne

108 **Furnishing option for hospitality box**
RheinEnergieStadion, Cologne

05 VIP areas & corporate hospitality

Regarding item a):
Functional layouts with direct connection are realized, for example, in the World Cup stadia of Cologne and Berlin.

In RheinEnergieStadion, Cologne, boxes are situated across two levels directly adjacent to the main business area. They are accessed directly via the 'FC Lounge' where all visitors of the VIP guest stand are welcomed and entertained.

Originally standard boxes, they were fitted out as conference boxes with seats placed directly in front of a fully glazed façade.

In the Olympic Stadium, Berlin, a continuous box ring between upper and lower tier is divided into segments containing several boxes by access tunnels leading regular supporters to the lower tier. These rampant ring segments are connected to VIP meeting points at the rear of the boxes.

Every group of boxes at the lower tier level +0 has five to seven units with a foyer each (around 150–250 m²). These open corridors are private meeting points with individual catering units. The main executive area is located at the main stand south, containing several special suites.

Regarding item b):
The concept of Veltins Arena in Gelsenkirchen was to realize two VIP box levels which are not directly linked to the main VIP lounge, the so-called 'La Ola Lounge'.

Guests in executive suites have no admission right to the lounge, up to a certain time after the match ends. A segregation of various VIP stands and catering zones is taken even further by providing 'event catering' in other special areas.

The Munich concept of a three-tier stadium enables a continuous box tier with separate VIP stand at the second tier. A corridor at the rear, circulating and generously spaced, serves as meeting point and recreational zone with a view of the stadium environs.

The planners of Fritz-Walter-Stadion in Kaiserslautern, as an example of an 'evolved' sports stadium, responded to a changed utilization by constructing further boxes within a multi-storey tower to the left and right of the main stand north. One wing became a media tower, the other a box tower, with boxes stacked over five levels.

109 **Panorama suite**
Allianz Arena, Munich
110 **Single-event hospitality box**
Commerzbank Arena, Frankfurt
111 **Conference space**
Frankenstadion, Nuremberg

page 61:
112 **System overview (lounge / box area)**
RheinEnergieStadion, Cologne

Concepts for private hospitality boxes 61

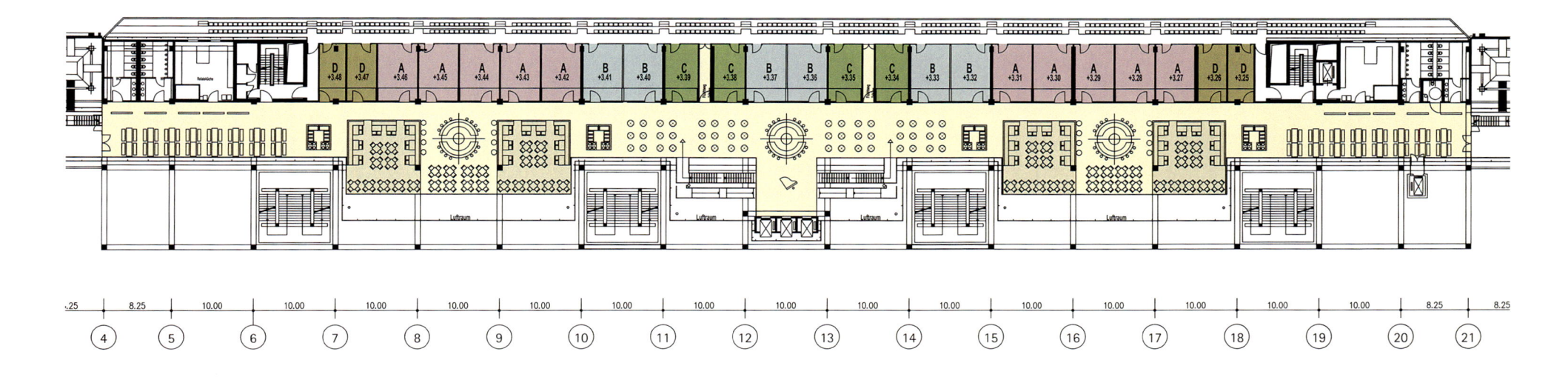

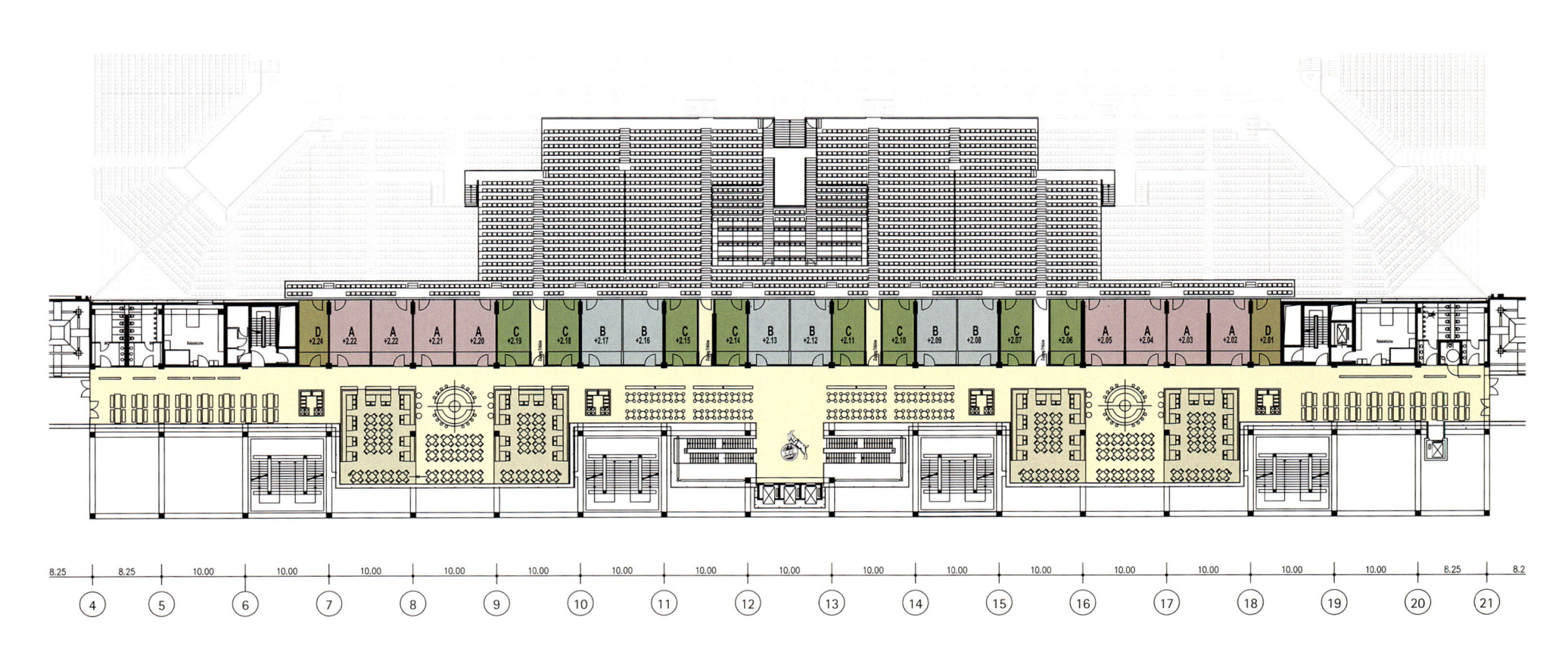

05 VIP areas & corporate hospitality

Definition of the term 'box tier'

Stands with large capacities are divided into several tiers. As a rule, we distinguish the classic 'earth-tier' stadium, which incorporates the natural or artificial topography into its construction plans.

Grandstands in built stadiums are constructed from the ground up, creating under-terrace areas which generally accommodate functional rooms. Incorporating the advantages of both systems, the combination stadium as building type was primarily realized in the two-tier stadia of the second generation during the 1970s.

Example:
Müngersdorfer Stadion, Cologne, 1975
World Cup Stadium Nuremberg, 1991

The geometric 'gap' created between upper and lower tier is situated on a circular artificial embankment with a framed-on terrace structure. It is used as main circulation area providing access to the lower tier from above and to the upper tier from central staircase vomitories. From a building-type viewpoint, the horizontal gap is a 'separation and access joint'.

Inserting a hospitality box tier into modern grandstands still serves the vertical separation of spectator blocks into two or three tiers, but in two-tier stadia this gap is taken over by boxes and special utilizations; mostly VIP boxes as part of the zones reserved for honorary guests.

The sketches on page 59 display diverse access systems. The decision on the number of tiers is based primarily on the planned holding capacity and utilizations.

Stadium planners have to decide on employing a height offset at the lower/central/upper tier for circulation purposes or for the provision of high-quality premises. However, a commercially successful utilization as hospitality boxes results in a pronounced height offset in elevational geometry. Depending on its dimensions, the successive upper tier will be characterized by a steeper slope.

Two parameters for the positioning of the box tier are paramount: the ground-plan position and clear room height (box offset).

Ground-plan position
The upmarket utilization units in this hospitality box tier should be within the 90-m best viewing radius, i.e. a maximum distance of 39 to 56 m from the touch line.

The further away from the pitch, the higher the hospitality box tier will be positioned.

Structural height beneath the box tier should match useful room height. The demand for clear height results from the maximum height of ancillary spaces located underneath the stand.

Heights
The German regulation on workplaces states on required room dimensions that rooms may only be used as working rooms if clear height is:[6]
for a surface area
below 50 m²: at least 2.50 m
of more than 50 m²: at least 2.75 m
of more than 100 m²: at least 3.00 m
of more than 2,000 m²: at least 3.50 m

---

113 **Wall circulation**
Müngersdorfer Stadion, Cologne (old)
114 **Two-tier stadium**
Frankenstadion, Nuremberg
115 **Grandstand with two box tiers**
RheinEnergieStadion, Cologne

Furthermore, the regulation on the construction and operation of garages states: 'Publicly accessible areas must have a clear height of at least 2,0 m, also underneath ventilation pipes, main beams and other building parts.'[7]

(Minimum headroom, according to *Guide to Safety* 5.5, is 2.0 m; if possible 2.4 m)

The World Cup Stadium Munich, with its 'grand promenade' forms a large gap toward the stadium interior with a height offset of 2.10 m. Here, the minimum value for the special scenario of cantilevered upper tier has most likely been approximated. This three-tier stadium uses the lower tier as free circulation area. The second circular box tier with around 3,500 executive seats is located at the upper end of the middle tier.

Minimum height (sight reference)

EN/DIN 13200-1: 'The height of roof coverings or overhanging structures such as balconies should be as far as possible designed so that the view of the centre and its height is unobstructed for all rows of spectator Viewing Area. Note: The information board or panel should be seen without obstructions from all places. For example in football facilities, the height from the centre of the playing field, is 15 m (for indoor international football competitions is 20 m) and the height at all points of the Activity Area is 7,5 m.'[8]

With a cantilever of 5.35 m (front edge of railing) up to the box façade and a clear height offset of just 1.50 m, the World Cup Stadium Nuremberg, exploits limit values. A person standing in the private box would be unable to view freely ahead (eye-point height = 1.65 m).

The corresponding sightline was constructed for seated spectators in the stand. The depth of the box makes it impossible to apply the sightline profile to the seating/standing places in the lounge interior. Standing guests should at any rate be able to view the 15-m point of reference above the centre of the pitch.

Height offset proposal for a standard box (< 50 m²):
2.50 m (height)
2 rows (offset)

Note: Each height offset has a negative effect on the slope of the adjacent terrace. Therefore, the advantages of a projecting upper tier should be examined closely as the inclination of the stand would have to be increased; a typical result for any decrease in distance to the pitch.

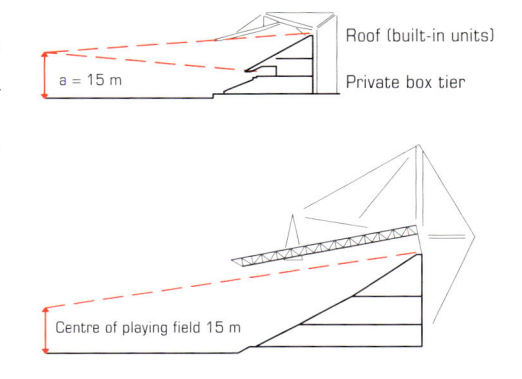

116 **Minimum height (sight reference), drawing**
Roof/terrace offset

117 **Private box seats (outside)**
Frankenstadion, Nuremberg

118 **Standard box tier, drawing**
with continuous 2.50 m gap

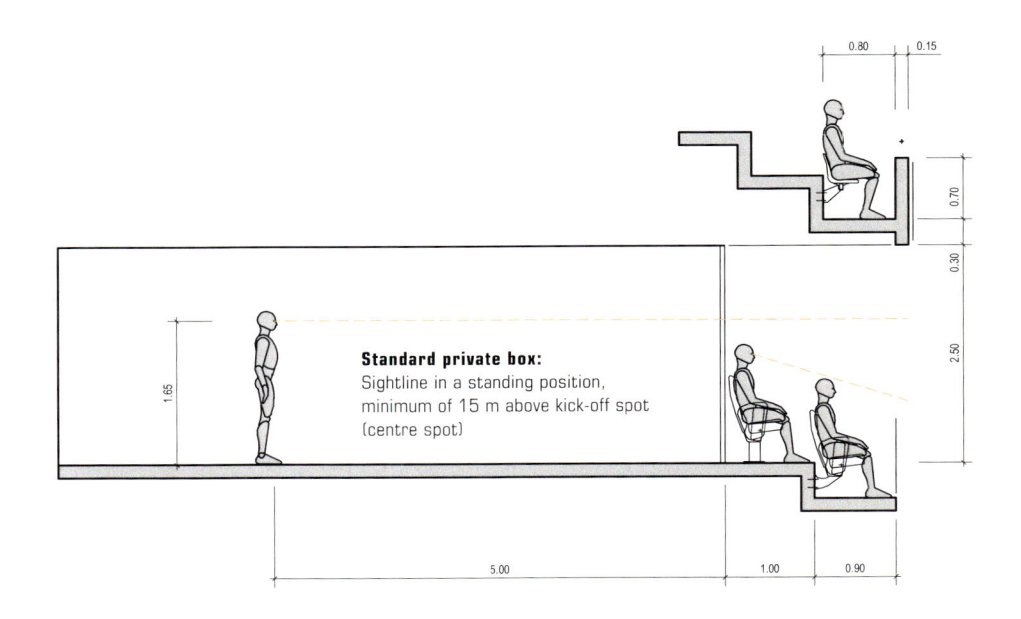

05 VIP areas & corporate hospitality

Lounge concepts

The layout of VIP areas strongly relies on the operating concept. Frequently, central lounges form the core of a multifunctional event space. The room plan itself is complemented by segregated premises, which are partly interconnected, partitionable and adaptable to all types of conference activities by providing various seating options.

In Olympic Stadium Berlin, a four-storey open space, where between 900 (seated) and 2,000 (standing) persons can be entertained, was created beneath the VIP stand, which is also linked to special rooms such as the 'Coubertin Hall' or the 'Jesse Owens Lounge':
South: 1 x VIP box: 270 m² – Level 1
South: 5 x VIP boxes: 21 m² – Level 2
South: 5 x VIP boxes: 21 m² – Level 3

Likewise, in Allianz Arena, Munich, the two executive lounges of the two main operators FC Bayern München and TSV 1860 München are supplemented by various event spaces.
(Level +5: 165–215 m² 2 x 4 party lounges with optional foyer 4 x 230 m².)

These areas are accessed via Stairwell 4 East as second VIP entrance. More special-use boxes with lobby are located in the transition area of both club lounges at the north/south curved corners (in total 2 x 180 m²). Lounge level 3 (west) is arranged as follows:
Telekom Sportsbar: 410 m²
Adidas-Audi Lounge: 245 m²
Champions League: 385 m²
FC Bayern Lounge: 1.840 m²
Allianz Lounge: 190 m²
Paulaner Lounge: 330 m²

The lounge concepts for the former Westfalenstadion Dortmund or AufSchalke arena are models for new builds or conversions of the third stadium generation.

New and modern gastronomy concepts were implemented which transformed the former special zone into a multifunctional executive area.

Borussia Park at level +3 is an event space for 100 – 1,200 guests located in the new-build section of the Westfalenstadion; *Förderverein* [sponsors' club] is situated one storey above, providing space for 100 – 500 guests. Varying in design, all these areas can be rented as single sections or entire levels.

Furnishings and exhibition pieces transform this lounge with direct terrace access into a 'museum lounge' and thus into an important point of identification for many visitors. The western part also accommodates two lounges for respectively 200 to 600 guests on individual storeys (levels +3/+4) which are connected via an internal flight of stairs. After completion of the four corner buildings, these areas can be structurally interconnected.

The example of Gelsenkirchen shows a differentiated grading of marketable areas. Apart from the main executive area with two box levels on top of it, other special areas are located in the curved corners on level +2:
La Ola Lounge: 2,180 m² – level +0 W
Gallery: 2 x 125 m² – level +1, +2
Schalker Markt: 600 m² – level +2 S/W
Incentive boxes: 8 x 45 m² – level +2 N,
 1 x 200 m² – level +2 N
Club rooms 04: 4 x 45 m² – level +2 N

---

119 **Lounge with exit to open area**
Borussia Park, Mönchengladbach
120 **Panorama Lounge with balcony**
Commerzbank Arena, Frankfurt
121 **Panorama Lounge**
Olympic Stadium, Berlin

In a second conversion phase from 1999 – 2003, the new 'Business Center' at Gottlieb-Daimler-Stadion in Stuttgart was completed, including a new upper tier with lounge and box area below.
Entrance foyer: 200 m² – level + 0
Foyer: 400 m² – level + 1
Reception: 175 m² – level + 2
VfB Club Room incl.: 120 m²
meeting room, kitchenette, WC: 70 m² – level + 3
Foyer with WC: 75 m² – level + 4

This building is also the new and representative stadium address. From the VIP multi-storey car park a glass bridge leads over Mercedes Street to the conference facilities. Premises for administration and press are also situated in this part of the building, complemented by a café for variety.

In a third construction phase and as multi-purpose space for around 290 people a 300-m² food court with an open buffet was installed below the second upper tier extension on level + 1 of the back straight. This area has no direct link to the stadium and is serviced from a passageway underneath the stand.

Levels accommodating hospitality lounges

Six of twelve German World Cup stadia have two 'stacked' single-storey lounge levels; only two stadia (Leipzig, Nuremberg) operate single-level lounges.

In Cologne, lounge space has been deliberately planned as a two-storey, large open space, glazed and with a built-in gallery.

For reasons of height, the business area in Gelsenkirchen was also realized as two-storey, but the boxes and event spaces of the upper level are only visually connected to the lounge.

Ancillary utilization for lounges

The effective depth of the lounges ranges from 20 to 25 m and is frequently adapted to the geometry of the stadium plan. The lounge at World Cup Stadium Leipzig, for example, follows the curved shape of the ground-plan with 5 to 18 m effective depth.

Useful depth and length of a lounge are generally determined by the architectural concept – depending on how many utilizations are to be accommodated on one side of the stadium and to what extent the available site allows an extension of the executive area and the stadium body. Another question would be to what extent the ancillary spaces on the main stand are to be distinguished as an independent structure.

The lounge lengths of the above examples vary from 27.5 m (Nuremberg) to 165 m (Cologne); for three of twelve stadia between 25 – 40 m, for three others between 50 – 90 m, and the remaining six main stands between 100 – 165 m.

World Cup Stadium Hamburg

2 x 25 boxes (4,0 x 7,5 m) are located at the main stand east and are directly connected to a lounge area at the rear (l = 120 m), which is separated into three sections by two ancillary utilization areas for catering, buffet and sanitary units. Thus, two 450 m² large restaurants are created to the left and right of a central reception area covering around 950 m².

This layout ensures short distances (here 60 m) to the sanitary facilities, during half-time.

The large effective depth of approx. 21 m with a clear room height of more than 3.0 m creates compact but also visually expansive rooms.

122 **Bar in FC Lounge**
RheinEnergieStadion, Cologne
123 **Casino in FC Lounge**
RheinEnergieStadion, Cologne
124 **'Brewery' in FC Lounge**
RheinEnergieStadion, Cologne

05 VIP areas & corporate hospitality

World Cup Stadium Cologne

Here, boxes are directly adjacent to the lounges. The effective depth of level +2 is only 14.5 m, while making maximum use of the 165-m lounge length since the total width of the stadium body is limited by two listed tree-lined avenues. Situated directly at level with the tree tops, the lounge opens up with generous glass fronts to the green environs outside.

The VIP area takes advantage of the full stadium length in order to accommodate around 3,000 m² of executive space on two lounge-/box levels.

Responding to the gauged length, quiet seating zones like the 'Casino' and such relaxed standing areas as 'Bar and Bistro' are provided.

125 **Group seating in the lounge**
Commerzbank Arena, Frankfurt

126 **Gallery in FC-Lounge**
RheinEnergieStadion, Cologne

127 **Table D: Lounge areas of World Cup stadia**
(Figures relate to lounge spaces with or without stadium bowl association.)

Table D Lounge	Location (stand)	Storeys	Level	Util. storeys	Catering area	Useful depth in m	Length in m	Location of WCs dist. in m
01 Berlin	North-south	4	−1 / −4	4	3,850	20 m	35.00	Central
02 Dortmund	N-E-W	2	+3 / +4	3x 1	7,150	24 m	97.00	Central
03 Frankfurt	East	3	+2 / +4	3x 1	3,300	24 / 9.5 m	130.00	80
04 Gelsenkirchen	West	2	+0 / +1	2x 1	3,300	21.5 m	105.00	105
05 Hamburg	East	2	+2 / +3	2x 1	3,650	21 m	120.00	60
06 Hanover	East	2	+1 / +2	2x 1	2,000	22.5 / 12.5 m	87 / 55	87 / 55
07 Kaiserslautern	North	3	−1 / +2	2x 1	2,800	21 / 26 / 13 m	40.00	Central
08 Cologne	West	2	+2 / +3	2	3,000	14.5 / 11.5 m	165.00	165
09 Leipzig	East	1	+5	1	2,000	5 to 18 m	90.00	90
10 Munich	Circulating	2	+3 / +4	2x 1	6,800	21 to 26 m	150.00	150
11 Nuremberg	West	1	+2	1	1,200	23 m	27.50	Central
12 Stuttgart	East-west	2	+2 / +3	2x 1	4,500	8 to 13.5 m	100.00	100

Commentary on table D
1) Figures relate to lounge space (total) with and without association to the stadium. The do not relate to the restaurant areas, unless business areas and stadium restaurant are identical (ex. World Cup Stadium Kaiserslautern).
2) In merely a third of the World Cup stadia, sanitary facilities are situated centrally; in more than two thirds, WC rooms in VIP areas are located for design reasons at a distance of between 50 to 90 m or 100 to 165 m.

Surface area demands for lounges

There are no direct regulations for seating and bar areas. For seating at tables, the relevant regulation merely sets a value per guest of *one* square metre surface area of congregation/exhibition space. Distance of a table seat to the next aisle should not exceed 10 m and distances between tables should not be smaller than 1.50 m.

Example: RheinEnergieStadion, Cologne
Required lounge dimension:[9]
Approx. 2,000 executive seats x 1 m² = 2.000 m²
Actual lounge dimension:
Level +2 = 1,600 m²
Level +3 = 1,400 m²
            3,000 m²
Thus, more than 1,000 m² of additional space are available.
3,000 m² : 2,000 persons = 1.5 m²/person

This figure reflects the generally increased demand on comfort in VIP areas in modern stadia. In comparison with other new plans for stadia and arenas, figures range from 1.50 – 2.0 m²/person. UEFA demands at least 2 m²/person for VIP reception areas.

Number of seats

Depending on the catering concept, seating capacities in canteens are calculated with a 2.5-fold occupancy: the number of potential guests at the table divided by 2.5 determines the number of seats.
Applied to the example of RheinEnergieStadion Cologne:
2,000 executive seats: 2.5 = 2,000 x 0.4 = 800 seats
In fact, there are 1,200 seats and 650 standing places available, that is around 50% additional seats.
3,000 m² lounge space (in total) divided by 1,200 seats = 2.5 m²

Equivalent calculation:
Veltins Arena, Gelsenkirchen
2,200 m² ('La Ola Lounge') divided by 750 seats
= 3.0 m²
Commerzbank Arena, Frankfurt
1,000 m² ('Business Club') divided by 500 seats
= 2.0 m²

First preliminary calculation: 1.5 m² per VIP

Second preliminary computation (depending on comfort and catering concept): 2.0 – 3.0 m² to calculate number of seating places

Values might increase during the very short catering period of 15 min at half-time during which 90% of all food and beverages are ordered.
Since there are as many furnished (seating and standing) places provided as executive seats on offer, occupancy rate is better calculated at 1:1. The ratio of standing to seating places is approximately 2/3 of the business capacity.
There are also no binding stipulations for bars and buffets. The length of bar/buffet systems depends on the utilized thermal facilities and equipment. Again, the required capacities are determined by the concept and organization of the operator.

Example: RheinEnergieStadion, Cologne
Lounge capacity: 2,000 people
Two-storey

Buffets (u-shaped layout, service from three sides):
– Number: 2 x 4
– Dimensions: 3.0 x 3.0 m
– Service length: 8.0 m
Applied to the number of guests: 30 persons per running metre of buffet length.

Bars (circular, service from all sides):
– Number: 5
– Dimensions: 5.0 m diameter
– Perimeter: 15 m
Applied to the number of guests: 25 people per running metre of bar length.

Note: These figures may be used in approximation as design calculation.

Chapter 06

128 **Buffet zone in FC Lounge**
RheinEnergieStadion, Cologne
129 **U-shaped layout (circa 3 x 3 m)**
RheinEnergieStadion, Cologne

Stadium catering
Concepts of culinary service for spectators

Stadium catering

Recreational activities are growing in significance in modern society, and there is clearly a high demand for services offered by leisure facilities. Due to increased wealth and spending power of consumers, modern stadium catering aims to respond to the changing and diversifying needs and demands on quality across the entire spectrum of potential guests.

Catering areas are generally divided into three types:
1) VIP lounge for business guests or prominent figures, partly independent of the event
2) Restaurant area for guests who come for the stadium experience, independent of the event
3) Kiosk or shop for stadium visitors or specific events

VIP lounge

High-quality meals are served in the executive lounge. Catering is generally limited to a self-service buffet, while drinks are often served at the table. VIP lounges tend to be the most upmarket utilization areas in a stadium. This type of catering is mostly financed by means of the high-priced tickets which are normally only available as season tickets.

Adjoining 'relay' kitchens guarantee a steady supply of meals for the high number of guests present. Fitted with regeneration technology, these facilities have a surface area of around 40 m², with spaces for cold storage and general storage.

Relay kitchens are only used during events. They are located on the same level as lounges and other VIP areas, in order to guarantee a fast and efficient service. Each relay kitchen can cater for on average 800 to 1,000 people. Buffets in VIP areas are fitted with thermal facilities and serve a comparable number of people.

Restaurant area

Modern operating concepts are geared to improving revenue by luring the public into the stadium much earlier than the actual start of a match or event. Convinced by entertainment and culinary offerings, people are enticed to remain longer and spend their money.

An executive lounge may be converted into a general restaurant and bar during extra time. People might remain into the late evening and use the opportunity to communicate and make business. Apart from the sports competition, the appeal of the executive areas is vital to the marketability of the respective seats and hence the commercial success of the venue.

The actual home of the club often still exists, since not all clubs have training grounds in the direct vicinity of the stadium new build, and the 'holy turf' is generally only played on match days. The location of *Geisbockheim* of 1. FC Köln, for example, is located only a few kilometres away in the Cologne green belt.

Half of all World Cup venues operate independent service restaurants outside of regular matches, promoting the stimulation of the stadium site during the week. Examples are the *Soccer Café* in Veltins Arena, Gelsenkirchen (230 m² on two storeys, level +0/+2 north) or *Der zwölfte Mann*, a sports bar in RheinEnergie Stadion, Cologne (550 m², single-storey, level +2 north).

Promoting their use as event area during a match, restaurants often provide a view into the stadium bowl through a wide panorama glass front.

In terms of multiple utilization, they are also suitable for other events like parties or private family celebrations (example: *Die Raute* in AOL Arena, Hamburg, with a 200 m² panorama restaurant on level +2, excluding ancillary rooms).

Allianz Arena Munich has an à-la-carte family restaurant named *Stube* at the south stand with 900 m² for around 400 people.

Outside of events, the stadium is generally locked up and monitored for reasons of safety. In case of operating restaurants during the week, accessibility must be clarified and should be provided by a separate entrance with external signage; uncontrolled access to the stadium and stands should be avoided. Catering in the restaurant can be buffet-style or with table service; the range of food and beverages available tends to be more extensive than that offered by kiosks.

For panorama restaurants with view into the stadium bowl, access to or even use of the viewing areas would be conceivable. In this case, block partitions should be provided as barriers.

Restaurant service during an entire event is feasible, in which case accessibility and checks need to be clarified with the organizer.

A separate and independent kitchen is provided for the restaurant area. Technical requirements being similar, a combination of food regeneration and production kitchen can be integrated into the restaurants. Kitchen areas can thus be limited in size and are dimensioned adequately with approx. 40 m², including storage and cold storage space. Reaction times for food preparation in a restaurant are much higher due to possible à la carte orders.

Shape and design of the restaurant ground plan and the underlying concepts may vary significantly; the dimensioning of service and kitchen areas is highly reliant on planned utilization.

The catering demand during an event day is notably higher than normal restaurant capacities would be able to cope with on non-match days. Self-service areas within a restaurant or independent self-service restaurants have proved to be practical in this respect.

During intervals at events the self-service concept may ensure an optimum use of the restaurant areas, e.g. in Munich Stadium with 2 x 1,250 m² fan restaurant for 1,000 persons each on level +3 in the north/south corner curves; or in Frankenstadion in Nuremberg with a fan meeting point as roofed beer garden (310 m²) with a 45 m² restaurant in the northern approach to the stadium.

This facility has an exterior link, which means that guest already have a reference point prior to admission to the stands. It is also much easier controlled than the bars, petrol stations or kiosks in the surroundings of the stadium, which are often used as meeting points prior to the match. Encounters of 'enemy' supporters in these areas are a major concern to security staff before kick-off.

130 **'La Ola Lounge'**
Veltins Arena, Gelsenkirchen

Kiosks or shops

Kiosks or snack shops cater for customers before, during breaks and after an event.

On the part of FIFA/UEFA, a general ban on alcohol exists before and during a game throughout the entire complex. Such a ban can be relaxed in agreement with safety authorities, for example by selling alcohol-reduced beer.[1] Kiosks are generally limited to selling alcohol-free drinks and hot or cold snacks.

Kiosks selling food and beverages are mostly based on a rectangular plan and are integrated into the stadium structure (at the under-terrace 'pendentive'). A kiosk may have several hatches (column grid) selling wares separately. Queuing customers should not impede circulation routes.

Kitchen technology is installed at the rear, and via a neutral 'feed-through' counter prepared meals are passed onto the service staff. No professional know-how is required on the part of the temporary staff, but rather logistic and organizational skills. Focus must be on supplies and restocking of the kiosk.

Kiosks or shops are located at strategically important positions and are evenly distributed throughout all stadium sectors.

At short distance from the terraces, large milling areas are designed for food consumption. Due to the short break periods at events, fast service is paramount. Catering management must carefully organize and coordinate the quantities of prepared food and actually sold food. The surface layout of these areas affects the logistics. Kiosks or shops require a minimum space of around 35–45 m² for preparing and selling food and drinks. Storage areas are located close to a central delivery point.

Supplied in advance, stocks are intended to outlast an entire event, based on the capacity of storage space within the kiosk.

Cooled drinks are supplied directly to the kiosk via a system of pipes from storage rooms. In Veltins Arena in Gelsenkirchen around 5 km of pipes deliver the beer of the main sponsor from large central tanks directly to the service counters.

Drinks are served in containers such as plastic or cardboard cups which cannot be used for dangerous activities.[2] Frequently, mobile vendors operate directly on the terraces (as in AWD Arena, Hanover; 60 vendors selling from rucksacks).

131 **Lower concourse with kiosks**
World Cup Stadium Nuremberg

132 **Table E: Kiosk surface areas at World Cup stadia**
(Figures rounded to standard dimensions)

Commentary on table E
Figures relate to kiosks, reduced to standard dimensions. Single measures vary significantly and are only reference values as system specifications.

Table E Kiosks	Kiosk type	Total number	Surface area in m² (mean value)	Total area in m²	Spectator capacity (national)	Customers per catering area (m²)	Useful depth in m	Shops (merchandise)
01 Berlin	Built-in	22	75	1,650	74,200	45	4.20	
02 Dortmund	Built-in	36	40	1,440	81,250	56	3.60	
03 Frankfurt	Built-in	22	40	880	52,100	59	4.85	
04 Gelsenkirchen	Built-in	32	60	1,920	61,500	32	3.80	5
05 Hamburg	Built-in	24	65	1,560	56,000	36	4.20	9
06 Hanover	Combination	12	70	840	49,500	59	3.70	
07 Kaiserslautern	Built-in	36	25	900	48,500	54	2.50	
08 Cologne	Combination	26	40	1,040	50,700	49	4.50	6
09 Leipzig	Combination	24	70	1,680	44,200	26	5.00	6
10 Munich	Built-in	28	50	1,400	66,000	47	4.75	15
11 Nuremberg	Stand-alone	24	45	1,080	46,800	43	3.90	9
12 Stuttgart	Stand-alone	19	50	950	56,000	59	5.00	

Kiosk concepts

Three structural system layouts are distinguished for kiosks or shops: built-in, stand-alone or a combination of both. Usually located at the general circulation routes to the stands, they also serve as supply and milling areas prior to and after the game, and during half-time.

Kiosks are located on the upper tier concourse in the so-called 'pendentives' created by the triangular elevation surfaces underneath the terraces. Therefore, useful depth and adequate headroom are decisive. An average kiosk in a World Cup stadium has a depth of 4.15 m, with single values ranging from 2.5 to 5.0 m. The measures of the respective surface areas are determined by stadium operators and set in drafts, ranging from 25 to 75 m² with an average of 50 m². Planning of kiosks and other vending points for merchandising should be geared to regular venue operation. The foregoing table is therefore based on spectator figures, including standing capacities.

In the mentioned twelve stadia there are 12 to 32 kiosks, on average 25 kiosks with 50 m² each; for a gross capacity of around 685,000 visitors, a total of ca. 15,000 m² is allotted for service outlets. Between 32 to 45 customers are estimated per square metre of vending area (average 45 persons/m²).

Requirements for doors and gates are stipulated as follows: F 90 (fire walls F 30) fire-retarding, smoke-proof and self-closing.[3] Therefore, openings are locked up with T 30 rolling shutters.

First-aid points

In agreement with local authorities, an appropriate number or a minimum of one medical service point is to be planned for every subdivided spectator area or sector.

DFB Guidelines (§ 16) propose at least one large room for medical and rescue services, including necessary outfits for initial medical treatment of spectators. An exact list of equipment should be worked out with the relevant paramedic services.

Premises should be easily accessible and clearly signed for spectators, inside or outside the stadium. Passageways are to be executed wide enough for transports on stretchers and wheelchairs; clear frame width is at least 90 cm which corresponds to minimum door widths for assembly rooms for less than 200 people.[4]

Earlier IAKS planning fundamentals quote the number of one station with around 20 m² for every 10,000 to 15,000 spectators. Rooms for trainers or referees (popular sports) may also be used as first-aid rooms.[5]

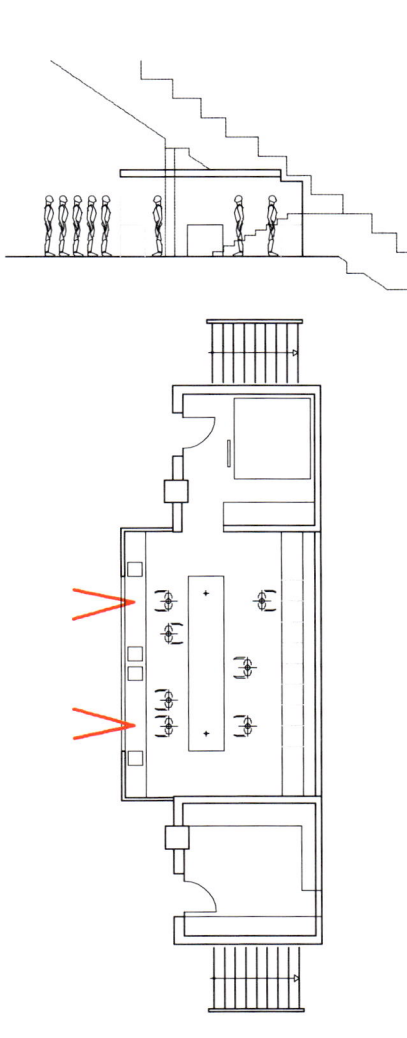

133 **Kiosk concept, three-part, with foyer**
World Cup Stadium Cologne (Level + 3/4)
1) Waiting zone in front of counter
2) Service behind till, in front of counter
3) Production at the rear;
   Day storage with cold storage cell

134 **Upper concourse with kiosks (Level + 3/4)**
RheinEnergieStadion, Cologne
135 **Counter situation 'Food & Beverage' (Level + 1)**
RheinEnergieStadion, Cologne
136 **Beer tanks for direct supply (5 km)**
Veltins Arena, Gelsenkirchen

137 Relay kitchens in lounge area

Production in the central kitchen

VIP areas are serviced from a central kitchen which also covers capacity peaks. In the interest of short supply routes, the central kitchen is located close to the storage areas. As a rule, these are planned at ground level to ensure a smooth delivery. The distribution of delivered goods may be organized by suppliers in advance.

In case of insufficient space at ground level, these utilizations may be relocated into the basement. Planning should consider that these areas have to be accessible to delivery vehicles. The clear height to be observed depends on expected vehicle sizes and has to be agreed with the operator/caterer. Ease of access needs to be ensured also for the large collection trucks of waste disposal services. The central kitchen has the largest share of thermal devices. Providing a broad range of high-quality meals for up to several thousand people affects the requirement specifications of modern stadium catering. Depending on the operating concept, the central kitchen can be used for regeneration or as production kitchen.

Size requirements (preliminary dimensioning)

An area of around 80 to 100 m² is required. The size of storage areas ranges from 25 m² for cold storage, 20 m² for the freezer room, 30 m² for food storage and 20 m² for non-food storage. Delivery intervals for goods are just in time. Stock turnover tends to be very high due to fast reaction times. Hence, not only can storage areas be minimized, but also the risk of spoiling goods. Supply from the central kitchen to the respective outlets may be done via goods lifts or tunnel systems; staircases should be avoided, if possible.

The central kitchen should cover all catering areas unless restaurant and VIP catering is operated by two different companies. In this case, ancillary utilization areas for production, staff and storage are doubled. Floor finishes must be shockproof, watertight, rot-proof, easy to clean and anti-skid. Up to a sufficient height, walls must be made from smooth, washable, watertight and anti-bacterial materials. A window, transparent door or wall should be provided at eye level with a view from the work area to the outside. [6]

Dishwashing

Dishwashing systems are also located in the kitchen environs; the unclean area strictly shielded from the clean area. Planning should always ensure a consistent routing between food production, service and returned dishes. Logistic requirements are paramount for scaled-down catering concepts. Further dishwashing areas may be planned for restaurants and in relay kitchens for VIP lounges. A total area of 150 to 200 m² should suffice to cover the dishwashing area for central kitchens.

Dishwashing areas are also buffer zones for the collection of returned dishes, with the dishwasher itself taking up the least space.

Staff and social rooms

Catering facilities in stadia require large numbers of employees and temporary help. Separate recreational and social areas are to be provided for break times. If special attire is required for the job, changing rooms are to be provided. Dressing rooms for kitchen staff should adjoin the work space unless the risk of catching cold is minimized by some other means (e.g. by heated passageways). Distance between a changing room and kitchen should, if possible, not exceed 100 m or one storey.

If personnel are exposed to smell nuisances, a separate storage space for work clothes (white) and street clothes (black) should be available. If detached, it seems practical to connect both areas with washrooms, since they must be accessible to one another.

Staff toilets should provide no direct access to areas where food is produced, processed or distributed. WCs are to be kept in an impeccable state and provided with a hand washing facility (warm and cold water), soap and disinfectant dispenser and an appropriate appliance for drying hands.

Outlook on future catering concepts

The technology for producing meals is constantly simplified and increasingly integrated into one single appliance. As production and output times decrease, performance demands on devices are growing, as is power demand. To improve economic parameters, the food industry is engaging in research on conservation and production processes; a development that will also affect stadium catering.

In order to ensure fast supply in a consistent quality, automation seems unavoidable. Logistically though-out access routes and increased self-service are necessary to achieve this. Future concepts such as locating production beneath output areas, normally one storey below, would allow service staff to serve customers in a matter of seconds. In a one-way movement, guests would leave the service area, thus avoiding counter-flow. Interior design can help to guide visitor streams by leading them efficiently to the vending and eating areas.

Simplifying production technology will result in minimized production time and reduced personnel demand. Capacities and spaces for storage can thus be reduced, depending on the degree of pre-fabrication and on improved supplier logistics. With an enhanced performance, in future only one or two devices will be necessary for the entire meal production.

Professional competence is exchanged for full convenience; automation of resource planning renders any storekeeping obsolete. Disposition of food and beverages can be optimized and made more economical. Quality and quantity are no longer reliant on the production staff, while hygienic risks are minimized to the greatest possible extent.

Receiving counters or machines enable a convenient and fast return of used glasses, etc. Waiting times will be reduced by technical and logistical systems to such a low level that, in future, service windows will be replaced by a continuous provision.

Cashless systems and machine operation are also being introduced to catering facilities. Reducing or abolishing cash payments sustains a fast and smooth service. In the wake of automation, cashless systems like money cards or credit systems are conceivable. Kiosks, shops or VIP areas may be optimized through innovative planning. The separation of payment and output of food and drinks simplifies handling and increases vending rates.

Transponder technology, for example, works touchless with a memory card that exchanges data with a debit terminal. Data carriers built into watches, armbands etc. would be recognized on passing a special barrier.

138 **Central kitchen with thermal production devices**

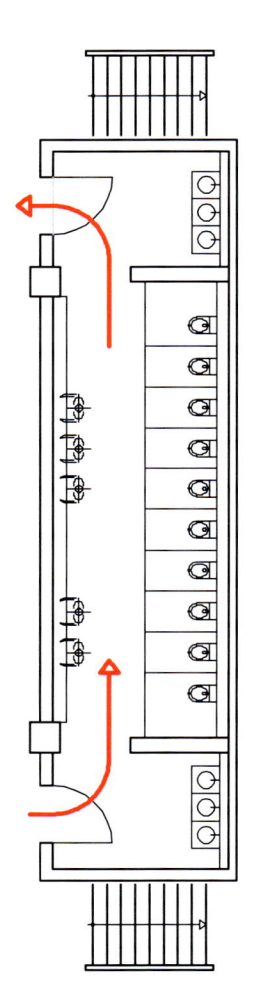

139 WC room concept
with separate entry and exit situation for two-directional traffic

General WC areas

The German regulation on places of assembly states the following requirements and minimum figures:[7]

1) Assembly places must provide separate toilet rooms for women and men.
2) Toilets should be provided on every storey.
3) There should be provided at least:
(per 100 persons)
– up to 1,000: WCs for women 1.2, WCs for men 0.8, urinals 1.2
– above 1,000: WCs for women 0.8, WCs for men 0.4, urinals 0.6
– above 20,000: WCs for women 0.4, WCs for men 0.3, urinals 0.6
4) The stated figures are to be rounded up.
5) If the regular allocation of toilets is not practical for a certain type of event, a different allocation may be employed for the duration, as long as facilities are marked accordingly.
6) Toilets available on-site or in the vicinity may be counted if they are accessible for visitors of the venue.
7) For wheelchair-users, a sufficient number of suitable barrier-free toilets, but at least one toilet for every ten wheelchair places, must be provided.
8) Every WC must have an anteroom with washbasin.

Note: Depending on the type of event, the regulation may be implemented accordingly, e.g. for trade fairs and comparable large events, in particular for temporary congregation spaces in the open. For the latter, mobile toilets must be installed.

Distance to an internal exit should not exceed 50 m and should ideally be on the same level (IAKS 1993).

Planning of WC areas should encourage a flexible allocation of toilets as different events may entail a different gender-specific quota, or only parts of the building might be used.

FIFA points out that the entrance situation for sanitary facilities must be realized jostle-proof. Especially during the short break periods, circulation routes should not be impaired.

Fittings and appliances of a WC system should be very robust, anti-theft and vandalism-proof since agitated spectators often cause major damage in sanitary areas.

The anonymity of a sanitary facility tends to promote a rather negligent behaviour on the part of many visitors. In the past, this phenomenon was met with fittings of a rather 'martial' appearance.

Stadium new builds of the last few years have instead started to respond by providing high-quality fittings and better hygiene. Experience has shown that higher WC standards are generally welcomed by stadium visitors.

Table F – WC facilities	WC type	Number	Male	Female	Useful depth [m]
01 Berlin	Double-plan	17	9	8	6.80/11.40
02 Dortmund	Single-plan	50	26	24	3.60
03 Frankfurt	Double-plan	34	17	17	4.80
04 Gelsenk.	Single/double-plan	66	36	30	3.0/4.5
05 Hamburg	Double-plan	44	22	22	4.20
06 Hanover	Single/double-plan	24	12	12	3.0/4.90
07 Kaisersl.	Double-plan	52	26	26	5.25
08 Cologne	Single-plan	46	22	24	3.30
09 Leipzig	Double-plan	22	11	11	5.00
10 Munich	Double-plan	38	20	18	4.10
11 Nuremberg	Single-plan	24	14	10	2.85
12 Stuttgart	Double-plan	36	18	18	4.10

140 **Table F: WC facilities at World Cup Stadia**

Commentary on table F
The figures do not relate to individual sanitary appliances or numbers of toilets since this is clearly stipulated in the foregoing regulation. Here, the number of facilities available on demand are considered.

141 **Single urinals (vandalism-proof)**
Estadio do Dragao, Porto, Portugal
142 **Urinal trough (vandalism-proof)**
Estadio do Braga, Portugal

Chapter 07

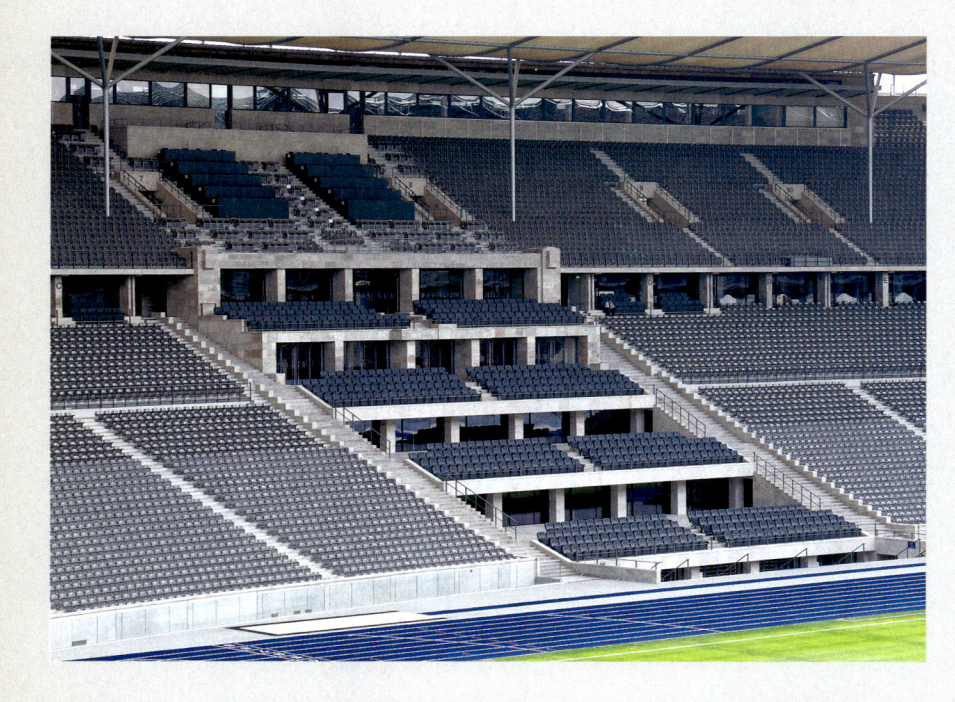

143 Press and honorary stand
Olympic Stadium, Berlin

Media facilities
Working areas and requirements of the media and press

Media

'The 21st century will be, even more than the last, an age of communication and, unfortunately, of manipulation.

Thanks to new technology, each individual will be able to hear, speak and see, and thus participate in, and be influenced by, public events from a distance in unprecedented ways.

When, this notwithstanding, stadia continue to be built at a great expense, this is because of the people themselves and of their natural need to be physically present with all their senses, in the middle of orchestrated life.

Thus it is the millions in front of the television that are the actual audience for the product advertisements for the consumer society. The public in the stadium acts as massed extras, a virtual bath in the masses that also becomes a vital feature of the orchestration' (V. Marg, 2005).[1]

Since the beginning of the 20th century, this type of orchestration has been taken over more and more by the media.

The Olympic games of 1936 took place before a fascinated and amazed global audience and featured a number of premieres: for the first time employing modern communications technologies, a 'live' radio/TV transmission and Leni Riefenstahl's film opus on the beauty of the human body.

In 1937, the first live-broadcast of an Arsenal London football game by the BBC took place, located in the vicinity of the studios and the station in Alexandra Palace, North London. With the majority of homes owning a TV set by the end of the 1950s, the view into the stadium became a reality for millions.

Franz Joachim Verspohl, in his 1974 dissertation on the direction and self-experience of the masses, approached this phenomenon in terms of building-type and in art-historical terms. In order to critically analyze the potential of architecture to influence the masses, he gives examples of sports and event venues from antiquity to the modern ages.

Note: Nowadays, the media hold a significant responsibility; journalistic freedom of reporting should not be reduced to mere orchestration for publicity purposes. Trademark, copyright or product placement as lucrative sources of revenue may augment the profitability of privately-operated sports venues, but they should never become more important than their original objective to make a sports competition or cultural event accessible to a larger audience.

With technological progress, certain areas of work and structural requirements are becoming more specialized. After the written press and photographers, commentators arrived on the scene; first as radio reporters, then as mediators for a global TV live broadcast reaching an audience of millions.

Working areas for the media

Three different types of working areas are generally distinguished according to location: on the stand, beneath the stand or external. Different structural prerequisites apply:

On the stand
different groups of people with different requirements are based:
1) TV and radio – Commentator positions and places for observers
2) Print media (written press) – Press seats
3) Photographers
4) Camera and presenter positions

Beneath the stand
the respective media workrooms for processing of sports news and reports are located:
5) Press conference
6) TV studios
7) Commentators' point and TV control room
8) Media centre and milling area

Outside the stadium
parking space must be provided for OB vehicles and technical equipment:
9) Parking space and set-up areas
10) Media transfer point

When for major sports events the number of media representatives multiplies, the main working areas are generally moved to so-called 'SMC – Smart Media Centres' outside the stadium perimeter. Adjacent, the technical TV compound is positioned.

Figures and approximations for this work group are highly reliant on the type of event. The number of media facilities may also vary from game to game or stadium to stadium.

A highly *flexible* approach is vital when it comes to installation of facilities. Technological developments cannot be foreseen, and comparing their pace with the time frame for planning and construction of venues clearly shows that even experts can only state approximation values.

Location of the media zone
With regard to camera positions, work places and TV studios, the provision of first-class media facilities is essential. The media area should be located in a central position on the main stand between the 16-m lines and in extension of the halfway line; on the same side of the stadium as VIP and players' area, below and above the honorary guest terrace, as the game can be watched best from there.

The optimum location for media stands at athletics competitions is between the halfway line and the finish line of the 400-m running track. It should be a controlled and secured zone with a dedicated access. All places should be covered, with good and unrestricted viewing of the whole playing field and with numbered single seats meeting the general requirements of safety and comfort. Journalists should have fast and intersection-free access to all media working areas.

144 **Media places on the upper tier**
AWD Arena, Hanover, during Bundesliga
145 **Photographers behind goal line**
ConFedCup 2005, Nuremberg

07 Media facilities

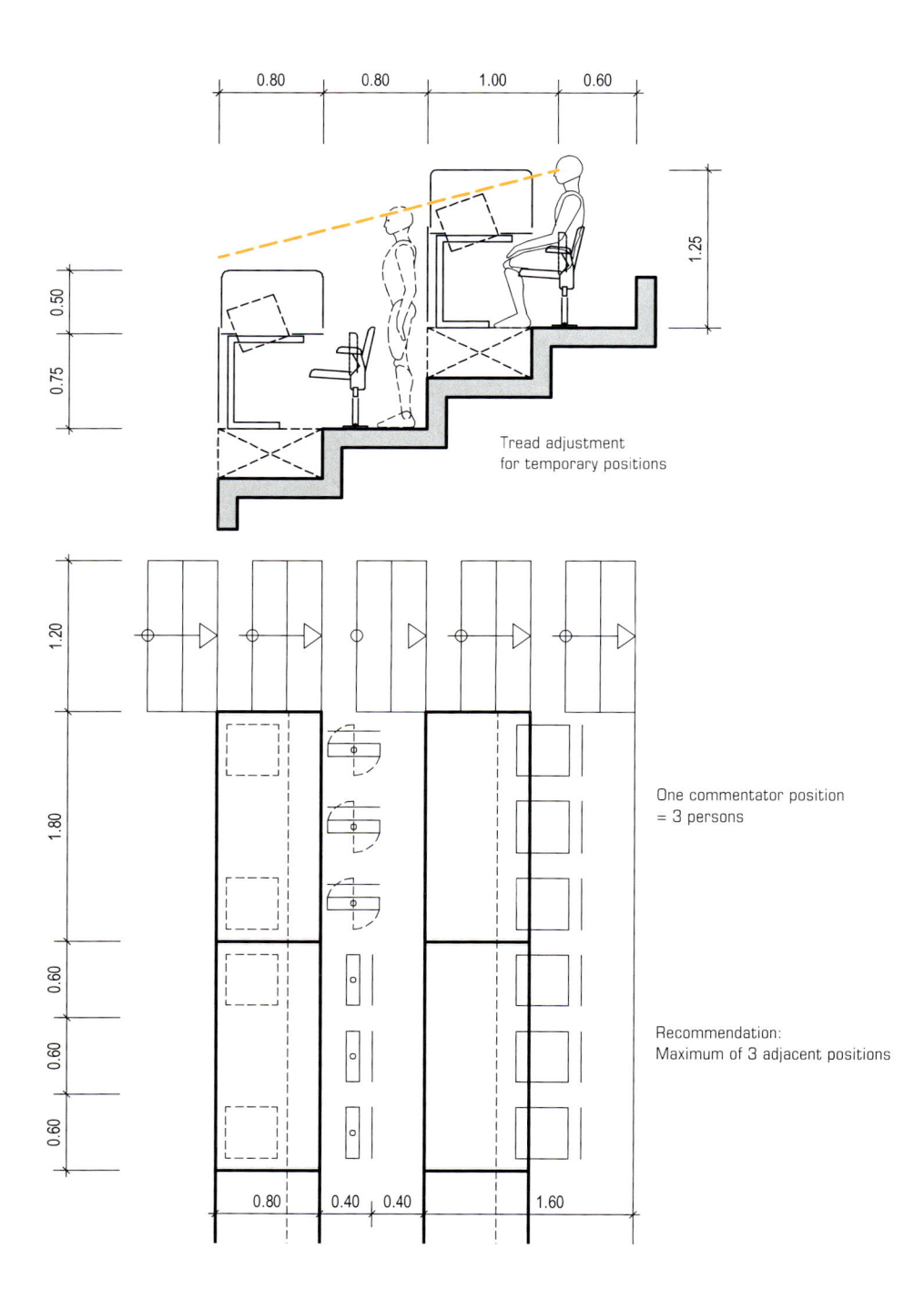

Commentators' positions

These are workplaces for radio or television reporters who comment on the games. Apart from observing the match or competition live, they simultaneously control stills, repeats and live images on 2 or 3 monitors (1 monitor for radio).

For special sporting events, the press stand is to be extended. In order to raise the regular media capacity to, for example, UEFA/FIFA requirements, regular spectator places are converted with the help of temporary measures. Depending on the event, commentator positions may have to be adapted to the new situation.

Location and fittings

On the main camera side, between 16-m lines: covered seats with unrestricted viewing of the entire pitch. Covered but otherwise open places are preferred to cabins, according to FIFA. Individual positions should be shielded by plexiscreens. One module consists of three working spaces at a worktop with slanted built-in monitor screens; a maximum of 3 adjacent positions with free access from the gangway. Presently, required dimensions of a commentator's position are given as follows:

Width: at least 180 cm, 3 x workplaces = 60 cm
Depth: at least 180 cm, new: 1.60 m
Height: at least 75 cm (table)
Clearway: at least 30 cm
Desk width: at least 70 cm

Three seating rows per position are cited as optimum by FIFA Technical Recommendations for World Cup 2006, with each position taking over nine seating places, based on the originally required minimum depth of 1.80 m. Regular seating depth is recommended at 80 cm, i.e. 2 x 0.80 m = 1.60 m.

VIP and media areas should be positioned on the same side of the stadium. During regular sports events the 'reduced' number of media seats may be accommodated in the press area on the same tier. Steppings for business seats are normally 90 cm deep, i.e. 2 x 0.90 m = 1.80 m; commentators' positions in this type of stand would meet the requirements of FIFA.

Minimum capacities for large events are much higher, resulting in press places taking over other sectors of a stand, for example the upper tier. For economic reasons, a comfortable seating tread depth of 90 cm cannot be generalized because media requirements on such a scale would be only short-term. The depth of the commentator's position must be adapted to the conditions of a regular stand. Adding the required individual measures confirms the adaptation of the new commentator's position to the regular measure for steppings with 2 x 80 cm = 1.60 m.

Worktop depth	0.80 m
Seat depth	+ 0.40 m
Clear width	+ 0.40 m
Total useful depth =	1.60 m

Figures clearly reveal that commentators' positions should require only two seat depths (2 x 0.80 m). Instead of nine, only six regular seats would be taken up for a conversion. With a required total capacity of 150 to 200 commentators' positions for World Cup 2006, the reduction would be around 30%, resulting in 450 to 600 additionally available spectator seats.

Photographers

Commentators' control room

FIFA expects the provision of a CCR – Commentary Control Room – of around 150 m² size in every stadium, with full technological equipment, near the commentators' press stand.

Observer seats

are regular seats for TV representatives without technical equipment. These should be close to and with access to the press stand.

FIFA (UEFA): minimum number of observer seats
Qualifying round / group games and
last sixteen: 200 (125)
Quarter final: 200 (140)
Opening, semi-final: (150)
Game for 3rd place, final: 375 (250)

Seats or photographers' platforms do not have to be located on the press stand, and provisions are made for each stadium individually.

The catering units should be located near the pitch since most press photographers work at the pitch perimeter and thus need direct access to the media working area. Their seats are placed between the advertising hoarding (pitch perimeter) and the stand. Most photographers work behind the goals up to the 16-m lines. The distance between the photographers' zone to the goal line is 5.0 m and to the touch line / corner 3.0 m.[2]

Digital image transfer has become standard. Extra room should be provided for communication outlets behind the photographers' zone or at the pitch perimeter.

As part of a complex telecommunications system, the provision of extensive WLAN networks within the entire stadium has become standard in recent plans.

FIFA: workroom for photographers
1) Dark room not requested.
2) Separate room (100 m² min) with sufficient ISDN cabling.
3) Swift and unrestricted access.
4) Technological outfit comparable to press stand.

FIFA (UEFA): minimum number of photographers
Group game:
– Stands: 50 to 100 (150)
– Pitch: 50 to 150 (20)
Final:
– Stands: 150 (200)
– Pitch: 100 to 200 (20)

page 78:
146 **System sketch of a (2 x 80 cm = 1.60 m) commentator's position**
Minimum space demand for one position
= 3 work places (taking over 6 spectator seats)

147 **Camera position and photographers**
1. FC Köln during Bundesliga
148 **Commentator's position**
1. FC Köln during Bundesliga
149 **TV centre**
Westfalen-Stadion Dortmund during Bundesliga (2006)

07 Media facilities

150 System sketch of a press seat
Proof of an 80 cm deep minimum seat space with desktop

Written press (print media)

Press seats are individual workplaces provided with a folding shelf fitting the size of a notebook.

FIFA measures for press places:
Width: at least 60 cm
Depth: at least 90 cm (new: 80 cm)
Height: at least 75 cm (table)
Seat depth: at least 40 cm
Desk width: at least 35 cm
Clearway: approx. 60 cm

Technical outfit (FIFA)
Workplaces with desks for notebooks and mounted or tip-up seats with backrests. Telecommunication points and power outlets, 3 electrical outlets, telephone point, ISDN multiple outlet and desk light. Monitors in a central position, one monitor for every 8 places.

Note: No extra light since good general lighting is assumed.

FIFA: Number of press seats
Group game / final: 600 / 2,000
– with desktop: 400 / 1,200
– without desktop: 200 / 800
– with telephone: 300 / 1,000
– during Bundesliga: 100

UEFA: Press stand
Group / quarter final / semi-final and opening / final: 500 / 600 / 800 / 1,000 seats;
of these provided with shelf, light, power supply, ISDN, telephone: 250 / 300 / 400 / 600

TV production

The host broadcaster is not only responsible for production and distribution of international TV signals, but also coordinates transmission of signals, facilities and services to stations. Depending on the broadcasting concept, two TV positions are distinguished:
– Multilateral: camera positions produce one TV signal for all (from Latin *multos* = many, *latus* = side, meaning 'multi-sided')
– Unilateral: camera positions produce their own TV signal with their own equipment.

To be FIFA-compatible, a stadium requires the provision of two fully equipped TV studios (and also a radio studio, according to UEFA Technical Recommendations 2008).

Individual size is at least 50 m² respectively, with a height of least 4.0 m, based on the media technology, including lighting, etc., which needs to be accommodated therein. Players, coaches and delegates should have easy access, and press rooms, commentators' positions and dressing rooms should be close by.

The FIFA requirement profile for World Cup 2006 also includes presenter positions as open platforms:
Dimensions: 5.0 x 5.0 m (approx.)
Group play: 3 x 25 m²
Final: 6 x 25 m²

Mobile glass studio with panoramic view of the stadium:
Number at least 3, size at least 2.0 x 4.0 m.

Presenter Studio

One of the above TV studios may be a glass studio, also called presenter studio. Typically, in television broadcasting a match is commented on before and after by one or several persons with an excellent view into the stadium.

Prior to the extensive measures of conversion and modernization implemented for World Cup 2006, television producers often installed simple glass containers near the athletics tracks or the marathon goal.

The camera perspective should be unrestricted and unhampered by spectators suddenly jumping up. Thus, an inclined perspective is preferable.

The glass front should be camera-compatible, AR-coated and inclined; positioning of the posts should not affect camera views.

The studio lies in a raised position above the playing field, e.g. on the same height as the box tier. It would seem rather uneconomical to have a clear room height of 4.0 m across the whole storey based on the requirements for just one room (see above TV studio).

Note: After consultation with the responsible TV producers and the LOC 2006, the room height of one of the studios may deviate in order to accommodate the box tier, for example.

UEFA requires 2 to 4 presentation studios with 6.0 x 4.0 m surface area and only 3.0 m room height.

Digression on storey heights

The regulation on workplaces requires the following clear room heights:[3]

'Rooms may only be used as work rooms if clear height is at least 2.50 m for a surface area of not more than 50 m²; at least 2.75 m for a surface area of more than 50 m²; at least 3.0 m for a surface area of more than 100 m²; at least 3.25 m for a surface area of more than 2,000 m². Dimensions stated in section 2 may be lowered by 0.25 m for shops, office rooms and other working rooms in which work is mostly light or executed while seated. Lower dimensions may be implemented for compelling structural reasons, provided there are no reservations relating to health issues. Clear height shall not be below 2.50 m.'[4]

Gathering places such as executive lounges or the press conference room must have a clear height of 3.0 m since they generally exceed 100 m² of surface area. Unnecessary storey heights are to be avoided for economic reasons. Nevertheless, rooms should always reflect the desired atmosphere of the architectural draft.

Note: Economical storey heights of 3.84 m (= 24 rises x 16 cm) are feasible within an 8.25 x 10.0 m supporting grid, including semi-prefabricated structures and fittings (air-conditioning).

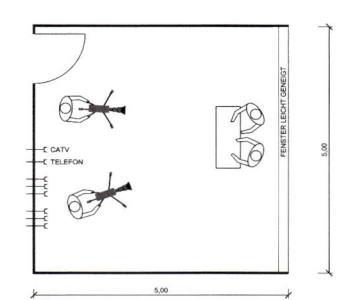

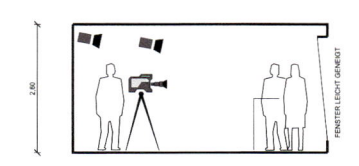

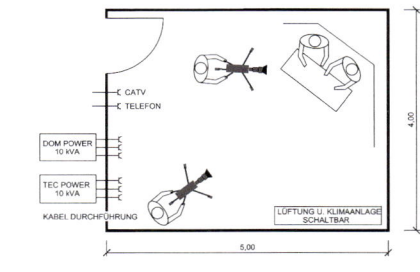

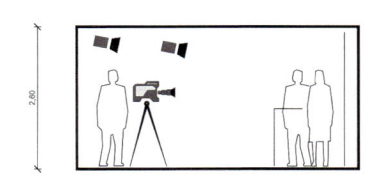

151 **System sketch of TV studios**
FIFA media specifications for World Cup 2006
a. Ground plan, standard studio 5 x 5 m
b. Section of studio with slanted façade
c. Ground plan, glass studio 4 x 5 m
d. Section, minimum clear height 2.60 m

152 **Temporary glass studio**
UEFA Technical Recommendations 2008
153 **Permanent TV studio**
Bundesliga, 1. FC Köln
154 **Flash-interview zone**
World Cup Stadium Cologne

Mixed Zone

Intermingling of the above mentioned groups of people (athletes and media) is to be avoided with the exception of the Mixed Zone, as mentioned in UEFA planning notes.

This area has a special function and represents an intersection of various user groups. Here, direct contact between otherwise strictly segregated players and media representatives is made possible for interviews. Via separate accesses, players and journalists may enter the mixed zone, particularly when teams are leaving the field after the game.

The routes on which players 'run the gauntlet' from the dressing room to the team buses generally meander by at least 1.50 m so as to provide the largest possible number of people with an opportunity for direct contact with players. Along these temporary barriers, journalists may be posted on a platform in the second row, depending on need. The area is screened off from the general public.

Partitions may be installed short-term (approx. 1.10 m) since the mixed zone is to be provided only for around 60 minutes. Before and after the match, this area can be used in other ways.

The mixed zone is mostly located between the teams' dressing rooms. For large sports events such as a Football World Cup, FIFA requires sizes of 600/750 m² for group games/finals. During Bundesliga, 100 to 200 m² would be sufficient.

Additional space demands are temporarily met by neighbouring areas such as underground car parks, warehouses and external tents.

UEFA grades the requirements of a mixed zone acc. to the significance of a game (group/quarter final/semi-final and opening/final):
– press representatives: 150/200/250/320
– surface area in m²: 250/300/400/600
– TV-ENG-Crews (2 persons each): 25/25/30/40

Flash interview zone

These types of interviews are taking place en route from the pitch to the changing rooms, either on the pitch or, shielded from the audience, in or next to the mixed zone.

For interviews with players or coaches during halftime (duration around 90 s) or after the end of a game, space requirements are 2 x 15 m for a team of around six people (FIFA).

UEFA gives the following spatial requirements for flash interviews for 2 – 8 people:
a) Interior space in the mixed zone: 3 x 3 m and 3 m height
b) Presentations at the pitch perimeter: 2 per 10 m²
c) Interviews at the main stand: 1 – 3 per 6 m²

155 **System drawing of the Mixed Zone**
World Cup Stadium Cologne

156 **Table G: Press area (during Bundesliga) in the twelve World Cup stadia**

Table G Press	SMC Media Center	Presenter studios World Cup 2006	TV studios World Cup 2006 in m²	Press conference area during Bundesliga in m²	Mixed Zone (Bundesliga/ World Cup) in m²	Location of tier	Location of stand	Commentators' positions	TV observers	Press seats with desk	Press seats without desk	Photographers
01 Berlin	5,500	6	3 x (35/55 m²)	./.	430 / 1,750	Lower tier	West	200	400	1,000	1,000	70
02 Dortmund	3,400	6	./.	./.	125 / 800	Lower tier	East	200	400	1,000	1,000	70
03 Frankfurt	1,850	6	4 x (50/65 m²)	250	630	Upper tier	East	150	200	500	100	50
04 Gelsenkirchen	3,000	6	2 x (45/50 m²)	230	200 / 530	Upper tier	West	150	200	500	100	50
05 Hamburg	2,500	6	3 x (25/50 m²)	265 – 390	380	Upper tier	East	150	200	500	100	50
06 Hanover	4,150	6	3 x (25/35 m²)	145	300	Upper tier	East	150	200	500	100	40
07 Kaiserslautern	1,600	6	6 x 40 m² + 240 m²	385	160 / 360	Upper tier	North	150	200	400	200	50
08 Cologne	3,000	6	2 x 60/75 m²	470	195 / 700	World Cup upper tier	West	150	200	400	200	50
09 Leipzig	3,300	6	./.	95	225 / 725	Upper tier	East	150	200	400	200	50
10 Munich	2,300	6	5 x 50 m²	./.	./.	Lower tier	West	200	400	1,000	1,000	70
11 Nuremberg	3,300	6	3 x (15/25 m²)	225	310	Upper tier	West	150	200	400	200	40
12 Stuttgart	3,500	6	4 x (35/40 m²)	190	380	Upper tier	West	150	400	700	700	50

Press centre in the stadium

During the regular season, media working areas are only activated on match day; the media working room with press catering, the supplementary media milling areas and the press conference room with adjacent press foyer and accreditation and information point.

UEFA divides the media area into a work and buffet zone with required dimensions of 150/250 m² for group games/finals.

Fitted with all the necessary sanitary facilities, this area is also perfectly suitable for large conferences outside of matches. The technical outfit usually meets the standards for international TV press conferences.

Premises are mostly located underneath the main stand, due to short distances to the pitch and VIP or players' areas. The media zone is only accessible for accredited personnel and should not be viewed into by the general public, thus, it could also be located in the unlit section of the stand.

Press conference room (FIFA)

The location within the stadium should be close to dressing rooms, for the benefit of participants: players/athletes, coaches, press officers/presenters. Lacking space, the press conference room (size for World Cup: 800 m) may be integrated into the press centre.

A podium is provided in the form of a cladded table with five chairs and three microphones, plus a background wall with a TV camera platform on the opposite side and a platform for at least 10 camera teams (UEFA 16 to 30 teams).

The technical infrastructure contains a split box which bundles the signal of several microphones and allows a centralised sound removal.

The room should be equipped with a first-rate sound system and, depending on the importance of the match, should be able to accommodate at least three cabins for simultaneous interpretation. Theatre-style seating for journalists has become obsolete.

UEFA has a different concept of a stadium press conference room. Dimensions are graded after match importance. A desktop is required for half of the number of seats (group/quarter final/semi-final and opening/final):
– seats: 125/150/300/600
– surface area in m²: 180/250/420/400

Note: A television-compatible room of this size should be, where possible, free of supports. The fact that the smallest press conference covers an area of 180 m² and a 10 x 10 m grid covers just 100 m² should be considered for structural analysis and for the column grid. The clear height of premises larger than 100 m² must be at least 3.0 m, according to the regulations on workplaces.[5]

Press centre (FIFA)

The press centre with communication facilities, information services, catering, transport office, etc. is activated for the entire duration of a tournament (access for the accredited only). Staggered multi-purpose furnishings represent a fitting-out option, possibly with false floor for technical communication equipment. A FIFA-compatible size would be at least 2,500 m² for existing buildings or for temporary facilities (tents).

Air-conditioning is requested for reasons of comfort, independent of the prevalent climate. The room plan is as follows:[6]

1) Entrance area:
Access control 100 m²
Information, reception 60 m²
Sponsors' area 180 m²
Catering area 220 m²
(with indoor and outdoor gallery, cloakroom, interview room and press conference room)

2) Administration area:
Accreditation office 150 m²
Information office 50 m²
Administration/press office/FIFA office 30 m²
LOC office 50 m²
Copy room 80 m²
Meeting room 80 m²
Telecommunication 100 m²
Fax zone 20 m²
Editorial room (ca. 300 work places)
800 to 1,000 m²
Agency zone – offices 200 m²
TV room 100 m²
Photographers' zone (ca. 80 – 100 work places)
200 m²
Photo partner (sponsor) 200 m²
Computer room 20 m²

UEFA gives details on the media centre only with variable capacity. Figures graded acc. to group/quarter final/semi-final and opening/final:
200/300/400/500.

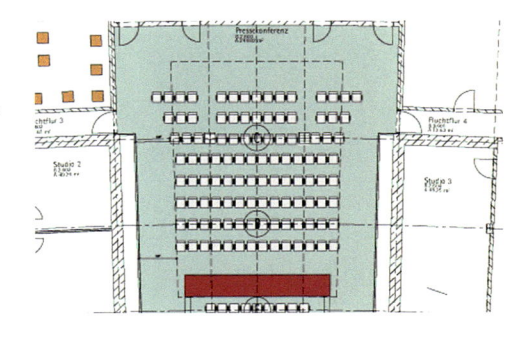

157 **Mixed zone, New York Road Runner**
Contact zone for press/athletes
158 **Press working area (Munich)**
Temporary tents (World Cup 2006)
159 **Press conference room**
Commerzbank Arena, Frankfurt

FIFA – TV compound

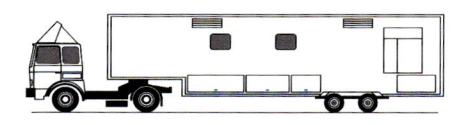

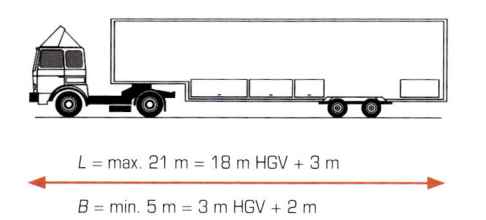

Demand for external media spaces varies from event to event, for multilateral or unilateral TV and radio broadcasts, including national and international satellite links (uplink/downlink). How many production companies will be involved in an event, how many broadcasting signals will be produced and what technical equipment will be used are decisive questions, as are the media marketing rights. During regular league, a pre-dimensioning of 1,200 m² would seem to be sufficient.

For all OB vehicles, FIFA requests a minimum set-up area, fenced and monitored, of 3,000 m² for group games and 5,000 m² for finals.

UEFA requires around 8,000 to 9,000 m² for fenced pitches, available at least 4 weeks prior to the tournament and positioned less than 300 m away from the stadium.

An earlier requirement catalogue situated parking spaces in the basement of the stadium so as to facilitate cabling; outdoor parking spaces were supposed to be covered. This prerequisite was relinquished in favour of capacities for restricted and VIP parking spaces within the stadium. Roofing is preferred but not obligatory for professional broadcasting vehicles.

Ducting and cable system

The transfer point between OB vehicle and stadium can be a permanent media connection point close to the media site. Since it holds moisture-sensitive terminal boxes, it should be roofed and encased. An efficient ducting system including adequate reserves ensures adaptability to coming technical innovations. Pre-installation of cable runs from the commentators' positions to the sites of OB vehicles may only be carried out after consultation with the official host broadcaster.

The structural connections between TV compound and stadium are event-related and hence temporary measures. Depending on demand, cabling can run through bridges or media ducts into the stadium. The bridge solution has the added advantage of bringing media representatives into the stadium, without intersecting other visitor streams.

Further requirements

The broadcasting zone for TES (transportable earth stations) is integrated into the site of the TV compound. Sites for satellite broadcasting vehicles should have an unrestricted view toward the southern horizon and an emergency power supply.

UEFA points out that around 300 m² are needed for media offices with air-conditioning and/or heating. Storage space for host broadcasters is approx. 400 m²/1,200 m³.

Dimensions of OB vehicles

(example: NDR, a German television channel)
a) Television OB vehicle (FÜ 1): 16.75 m long, 4 m high, 31 tons, equipment truck for reports, shows, theatre
b) Television OB vehicle (FÜ 2): 9.20 m long, 4 m high, 14.3 tons, equipment truck for sports broadcasts and live reporting
c) Television OB vehicle (FÜ 3): 10.65 m long, 4 m high, 20 tons, equipment truck for concert recordings and talk shows
(All vehicles contain complete equipment such as cameras, microphones, cable drums, etc.)
d) Digital satellite camera truck: (SNG – Satellite News Gathering)
e) Mobile technology/editing

L = max. 21 m = 18 m HGV + 3 m
B = min. 5 m = 3 m HGV + 2 m

160 **Parking spaces for media vehicles (Bundesliga)**
Tree-lined avenue (west stand)
161 **Maximum dimensions of OB vehicle**

Temporary media facilities

Technical equipment (FIFA)
Mounted desktops with 2 slanting monitors built into the shelf (plus desk light); telecommunication and electrical points (6 to 8 socket outlets, telephone, ISDN multiple socket); cable duct and two separate cable runs from the front face for high- and low-voltage current; 20% of commentators seats provided with a 'ComCam' in the upper rows for TV images by commentators.

The technical equipment for commentators is constantly evolving. Flatscreen monitors in the commentators' zone require less space compared with regular picture-tube sets.

FIFA (UEFA): figures for commentators' positions
Qualifying round/group games: (85)
Last sixteen: 200
Quarter final: 250 (115)
Opening, semi-final: (115)
Game for 3rd place/final: 300 (125)
Bundesliga – at least five radio and TV positions: 10

Media capacities for a large sports event are significantly higher than for regular matches. Hence, these are temporarily erected or supplemented. As a rule, temporary and permanent cabling are distinguished.

A permanent installation of cable ducts and access points makes sense only for directly required media work spaces. If the stadium has a fan television programme, permanent cabling seems practical since personnel and technology are established.

This is frequently not the case when planning of a stadium new build takes place prior to large sports competitions. Further provision would not be advisable due to fast technological development.

In order to provide sufficient openings and structural cable shafts, the location of media facilities and the respective cable runs should be considered during final planning. To maintain flexibility, a parallel system of fixed and loose cabling is conceivable.

For temporary large events, cabling should not pass through public spectator areas. In principle, three distribution systems are feasible.
a) *from above/vertical:* All cables are lead from the external media site to the upper edge of the respective media stand and distributed down with a fan system through mounted ducts between commentators' positions.
b) *from below/horizontal:* The main supply lines are running from under the stand through openings in the riser area. An adequate transverse distribution is arranged horizontally and integrated into the media furnishings.
c) *combination system:* In any case, transfer points within the viewing area should be structurally and statically provided for. Dimensions are to be agreed with the host broadcaster and media planner.

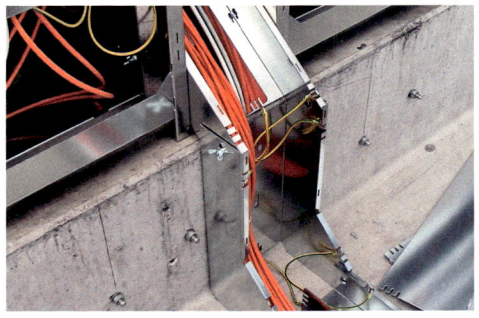

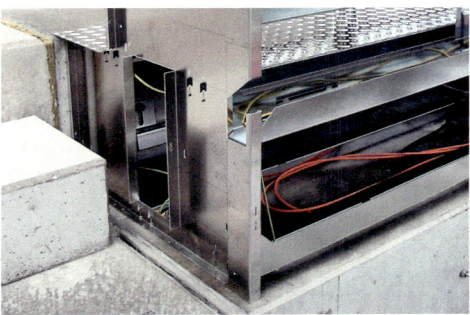

162–164 **Commentators' positions (details)**
Temporary set-up for World Cup 2006

page 86:
165 **Site plan, World Cup Stadium Cologne**
With sponsors' village (marked in yellow) and TV compound (marked in orange)

page 87:
166 **Tier plan, World Cup Stadium Cologne**
With camera/commentators' positions (marked in red)

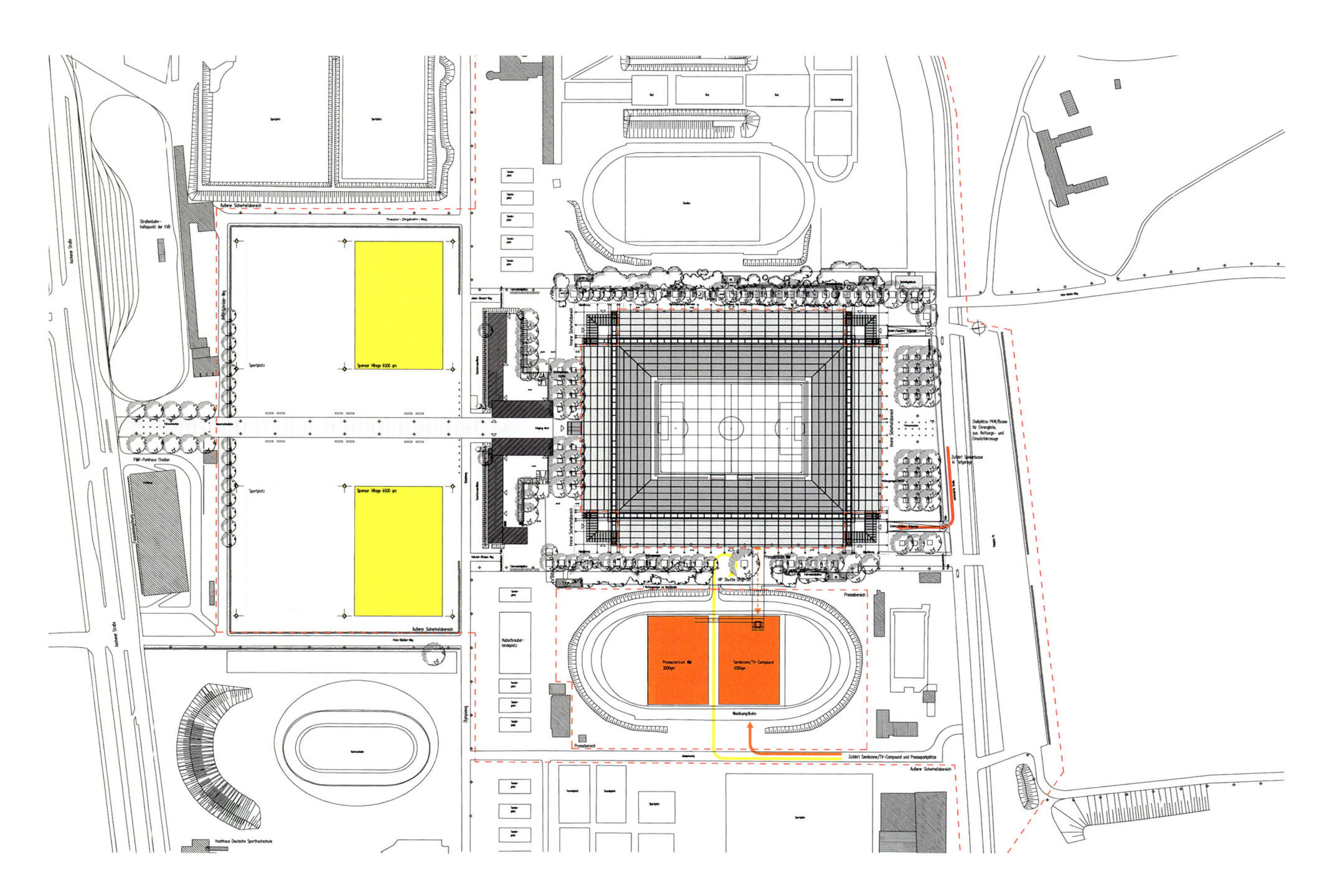

07 Media facilities

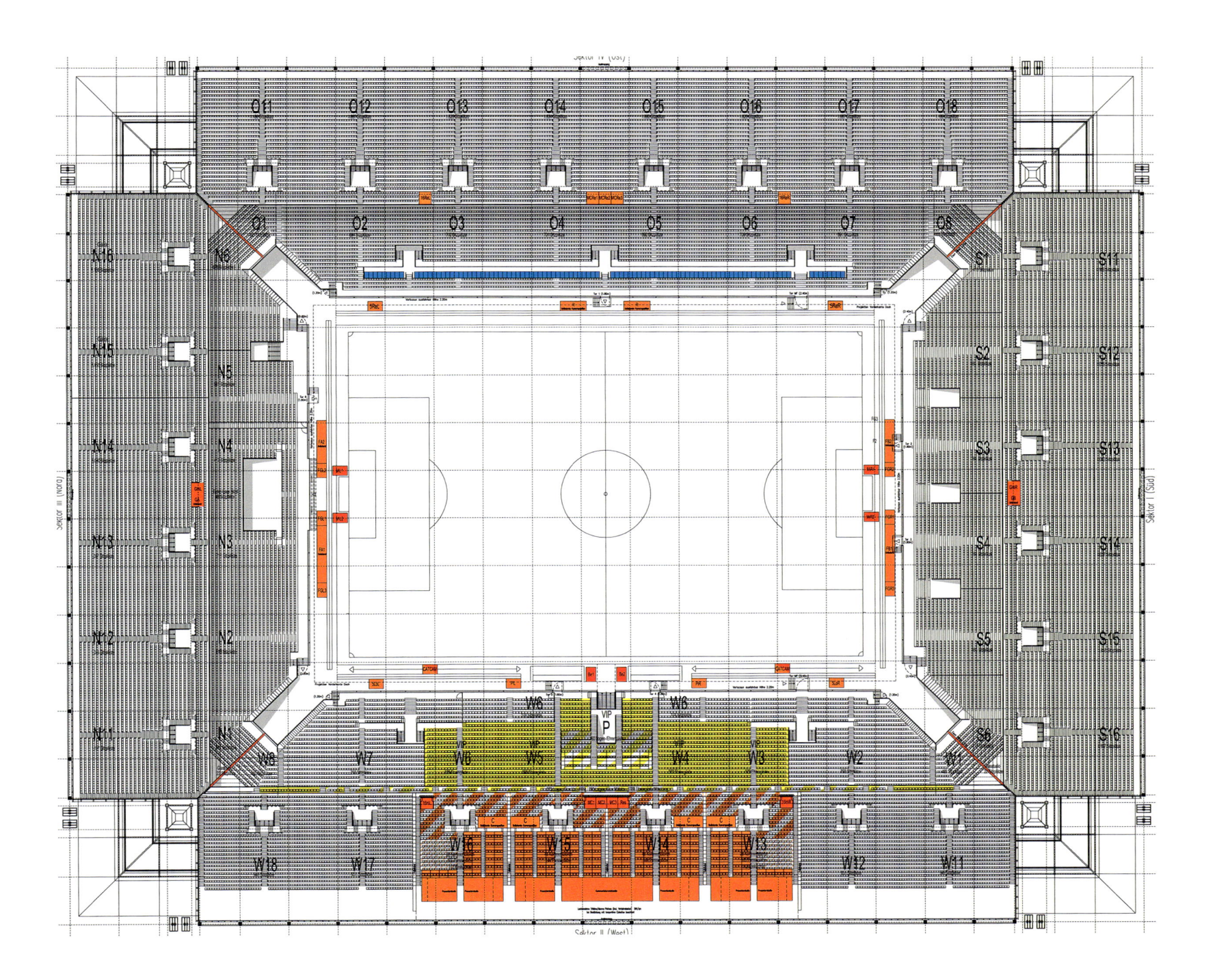

Temporary media facilities 87

Camera positions

All positions have to be agreed between FIFA, the host broadcaster and the LOC. The host broadcaster is authorized to determine the number and locations of additional cameras. Positioning and technical equipment are highly important factors, as is observing the minimum dimensions. Also, cameras should always be facing away from the sun during afternoon and evening games. The main or lead camera is situated on the main stand; its fitting height depends on the camera angle and on the distance and elevational geometry of the stand. According to the FIFA Technical Recommendations, camera angles are measured to the horizontal:
First touchline: 27° to 36°
Centre spot: 16° to 20° (UEFA 15°)
Penalty spot to be seen above crossbar: 12° to 15°

FIFA special features

Atmospheric cameras at the pitch are portable (handheld cameras, steady cam, globe or catcam track cameras). Generally, for each multilateral camera a second space for a unilateral camera is provided. Touch-line cameras are padded by the host broadcaster. A camera crane may be installed behind the goal, as long as viewing restrictions are kept to a minimum. Except for the manager's bench, microphone positions are distributed throughout the entire stadium bowl.[7]

Innovations

Media technology is developing rapidly, especially for reproduction and recording equipment (HDTV, etc.). Consequently, the foregoing details given by FIFA/UEFA represent the current status as of 2002. Sports associations' guidelines can only be stipulated with reservations and are subject to further progress.

The presence of circulating roof structures has created a new camera type ('spider cam'). Suspended from four diagonal ropes, which can be controlled independently, it can be moved to almost all pitch positions and all heights.[8]

Space requirements of camerapersons' positions: Two positions for camerapersons are distinguished: *standing* or *seated*.

Note: None of these two operating modes is deemed obligatory by FIFA/UEFA/DFB. Viewing restrictions for spectators placed behind these positions can therefore not be ruled out.

All following figures in brackets are requirements by UEFA. Fewer cameras are employed for broadcasting Bundesliga matches.

---

167 **Leading camera position (fish-eye)**
During Bundesliga, 1. FC Köln
168 **Camera position**
Main camera at centre line/lower edge of upper tier
169 **16-m camera**
During Bundesliga, 1. FC Köln, position 16 seated, low right or low left

01. Multilateral lead camera
Number: minimum of 3 (4)
Type: MC 1 MC 2 MC 3
Location: main stand at halfway line
Platform size: 3.0 x 8.0 m (same side as commentators, no obstruction to spectators)
First main camera: wide angle
Second main camera: close-ups

02. Unilateral main cameras
Number: minimum of 8 (6–10)
Type: C 1–8
Location: to the left and right of the main camera
Platform size: 2.5 x 10.0 m

03. Multilateral cameras behind the goal
Number: minimum of 2 (2–4)
Type: GHIL GHIR
Location: longitudinal axis of the pitch, behind the goals
Position: behind the hoardings
Platform size: 2.5 x 2.5 m

04. Unilateral cameras behind the goal
Number: minimum of 4 (6)
Type: GA GB
Location: next to goal cameras
Platform size: 2.5 x 5.0 m

05. Sixteen-metre cameras
Number: minimum of 2
Type: 16 HIL 16 HIR 16 LoL 16 LoR
Location: main stand (penalty area), samel level as main camera
Platform size: 2.5 x 2.5 m

06. Five-metre cameras (Slow motion images)
Number: minimum of 2
Type: 5 LoL 5 LoR
Location: main stand, slightly raised at the goal lines
Platform size: 2.5 x 2.5 m

07. Multilateral pitch cameras ('Steady-Cams')
Number: minimum of 4 (2–4)
Type: PIL PIR StL StR
Location: touch line main stand
Position: at least 4 m away from the touch line (UEFA)
Number: minimum of 2 x 3
Type: FGL 1–3 FGR 1–3
Location: behind every goal line
Number: minimum of 2 x 2
Type: miL 1–2 miR 1–2
Location: behind every goal line two mini cameras to either side of the goal

08. Unilateral pitch cameras
Number: minimum of 4
Type: FA 1–2 FB 1–2
Location: behind every goal line
Main camera side: 12 m from the goal post
Opposite side: 9 m from goal post

09. Players' bench cameras
Number: minimum of 4
Type: Be 1 Be 2
Location: manager's bench area

10. Reverse angle – opposite side
Number: minimum of 5 (6–8)
Type: MCRe 1–3 16 ReL/R
Location: opposite main camera and close to penalty areas
Platform size: 3.0 m x 8.0 m or 2 x 2.5 x 2.5 m

11. Unilateral reverse angle
Number: minimum of 4
Type: R 1–2
Location: opposite main camera, close to penalty areas
Platform size: 2.5 x 5.0 m

12. Further camera positions
a) Wireless camera in flash-interview zone or players' tunnel camera (fixed)
b) Catcam-track camera: along the main stand between the side hoarding and the spectator tiers
c) 'Beauty Camera': view of the entire stadium either from the stadium roof, unmanned and remote-controlled, or manned on an external mobile platform, BLIMP camera, helicopter or 75-m crane/platform

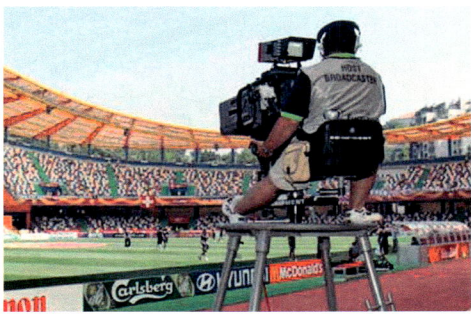

170 **'Steady Cam'**
Portable equilibrium camera
171 **Minimum space requirement – EURO 2005 Portugal** (exceptional situation, not representative)
172 **Temporary camera position**
American Football, RheinEnergieStadion, Cologne

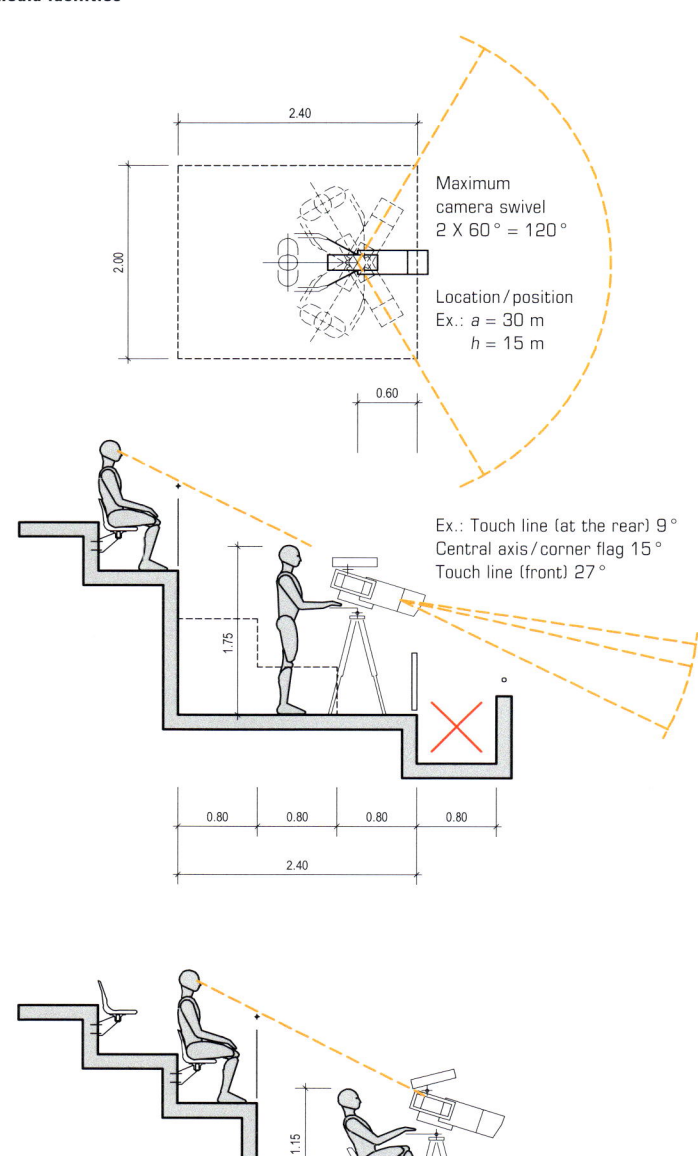

Space requirements of camera positions

The relevant sports associations stipulate significantly diverging space requirements for a standard camera position. Therefore, measures should be adjusted and adapted to the conditions of a regular stand:
FIFA Technical Recommendations 2006:
2.5 x 2.5 m = 6.25 m² per camera
UEFA Technical Recommendations:
2.0 x 2.0 m = 4.00 m² per camera

Since this roughly 30 % difference in measure has a strong impact on planning, the position of a camera, standing or seated, shall be examined in more detail.

Minimum space requirements of a camera

Seated, the cameraperson is surrounded by a half-circle track up to the corner flag ('spider' width = 2.40 m, or as special construction $w = 1.60$ m; compare to Stuttgart stadium).

When standing, the cameraperson works with a height-adjustable tripod or telescope base, depending on the device. The distance to the pitch of a leading camera requires a maximum camera swivel of 2 x 60° = 120°.

Note: Because up to now production companies have issued no clear guidelines, relevant conclusions can be drawn only from empirical surveys of several TV channels and from the experiences of the ConFedCup 2005.

The sectional geometry of a regular stand is generally based on an 80-cm step grid to which space requirements should be geared; structurally, this results in a multiple set-up depth:
from 2 x 0.80 m = 1.60 m
to 3 x 0.80 m = 2.40 m (minimum 2 m)

Step 1: FIFA standard

Translation of 2.50-m FIFA space requirement to a standard step measure of 3 x 80 cm = 2.40 m.

173 **Example for leading camera (standing)** Standard situation (permanent at lower edge, upper tier)
 a) Camera position is integrated into the regular stand profile.
 b) First seating row in front of camera must remain clear (spectators rising to their feet).
 c) 90-cm barrier remains without additional safety measures.
 d) Camera position in the second row, taking advantage of barrier/advertising hoarding (90 cm).
 e) First to third successive seating rows are viewing restricted due to camera in place.
 (point of focus: distance = 30 m, eye-point height = 15 m)

174 **Example for leading camera (seated)** Standard situation (temporary, lower edge, upper tier)
 a) Camera position is mounted onto the regular terrace front.
 b) First seating row in front of the camera is omitted.
 c) First seating row behind is highly viewing restricted due to the cameraperson.
 d) First to tenth seating row behind are highly viewing restricted due to the cameraperson (point of focus: distance = 30 m, eye-point height = 15 m).

Space requirements of camera positions 91

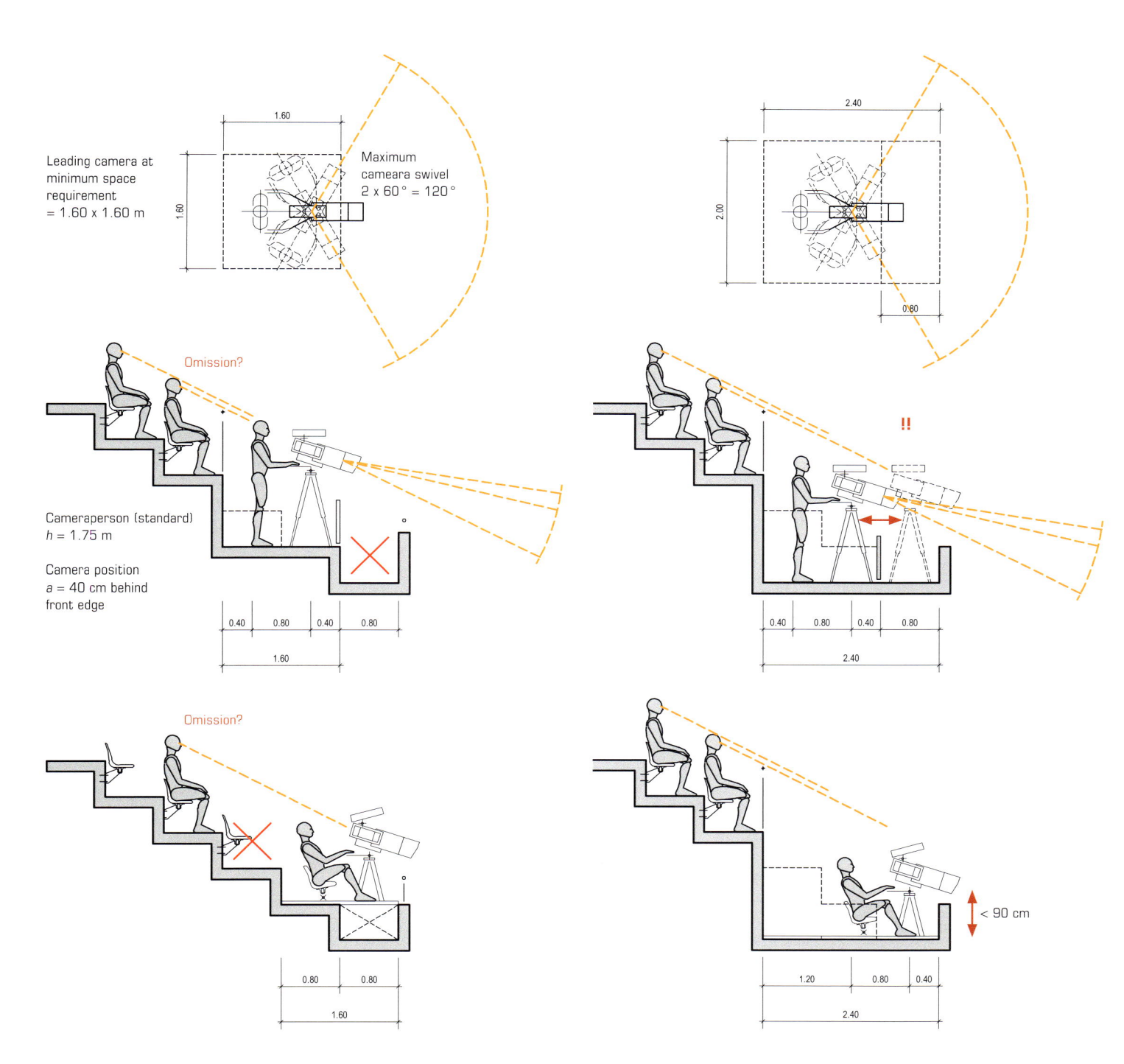

Step 2: Minimal variant 1.60 m (new)

left column:

175 **Reduction of camera space requirements to the minimum measure specified for planned use**
Standing camera position at minimum space requirement but with viewing restrictions for spectators.

176 **Seated position as opposed to standing operation (clearly less viewing restrictions)**
Seated camera position at minimum space demand, but forward movement is required.

Leading camera at minimum space requirement = 1.60 x 1.60 m

Maximum camera swivel 2 x 60° = 120°

Cameraperson (standard) h = 1.75 m

Camera position a = 40 cm behind front edge

Step 3: Special scenarios

right column:

177 **Rear position camera axis minus 1.20 m (in the centre of the camera pitch 2.40 x 2.40 m)**
Optimization of ground plan position due to sight impairment for spectators.

178 **Seated position in the same set-up space (clearly less viewing restriction, as opposed to standing operation)**
Ground plan definition redundant. Temporary omission of barrier to allow camera swivel.

Space requirements of a camera position
(Tread depth x set-up width)

Standard   2.40 x 2.00 m
Exception 1.60 x 1.60 m

According to the rules of sightline calculation, the eye point of the cameraperson should be complying with sightline geometry. This means that the most projected point, or head of the camera, might be restricting the field of vision of people sitting behind.

Therefore, the fixed camera position normally cuts into the terrace body in order to restrict viewing for as few seats as possible.

The position is taken up by the apparatus itself including cables, obligatory equipment and the cameraperson. The set-up situation at ConFed-Cup 2005 serves to show that depths of 1.60 m are sufficient if the camera head is allowed to jut out of the double row (2 x 0.80 m).

No people are allowed to sit in the immediate area of the camera movement. Spectators standing or jumping up with arms stretched out should not restrict the sightline of a slanted camera. The advantage of a position in the first row is no seat loss, but a low pitch perimeter hoarding; the location of the camera in the second row means a circulating barrier, but also loss of seats.

Minimal variant (1.60 m)

The diverse set ups of the camera positions have clearly demonstrated that installation surfaces may be reduced in size. This would result in the loss of the first seating row behind. For spectators in rows 2 to 7, cameraperson and camera are in the field of vision toward the point of focus.

Hence, this type of installation seems suitable only for temporary use. Additional positions for large sports events do not have to be provided for structurally but can be realized as temporary measure within the standard cross-section of the 80-cm step.

FIFA variant (2.40 m)

The size of a single camera space is adapted to the standard tread depth of 80 cm, resulting in a 2.40 m set-up space. This dimension is for permanently fitted camera positions which are integrated into the terrace front but not positioned above the sightline curve. For the ground plan location of a camera, a rear position is urgently recommended (camera axis = centre space; –1.20 m to the front edge) as the 'front' position restricts viewing for several seating rows behind the camera.

Note: Experience has demonstrated that mere oral instructions tend to have little effect. Hence, a small railing should enforce the rear position for camerapersons; controlling the unrestricted swivel toward the corner flag is a basic prerequisite. (Distance between point of focus and corner flag is almost double; thus, the vertical visual angle is halved.)

179 **Mounted permanent camera**
     World Cup Stadium Nuremberg
180 **'Spider Cam'**
     (during Bundesliga)
181 **Rope-guided cable camera**
     FIFA World Cup Korea/Japan 2002

Super slow motion

Around seven regular cameras plus one special camera are employed for super slow motion at Bundesliga matches. 'Superslomos' produce triple the amount of images compared to regular cameras, to avoid jerking frames. Hence, significantly less time is available for every individual image. On the other hand, reduced shutter speeds require much better light. Since the super slow motion camera is installed low down at the pitch perimeter, it can only make use of the vertical illuminance E_v.

'Camera technology is progressing toward requiring less light for ever improving images. We may assume that, in future, TV technology will be less demanding concerning the brightness of floodlights.

Whereas both cameras would produce exactly the same picture quality, light penetration could be one f-stop lower for a camera of the latest generation than for a camera manufactured 7 years ago. With regards to brightness, modern cameras have higher reserves at their disposal.

In future, TV producers will not be demanding brighter floodlights; only the colour temperature would leave room for improvement,' as stated by the head of technology at WIGE Media AG.[9]

High Definition TV (16:9)

Apart from plasma TV, HDTV is the term to capture the imagination in the coming years.

The abbreviation HDTV stands for High Definition Television: a global digital TV standard with significantly improved image quality, in wide screen format with sharp outlines, rich colour and high depth of field.

The PAL standard is widespread in Europe and has a resolution of 576 visible lines (vertical) and a maximum of 720 lines (horizontal) with a 50-Hz frame rate and interlace scan.

Since HDTV will streamline television standards, not the least as a consequence of global sports broadcasts, we should mention the US-American counterpart to PAL, the NTSC standard. It has a resolution of 480 visible lines with the same maximum of 720 lines, frame rate is at 60 Hz, interlacing. Since resolution defines the richness in detail of an image, the frame rate and broadcast system (PAL or NTSC) guarantee a flicker-free TV image.

Frame frequency signifies how often per second an image signal is transmitted; 50 times per second for 50 Hz. European HDTV development operates with 1,250 lines, a frame rate of 50 Hz and a new aspect ratio (16:9).

The Olympic Games in Athens were already recorded and broadcast to large parts of the globe in HDTV (720 p). 16:9 relates to image reproduction equipment (television, video projectors) or image recording equipment (video cameras) with a corresponding visual field format. Classic formats for video technology would be 4:3 and for photography 3:2; today, standard page format for SDTV is still 4:3.

As today almost all cinema formats range above 1,78:1, endeavours have been focusing, since the beginning of the 1990s, on changing the standard format of television from 4:3 to 16:9. Some of those attempts have included the introduction of PALplus, the anamorph codification on DVDs and for DVB (Digital Video Broadcasting) and the determination of 16:9 as standard format for HDTV.

DVB uses one satellite channel to broadcast data-reduced image and sound signals. Depending on the quality standard, 4 to 10 television broadcasts may be transmitted on one channel. Video signals are data-reduced, i.e. only changes in active image data are transmitted. Reception of DVB requires a digital decoder, a so-called set-top box. The decoder is connected via a Scart point with the TV set.

Note: New TV sets, with LCD or plasma monitors, are needed to fully appreciate this new technology. HDTV affects floodlighting. The higher resolution of the image requires a higher TV-compatible illuminance. Instead of the present 1,500 Lx, future requirements will be 1,800 to 2,000 Lx.

Chapter 08

182 Players' changing room
RheinEnergieStadion, Cologne
183 Players' changing room
Ibrox Stadium, Rangers Football Club, Glasgow
Scotland, capacity: 50,500 covered seats

Players' area and changing rooms
Facilities for players, referees and team officials

Players' area

Changing rooms for the playing teams clearly exemplify the usability of a sports stadium; they are mostly located at the side of the main stand, close to the field. Originating from the catacombs of Roman amphitheatres, they belong to the extensive ancillary rooms of a stadium and are obligatory for matches and competitions in modern sports venues. Rarely will regular spectators be allowed access to these areas since some sort of accreditation would be required.

Access in order of priority: Players and team officials, referees and UEFA officials, VIP and UEFA guests, accredited media and official partners.

To the extended group of functional personnel belong, apart from medical staff and suppliers, all persons participating in the further entertainment programme (ball boys, musicians and artists). Due to the multitude of people present, the main corridors within the players' area should be clearly routed and sufficiently wide ($w = 1.50$ m). Doorways should have at least 1.0 m clear width to allow passage of material (roller container of the kit manager) and first-aid transports (stretchers). Sports stadia which are also used for popular sports or training purposes should provide shoe-cleaning facilities and information boards at the entrance.

Team rooms

First-class changing rooms for teams and referees have become standard; located generally on the same side of the stadium as VIP stand and media facilities, inaccessible for spectators and media representatives.

FIFA/UEFA attach value to the fact that the fitting-out standard should be equal for resident and guest teams. Installation of cupboards or lockers for safekeeping valuables is recommended, particularly for athletics competitions lasting all day:
– Number: at least two, preferably four rooms
– Size: 150 m² each (FIFA), 100 m² each (UEFA)
– Number of persons: at least 25
– Fittings: clothes' hook or lockers, refrigerator, telephone, television, tactics board
– Sanitary area: showers and toilets, directly adjacent
– Number: 10 showers, 5 washbasins,
3 WCs and 3 urinals, 5 hairdryers

Relaxation pool

is required according to FIFA and stipulated by IAKS (1993) with 2.0 x 0.60 m reclining area per user and with resting/seating spaces ca. 60 cm under water. Circulation routes shall be at least 1.20 m wide.

Massage room

According to FIFA, two massage benches (perhaps a room) and, according to IAKS, at least one bench, 1.20 x 2.50 m, accessible from three sides at 60 cm; stool, cupboard, shelf, washbasin (wardrobe) are to be provided.

Team coaches' changing rooms

are dedicated rooms with wardrobe, sanitary area with shower and washbasin, close to the players' changing rooms. These rooms should be provided with a desktop (PC) and chairs.

Referees' room

According to FIFA: two changing rooms for fours persons each, a room with sanitary facilities, modern infrastructure and direct covered access to the pitch. Location is separate from players' area, but close to the changing rooms.
– Size (total): at least 40 m²
– Fittings: seats, table, massage bench, refrigerator, 2 showers, 1 WC, 1 urinal, hairdryer and telephone

Dope testing area

A room close to the players' rooms, not generally accessible.
– Size: at least 20 m²
– Fittings: desk top with two chairs, fax/copier, lockable wardrobe (storage of samples), telephone

A waiting area with seating for eight people, a shower and WC with monitored toilet access.

Medical care

A medical examination room for players and referees, in emergency also for injured persons, close to the dressing rooms, respectively playing field (DFB Safety Guidelines § 16). Access should be wide enough for medical transports (stretchers).
– Size: at least 24 m²
– Fittings: bright and hygienic (FIFA)

An examination table, appropriate medical equipment (oxygen tank with mask, blood pressure gauge, hotplate for instruments), two portable stretchers, writing desk and chairs, lockable glass cabin for medication and devices, washbasin and telephone.

Other rooms

Room for ball boys/girls (recreational space and seating for 15), musicians and instruments, organizer's offices, storage rooms, special purpose room (WC nearby).

Warm-up area

FIFA requires a warm-up area for each team near the dressing rooms:[1]
– Size: at least 100 m² each
– Fittings within the building: ball-proof and air-conditioned
– Fittings out in the open: grass or artificial turf (separated)

UEFA also demands two private and separated warm-up areas.[2] Internal statements by FIFA/UEFA have revealed, though, that this area is hardly ever used by coaches. Consequently, space reserves of 2 x 100 m², erected at high structural expense, with sports floor or ball-proofing for installations (air-conditioning), are remaining largely inactive. Hence, this requirement should be analyzed and, if justified, cancelled.

Training grounds

FIFA requires for each venue two excellent natural-grass playing fields in an enclosed, secured complex with a roofed section and separate access for media representatives. UEFA demands that the best possible training facility shall be less than 20 min away from the team hotel.

184 **Massage area**
Veltins Arena, FC Schalke 04
185 **Relaxation pool**
RheinEnergieStadion, 1. FC Köln
186 **Warm-up area (ball-proof)**
RheinEnergieStadion, 1. FC Köln

Players' bench and tunnel

On their way from the changing rooms to the pitch, players should pass through (ideally separate) corridors opening to the field near the centreline. When entering and leaving the field, players are to be protected by suitable structures from being molested or threatened by spectators.

Exception: If the first row is not placed significantly higher, players' benches are to be integrated into the stand and secured.

Players' tunnel
A closed, fireproof steel structure, with 'cabriolet' roof or as extendable telescopic tunnel.

The mouth must be at a safe distance to the spectator area with a pitch distance of 1.50 to 6.0 m (FIFA). Since the tunnel is usually retracted after players have emerged, there should be a fast opening mechanism, in case of substitutions.

Clearance is $h = 2.40$ m (FIFA), flooring should be skid-proof, and ideally a small WC should be fitted near the entry.

Players' benches for two teams
Each bench should be able to accommodate 25 persons (World Cup 2006) or at least 20 persons (World Cup 2002). Seating benches should have backrests; fittings comparable to regular single-seat shells or more comfortable as single chairs. Both benches should be equal in finish and heated in winter; cold feet are to be avoided by suitable measures. Frequently unavoidable on account of one-site conditions, the manager's bench is placed within viewing of the spectator stand behind. Therefore, a transparent roof protecting from weather influences and hurled missiles should be installed.

Lowering the manager's bench by one or two steps is feasible but the floor indentations often impair the servicing and supply of the stadium bowl for other events. It also restricts the flexible response and event-related repositioning of the bench, e.g. through addition of a central camera position.

A manager's bench is ideally 2.0 m high to allow quick entry and exit, but a lower total height (standard height approx. 1.65 m) means less sight restrictions for expensive spectator seats.

In any case, all edges must be padded. Players' benches should be positioned at field level and at least 5.0 m away from the halfway line (FIFA), i.e. 2 positions with at least 10 m distance, parallel to the pitch. An enclosure for match supervision (4 persons) is to be positioned between the teams. The Technical Zone (players' area) is at a distance of 1.0 m from the pitch and 1.0 m to the left and right of the team bench.

Flagpoles
For representative purposes of the playing teams, FIFA demands five flagpoles or alternatives for presentation of the national flags.

187 **Players' tunnel, prismatic shell**
Estadio do Braga, Portugal
188 **Players' tunnel (telescopic)**
World Cup Stadium Nuremberg
189 **Sunken manager's bench**
Bundesliga stadium, Bayer Leverkusen
190 **FIFA-protocol flags at the roof edge**
World Cup Stadium Cologne, CZE vs. GHA
(17 June 2006)

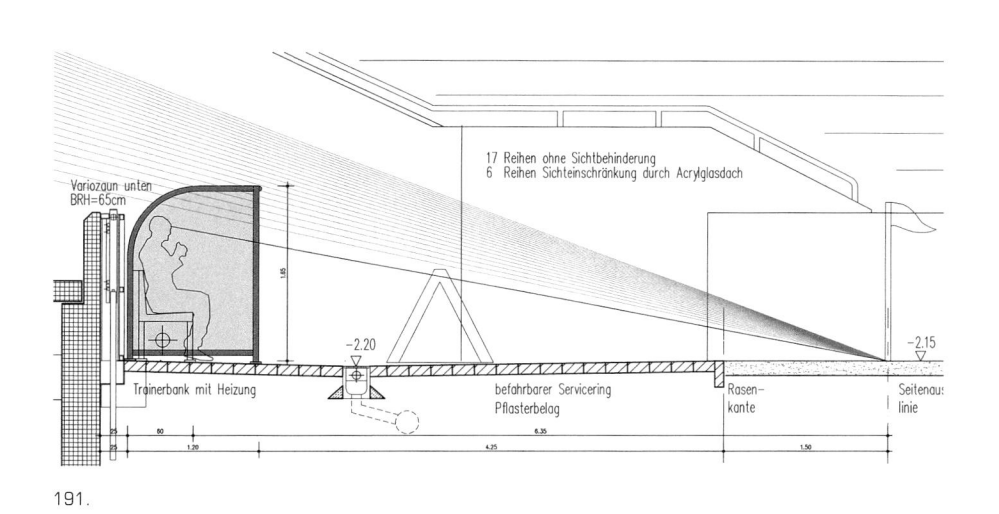

top left:
191 **Players'/Coach's bench, classic shape**
RheinEnergieStadion, 1. FC Köln
Rear façade, transparent and rounded
Note: Material must be reflection-free. In a position close to the pitch, viewing restrictions are to be expected for the first seating rows.

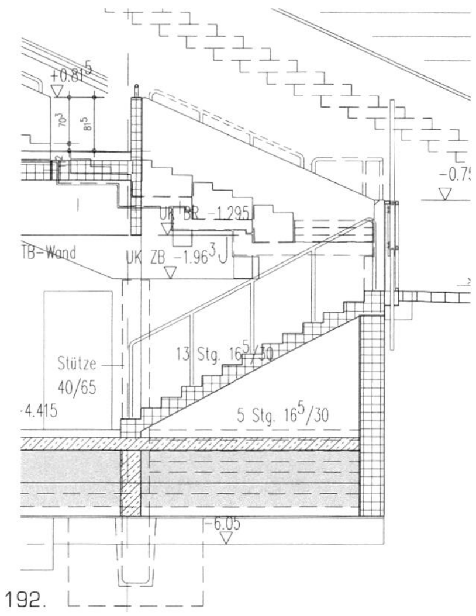

top right:
192 **Access to players' tunnel**
Note: Tread covering of stepped players' tunnels should be stud-proof. If the exit is within the service ring, the stand vomitory is omitted in favour of a trafficable lift slab which renders the stadium bowl fully operational outside of matches (cf. Allianz Arena, Munich).

193 **Position of coach's bench**
Note: The substitutes' benches should ideally be modular as they need to accommodate varying capacities. Location/position should be changeable and for example accommodate the seating bench of FIFA/UEFA game supervisors between the two players' benches.

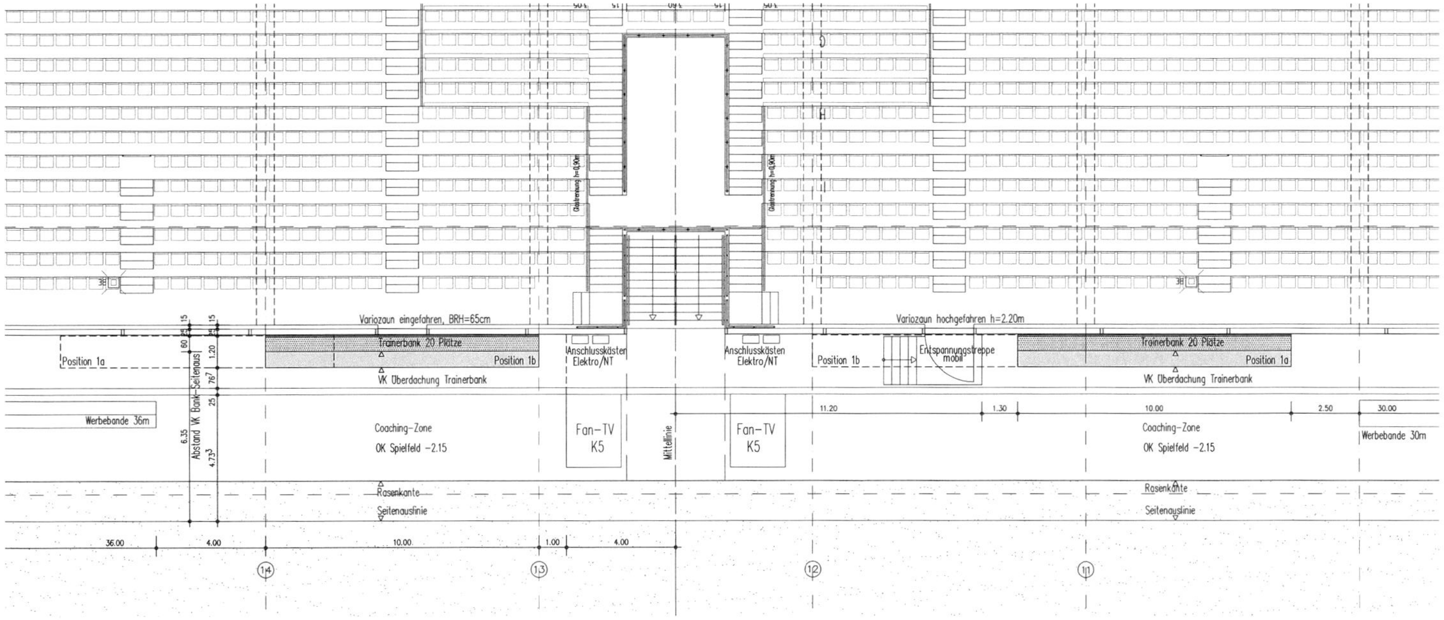

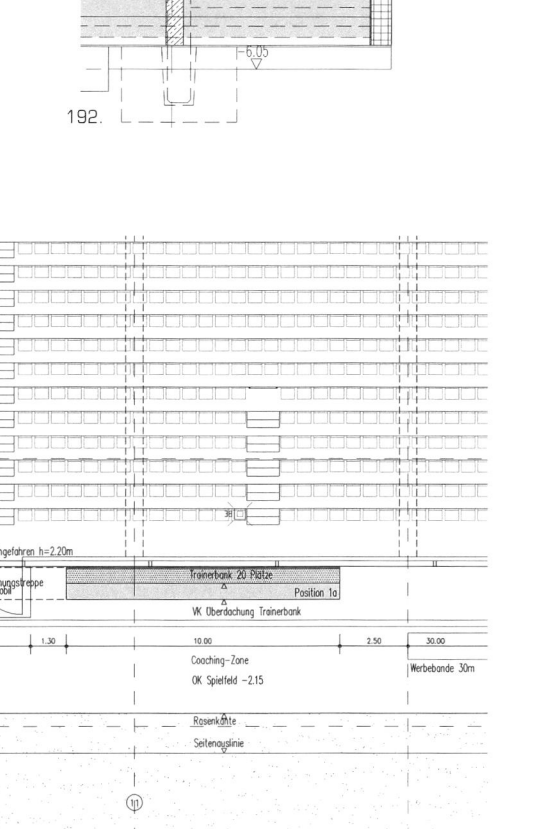

Chapter 09

194 Stadium control room
Estadio do Dragao, Porto (Portugal 2004)

Stadium administration
Essential functional rooms such as stadium control room and other ancillary rooms

Stadium control room

Given that operation of a modern stadium is a technologically complex and event-specific undertaking, a control centre with clear and unrestricted viewing of the entire stadium bowl is requested by FIFA and UEFA. Located in the direct vicinity of the centre of operations for security staff, it should be realized as a secured zone with designated entrances, separate from the general audience.

DFB Safety Guidelines (§ 10) require provision of rooms for police, fire service, stewards and medics. These premises have to be adequate in size and number, as separate but adjacent rooms with appropriate technological outfit.

Note: Specifications for installations are not further detailed in this context as domestic engineering is progressing at a rapid pace, and on-site technology should be constantly adapted.

Spatial requirements are to be clarified in advance with the persons in charge. Additionally, an extensive safety concept geared to special buildings of this type must be agreed upon with local authorities.

In the run-up of a large sports event, sports associations demand a so-called 'safety certificate' which must be confirmed by local authorities and must be renewed annually.

Sound, floodlight, lighting systems and scoreboard are operated directly from a technical stadium control room. Therefore, a public address system control room overlooking spectator accommodations and central areas is obligatory. It must have a priority link (override switch) to police, and the centre of operations for police and fire service must be provided with a direct (structural) link to the stadium speaker.

Video monitoring system

Mounted, remote-controlled colour video cameras, revolving and with zoom function, should be installed in all relevant public areas inside and outside the stadium. Areas to be monitored are concourses, spectator routes, approaches, entrances, squares as well as external areas. The police operations centre should be capable of producing still-shots in portrait size (plus print-out) for identification purposes. The video monitoring system, which is used by security services outside of matches, must be fitted with a priority interrupt for police.

Loud-speaker facilities

As multi-purpose event venues, modern sports stadia require an excellent quality of sound for speech and music with individual sector alignment. Security staff suggest an either totally or selectively switchable PA system, e.g. for intermediate areas behind the goals, straight and back straight (esp. home and away fans) or the playing field.

UEFA states the following requirements: a modern and efficient system which produces clearly audible announcements during the game, inside and outside the stadium.

Intelligibility has to be maintained in spite of unfavourable conditions such as spectator noise, sudden rises in noise levels or loss of main power supply.

According to Technical Recommendations of FIFA/DFB, a police priority interrupt with dedicated microphone should be installed, and an automatic regulation of sound volume should be feasible for emergency announcements. Rooms for loud-speaker technology are to be located near the stadium control room. FIFA determines the sound pressure level with 105 dB (A), 10 dB (A) above the interference level.

Other ancillary rooms

Stadium police station
According to FIFA and DFB Safety Guidelines, a stadium police station must be provided as easily accessible rooms on the premises with detention and arrest cells for up to 20 people.

Rescue routes
Accesses must be trafficable and barrier-free. Set-up and circulation areas for police forces and rescue services must be secured, clearly identified and accessible from separate entrances. Security and rescue personnel must be able to drive into the stadium bowl with large emergency vehicles. Off public access roads, shortest possible distances to rescue centres (fire brigade) and hospitals should be guaranteed; emergency evacuation routes within the complex should be two-lane.

For emergency rescue helicopters, at least one clear landing space should be provided at short distance to the stadium (areas of operation in case of disaster).

Organizing a tournament requires extensive preliminary planning, and assistance during implementation is obligatory. Therefore, FIFA expects the provision of dedicated office space at the venue. Rooms should be adjacent to the playing field and changing rooms. Fittings comprise a TV set, furnishings and telecommunication points, washbasins and WC (except for group games, plus three rooms with 10 m² each). Two years prior to the tournament, at least two operational LOC offices should be provided.

2 x FIFA office (approx. 10 to 16 m²)
2 x LOC office (approx. 8 to 10 m²)
1 x stadium operator
1 x police (not operations centre)
1 x stewards
1 x medical service
1 x meeting room (approx. 30 m²)

UEFA offices:
– group games 5
– opening game 12
– at least 12 workplaces inside the stadium,
– two conference rooms (80 / 30 m²).

IAKS – Competition offices:
For every 50 persons one m² in two separate rooms, conference room, appeals room ('Jury d'Appell') ca. 20–30 m². For purposes of event organization, all rooms are to be adequately equipped.

Storage rooms:
Storage capacities have to comply with the size and utilization of the sports ground and the operating structure.[1]

A translated quote from the relevant German regulation for spectator facilities:[2]
1) Adequate workshops must be provided for carrying out fire-hazardous works such as welding, soldering or sticking.
2) Dedicated storage rooms must be provided for storage of decorations, props or other inflammable materials.
3) For collection of waste and recycling material, adequate containers must be provided out in the open or in special storage rooms.
4) Workshops and storage rooms should not be directly connected to essential stairwells.

When hosting a World Cup tournament, FIFA requires a warehouse of ca. 650 m² plus an LOC 2006 warehouse of ca. 300 m², dry and lighted.

IAKS reference value for equipment room: one square metre per 500 – 700 m² of sports ground. Access should be at least 2.20 x 2.20 m (smaller doorways at least 1.50 m), adapted to relevant vehicles for ground maintenance. Clear height should be at least 3.0 m throughout.

UEFA storage rooms:
Two storage rooms with direct access to the field and two additional storage rooms for the opening game and final (800 and 200 m²) are to be provided.

IAKS fleet:
Approximate value: one square metre per 400 – 500 m² sporting area for dedicated fleet, one square metre per 700 – 900 m² sporting ground for external fleet.

195 **Headquarters (police)**
World Cup Stadium Dortmund (2006)
196 **Stadium control room**
World Cup Stadium Nuremberg

Chapter 10

197 **The FIFA World Cup** (impression taken from the ISE Hospitality Programme)

Planning principles
A listing of essential regulations and recommendations for the construction of spectator facilities

Regulations and Recommendations

MVStättV
Title: Regulation on the Construction and Operation of Places of Assembly
Version: June 2005
Status: special building regulation – legally binding

FIFA
Title 01: FIFA Technical Recommendations Profiles and requirements for cities and stadia for hosting the FIFA World Cup 2006
Version: November 2002
Title 02: Technical Recommendations and requirements for the new build or modernization of football stadia
Version: 3rd edition 1995
Title 03: Safety Guidelines
Version: January 2004
Status: urgent recommendation for hosting of FIFA tournaments

UEFA
Title: 'Stadia List' and UEFA Technical Recommendations 2008, requirements for host venues
Version: 2004
Status: urgent recommendation for hosting of UEFA tournaments

DFB
Title: Guidelines for the improvement of safety at Bundesliga matches
Version: January 2004
Status: urgent recommendation for hosting of UEFA tournaments

DIN EN 13200 – Spectator Facilities
Part 1: Criteria for the spatial layout of spectator accommodations
Part 3: Barriers (draft)
Part 4: Seats and product features
Version: September 2003
Status: Part 1 – valid throughout Europe; expected to be implemented as National Law.

British Standard
Title: Guide to Safety at Sports Grounds
Version: 1997 (2005/4th)
Status: blueprint for EN/DIN 13200-1; the Scottish Office, Department of Culture, Media and Sport

DIN 18035 on sports grounds
Part 1: Open-air spaces for games and athletics, planning and dimensions
Version: February 2003
Status: valid in Germany

DIN 18065 on stairs in buildings
Title: Definitions, rules and principal dimensions
Version: January 2000
Status: valid in Germany

IAKS / IOC
Title: Sports grounds / Stadia Planning principles No. 33
Version: 1993
Status: update required

IAAF
Title: Track and Field Facilities – Manual
Version: 1995
Status: urgent recommendation for hosting competitions of the International Amateur Athletic Federation

Mandatory construction specifications

Various types of utilization present special requirements to a modern sports and event venue. On the opposite page, only the essential and most current regulations issued by lawmakers and by the large sports associations are listed; all other effective building regulations and executive orders (national law) and the relevant commercial guidelines or other EN/DIN standards remain valid.

In Germany, sports grounds are subject to the building regulations of the federal states [*Landesbauordnungen*] on the condition that 'generally recognized codes of practice' are to be observed.[1] In the build-up of planning a large project, such as the special building-type of a stadium, it seems recommendable to determine the legal foundations in cooperation with local authorities. Planning and construction often take up an extended period of time during which the effective legal situation might be up-dated or altered. Normally laid down in contract before planning begins, this is particularly relevant for international projects which might not be regulated by national law. In this case, comparable standards and regulations of other countries might apply. The building code law [*Bauordnungsrecht*] of the Federal Republic of Germany falls within the jurisdiction of the federal states [*Bundesländer*]. According to the German civil code, the federal legislator may issue special federal building regulations which are effective throughout Germany.[2] However, a uniform and comprehensive regulation is effectively left to the federal states.

For this purpose, the Bundesländer have founded a consortium called ARGEBAU, where expert committees and project groups prepare master drafts, which in turn are recommended to the state ministries for introduction.

Since not all states implement these proposals, regulations may diverge. In future, these master drafts, as well as the DIN regulations of the Standards Committee, will be examined, notified and recognized as 'National Law'. Appeals by other EU countries may only be lodged against aspects regarding competition law.

Since 1991, Germany has a national concept for sports and safety which was initiated by the federal state ministers of the interior.[3] DSB and DFB were integrated in setting up these safety recommendations. Since 1988, DFB has been updating its guidelines for improving safety at Bundesliga matches, which can be seen as a continuation of the National Concept.[4]

Versammlungsstätten-Verordnung [Regulation on the Construction and Operation of Places of Assembly]

The legal foundation for such building types is summarized in this regulation on the construction and operation of spectator facilities.

Stipulated first in 1978, it was introduced later (in a 90s' version) within the framework of German reunification throughout Germany. A revision of the design draft was concluded in the process of deregulation in 2002, reducing the number of paragraphs from 137 to only 48. The purview of this regulation was extended (e.g. to the catering trade).

Focus is currently shifting from theatre and event halls to multi-purpose sports stadia and their diverging requirement profiles.

The soon to be introduced legally binding regulation throughout Germany, the *Muster-Versammlungsstätten-Verordnung* (MVStättV 2005), shall be used as the basis for all further considerations and examinations.

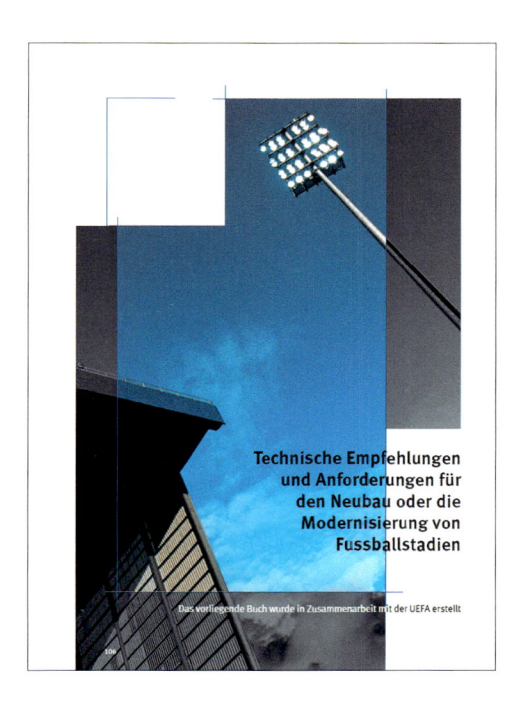

198 **FIFA / UEFA 'Pflichtenheft' 1995**
(cover of Technical Recommendations)

Three types are distinguished:
- Regulations [*Verordnungen*]
- Guidelines [*Richtlinien*]
- Recommendations [*Empfehlungen*].

Regulations are to be observed by the construction authorities and by all relevant natural persons and legal entities (clients, operators, organizers, etc.) after its publication in the law gazette of a federal state.

Administrative guidelines are binding for all administrative authorities after publication.

In the framework of the relevant competence, these are put into practice through single administrative orders (under the relevant conditions of approval notification).

Recommendations are comparable to guidelines; breaches are only counted as legal infringements if acting against regulations or guidelines.

German Civil code on 'special building types':[5]
(1) For structural complexes and rooms of a special type or utilization, special prerequisites might apply in individual cases for implementation of general requirements, according to §3 (section 1, clause 1). Simplification/deviation might be admissible in individual cases, inasmuch as keeping to regulations is not obligatory
a) due to the special type or utilization of buildings and rooms, or
b) due to the special requirements according to clause 1.

(2) Requirements and deviations may apply to […]
4) the construction type and layout of all building components essential for stability, traffic safety, fire protection, heat and noise protection or health considerations,
5) fire protection and precautions, […]
7) the layout and fabrication of lifts as well as stairs, staircases, corridors, exits, other escape routes and their signage,
8) the admissible number of users, layout and number of admissible seating and standing accommodations for places of assembly, restaurants, entertainment centres, spectator stands and temporary structures.

Sports venues are classified as 'special builds', according to the simplified approval procedure of the civil code.[6]
'The simplified approval procedure is not valid for […]
7) places of assembly with room for more than 200 persons,
8) sports venues with more than 1,600 m² surface area or more than 200 spectator places, uncovered sports grounds with more than 400 seats.'

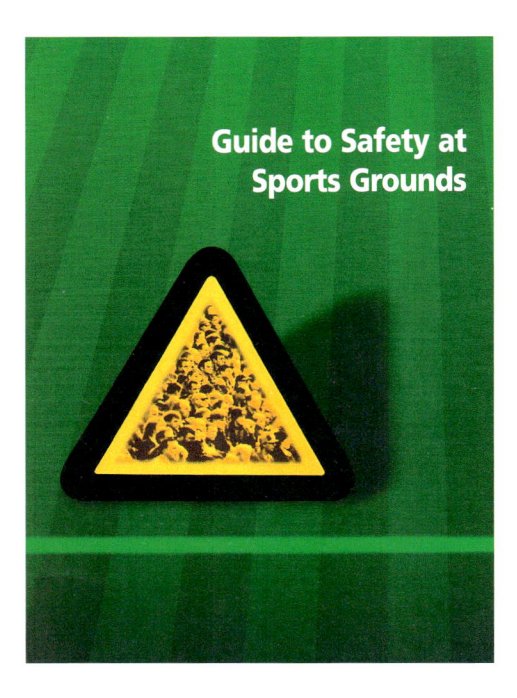

199 **Technical Guidelines**
SCOTTISH OFFICE (1997) Fourth Edition, *Green Guide* (The Stationary Office, London)

Regulations / Recommendations

The hosting sports associations are to be contacted urgently in the build-up of construction, or rather their respective regulations and recommendations are to be considered.
Conditions for keeping to these regulations are based on a suitability test for the hosting of such an event. Example:
– FIFA / UEFA-suitable or
– according to IAFF-standard or
– corresponding to *Green Guide*.

Note: The general building regulations are to be considered (cf. *Green Guide to Safety at Sports Grounds*), and the necessary assessment and evaluation should be executed by experts.

The European Committee for Standardization, or C.E.N., was established by the European Union to set up 'TCs – Technical Committees' with the aim of achieving far-reaching standardization in the member states of the EU. In the standards contract [Normenvertrag] of 1975, the German Standards Institution [Deutsches Institut für Normung (DIN) e.V.] in Berlin was recognized by the German government as the competent authority for standardization. In return, the DIN committed itself to take into account the public interest when formulating a standard.

In the framework of a conference on sports venues that took place in March 2004, the following definition of the DIN term was formulated:[7]

DIN standards reflect the generally recognized codes of practice, unless materials and construction methods have considerably progressed, in which case the up-to-date technological development as published in the relevant specialist literature and discussed in expert seminars becomes decisive.

If the contractor's performance corresponds to the relevant DIN standards, then proof has been delivered (prima facie) of works executed in accordance with regulations and free of faults. Such is current practice because DIN standards are claimed – refutably – to correspond to the acknowledged rules of technology.

This proof of can be contested, for example, in the event of a court case, when a professional expert can either prove that there is a generally acknowledged code of practice which deviates from the existing DIN standard but is more pertinent to the project; that in the meantime significant progress has occurred; or that new findings, e.g. on the building material properties, have been obtained which were unknown at the time of the standards consultation.

At the same time, not every breach of a contractor against individual guidelines necessarily represents a defect; success relies on the executed contract work. If sound in this respect, it remains so even if during execution individual aspects of the acknowledged rules of technology were disregarded. Observing these rules is intended to protect clients from disadvantages caused by faulty works. Standardization is not an end in itself though. When the individual components of a performance meet the requirements of DIN standards, but the building as a whole is not operational, legal claims can be made. The contractor is responsible for the success of the operation. Visual defects and deviations in quantity or thickness may lead to claims for defects on the client's part. Legal claims are ruled out for regular wear and tear.

A special DIN committee determined the following principle for the implementation of DIN standards:

'Every legally responsible person is also liable for his/her actions. Users of a DIN standard are not exempted.' DIN standards are published technical guidelines intended to create generally recognized codes of practice, so that these may be used as reference in legal regulations or private-law contracts (such as work contracts).

Verordnungen
[regulation, ordinances, orders]
are binding for all natural and judicial persons.

Richtlinien
[directives or administrative guidelines]
are binding for all administrative authorities after publication.

Empfehlungen
[recommendations]
are comparable to guidelines; breaches are only counted as legal infringements if acting against regulations or guidelines.

200 **DIN EN 13200-1**
'Spectator facilities Part 1: Layout criteria for spectator viewing area – Specification'
(cover of Technical Recommendations)

The German MVStättV 2005
Model regulation for places of assembly

§ 1 Scope of application
a) Assembly rooms for more than 200 visitors.
b) Open-air venues with stage surfaces for more than 1,000 spectators.
c) Sports venues with a holding capacity of over 5,000.

Note: The licensing procedure of DFB for regional-league clubs requests a holding capacity of at least 10,000, or provisionally 5,000 spectator seats. MVStättV 2005 (section 2 in §§ 26–30) also refers to stadia with a capacity of over 5,000. Furthermore, it gives details (§ 27) on the formation of blocks in stadia with over 10,000 visitors.

§ 2 The German term 'Versammlungsstätten'
[places of assembly] is used for
Clause 1) Building complexes for large numbers of people simultaneously present at events hosted 'for educative, economic, social, cultural, artistic, political, sports or entertainment purposes'.
Clause 3) Areas for consumption of food and drinks, halls and foyers, lecture halls and studios.
Clause 4) Stage surfaces (less than 20 m² is not considered a stage surface).

This regulation does not apply to places of congregation for church service, class rooms in general and vocational schools, exhibition spaces in museums and temporary structures.

Further definitions are intended to delineate areas to which the foregoing safety goals apply:
– Spectator facilities (as stated in EN/DIN 13200-1, item 3) comprise all areas of constant or temporary congregation of audiences at a sports or other event.
– Playing and sporting surfaces are all areas suitable in their construction and outfit.
– Supplementary areas are surfaces which are situated on the stadium site but not directly used for sports activities. However, they are necessary for operating a sports complex (routes, green areas, etc.).

MVStättV also defines as follows (§ 2, clauses 6–12):
– Multi-purpose halls or arenas: are covered venues for diverse types of events.
– Foyers: are reception areas and recreational spaces frequented by visitors during intervals.
– Sports stadia: are spaces of congregation with spectator stands and unroofed sports surfaces.

§ 6 Principles of emergency evacuation routing
All rescue routes must lead out in the open to public circulation areas and are to be kept clear.
1) passageways and gangways
2) exits from spaces of assembly
3) essential corridors and stairways
4) exits into the open, galleries
5) roof terraces and exterior stairs
6) rescue routes in open grounds

Skid-proofness must be guaranteed under wet or dry conditions.[8] Evacuation routes should be straight, easily discernible and clear of obstructions. Entry and exit gates are to be permanently staffed by stewards.[9] Every floor accomodating recreation/assembly rooms requires two independent escape routes, which also applies to stands (compare to § 9 DFB for 'all blocks').

Joining up both escape routes into one mutual corridor is permitted; routing through foyers and halls is admissible if for every level a further independent structural escape route is provided. Assembly rooms with more than 800 occupants are to be assigned only to this particular storey.

For milling spaces with more than 100 m² surface area, two far-apart exits into the open or escape routes in opposite directions are to be provided (MVStättV § 6).

Signage of emergency evacuation routes
Guiding to the closest exit at hand, safety signage must be permanent and clearly visible. In accordance with MVStättV 2005 § 10 Section 8, escape routes in multi-purpose halls with more than 5,000 places and in sports stadia must stand out clearly in colour. According to DFB/FIFA and DIN 4844, radial gangways are to be marked by signal colours.[10] In all spaces of congregation, escape routes must be discernible during blackouts, even without safety lighting. In sports stadia with emergency lighting, an extra lighting of rows is not obligatory.[11]

Dimensioning of escape routes

Distance and walking length
(MVStättV §7, Section 1)
'The distance of each spectator seat to the next exit leading out of the assembly space or the spectator stand may not exceed 30 m. Provided clear height is above 5 m, for every additional 2.50 m of clear height an extension of the maximum distance by 5.0 m is admissible. A distance of 60 m to the next exit may not be exceeded.'

Within corridors or the foyer, the distance to the next exit or staircase must be less than 30 m. Distance is measured in walking line along the rows and gangways. EN/DIN 13200-1 states in item 8.0 on walking length/distance: exit system inside buildings at a maximum of 30 m and in the open at a maximum of 60 m.

Dimensioning
(MVStättV §7, Section 4)
Width of flight routes is to be proportionate to the largest possible number of people that might have to use them and is measured as follows.

Minimum $w = 1.20$ m and maximum $w = 2.40$ m (straight flight width for two people using the hand rail, plus one other person):
– for assembly places in the open and sports stadia: $w = 1.20$ m for every 600 persons;
– in other assembly areas: $w = 1.20$ m for every 200 persons.

According to item 7.5 in the *Guide to Safety*: $w = $ max. 1.80 m with a maximum of 16 risers per flight; change of direction after 36 risers max. if slope is $< 30°$.

Grading of escape route widths
(60-cm intervals)
For milling spaces with less than 200 m² ground area $w = 0.90$ m and for working galleries $w = 0.80$ m.

The explanation to MVStättV (June 2005) of the ARGEBAU expert commission on construction supervision states on the grading of flight route widths as follows:

VStättVO (1978 §19, section 4) assumed that emergency evacuation could be organized without having to accommodate streams of people from various storeys.

In fact, it has been shown that congestion and panic behaviour is frequently the result of different spectator flows encountering on staircase landings.

Since the character of events has changed and assembly places need to be at capacity for economic reasons, reducing the numbers of occupants seems unjustified.

Modules
The exit module is based on a clear width of 0.60 m for a person and that two people at a time may use an escape route without obstructing each other.

The exit module width is therefore at least 2 x 0.60 m; a minimum width of 0.90 m would be required in any case, according to DIN 18024.

Two people at a time may exit through a door opening with an exit module width of 1.20 m; 100 people would need around one minute to do this.

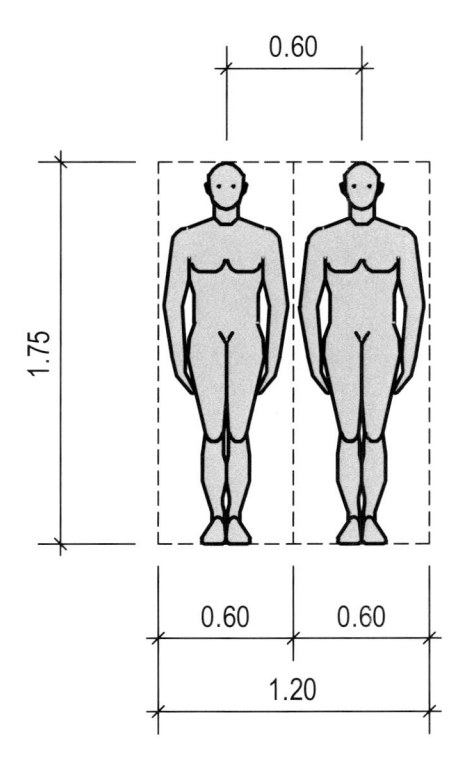

201 **Person module 60 cm**
System drawing to explain minimum flight route width (MVStättV 2005)

202 Opening celebration of Olympic Stadium Berlin
Stadium bowl view on 31 June 2004
Capacity: 74,600 spectators (two tiers)

To leave a space through a 0.60-m module, 50 people would require one minute.

Clearing time does not change for door widths of 0.80 m, 0.90 m or 1.0 m, as these can also be passed through by only one person at a time.

The basic concept of the module corresponds to current scientific research and hydraulic models, e.g. by Predtetschenski and Milinski, and has been confirmed by various trials in practice.[12]

The regulations of VStättVO 2002 have rendered obsolete evacuation scenario calculations for each individual case. The new dimensioning formula based on a 1.20-m width for every 200 persons does not constitute any complication compared to the existing dimensioning rule of 1.0 m per 150 persons.

Up to now, 8 m of exit width were required for 1,200 people, whereas in future an exit width of only 7.20 m will be obligatory. This reduction in width is possible because escape routes can be used more effectively with the module system and, hence, flow capacity can be increased.

Assembly spaces in the open and non-roofed sports stadia are clearly favoured in the dimensioning regulation (clause 3, no. 1).[13]

In sports stadia, this applies only to escape routes from the stands and from the central area. Evacuation routes from the interior milling areas are covered by no. 2.

New structural and architectural developments have lead to covering sports stadia with moveable roofs and turning them into multi-use arenas. Arena AufSchalke in Gelsenkirchen would be a typical example of this.

'If the roof above the pitch is completely retracted for an event and only the terraces are covered, dimensioning rule no. 1 applies; if events take place in a covered arena, dimensioning rule no. 2 takes effect.'[14]

Flow capacity

According to Neufert, C. von Eestern gives the following egress figures for Amsterdam Stadium:[15]

7 minutes per 5,000 spectators = 420 seconds for 9.5 m exit width (Los Angeles 12 min. and Turin 9 min.):

$$\frac{9.5 \text{ m} \cdot 420 \text{ s}}{5,000 \text{ persons}} = 0.8 \text{ s}$$

One person needs 0.8 s in relation to 1 m exit width, resulting in the following approximation formula for the calculation of egress times (old).

Egress times

$$\text{Calculation formula} = \frac{\text{number of spectators}}{\text{exit width} \cdot 1.25}$$

When comparing this formula to the regulations of MVStättV 2005, the following egress times arise:
50 persons = 1 minute per 0.60-m module
100 persons = 1 minute per 1.20-m min. width

Thus, two people require 0.60 s for approx. 1.20 m per exit module.
0.8 s = 1.0 m (old) → 1.0 s = 1.25 pers./min. width
0.6 s = 1.2 m → 1.0 s = 1.65 pers./min. width

According to MVStättV 2005, the minimum exit width is 1.2 m = 2 x 0.60 m:
1.65 > 1.25 (old) → 25 % increase in rate of passage

$$\substack{\text{New calculation formula} \\ \text{(for egress time)}} = \frac{\text{number of spectators}}{\text{exit width} \cdot 1.65}$$

Compare to egress times calculated for a 1.20 m exit module width (see above, C. van Eestern, Amsterdam Stadium) 420 s > 20 % > 315 s

Example:
$$\frac{40{,}000 \text{ spectators}}{600 \text{ people}} = 66.7 \text{ m} \cdot 1.2 \text{ m}$$
$$= 80 \text{ m exit width}$$

$$\text{Egress} = \frac{40{,}000 \text{ people}}{80 \text{ m} \cdot 1.65} = 300 \text{ s}$$
$$= \text{circa 5 min}$$

As stated in DIN 13200-1 'Flow capacity' in item 8.0 and Annex E (informative), 100 people can reasonably pass through an exit of 1.20 m width on a level surface.

For stepped surfaces, the number is reduced to 79 people passing an exit of 1.20 m in one minute (compare to *Green Guide* = 73).

Times for the evacuation of spectator areas and reaching a place of safety are approximately:
– in the open: a maximum of 8 minutes (15)
– inside buildings: a maximum of 2 minutes

Note: In the mid-1990s, a period of ca. 10 min was deemed sufficient for egress; around 40 people passing per minute through a width of 60 cm (IAKS Planning guidelines 2.3.3.6, 1992).

FIFA recommended a so-called 'funnel system': arrival around one hour before kick-off, but egress in a very short period of time.

After consultation with HHP West Engineering Consultants (fire protection), a realistic value of 15 min may be estimated for evacuation into the open; unless the area to be reached within 8 min is, as stated in the *Green Guide*, a 'place of comparative safety' (intended for a sojourn of 30 min) similar to the pitch or circulation areas with free external stairwells. An egress period of 15 min is proposed since DIN stipulates very positive values at 79/100 persons/min.

Egress or emergency evacuation time
Guide to Safety at Sports Grounds
Chapter 9.6 – 9.9 (B028) as comparison/information

For all routes within seated accommodation (incl. gangways, concourses and under-terrace staircases) flow capacity is: 73 people per metre/minute

For standing areas: 109 persons per metre/minute

Normal egress time: max. 8 minutes
Evacuation time: 2.5 to 8 minutes

... the emergency exit system from the viewing accommodation to a place of safety [p. 82].

A place of safety is a place which people can reach safely via the escape route, and which will be safe from the effects of fire. It may be a road, walkway or open space adjacent to, or even within the boundaries of the sports ground [p. 165].

Places of comparative safety = safe for min. 30 minutes

Only in exceptional cases and after consultation with a fire protection expert, the playing field may be used as an emergency evacuation route, on condition that it provides a direct access to a place of safety.

This option does not exist in MVStättV.[16]

Dimensioning of escape routes

Places of congregation in the open, such as sports stadia
w = 1.20 m   for every 600 people
In other assembly places
w = 1.20 m   for every 200 people

Stair flight width measured between two handrails
w = 2.40 m   max.
w = 1.80 m   max. (acc. to *Green Guide*)

10 Planning principles

Table H Capacities	International (seats only)	Standing places	National total (incl. standing places)
01 Berlin	74,200	./.	74,200
02 Dortmund	65,900	27,500	81,250
03 Frankfurt	48,400	9,300	52,300
04 Gelsenkirchen	53,600	16,300	61,500
05 Hamburg	51,300	8,900	55,800
06 Hanover	44,900	7,200	49,800
07 Kaiserslautern	47,700	12,000	48,500
08 Cologne	46,300	9,200	50,300
09 Leipzig	44,300	./.	44,300
10 Munich	65,600	20,000	66,000
11 Nuremberg	43,800	7,800	46,800
12 Stuttgart	53,100	4,300	57,000
Total	**639,100**	**122,500**	**687,750**
Average number	53,258	12,250	57,313
Minimum	43,800	4,300	44,300
Maximum	74,200	27,500	81,250

203 **Table H (summary of table A): Holding capacity of the twelve World Cup stadia**
(information by LOC 2006, Frankfurt)

'Green Guide' (B028) to Safety at Sports Grounds

A final capacity is restricted by four individual considerations, respectively limited by the smallest value:
a) the holding capacity resulting from the available viewing area (A) reduced by the appropriate spectator density (D)
b) the entry capacities/widths
c) the exit capacities/widths
d) the emergency evacuation capacity

Concluding, two independent reduction factors are set for all assessed capacities:

(P) factor
'physical condition or technical standard of the viewing accommodation'

(S) factor
'safety management of the viewing accommodation'

Exact values for (P) or (S) are not stipulated. A determination between 0.0 and 1.0 is left to the responsible parties [p. 19].

Holding capacity

Spectator stands are staggered rows on top of an embankment or simple structural framework or in multi-tier buildings. Combination types are widespread and should strive to exploit the topography in the interest of cost reduction.

Capacity is geared to the types of events to be hosted in the stadium. As gross capacity it refers to all available seating or standing places within the stadium bowl, places for the disabled, VIPs, media as well as boxes/suites.

Representing the economic foundations of a venue, the net capacity figures relate to paying spectators after deduction of media spaces, viewing-restricted seats, VIP seats, etc.

DFB/DFL determine 10,000 places as minimum licensing standard for the regional divisions.

FIFA minimum capacities:
30,000 seats for international tournaments
60,000 seats for group games, last sixteen and quarter finals
60,000 seats for opening matches, semi-finals and finals
50,000 seats for the Confederations Cup

UEFA minimum capacities:
30,000 seats for group games, last sixteen and quarter finals
40,000 seats for quarter and semi-finals (four-star stadia)
50,000 seats for opening matches/finals (five-star stadia)

Ratio standing/seating places

According to DFB Safety Guidelines §9, all Bundesliga stadia are to be successively converted into all-seater stadia; a quota of max. 20% for standing places is permitted.

FIFA/UEFA require that all spectators should have a seat at their disposal; a stadium is also considered an all-seater if standing accommodation remains closed. This applies to qualifying matches, the Confederations Cup, World Cups and Olympic Games.

Determining the ratio between standing and seating places depends also on the duration of the event (IAKS), e.g. all-day athletics competitions.

Ticket information by FIFA regarding World Cup 2006

Seats means all the seating places available at FIFA World Cup 2006, which are calculated from gross capacity minus those seats which are unavailable for FIFA World Cup 2006 (due to viewing restrictions and safety reserves). This capacity is based on the current estimates and may change after final stipulation of requirements.

Available (selling) tickets means all admission tickets minus those tickets reserved by the LOC for the media, written press, VIPs or made available free of charge. This capacity is based on estimates (maximum demand of the media) and may change after final stipulation of requirements.

Holding capacity 109

204 **Berlin** stadium plan
during Bundesliga season

205 **Dortmund** stadium plan
during Bundesliga season

206 **Frankfurt** stadium plan
during Bundesliga season

204 **Gelsenkirchen** stadium plan
during Bundesliga season

208 **Hamburg** stadium plan
during Bundesliga season

209 **Hanover** stadium plan
during Bundesliga season

210 **Kaiserslautern** stadium plan
during Bundesliga season

211 **Cologne** stadium plan
during Bundesliga season

212 **Leipzig** stadium plan
during Bundesliga season

213 **Munich** stadium plan
during Bundesliga season

214 **Nuremberg** stadium plan
during Bundesliga season

215 **Stuttgart** stadium plan
during Bundesliga season

216 **Main stand west (below tent roof)**
Olympic Stadium, Berlin
217 **Reverse stand – (Estadio Azteca)**
Guillermo Cañedo, Mexico City
Capacity 105,060

Pre-dimensioning and circulation
Shedding light on the relationship between circulation systems and holding capacity

External circulation systems

A fundamental planning principle for all circulation systems is to provide easy orientation to visitors of the stadium. This should preferably be achieved through structural provisions, i.e. visibility of routing through decreasing hierarchy of space proportions, rather than through additional visual measures such as signage, etc. Entrances should ideally be visible from afar and directly from the general circulation routes around the stadium. Colour coding on tickets designating each sector (blue, green, yellow, red) should also be clearly reproduced at the relevant entrances. Segregation of spectator streams is part of the safety concept of guiding different groups to their sectors along intersection-free routes:
– Home fans/Visitors
– Regular spectators/VIPs
– Press/VIPs
– Players/Service personnel.

Zone 1 – Playing field   central area
Zone 2 – Spectator stand auditorium
Zone 3 – Stadium   circulation area
Zone 4 – Sports complex   external area

1st check   at the outer safety ring
2nd check   at the stadium entrances
3rd check   at the concourse access to the stand
4th check   at the stand vomitories

Corresponding to the above mentioned areas, *external* and *internal* circulation are distinguished. All routes beyond or below the stand belong to the external group. All routes to seating/standing places on the stand (from access/vomitory) belong to the internal group.

Number of tiers

Depending on the concept and capacity, the stand structure is divided into *tiers*. The classical earth stadium with just one lower tier is either created by lowering of the pitch into the ground (or topographical trough/hillside) or by constructing a circulating embankment.

The concept of the built stadium either replaces the terrain elevation of the Greek or Roman theatre, or complements the spectator embankment by a further structural upper tier (as in Müngersdorfer Stadion, Cologne, 1975). The number of consecutive tiers depends on the required capacity and the utilizations designated to the ancillary rooms beneath the terrace.

World Cup Stadium Stuttgart: one-tier stadium
World Cup Stadium Cologne: two-tier stadium
World Cup Stadium Munich: three-tier stadium

Note: Mistakes may occur when determining the number of tiers of a grandstand. For example, a two-tier stadium from a visual or structural viewpoint, the Aztec Stadium in Mexico City (Guillermo Cañedo) is a three-tier stadium with regards to circulation. Ingress and egress at the upper tier of World Cup Stadium Hamburg, takes place on two access levels. Technically a two-tier stadium, the size of the spectator blocks at the upper tier forms two tiers (middle and upper tier) which are fed from a lateral gangway. To avoid confusion, terraces should be distinguished by clear visual separations such as 'box tiers' (one joint = two tiers).

To reach their seats, spectators follow two directions of movement: *vertical* and *horizontal*.

A stadium with more than one tier should guide spectator streams vertically to the different lateral gangways/concourses.

Generally, the routes leading there and back are the same. Determination of necessary exit widths applies only to emergency evacuation. Regular routing may be complemented by additional escape routes (e.g. additional evacuation gates). The following vertical systems of circulation are distinguished:

Stairs
This is the main and most frequently implemented option of leading spectators to the respective circulation areas. For the external area, a design value of 1.20 m per 600 persons and max. 2.40 m clear flight width applies.[1]

Split / staggered stairs seem especially suitable for confined spaces; straight-line stairs, however, are better as short stairs for terrace access along the vomitory axis (compare to middle tier Allianz Arena, Munich, or Stade de Suisse, Berne / Switzerland)

Cascading stairs covering several storeys are relatively long and are located as near the façade (as in Köln-Arena, architects G. Böhm). If necessary, they intersect under-terrace utilizations (as in Allianz Arena).

Ramps
This circulation type is becoming increasingly rare. There are, however, prominent examples, such as the Estadio Santiago de Bernabeu in Madrid, Spain, or the Guiseppe Meazza Stadio (San Siro) in Milan, Italy.

Since the shape of such systems tends to be prominent, they possess a significant recognition value. The relevant DIN standard describes the structural requirements for ramps as follows:[2]

'Circulation areas for buildings with a longitudinal gradient of more than 3% are to be realized as ramps, the rake of which may be 6% maximum. A transverse slope is inadmissible.

The length of each ramp flight is limited to 600 cm maximum. Platforms are to be installed between the flights. No descending stairs may be placed at the prolongation of a ramp.'

Lifts and escalators
Passenger and goods lifts are planned for almost every large sports or event venue. Goods lifts connect the low-lying storage rooms with the kitchens and kiosks in the higher storeys. In case of vertical stacking of catering service units, such as production kitchen, dishwashing facility or storage space, at least one lift system is obligatory. Provision of further goods lifts depends on the catering concept or stadium operation.

Passenger lifts in the VIP area facilitate convenient access to VIP suites as these are mostly located on the higher levels of the box tier, at the separation between upper and lower tier. Their number and outfit depends on the wishes of the stadium operator, or rather on the number of people intended to be transported up or down per hour.

Escalators, as in Gelsenkirchen or Munich, are employed less frequently.

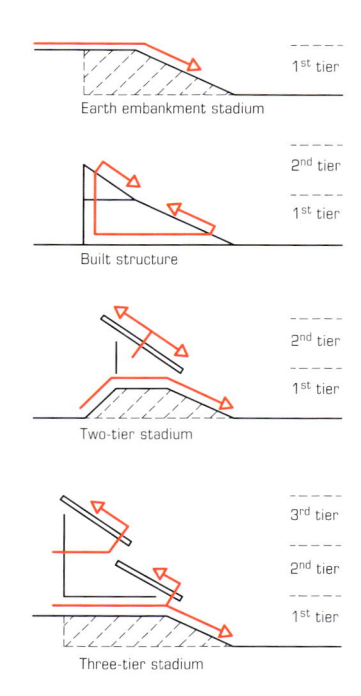

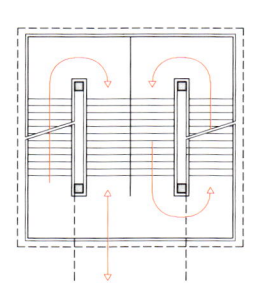

218 **Principles of circulation / tier sections**
(4 system drawings)
219 **Staggered stairs (drawing ca. 10 x 10 cm)**
Type RheinEnergieStadion, Cologne
220 **Circular ramp tower**   
Stadium Faro, Portugal, EURO 2004
221 **Short stairs between twin joggle beams**
Stadium Stade de Suisse, Berne, EURO 2008
222 **Wall stairs**
World Cup Stadium Nuremberg 2006

Internal circulation systems

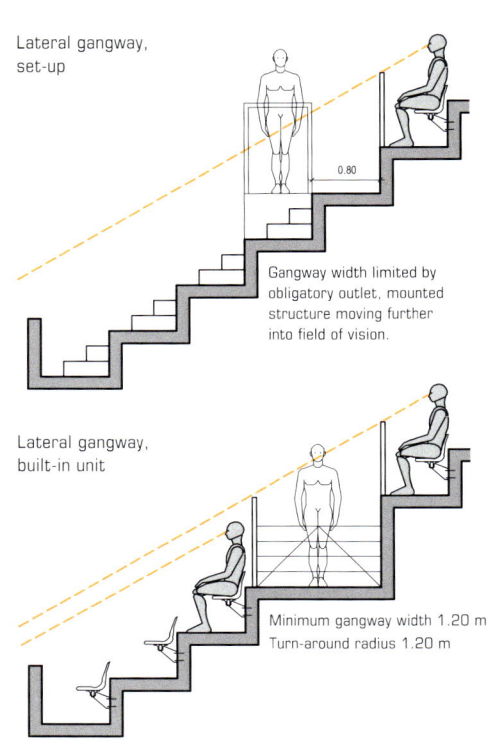

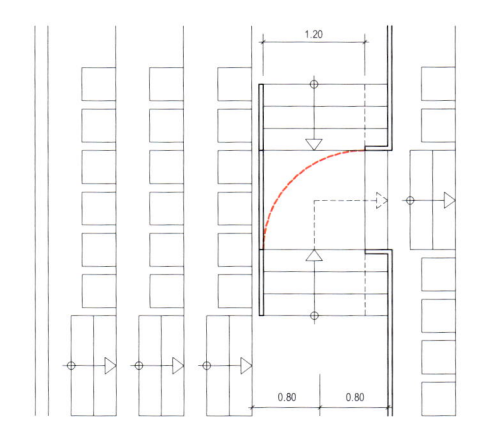

First: Terrace access

Two options of internal terrace access are common:
a) from a vomitory (at the terrace), same-level or with short stairs
b) from the concourse (behind the tier)

A concourse or tier access is the most efficient option, as the terrace is accessed directly from the rear without loss of seating caused by vomitories. This type of access is normally integrated into the gap between lower and upper tier (as in World Cup Stadium Nuremberg).

In some cases though, this joint is already taken up by high-quality hospitality suites, mostly with a large VIP area adjoining at the rear. An exception would be the Olympic Stadium in Berlin which has corridors from the lower circulation levels passing through between individual groups of boxes.

Access to the upper tier accommodation is mostly via vomitories, as it seems rather unpractical to bring spectators to the height of the upper edge/upper tier in order to then access the tier from the rear. (Exception: AWD Arena, Hanover, where the upper tier of the west stand is accessed via the wall system and the lower tier via a lateral gangway called 'Avus').

Second: Radial gangway system

As MVStättV 2005 limits block width to a maximum of 2 x 20 places, the distribution of gangways becomes essential:
Maximum block size: 30 rows x 40 persons
Maximum gangway grid: 22.40 m
30 rows x 40 places = 1,200 persons
per 600 pers. = 1.20 m escape route width
40 x 0.50 seating grid = 20 m + 2.40 m

Regular distance

The intervals at which vomitories are installed vary significantly in all German World Cup stadia (from 13.5 m to 37.0 m) and are reliant on the structural and circulation grid. The number of vomitories is essentially based on the underlying circulation principle, i.e. the way spectators reach the gangway:
a) vertical circulation (radial gangways)
b) horizontal circulation (lateral gangways)

Either the vertical circulation is directly adjoined to the terrace circulation (a) or the spectators are distributed horizontally up to the respective gangway.

Two different principles are distinguished: *one vomitory – one gangway* or *one vomitory – two gangways*. In the latter case, the gangways are located along the field axis, i.e. one vomitory for every other planning grid.

In the case of horizontal distribution of spectators (as for the upper tier of World Cup Stadium Hamburg), the gangway width should be integrated into the terrace elevation to keep it from affecting the sightline profile of people sitting behind. Depending on the number of people it has to accommodate, a larger width requires a short-stair gangway leading the evacuation route from the vertical to the horizontal, without diminution. The stairway itself may cause viewing restrictions.

223 **Vomitory circulation**
Stade de France, Paris
Capacity 80,000

224 **Access situation through vomitory**
Olympic Stadium, Berlin

225 **Gangway landing with turn-around radius 1.20 m**
System drawing with change of direction horizontal on vertical

Radial gangway types

There are three principal systems for a vomitory circulation:
a) axial gangway
b) single-plan gangway
c) double-plan gangway

The rampant gangway represents a very efficient type of circulation as it circumnavigates the problem of joining the double-plan gangway to the right and the left of the vomitory into an axial gangway. The axial route to the vomitory requires a safety barrier (to prevent people from falling) of 1.10 m at the turn-around platform of the vomitory lintel. This railing (also called 'cow catcher') often results in viewing restrictions towards the corner flag. Depending on the visual angle zone, the number of affected spectators may be excessively high.

Note: Glass railings or fillings are urgently advised for all components of the stand which might be located within the sightline profile of spectators (barriers, etc.).

Escape routes are not to be tapered along their course. Their exit width may be combined and added in the direction of flight, but under no circumstances may the width be reduced.

This precondition creates an additional problem for the stands: a seating row must be at least 40 cm wide (clearance); but exit width is calculated only for the gangway. Here, the number of people reliant on this particular gangway becomes crucial. Minimum width is 1.20 m, which is 30% wider than the recommended tread depth of 80 cm.

Therefore, more than one riser is needed for the horizontal distribution of spectators and the evacuation route (width) frequently leads directly around the vomitory and upward.

Viewing obstructions

Lateral gangway
A horizontal distribution of spectators may result in sight restrictions for occupants on the stand and on the turn-around platform to the gangway. Planning should ensure that the number of people allocated to it and the resulting sight restrictions are kept to a minimum (compare to *Green Guide*, B028, p. 110).

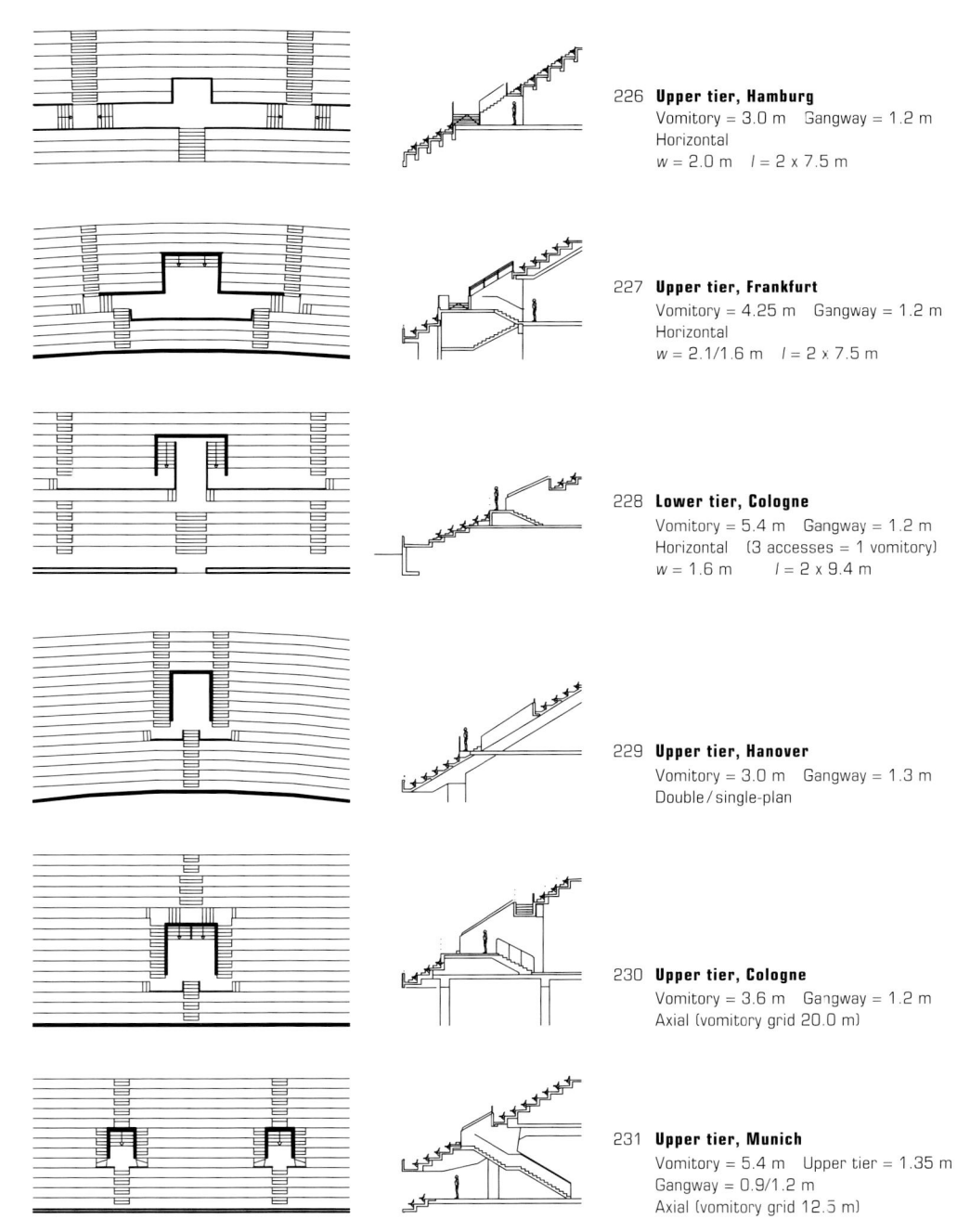

226–231 **Vomitory typologies with section sketch**
System drawings of built examples for World Cup 2006

226 **Upper tier, Hamburg**
Vomitory = 3.0 m   Gangway = 1.2 m
Horizontal
w = 2.0 m   l = 2 x 7.5 m

227 **Upper tier, Frankfurt**
Vomitory = 4.25 m   Gangway = 1.2 m
Horizontal
w = 2.1/1.6 m   l = 2 x 7.5 m

228 **Lower tier, Cologne**
Vomitory = 5.4 m   Gangway = 1.2 m
Horizontal   (3 accesses = 1 vomitory)
w = 1.6 m   l = 2 x 9.4 m

229 **Upper tier, Hanover**
Vomitory = 3.0 m   Gangway = 1.3 m
Double / single-plan

230 **Upper tier, Cologne**
Vomitory = 3.6 m   Gangway = 1.2 m
Axial (vomitory grid 20.0 m)

231 **Upper tier, Munich**
Vomitory = 5.4 m   Upper tier = 1.35 m
Gangway = 0.9/1.2 m
Axial (vomitory grid 12.5 m)

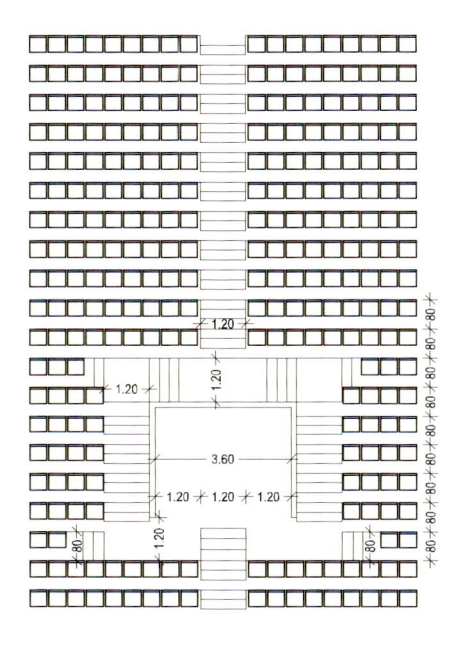

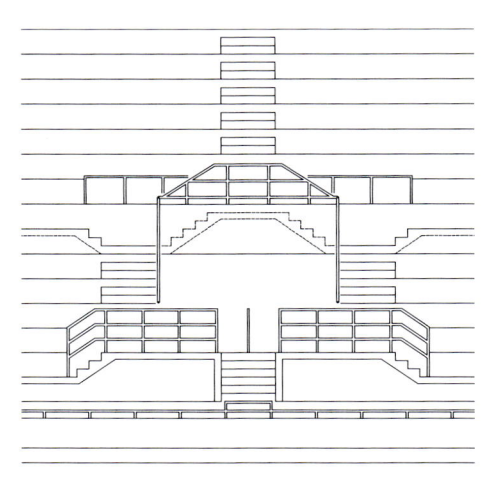

RheinEnergieStadion, Cologne
232 **Vomitory type (system 3.60 m)** (left)
233 **Vomitory type (elevation)** (right)

Allianz Arena, Munich
234 **Vomitory type (system 1.20 m)** (left)
(scale of execution, clear width = 1.35 m)
235 **Turn-around radius 90 cm** (right)

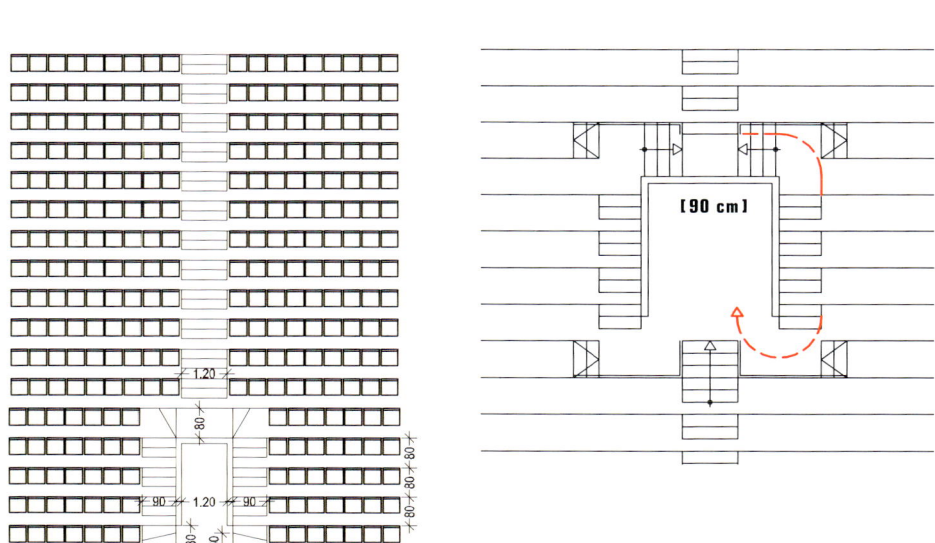

The vomitory type at the upper tier in World Cup stadium *Cologne* shows that the necessary radius of the escape route can only be achieved by placing a double riser at the vomitory exit.

The vomitory type in *Munich* has a gangway of just 0.90 m to the left and right which is joined up axially above the vomitory into a 1.20-m gangway.

According to the relevant regulation for places of assembly, a clearway of 0.90 m is permissible for a limited number of people (interior areas < 200 persons).[3]

There are around 20 seating rows above the vomitory at the middle tier of World Cup Stadium Munich. The axial spacing of vomitories is circa 12.5 m, so with a seating grid of 50 cm each area to the side of the gangway has to accommodate around 200 people. The 90-cm platform on both sides of the vomitory is executed as a 'spiral stair'; the advantage being that no second double riser (see above) is necessary for the 180° change of direction at the vomitory.[4] Before the entire stand geometry is based on this principle, clarification with the respective approval body prior to planning should be sought.

Note: The low number of two risers surely suggests a deviation from the planning regulations, especially since the upper riser (external w = 40 cm) is covered completely by the next seat. (To the author, the double riser at the spiral point appears problematic.)

The axial gangway, which separates (descending) to the left and right above the vomitory, promotes the flow of spectators in flight. Therefore, the 1.20 m axial gangway could be replaced, if need be and in coordination with approving authorities, by two 90-cm vomitory passageways. If more than 600 people are sitting above the vomitory (example: 1.80 m gangway = max. 900 people), and are not evenly distributed to the left and right, the vomitory passageways should not be reduced to 90 cm.

In this case, the number of people positioned above the vomitory may be balanced by the vomitory height in the terrace elevation, in order to adjust numbers in this gangway section to below 600 persons.

Note: The 90-cm rule is not yet represented in the official regulations (MVStättV 2005).

Reduction factor

The effects a chosen vomitory type might have in terms of seat loss will be systematically clarified for four typical angles of slope (20°/25°/30°/35°) without taking into account horizontal distribution.

By means of preliminary dimensioning, capacities may be approximately calculated in the early planning stages and verified during the ensuing design process. The principle of capacity determination in relation to space is based on tier access, that is, without vomitories.

Variant A represents the classic radial gangway as axial continuation of the vomitory. Regarding the ground-plan graphics, this is a very 'clean' type as the symmetry of the supporting and circulation grid may overlap. Seat loss is relatively high at minimum evacuation route widths of 1.20 m.[5] Efforts to reduce the vomitory width, or rather the loss of seats, of around 6.0 m seem reasonable enough.

Proposal D works, for this reason, with a 2 x 90 cm division of the axial 1.20-m gangway, reducing the number of reliant persons to 2 x 300.[6] For the vomitory grid this means: for 30 rows (block size) and a maximum of 600 people on 1.20 m gangway width = 2 x 10 people to each side of the gangway. This roughly corresponds to the 12.50 grid implemented at World Cup Stadium Munich. Variant D thus produces a denser vomitory grid than variants A – C because no further restrictions on the number of people, other than 600/1.20 m, are necessary.

Interpolation of both limit values should be possible; the other extreme would be to exploit the capacity of 2 x 20 persons for each side of the gangway. The maximum vomitory grid (22.40 m) can be exhausted, but the number of rows is reduced by the 90-cm limit to the left and right of the vomitory to a total of 600 people. Check calculation: 600/(2 x 20 people) = 15 rows.

In exceptional cases the required minimum width of 1.20 m may be reduced by around 5%; a request for deviation would have to be submitted as part of a fire protection survey.

Possible reasons as given in a relevant fire-protection survey: '… since the minimum value, according to VStättVO, from the upper edge of the concrete barrier or above the safety barrier is observed and the increased space demand for a person's upper body is taken into account.'[7]

236 **System sketch to calculate a reduction coefficient for seats loss due to vomitory instead of open circulation via radial gangway**
 a. Estimation table (reduction)
 b. Sectional sketch of the tier 20°/25°/30°/35°
 c. Ground-plan system types A – D

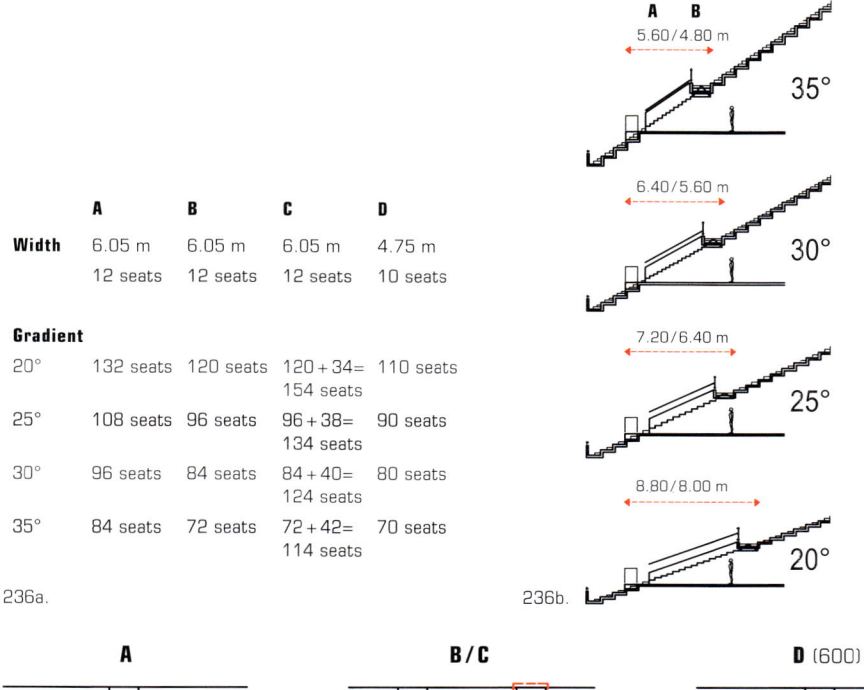

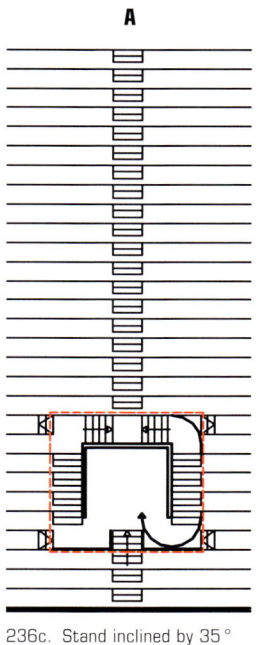

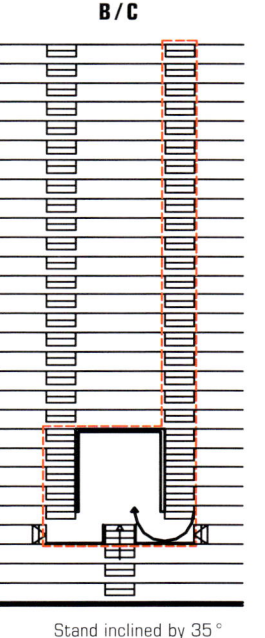

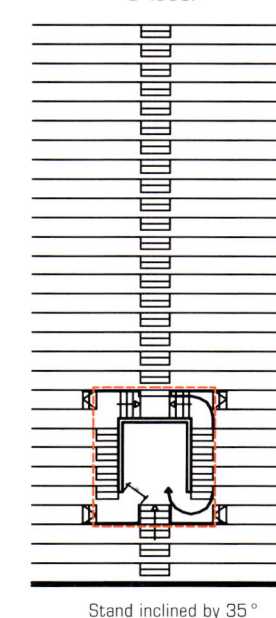

	A	B	C	D
Width	6.05 m	6.05 m	6.05 m	4.75 m
	12 seats	12 seats	12 seats	10 seats
Gradient				
20°	132 seats	120 seats	120+34= 154 seats	110 seats
25°	108 seats	96 seats	96+38= 134 seats	90 seats
30°	96 seats	84 seats	84+40= 124 seats	80 seats
35°	84 seats	72 seats	72+42= 114 seats	70 seats

236a.

236b.

236c. Stand inclined by 35°   Stand inclined by 35°   Stand inclined by 35°

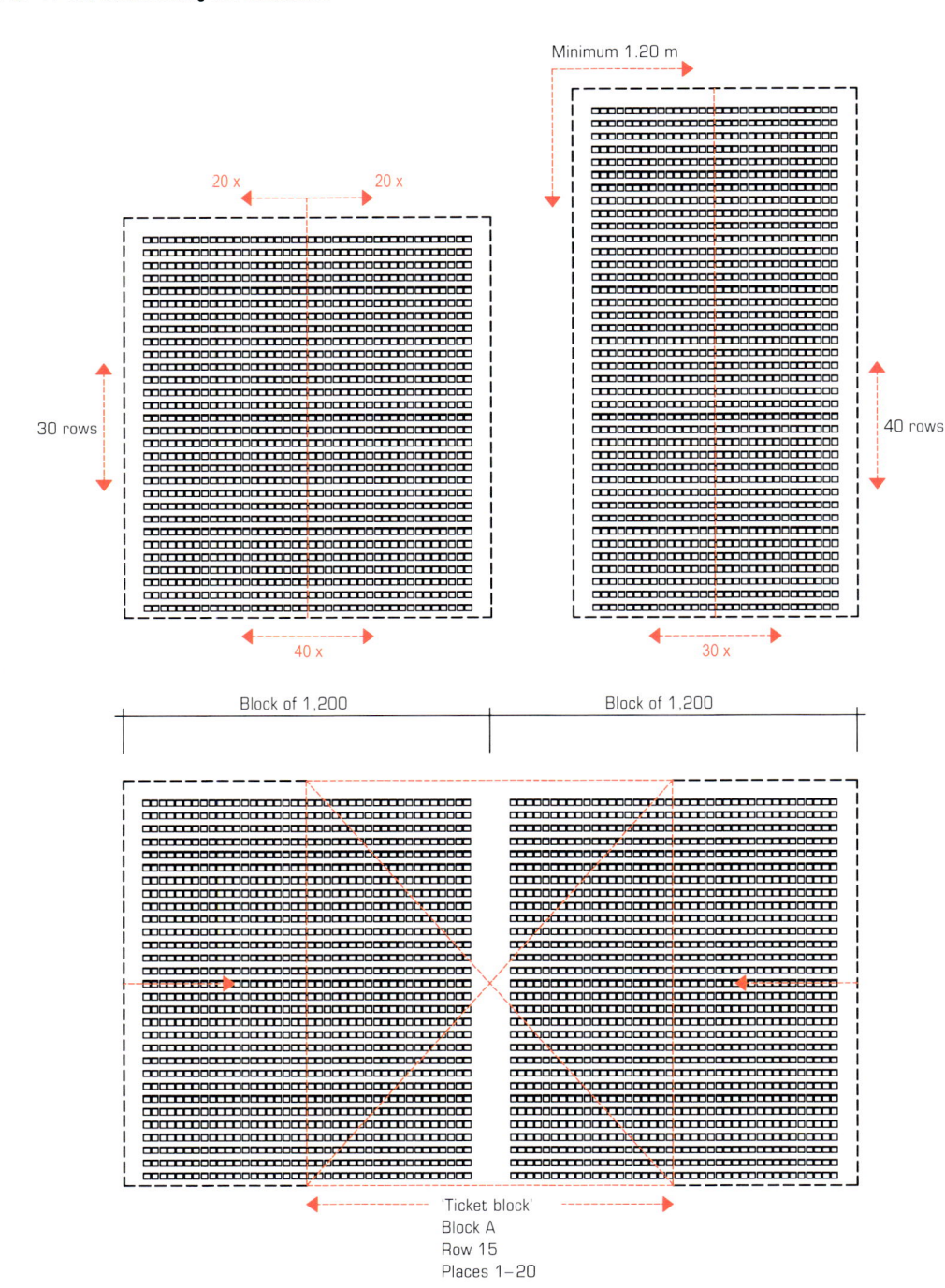

Block definition for seating accommodation

The following block definition is stated in the relevant German regulation: 'Seating accommodation must be arranged in blocks of 30 seating rows maximum. Behind and between the blocks gangways of at least 1.20 m length have to be arranged, leading at the shortest distance to the next exit.'[8]

The new '60-cm staggering' method requires a three-fold exit width for modern event arenas with convertible roofs as compared to open-air stadia. This far-reaching regulation for covered stadia also applies to theatres and cinemas.

In former legal regulations, between two gangways up to 32 people (2 x 16) are allowed to sit in rows without steep slopes.[9] For steeply inclined rows the number is reduced to 24 persons (2 x 12) and a maximum of 50 if flight doors at least 1.50 m wide are provided for every four rows.

Today, MVStättV 2005 states a maximum of 20 seats to each side of a gangway in uncovered areas, i.e. a maximum of 40 seats between two gangways. The values are reduced to 10/20 for indoor venues. A maximum of 50 successive seats is the exception, and only if a door of 1.20 m width is provided for every four rows.

Note: Deviating from this, EN/DIN 13200-1 states: seating between two radial or parallel passageways indoors 2 x 14 = max. 28 places.[10] Thus, the European standard follows the guidelines of *Green Guide* 1997 (chapter 11.14).

In Germany (status as of 2004), the project group 'Muster - Versammlungsstätten - Verordnung' is looking into defusing the regulation 1.20 m/200 persons for indoor assembly places (theatres, cinemas, etc.). Alterations would be conceivable if fire-protection aspects are considered (fire load, room height and smoke exhaust systems).

Revision to block definition [11]

The maximum size of a seating accommodation block is set at 1,200 people. The combination of row/place can be interpolated, inasmuch as the maximum length of emergency evacuation routes is not exceeded:

30 rows x 40 persons = 1,200 persons
40 rows x 30 persons = 1,200 persons

Note: This new definition of the term 'block' has so far not been entered into MVStättV 2005. However, it is deemed a viable planning base among experts.

Two definitions of blocks are distinguished. Depending on the dimensioning of spectator blocks, the block boundaries are shifted toward the centre of the block or field:
a) spectator blocks, i.e. standing/seating accommodation between two gangways, and
b) ticket blocks, i.e. the definition and numbering of single seats, generally complying with the circulation system. In this case, the term block applies to the gangway and the adjoining seating to the left and right. Hence, only spectators reliant on this gangway are counted.

237 **Block definition**
  a. Sketch according to MVStättV 2005 (top)
  b. Definition 'ticket block', according to *Stadium Atlas* (dissertation)

Circulation areas

Due to the minimum gangway width of 1.20 m between spectator blocks and the evacuation route of 1.20 m running behind each block, the relation between required total surface and necessary service area changes, depending on the layout. Following, three gangway widths are introduced for the external area, i.e. for the building type of stadium with open playing field (uncovered) and for the auditorium of an arena, theatre or other type of indoor venue.[12]

The more people are seated in one row, the more efficient the layout of seating accommodation.

The share of circulation areas increases with interpolation from 30 x 40 to 40 x 30 by around 2%. At the same time, exploitation of the surface decreases from 2.12 to 2.09 seats/m². The more people are accommodated within one block and the more balanced the selected proportions are, the more efficient capacities can be organized. Based on the evacuation module width of 0.60 m, three main groups of gangways emerge.

External area:	2.40 m	1,200 persons
	1.80 m	900 persons
	1.20 m	600 persons
Interior area:	3.60 m	600 persons
	2.40 m	400 persons
	1.80 m	300 persons
	1.20 m	200 persons

The share of circulation areas (e.g. 600 people) is doubled for the interior area. Block expansion, or rather the general possibility for interpolation, might play an important role in the course of design and final planning. This would allow planners to adjust the number of rows more easily to the stadium geometry, or, space admitting, to extend capacity by providing several additional rows. The admissible minimum escape route width remains a fundamental precondition, in relation to the number of people reliant on it within one gangway section.

The decision on the block size and the corresponding width of the gangway complies with the structural grid of the stadium geometry. In the case of a vomitory circulation, the position of the terrace access, similar to a stair opening, is arranged in alternation to the main supporting grid.

The more rounded the overall shape, the shorter the polygonal sections of the prefabricated parts. The tighter the grid and thus lower the span of indented beam/support beam, the more dense the distribution of vomitories and the smaller the dimensions of spectator blocks.

The standard block '1,200 persons – 2.40 m' occupies at maximum dimension a circulation surface share of 15% at 2.12 seats/m². Without an additional lateral gangway in the stand, the axial distance of the vomitory is 22.40 m.

For a block of 600 with narrow 1.20 m gangways, the interval grid is halved. Thus, the share of circulation areas increases to 23% and only 1.92 seats/m² can be installed.

A general recommendation is not given here; arrangements should be made for each individual case.

238 **Gangway definition (uncovered) and capacity consideration for maximum block sizes**

Block of 1,200 with 2.40-m gangway

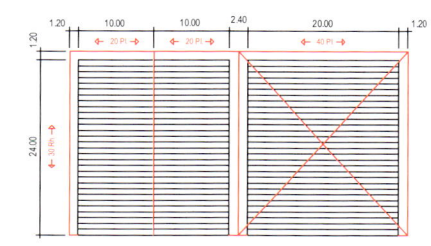

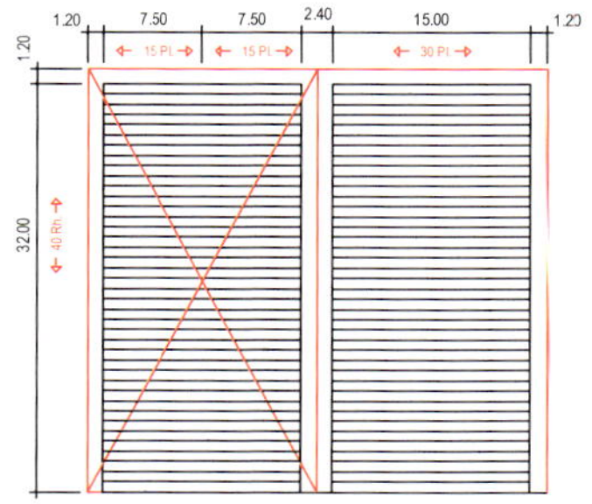

30 rows x 40 seats = 1,200 persons
Gangway width: 1.20 m per 600 persons, three-sided circulating gangway
Total surface area: (30 x 0.80 m + 1.20 m) x (40 x 0.50 m + 2 x 1.20 m) = 565 m²
Circulation area: (2 x 24.0 m + 20.0 m) x 1.20 m = 85 m² = 15%
Surface area per seat: 565 m²/1,200 persons = 0.47 m²/seat
Seats per m²: reciprocal value 0.47 = 2.12 seats/m²

40 rows x 30 seats = 1,200 persons
Gangway width: 1.20 m per 600 persons, three-sided circulating gangway
Total surface area: (40 x 0.80 m + 1.20 m) x (30 x 0.50 m + 2 x 1.20 m) = 575 m²
Circulation area: (2 x 32.0 m + 15.0 m + 2 x 1.20 m) x 1.20 m = 97 m² = 17%
Surface area per seat: 575 m²/1,200 persons = 0.48 m²/seat
Seats per m²: reciprocal value 0.48 = 2.09 seats/m²

Block of 900 with 1.80-m gangway

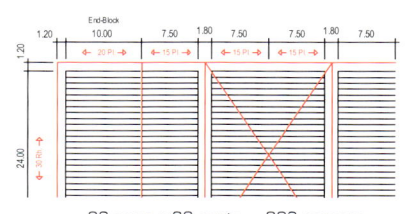

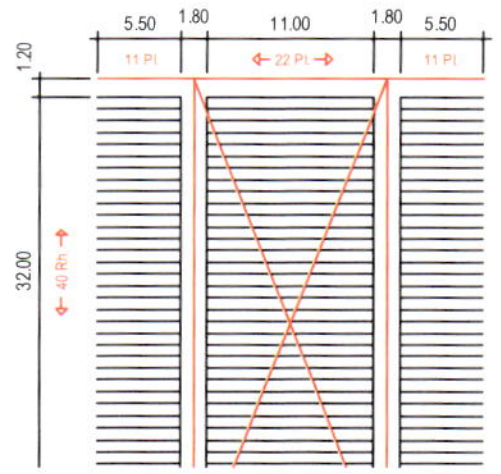

30 rows x 30 seats = 900 persons
Gangway width: 1.20 m per 600 persons, three-sided circulating gangway
Total surface area: (30 x 0.80 m + 1.20 m) x (30 x 0.50 m + 2 x 0.90 m) = 425 m²
Circulation area: (2 x 24.0 m) x 0.90 m + (15.0 m + 2 x 0.90 m) x 1.20 m = 85 m² = 15%
Surface area per seat: 425 m²/900 persons = 0.472 m²/seat
Seats per m²: reciprocal value 0.472 = 2.12 seats/m²

40 rows x 22 seats = 880 persons
alternatively: 41 rows x 22 seats = 902 persons or 39 rows x 23 = 897 seats
1.20 m per 600 persons, three-sided circulating gangway
(40 x 0.80 m + 1.20 m) x (22 x 0.50 m + 2 x 0.90 m) = 425 m²
(2 x 32.0 m) x 0.90 m + (11.0 m + 2 x 0.90 m) x 1.20 m = 73 m² = 17%
425 m²/900 persons = 425 m²/880 persons = 0.483 m²/seat
reciprocal value 0.483 = 2.08 seats/m²

Block of 600 with 1.20-m gangway

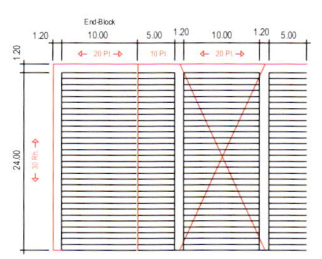

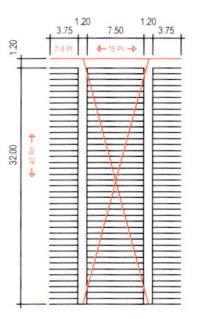

30 rows x 30 seats = 900 persons
Gangway width: 1.20 m per 600 persons, three-sided circulating gangway
Total surface area: (30 x 0.80 m + 1.20 m) x (20 x 0.50 m + 2 x 1.20 m) = 313 m²
Circulation area: (2 x 24.0 m + 1 x 10.0 m + 2 x 1.20 m) x 1.20 m = 73 m² = 23%
Surface area per seat: 313 m²/600 persons = 0.522 m²/seat
Seats per m²: reciprocal value 0.522 = 1.92 seats/m²

40 rows x 15 seats = 600 persons
1.20 m per 600 persons, three-sided circulating gangway
(40 x 0.80 m + 1.20 m) x (15 x 0.50 m + 2 x 1.20 m) = 330 m²
(2 x 32.0 m + 1 x 7.5 m + 2 x 1.20 m) x 1.20 m = 88 m² = 35%
330 m²/600 persons = 0.55 m²/seat
reciprocal value 0.55 = 1.82 seats/m²

11 Pre-dimensioning and circulation

Capacity pre-dimensioning

Block of 600 with 3.60-m gangway

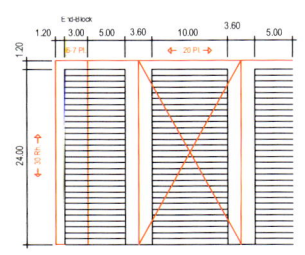

30 rows x 20 seats = 600 persons
Gangway width: 1.20 m per 200 persons, three-sided circulating gangway
Total surface area: (30 x 0.80 m + 1.20 m) x (20 x 0.50 m + 2 x 1.80 m) = 343 m²
Circulation area: (2 x 24.0 m) x 1.80 m + (10.0 m + 2 x 1.80 m) x 1.20 m = 103 m² = 30%
Surface area per seat: 343 m²/600 persons = 0.572 m²/seat
Seats per m²: reciprocal value 0.572 = 1.75 seats/m²

40 rows x 15 seats = 600 persons
Gangway width: 1.20 m per 200 persons, three-sided circulating gangway
Total surface area: (40 x 0.80 m + 1.20 m) x (15 x 0.50 m + 2 x 1.80 m) = 370 m²
Circulation area: (2 x 32.0 m) x 1.80 m + (7.50 m + 2 x 1.80 m) x 1.20 m = 128 m² = 35%
Surface area per seat: 370 m²/600 persons = 0.617 m²/seat
Seats per m²: reciprocal value 0.617 = 1.62 seats/m²

Block of 400 with 2.40-m gangway

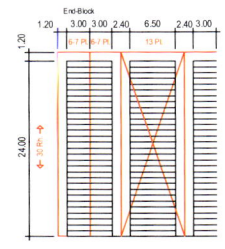

30 rows x 13 seats = 390 persons
Gangway width: 1.20 m per 200 persons, three-sided circulating gangway
Total surface area: (30 x 0.80 m + 1.20 m) x (13 x 0.50 m + 2 x 1.20 m) = 225 m²
Circulation area: (2 x 24.0 m + 6.50 m + 2 x 1.20 m) x 1.20 m = 68 m² = 30%
Surface area per seat: 225 m²/400 persons = 0.56 m²/seat
Seats per m²: reciprocal value 0.56 = 1.78 seats/m²

20 rows x 20 seats = 400 persons
Gangway width: 1.20 m per 200 persons, three-sided circulating gangway
Total surface area: (20 x 0.80 m + 1.20 m) x (20 x 0.50 m + 2 x 1.20 m) = 198 m²
Circulation area: (2 x 16.0 m + 10.0 m + 2 x 1.20 m) x 1.20 m = 53 m² = 27%
Surface area per seat: 198 m²/400 persons = 0.495 m²/seat
Seats per m²: reciprocal value 0.495 = 2.02 seats/m²

Block of 300 with 1.80-m gangway

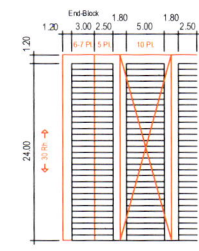

30 rows x 10 seats = 300 persons
Gangway width: 1.20 m per 200 persons, three-sided circulating gangway
Total surface area: (30 x 0.80 m + 1.20 m) x (10 x 0.50 m + 2 x 0.90 m) = 171 m²
Circulation area: (2 x 24.0 m) x 0.90 m + (5.0 m + 2 x 0.90 m) x 1.20 m = 51 m² = 30%
Surface area per seat: 171 m²/300 persons = 0.57 m²/seat
Seats per m²: reciprocal value 0.57 = 1.75 seats/m²

20 rows x 15 seats = 300 persons
Gangway width: 1.20 m per 200 persons, three-sided circulating gangway
Total surface area: (20 x 0.80 m + 1.20 m) x (15 x 0.50 m + 2 x 0.90 m) = 160 m²
Circulation area: (2 x 16.0 m) x 0.90 m + (7.50 m + 2 x 0.90 m) x 1.20 m = 40 m² = 25%
Surface area per seat: 160 m²/300 persons = 0.53 m²/seat
Seats per m²: reciprocal value 0.53 = 1.87 seats/m²

239 Gangway definition (indoors) and capacity consideration for maximum block sizes

A surface-related capacity guideline is given with two persons/m².[13] In 1993, IAKS starts out from 2.5 seats/m² of terrace accommodation.[14] In fact, pre-dimensioning is geared to the individual place without consideration of the necessary gangway surface.

Due to complex factors of influence, an early and exact capacity determination can hardly be made, or only in approximation and in the further course of planning. Actual total capacity is also affected by the circulation system, the block definition or the number of tiers of the stand. Partial capacities such as business or media seats, commentator or camera positions and the desired number of special seats also play a role. Pre-dimensioning should provide for sufficient 'buffering' in order not to fall below the required minimum capacity in the final planning stage. Two capacity-related considerations are distinguished: *seat-specific* or *block-specific*. Examination of simple stadium geometries (one-tier stadia) generates an overview which enables an approximate pre-dimensioning of a grandstand and a pre-determination of tiers.

The table is based on the standard scenario of 80-cm tread depth and a 50 cm seating grid. Depending on tread depth B (incl. the gangway surface), a seat takes up a certain gross surface share per m².

Seat/0.80 m = 1.25 pers./m² (see above and compare to IAKS 2 x 1.25 = 2.5 pers./m².)

For uncovered venues, regular gangway width is 1.20 m with a maximum of 600 persons assigned to it, or 200 persons for indoor spaces.

At a maximum block width of presently 30 rows this result in the following number of seats fitting in-line:
600/30 rows = 20 people = 10 people per gangway side = 10 running-metre seating step
$\rightarrow$ 10 m + 1.20 m (gangway)
$\rightarrow$ 20 seats/11.20 m = 1.785 running metres.

For the above mentioned scenario (80/50) the following capacity value (estimation) emerges: 1.25 m x 1.758 = 2.23 seats/m².

In a preliminary examination, layout types are arranged in blocks, and all grandstand geometries are accessed from the concourse/tier and without vomitories.

Depending on the length of the row, calculated values fluctuate between 1.78 to 1.67 running metre $\rightarrow$ 6,5%; and depending on surface share (at 80/50) between 2.20 to 2.09 $\rightarrow$ 5.0%.

A closer examination, excluding the corner areas, shows a fluctuation of 20%. Hence, deviations arising from the layout types are considerably reliant on the structural execution of the corner areas.

Further, test geometries of various layout types for 10,000/20,000/25,000 people are examined seat-specific instead of block-specific manner, in order to achieve more precise results. With a 5% deviation the calculated values are so similar that the characteristic value is simplified as pre-dimensioning tool, in approximation and independent of the shape. The result is a possible reduction factor of approximately 98,5% of the theoretical estimate (2.23), which generalizes the capacity values for all layout types.
$\rightarrow$ 2.23 x 0.985 = 2.20 seats/m²

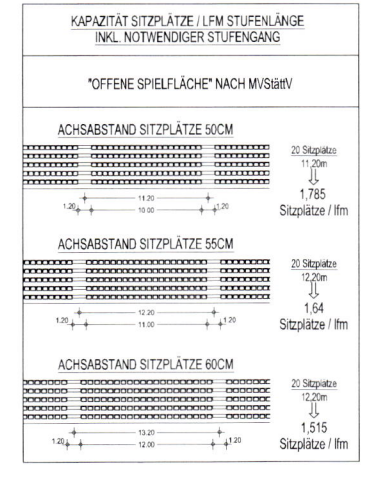

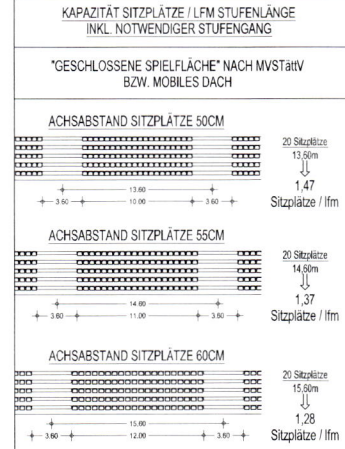

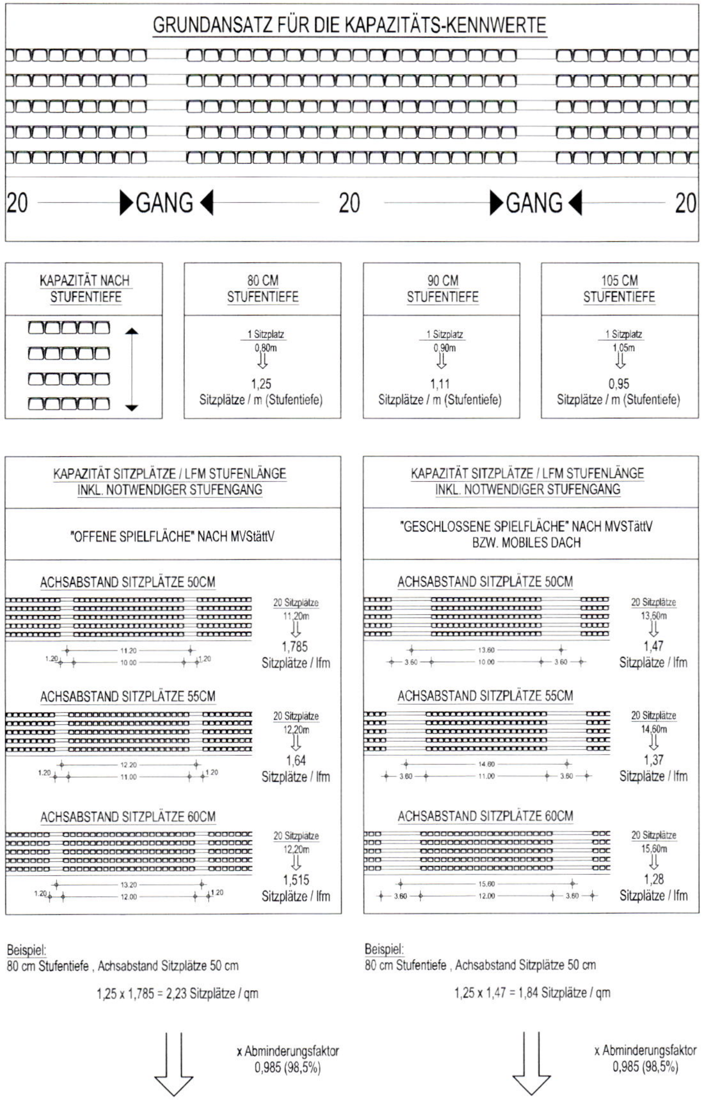

Capacity pre-dimensioning 119

Geometrically open rectangle – 29 rows

Stand geometry

Corner, section

Doubly curved rectangle – 25 rows

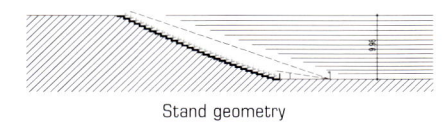

Stand geometry

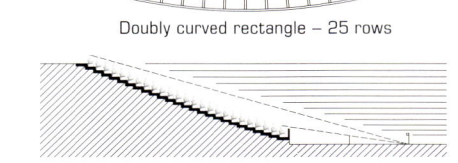

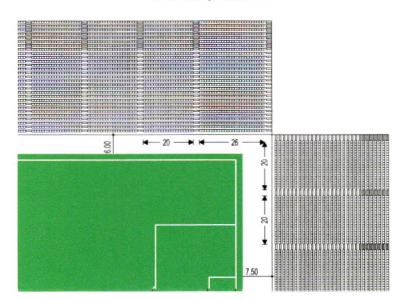

Corner, section

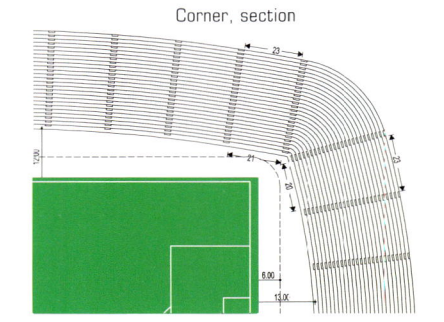

left:
240 **Capacity value**
Graphical explanation of the surface area assumption (excerpt from the study of RWTH summer term 06)

241 **System drawing**
20,000-cap. lower tier stand with rectangular geometry

242 **System drawing**
20,000-cap. lower tier stand with circular geometry

page 120/121:
243–244 **Table J: Pre-dimensioning overview**
after theoretical-geometric pre-measuring

11 Pre-dimensioning and circulation

Commentary to comparative table J:

The following overview shows a theoretical result, founded on the surface assumptions for seating places, including the necessary circulation areas per block. Internal circulation systems (level access, vomitory or lateral gangway) are not taken into account. The figures are based on a simple one-tier geometry, since the circulation type generally changes in the next tier above the main concourse level (e.g. to vomitories).

The number of tiers is divided, according to currently valid maximum limits for seating rows, into blocks with 30 rows. If a block (row number) has reached this size, theoretically another tier is adjoined. This can be balanced by alteration of the block proportion (interpolation).

Higher capacities can be achieved by adding of rows, e.g. by projecting terraces. In this case, the respective seating rows are located above one another, resulting in a doubling of individual rows.

Assuming that after 30–40 rows a new block begins, and with it a new tier, a total number of seats may be approximately calculated by adding individual tier segments (tier capacities).

The stepped surface of an upper tier is always larger than the tier below (exception: a geometrically open rectangle). It will continue to grow if the perimeter of a form is expanded. This increase in capacity should be considered during pre-dimensioning by rounding up the values presented in the table, as the exact spectator capacity during the course of final planning tends to decrease.

The stated number of rows is mathematically projected from the empirical test geometries and must be viewed only as an approximate value since it strongly depends on the circulation system (vomitory, block dimension, etc.).

In the present overview, after determining a desired capacity via a 'form-neutral' 2.2 parameter, a theoretical space demand for grandstands is approximately calculated.

After consideration of special areas such as VIP, media etc., and after conclusion of approval and final planning, deviations (gross/net) of up to 15% are revealed. This difference may not be counted as a general reduction factor since the individual requirements (number and size of special areas) vary from project to project and no stadium/arena truly matches a particular layout type but rather appears to be a combination or hybrid form.

Table J: Comparison of form and capacity

Capacity and theoretical space requirements

Step 80/50 cm 2.20 seats/m²	Basket arch geometry Rows m²	Semicircular geometry Rows m²	Oval geometry Rows m²	Octagonal geometry Rows m²	Step 80/50 cm 2.20 seats/m²
10,000 4,545 m²	10.86 –> **11** 4,604 m²	10.72 –> **11** 4,223 m²	11.37 –> **12** 4,389 m²	13.04 –> **13** 4,531 m²	**10,000** 4,545 m²
15,000 6,818 m²	15.91 –> **16** 6,860 m²	15.71 –> **16** 6,949 m²	16.61 –> **17** 6,990 m²	18.88 –> **19** 6,866 m²	**15,000** 6,818 m²
20,000 9,090 m²	20.74 –> **21** 9,215 m²	20.50 –> **21** 9,332 m²	21.62 –> **22** 9,266 m²	24.38 –> **25** 9,352 m²	**20,000** 9,090 m²
25,000 11,364 m²	25.39 –> **26** 11,670 m²	25.10 –> **25** 11,311 m²	26.42 –> **27** 11,643 m²	29.61 –> **30** 11,541 m²	**25,000** 11,364 m²
30,000 13,636 m²	29.86 –> **30** 13,706 m²	29.54 –> **30** 13,875 m²	31.03 –> **31** 13,618 m²	34.57 –> **35** 13,835 m²	**30,000** 13,636 m²
35,000 15,909 m²	34.19 –> **35** 16,343 m²	33.84 –> **34** 15,998 m²	35.49 –> **36** 16,176 m²	39.32 –> **40** 16,241 m²	**35,000** 15,909 m²
40,000 18,182 m²	38.38 –> **39** 18,524 m²	37.99 –> **38** 18,185 m²	39.79 –> **40** 18,295 m²	43.89 –> **44** 18,238 m²	**40,000** 18,182 m²
45,000 20,455 m²	42.45 –> **43** 20,770 m²	42.03 –> **42** 20,437 m²	43.96 –> **44** 20,478 m²	48.29 –> **49** 20,829 m²	**45,000** 20,455 m²
50,000 22,727 m²	46.40 –> **47** 23,080 m²	45.95 –> **46** 22,754 m²	**48** 22,726 m²	52.54 –> **53** 22,980 m²	**50,000** 22,727 m²
55,000 25,000 m²	50.24 –> **51** 25,454 m²	49.78 –> **50** 25,134 m²	51.93 –> **52** 25,038 m²	56.65 –> **57** 25,195 m²	**55,000** 25,000 m²
60,000 27,273 m²	53.99 –> **54** 27,273 m²	53.51 –> **54** 27,578 m²	55.77 –> **56** 27,414 m²	60.64 –> **61** 27,483 m²	**60,000** 27,273 m²
65,000 29,545 m²	57.65 –> **58** 29,763 m²	–	59.50 –> **60** 29,238 m²	64.51 –> **65** 29,835 m²	**65,000** 29,545 m²
70,000 31,818 m²	61.23 –> **62*** 32,315 m²	–	63.14 –> **63*** 31,732 m²	–	**70,000** 31,818 m²

Note: Figures marked with * refer to seats situated beyond the maximum admissible 150-/190-m viewing distance.

Capacity and theoretical space requirements

	Rounded rectangle			Circle-segment geometry			Circular geometry			'Geometrically open' rectangle		
Step 80/50 cm 2.20 seats/m²	Rows m²		Tiers	Rows m²		Tiers	Rows m²		Tiers	Rows m²		Tiers	Step 80/50 cm 2.20 seats/m²
10,000 4,545 m²	13.07 –> **13** 4,517 m²			12.99 –> **13** 4,547 m²			– –			14.20 –> **14** 4,480 m²			**10,000** 4,545 m²
15,000 6,818 m²	18.96 –> **19** 6,835 m²			18.85 –> **19** 6,877 m²			**15–40** 6,913 m²			21.30 –> **21** 6,720 m²			**15,000** 6,818 m²
20,000 9,090 m²	24.51 –> **25** 9,299 m²			24.39 –> **25** 9,350 m²			**21–46** 9,269 m²			28.41 –> **29** 9,280 m²			**20,000** 9,090 m²
25,000 11,364 m²	29.77 –> **30** 11,210 m²			29.64 –> **30** 11,521 m²			**27–52** 11,524 m²			35.51 –> **36** 11,520 m²			**25,000** 11,364 m²
30,000 13,636 m²	34.79 –> **35** 13,452 m²			34.66 –> **35** 13,793 m²			**32–57** 13,635 m²			42.62 –> **43** 13,760 m²			**30,000** 13,636 m²
35,000 15,909 m²	39.61 –> **40** 16,100 m²			39.47 –> **40** 16,166 m²			**38–63** 16,130 m²			49.72 –> **50** 16,000 m²			**35,000** 15,909 m²
40,000 18,182 m²	44.23 –> **45** 18,571 m²			44.09 –> **44** 18,136 m²			**42–67** 18,159 m²			56.82 –> **57** 18,240 m²			**40,000** 18,182 m²
45,000 20,455 m²	48.68 –> **49** 20,620 m²			48.55 –> **49** 20,690 m²			**47–72** 20,572 m²			63.92 –> **64** 20,480 m²			**45,000** 20,455 m²
50,000 22,727 m²	52.98 –> **53** 22,736 m²			52.86 –> **53** 22,805 m²			**51–76** 22,765 m²			71.02 –> **71** 22,720 m²			**50,000** 22,727 m²
55,000 25,000 m²	57.15 –> **58** 24,915 m²			57.03 –> **58** 25,539 m²			**56–81** 25,036 m²			78.13 –> **79*** 25,280 m²			**55,000** 25,000 m²
60,000 27,273 m²	61.20 –> **62** 27,731 m²			61.08 –> **62** 27,780 m²			**60–85** 27,383 m²			– –			**60,000** 27,273 m²
65,000 29,545 m²	65.12 –> **66** 30,062 m²			65.02 –> **66** 29,356 m²			**65–90** 29,808 m²			– –			**65,000** 29,545 m²
70,000 31,818 m²	68.95 –> **69*** 31,852 m²			68.85 –> **69** 31,908 m²			**68–93** 31,947 m²			– –			**70,000** 31,818 m²

Note: Figures marked with * refer to seats situated beyond the maximum admissible 150-/190-m viewing distance.

Comparing the built examples of World Cup 2006 with the table above and calculating the theoretical space demand of the existing World Cup capacities with the parameter of 2.2 seats/m², the following deviations of 5–15% arise.

Example World Cup Stadium, Cologne:
46,300 spectators
24,100 m²

Pre-dimensioning with parameter
2.2 seats/m² = ca. 21,000 m² → –13%

As a result, the actual space demand after execution planning is around 15% higher than after pre-dimensioning; or rather, the final capacity of the pre-draft is actually around 15% lower. These deviations, in the case of Cologne Stadium, have the following reasons:

Hybrid form: Lower tier, elongated octagon
Upper tier, single stand
(165 m instead of 120 m)
Stair run: B is 80/85/95 deep for reasons of sightline optimization
Circulation: Lateral gangway in front of first row
Gateways in every corner

As a rule, the empirically determined space demand of a stand is higher than the undifferentiated pre-dimensioning. Terraced surface (top view, cantilever not considered):

		existing	compared with theoretical pre-dim.
01	Berlin	35,300 m²	105%
02	Dortmund	33,400 m²	112%
03	Frankfurt	26,600 m²	121%
04	Gelsenk.	27,850 m²	114%
05	Hamburg	28,000 m²	120%
06	Hanover	23,800 m²	117%
07	Kaisersl.	24,800 m²	114%
08	Cologne	24,100 m²	115%
09	Leipzig	19,000 m²	94%
10	Munich	31,500 m²	106%
11	Nuremberg	18,500 m²	93%
12	Stuttgart	29,000 m²	120%

Note: Demands on the individual stadium do overlap in the requirement profiles of certain event types (FIFA-suitable, Technical Recommendations), but every project differs from the next. Therefore, a generally valid pre-dimensioning approach cannot be selected. The approach described above takes into consideration only the conditions stated in MVStättV 2005 and EN/DIN 13200-1 for the dimensioning of a block.

Chapter 12

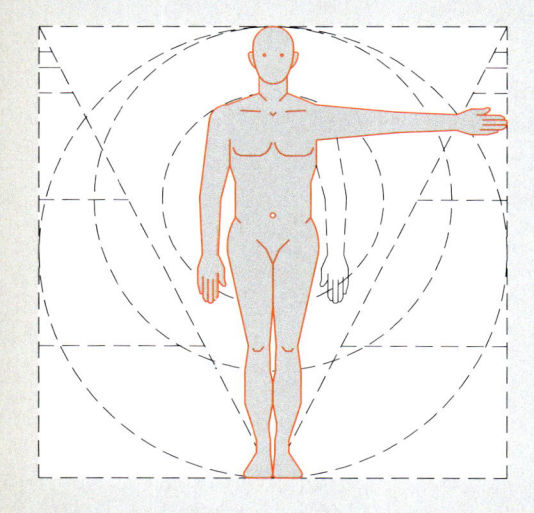

245 The Modulor 'EN'
application of the golden section

The Modulor 'EN'
The definition of a typical stadium spectator as an anthropometric minimum standard and contractually binding reference size for the planning and construction of spectator stands

The Modulor 'EN' (European Norm)

The design, planning and construction of a sports or events stadium has to accommodate the needs of diverse user groups.

This is true not only for the athletes on the playing field who engage in competitive sports or for performers on the stage, but also for the spectators who observe the competition, game or cultural event.

It is therefore indispensable to analyze more closely the dimensions of the typical human individual: a direct relationship exists between these dimensions and the specific function of the spectator stands, which must ensure the comfortable and safe circulation of spectators and their unimpeded view of the action.

The stadium management is responsible for the facility's economic success, which is the ultimate purpose of such a building. It has an interest not just in ensuring security and safety for visitors, but also in providing appropriate levels of comfort.

Such ergonomic standards are not only vital to human health and well-being while sitting and working, but are also decisive for the construction and inspection of the sightlines present within a given stadium or place of assembly.

The adaptation of a building to the anthropometric givens of the human user applies both to the width of circulation routes leading to and from visitors' seats in the stands, as well as to the dimensions of exits and escape routes to be used in case of evacuation.

When defining the required width of escape routes, the latest German regulations on places of assembly, for example, works with multiples of the shoulder width of a typical human individual, namely a module of 60 cm.

The precise depiction of the human body as a central criterion for the planning of grandstands must be regarded as indispensable; an inspection of merely the finished architectural results would present excessive economic risks.

The ground plan geometry of a grandstand is generally fixed quite early during the planning stages for the shell construction and the prefabricated parts, and is hence seldom reversible. Since the primary use of the facility is to provide secure and unimpeded views of the game or event taking place below, the most precise possible preliminary definition of sightlines is an absolutely decisive economic factor affecting the planning process as a whole.

The graphic translation of historical systems of proportion in combination with reference values provided by the German Federal Office of Statistics into a 'model person' can be an important planning instrument providing a common basis for decision-making on the part of client, architect and the responsible authorities. The structure of the Modulor 'EN' is based on empirical data concerning human bodily dimensions, as prescribed in the EN/DIN 33 402-2. Soon to become valid throughout Europe, these can therefore become generally applicable and specified contractually.

In this way, the architect is provided with a useful tool in the detailed planning and adaptation of a stadium: the previously defined reference dimensions can be used in order to avoid later problems, for example viewing restrictions for seats set behind camera positions, resulting in seat loss for the originally contracted capacities.

The canon of proportions

With the help of the powers of reason, the human individual constantly attempts to define his/her immediate environment by means of explanation and investigation. Throughout millennia of intellectual history, the striving of human-

Architecture and harmony

ity toward an ordering principle has led to numerous attempts to set out a canon of the proportions of the human figure. In fact, virtually all human societies have attempted to come to terms with the question of human proportion.

Investigations concerning questions of human proportion, however, are far from uncontroversial. 'A modern spectator, still under the influence of this Romantic interpretation of art, finds it uninteresting, if not distressing, when the historian tells him that a rational system of proportion, or even a definite geometrical scheme, underlies this or that representation', as E. Panofsky remarked in 1955.[1]

'Man as the measure of all things' becomes the standard for his own environment. Right up to our own times, prescribed dimensions such as feet and ells are customary units for measuring lengths.

The globally-diffused sport types are oriented toward two different systems of measurement: The two basic units of length are the 'meter' and the 'foot / yard'.

The basic unit is the 'inch', whose relationship to the meter has been defined internationally since 1 July 1959 as precisely 2.54 cm. A foot has 12 inches and is exactly 30.48 cm in length.

1 inch [in] = 2.54 cm
1 foot [ft] = 30.48 cm
1 yard [yd] = 91.44 cm
1 mile [mi] = 1.61 km
               = 1,760 yd = 5,280 ft

1 yd = 3 ft = 36 in

In describing the dimensions of prescribed architectural elements, the theorists Vitruvius and Alberti were oriented towards the basic principles of proportion, order and beauty. Paul v. Naredi-Rainer refers to Vitruvius' 'doctrine of beauty' as being reflected in the two leitmotifs of classical aesthetics.[2]

The first leitmotif is the Pythagorean-Platonic idea, which rests upon a definition of the beautiful as being 'objective and regular, as being based on number and proportion, and which ultimately regards beauty as being graspable by reason.' The second leitmotif is the Hellenistic notion of the beautiful as being 'never fully independent of taste, and as being accessible only through the senses'.

'You have ordered everything according to number and measure' [Omnia in mensura et numero at poneri dispuiti], is an Old Testament proverb from the *Liber sapientiae* (11,21). Saint Augustine (354 – 430), strongly influenced by Neoplatonism, and promoting the early-Medieval idea of 'ordo'.

'Order is the means by which everything is defined which has been determined by God', writes Saint Augustine. The concept of 'ordinatio', as used already by Vitruvius (vol. 2 / bk. 2), finds its corollary in the to some extent non-uniform definitions of the significance of the classical orders (see the chapter entitled 'The theatre of Alberti'), which is on the one hand derived from nature and from the human figure, and on the other derives its proportional system from the numerical relationships found in music theory. In the 6th century BC, the association of the Pythagoreans constructed an all-encompassing world view from the unification of harmony and the cosmos. They believed that the principles of mathematics were also those which governed all of existence, even guiding the human soul.

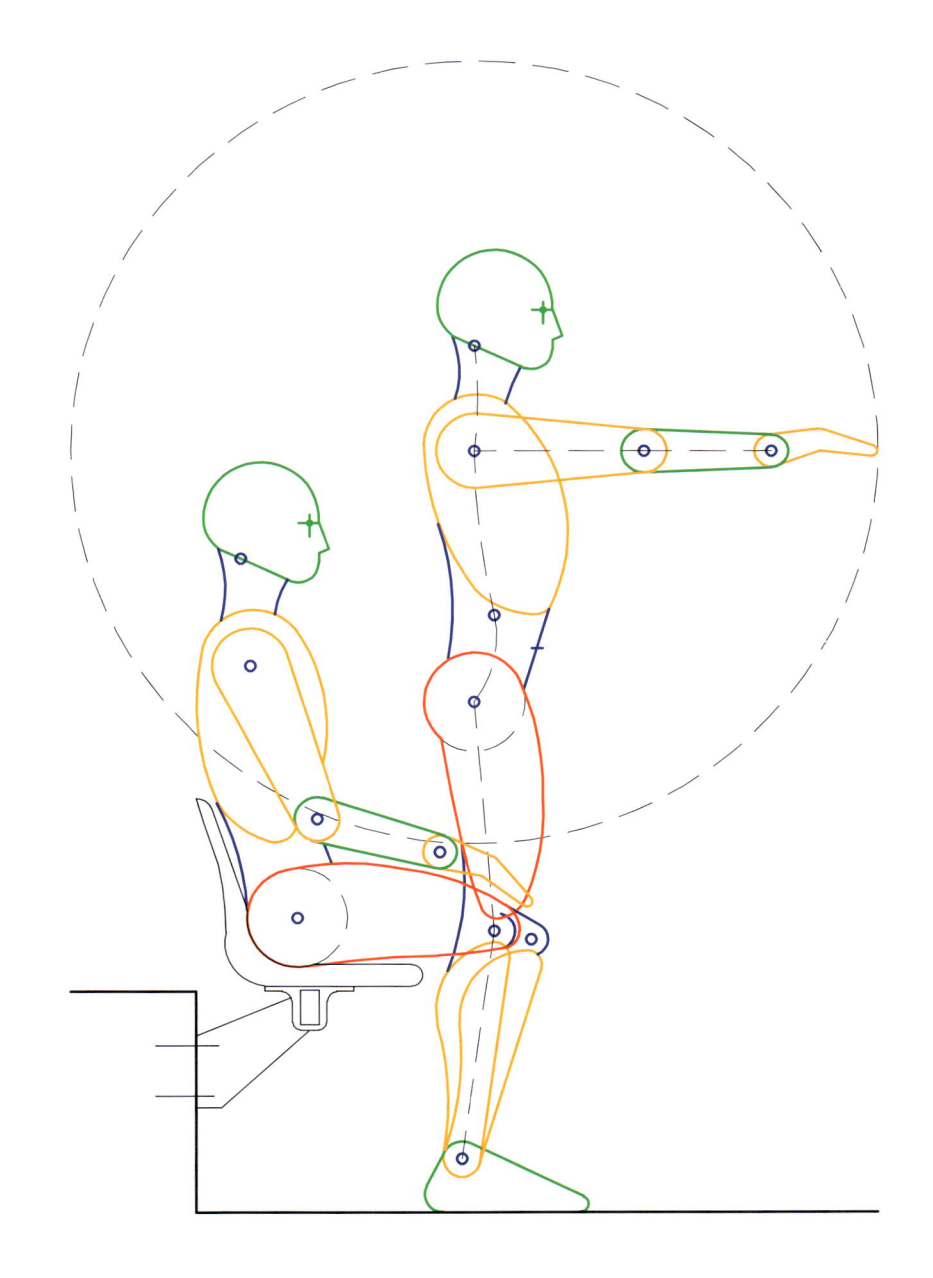

246 **Proportion study of the Modulor 'EN' seated / standing**

The 'Golden Section'

For centuries, the concept of 'harmony' (A. Hausmann) – translated freely from the Greek and meaning the combination of opposed entities to create an ordered whole – formed the foundation of both Classical and Christian world views. Hausmann continues: 'The beauty and perfection of the world reveal themselves through mathematical laws which correspond to the relationships of tones and numbers.'[3]

Leon Battista Alberti's definition of beauty was as follows: 'Beauty is a type of correspondence and a concordance of separate parts to form a whole, an arrangement that is disposed according to a specific number, a certain relationship, and in such a way that it promotes the most perfect and the highest natural law (*concinnitas*).'[4]

Following from this, *concinnitas*, as the regular correspondence of all parts, finds its clearest expression in the art of music and its numerical relationships. *Concinnitas*, then, is the visual equivalent, so to speak, of acoustic *consonantia*. Nonetheless, architecture does not simply understand itself as transposed music, but instead sets itself up, with its own lawfulness, as an equal alongside music.

'By itself, even the most superior knowledge of proportional relationships will not produce good architecture – just as a profound knowledge of the teachings of harmony will not in and of itself create good music. A thorough knowledge, however, can be a precise tool for evaluating and defining the results of an intuitive feeling for form.'[5]

The 'Golden Section' (Lat. *sectio aurea*) is a specific relationship between two measures or numbers. It represents a principle of order in art and architecture that is frequently applied in the form of lengths or measurements of structural components, generating unified formal groupings. The golden section is often regarded as an ideal proportion, as the essence of aesthetics and harmony. The repetition of the same element is immediately recognized by the eye as an ordering principle, whereas the eye registers minimal deviations less readily than does the ear (i.e. in music).

Moreover, the golden section appears regularly in nature and is characterized by a series of interesting mathematical features.

Two lengths conform to golden section criteria when the larger, or 'major' is to the smaller or 'minor' as the sum of both is to the larger of the two.

This relationship is generally referred to with the Greek letter Φ and goes back to Euclid, who describes the phenomenon for the first time in the second book of his *Elements*. If the larger element is designated a and the shorter b, then the relevant formula can be set out as:

$$\frac{a}{b} = \frac{a+b}{a}$$

If the shorter length is subtracted from the longer one, then (an even smaller) length is generated, to which the shorter, in turn, now stands in the relationship also defined by the golden section.

The term 'continuous subdivision' refers to the circumstance that this procedure can be repeated ad libitum, producing the same relationship each time. If a square is removed from a 'golden rectangle', the result is a rectangle also featuring golden section proportions.

An equilateral triangle two of whose sides feature golden section proportions is referred to as a 'golden triangle'. Geometry considers only constructive procedures involving circles and lines. There are an abundance of constructive procedures designed to divide a length according to the *proportio divina* [Lat.]. Because of its simplicity, the following triangle method is widely employed:
1) Along the length AB, mark a vertical line beginning from B which has half the length of AB, and whose endpoint is C.
2) The circle revolving around C cuts through the connection AC at point D.
3) The circle around A, with the radius AD, divides the length AB into a golden section proportion.

Note: This Euclidean procedure encompasses an 'external and internal division' which generates a series of numbers into progressively smaller or progressively larger divisions on the basis of an initial or reference length (see Modulor).

A rectangle whose side lengths are a and b lies exactly in the golden section if the relationship between the enlarged rectangular a plus b has a golden section relationship to a.

Fibonacci Numbers

Closely associated with the golden section is the infinite sequence of Fibonacci numbers, which goes back to the mathematician Leonardo Pisano in 1202 AD and his *Liber abaci*. Pisano was a calculation master in Pisa and counts as one of the most important medieval mathematicians.

Each number in the series is calculated as the sum of the two preceding numbers. The relationship of two succeeding sequences of

247 Classical constructive procedure Golden section

Fibonacci numbers will progressively approach the golden section proportion.

0 1 1 2 3 5 8 13 21 34 55 89 144 …

The definition of a Fibonacci sequence is defined by the recursive law of formation with the initial values being 0 and 1. In other words: for the first two numbers, the values 0 and 1 are pre-established. Each additional number is the sum of two preceding numbers.

The relationship between the golden section and the Fibonacci sequence a_n is immediately evident through the recursive law of formation.

$$\frac{a_{n+1}}{a_n} = \frac{a_n + a_{n-1}}{a_n} = 1 + \frac{a_{n-1}}{a_n}$$

This means that the relationships of succeeding pairs of Fibonacci numbers increasingly approach the golden section as the sequence proceeds. To the extent that the series of these relationships converge toward a limit value Φ, the following must be true:

$$\Phi = 1 + \frac{1}{\Phi}$$

$$\Phi = \frac{a}{b} = \frac{a+b}{a}$$

$$\Phi = 1 + \frac{a}{b} = 1 + \frac{1}{\Phi}$$

$$\Phi = \frac{1 + \sqrt{5}}{2} = 1.618033998 \ldots$$

This relationship, however, is true precisely for the golden section. There exists, for example, the following association between adjacent members of the first 12 members:

1 0.5 0.67 0.625 0.6154 0.619 0.6176 0.6182 0.61806 …

In this way, the inner subdivisions of the golden section are determined precisely by the four decimal positions coming after the period.

$$\Phi - 1 = 0.61803 \ldots$$

As determined by the music theoretician Johannes Kepler (1571–1630)[6], the quotient of two succeeding Fibonacci numbers approaches the golden section Φ. (Drawing the squares of a series of quarter circles, then the result is an approximation of the 'golden spiral'.)

$$\frac{f_{n+1}}{f_n} \approx \frac{\Phi^{n+1}}{\Phi^n} = \Phi \approx 1.618 \ldots$$

Remarkably, the quotient of pairs of succeeding Fibonacci numbers present themselves as continued fractions.

$$\frac{1}{1} = 1 \qquad \frac{2}{1} = 1 + \frac{1}{1} \qquad \frac{3}{2} = 1 + \frac{1}{1 + \frac{1}{1}}$$

$$\frac{5}{3} = 1 + \frac{1}{1 + \frac{1}{1 + \frac{1}{1}}} \qquad \frac{8}{5} = 1 + \frac{1}{1 + \frac{1}{1 + \frac{1}{1 + \frac{1}{1}}}}$$

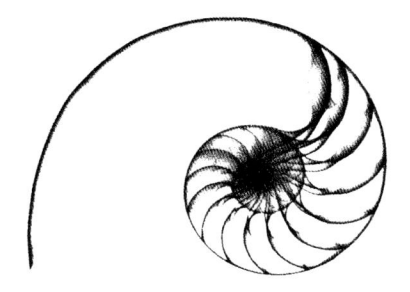

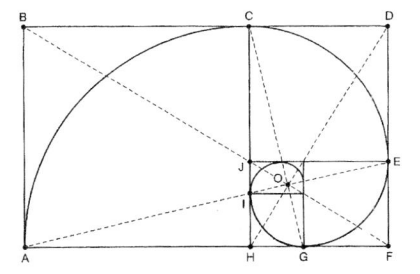

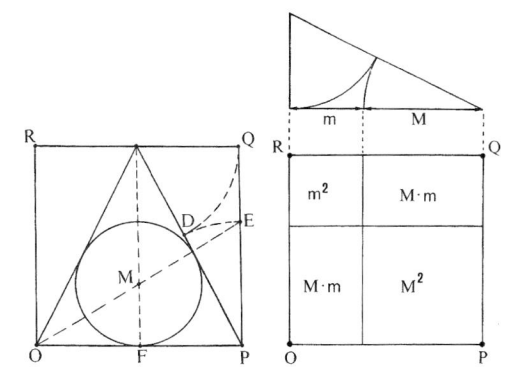

248 **Calcium skeleton of a nautilus pompilus (x-ray photograph)**
249 **Construction of the golden quadrant spiral**
250 **Geometrical relationships of square, triangle and circle vis-à-vis the golden section**

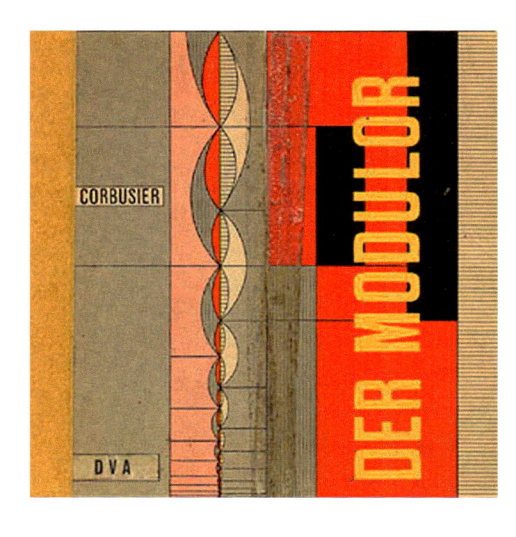

Le Corbusier's Modulor

'Blue series'	'Red series'
	774.7
957.6	478.8
591.8	295.9
365.8	182.9
226.0	113.0
139.7	69.8
86.3	43.2
53.4	26.7
33.0	16.5
20.4	10.2
7.8	6.3
4.8	2.4
3.0	1.5
1.8	0.9
1.1	

251 **Le Corbusier: title page of the German translation 'Der Modulor' (1953)**
252 **Drawing of 'Le modulor II' (1950)**

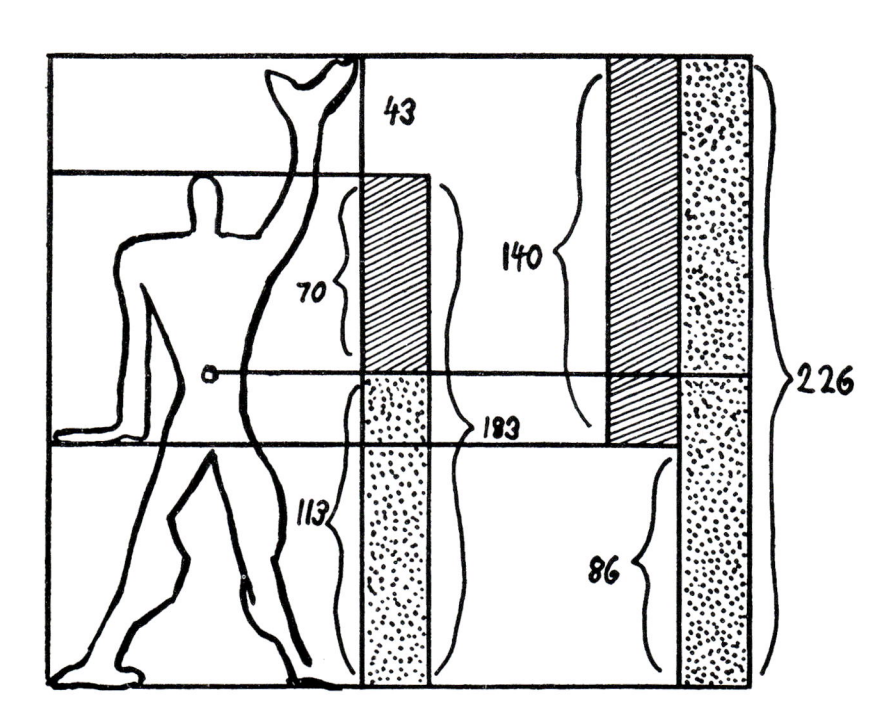

In 1948, the architect Le Corbusier (1887–1965) published his treatise *Le Modulor*. The subject of this volume is 'a harmonious measure to the human scale universally applicable to architecture and mechanics'.[7]

The Modulor, as v. Naredi-Rainer puts it, ranks as 'the most important modern attempt to endow architecture with a mathematical order oriented to the dimensions of the human being'.[8]

For the first version of the Modulor, Le Corbusier used a reference or body size of 175 cm. With the foot, the solar plexus, the head and the fingers of an up-praised hand, he marked the principal intervals of the human form. Through the application of a principle of subdivision based on the golden section, Le Corbusier arrived at bodily dimensions of 175 cm / 108.2 cm / 66.8 cm / 41.45 cm / 25.4 cm. The last dimension corresponds to exactly 10 inches = 25.4 cm according to the English system of measurement, whereby he discovered a method of transition to the English system.

The unit	A = 108
The doubling	B = 216
The 'extension'	A = C = 175 (108 + 67)
The 'shortening'	B = D = 83 (143 + 83)

In Le Corbusier's opinion, however, this system revealed itself as not useful, since at larger values, no superimposition is possible with the English unit of measurement, i.e. the inch. Therefore, in his *Modulor 2* from 1955, he developed a new proportional series specifying a body size of 6 feet:

6 ft = 6 x 30.48 cm = 182.88 cm

Based on the standard size for the human body (= 1.83 m), he again marked three intervals whose interrelations are consistent with the golden section; forming a golden section series.

1.13 m as the height of the solar plexus. (Despite frequent confusion on this point, this point lies below the sternum and not at navel height.)

1.83 m as the new standard body size (corresponds to exactly 6 feet)

2.26 m total height of the human form with outstretched arm (corresponds to 2 x 1.13 m, the height of the solar plexus)

By means of constant division, there emerges a red series (standing man with outstretched arm and fingers). After doubling, we arrive at the height of the 'blue series'. This is derived from the height of the solar plexus: 2 x 1.13 m equals 2.26 m (doubled square).

Calculation formula for ascertaining the levels of division with help of the golden section.

$$\Phi = \frac{1 + \sqrt{5}}{2} = 1.618033998\ldots$$

Application example:
Minor x Φ = major
Solar plexus x Φ = body size
1.13 m x Φ = 1.828 m

'Homo bene figuratus'

In the first century, Vitruvius, the Roman architect, whose full name was Marcus Vitruvius Pollio, composed the 10-volume treatise *De architectura Decem*, in which he sought to present a normative architectural system.

Also of historical significance is his doctrine of anthropometric proportions, a theory concerning the well-formed human body, the so-called 'homo bene figuratus', whose outstretched limbs were held to define either a square or a circle.

During the Renaissance, numerous artists illustrated the relevant passage in Vitruvius (Francesco di Giorgio, Albrecht Dürer, Cesare Cesariano). The text, following Vitruvius, found on Leonardo da Vinci's graphic studies of proportion above the circle, reads as follows:

Vitruvius said that proportions of the human form should be laid out in the following manner:

'4 fingers make 1 hand width, 4 hand widths make 1 foot, and 6 hand widths make 1 ell. 4 ells give you 1 fathom, and 24 hand widths is equal to the height of a man; and these proportions are also found in his buildings. If you spread out your legs, then your height, measured from the head, is reduced by 1/14th. And if you spread your arms out as far as possible and raise them so that you touch a line at the height of the crown of your head with your middle finger, then you should know that the centre of the furthermost point of the outstretched limb, the navel, and the space between the legs together form an equilateral triangle.'[9]

Before the assumptions of Vitruvius are united with those of le Corbusier, the historical situation of the applied proportional systems must be more closely illuminated, since Vitruvius once remarked that the leading system is based on two divergent basic units. First, there is the width of a finger (Lat. *digitus*) or of a thumb (Lat. *pollex*), and the unit of length known as a foot can be divided into 16 finger widths or 12 ounces (Italic system).

In the Greek colonies of Sicily, the width of a plot of land was defined as 10,000 finger widths, and one digitus as 1.85 cm. According to the system, a stadium = 10,000 digitus = 185 m. The Italic system was based on the foot [Lat. *pes*] as a unit of measure, which was subdivided into 12 ounces (thumb widths).

1 foot corresponds to:
a) subdivided by 12 = 2.47 cm
   1 *pollex* (thumb width) x 12 = 29.6 cm
b) subdivided by 16 = 1.85 cm
   1 *digitus* (finger with) x 12 = 29.6 cm

The proportional system notated by Leonardo da Vinci according to the Vitruvius' doctrine of proportions is as follows:

	1.85 cm	1 finger width
4 x	1.85 cm = 7.40 cm	1 hand width
6 x	7.40 cm = 44.40 cm	1 ell
4 x	44.40 cm = 177.60 cm	1 fathom
24 x	7.40 cm = 177.60 cm	1 man height

The English unit in use today is the foot: 1 ft. = 30.48 cm instead of 29.60 cm, thereby lying 88 mm above the standard used in classical antiquity.

At four hand widths = 1 ft. = 30.48 cm, Vitruvius would today base his system on a hand width of +3% = 7.62 cm instead of 7.4 cm. If we increase the size of the human figure in use in antiquity from 177.60 cm by 3%, the result is 1.83 cm. With his *Modulor 1* of 1950, Le Corbusier based his assumption on a human size of 1.75 m (derived from the octometric proportional system, assuming this system was brought into consonance with the human size of 177.6, the standard used in classical antiquity).

Note: In his reworking of the *Modulor*, Le Corbusier raised this height to 1.83 m as an adaptation to the Anglo Saxon system of measurement. The modern measure of 30.48 cm for 1 foot attempts to respond to the fact that the average size of the human being has increased.

If we check now by transposing *Modulor 2* to the Vitruvian 24-hand widths, the result is that the height of one man = 24 x 7.62 cm, (incl. 3%) = 1.8288 m = 1.83 m.

253 **Leonardo to Vinci – manual sketch showing the proportions of the human head**
254 **Leonardo to Vinci – proportional scheme of the human body, acc. to Vitruvius, 1485–1490, Venice, Galleria dell' Accademia**

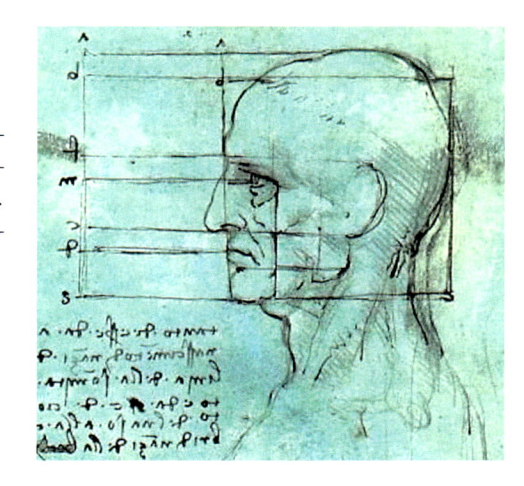

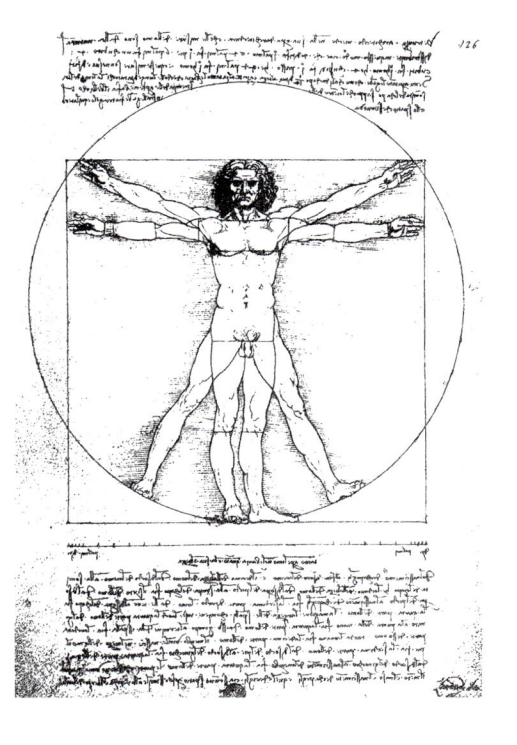

Anthropometrics

Table K: Males Dimensions [mm]	Percentile 50%	Percentile 95%	Registered Ages 18/65	Notes
Standing:				
Body height	1,750	1,855	1,750	(Women: 1,625 / 1,720 mm)
Eyes height	1,630	1,735	1,640	(System norm sightline = 1,650 mm)
Shoulder height	145	155	146	
Shoulder width	480	525	480	(bi deltoid = outer)
	405	435	325	(bi acrominal)
Hip width (standing)	360	385	340	According to Golden Section: 6th level = 1 x 0.16 m
Knee height (standing)	460	480	510	(Tibal height = floor to knee) according to Golden Section: 2 x knee height = navel height
Solar plexus height	1,100	1,175	1,080	Elbow height
Navel height			1,020	According to Albrecht Dürer = 1/4 h and chin/navel = ankle/knee
Crotch height	830	905	826	
Seated:				
Body height	910	965	917	Length of torso
Eyes height	795	855	807	
Shoulder height	625	670	625	
Hip width (seated)	375	420		
Knee height (seated)	535	585	545	
Height of seat base	450	490	400	Length of lower thigh to floor (women: 415 / 450 mm)
Seating depth	495	540	540	
Head length	220	235	220	According to Golden Section: eyes = 1/2 head height
Head depth	195	205	205	
Head width			200	Including ears
Face width	155	165	155	
Morphological face length	115	130	120	Chin to eyes
Distance between pupils	61	69	65	
Ankle height			100	According to Golden Section: 7th level = 0.10 m
Depth of rib cage	225	270	255	
Length of upper arms	365	400	400	
Length of lower arms	475	510	453	To tips of the fingers
Hand length	189	207	190	According to Albrecht Dürer = 1/10 h
Hand width	107	117	100	With thumbs
Finger length	84	93	90	Middle finger
Finger widths	17	19	19	Middle finger
Thumb width	22	24	24	Extended
Length of feet	265	285	290	According to Albrecht Dürer = 1/6 h = 292 mm
Width of feet	101	111	110	
Bodily extension:				
Extension of hands 1.75 m (assuming a body size of 1.83 m)	2,075	2,205	2,095 / 2,190	Both arms extended upward
Extension of hand 1.75 m (assuming a body size of 1.83 m)			2,160 / 2,260	One arm extended upward (see Modulor 2)

Anthropometrics is the doctrine of the dimensions of the human body. It is used principally in ergonomics in order to design workplaces, tools and furniture as well as in the field of work safety. It is consulted in terms of security measures, for example for determining the measurements of protective coverings or setting the necessary distances to be maintained between workers and dangerous equipment. Experts in the field of economic history have been using statistics on body sizes as indicators for living standards.

Every year, the German Federal Office of Statistics publishes the results of a micro-census on the body sizes and body weights of the general population and derives from it the so-called 'body mass index'. The body mass index, or BMI, is a measure used to evaluate the body weight of a human individual in relationship to the square of his or her height.

Not uncontroversial, the body mass index offers only a rough approximation since it obviously does not take into account the individual's stature, his or her specific composition of body weight from fat and muscle tissue. The result of these anthropometric surveys are not used directly, but are instead stored in data tables or norms in order to serve, for example, the classification of developmental tendencies within a given population. The European standard ISO 7250 defines essential standards for industrial design, and the DIN 33402-2:2005-01 (draft proposal) on ergonomic values comprises the dimensions of the human body:

'Consistent with a procedure that has, in the meantime, become the European standard, the population under investigation is defined not according to citizenship, but instead to encompass all individuals residing in the area under investigation, namely the Federal Republic of Germany (the resident population).

In the light of increasing levels of migration to and within Europe, and of non-uniform patterns about naturalization, it is clearly appropriate to orient the gathering of anthropometric statistics to the criteria of residency, in order to respond to the realities of working life and to the requirements of the user population. In contrast to earlier methodologies, this approach takes into account transformation of variability in all age groups.'[10]

'This norm determines the value of the bodily dimensions of undressed individuals. Clothed individuals are not considered, since the details of clothing and other accessories worn on the body undergo rapid changes as a result of fashion styles or technical developments. The function of these definitions is to convey knowledge concerning bodily dimensions and their variability.'[11]

Note: Found in the table appearing on the left are comparative standards for the graphic representation of a model human individual conveyed by the Modulor 'EN' and the entries into the database of the EN/DIN 33402 concerning bodily dimensions. The basis of these considerations is a body size with a reference measure of 1.75 m for a neutral stadium visitor. The registered values (green) lie within the stated spectrum of EN/DIN bodily dimensions. The 'deviations' serve an idealized adaptation of the model person to the ideal proportional principles of the golden section.

Data on bodily dimensions (DIN 33402-2)

If we follow le Corbusier's deliberations of the 1950s, as discussed above, then the size of the human body would be set at 1.83 m (as opposed to the one set out in *Le Modulor*, namely 1.75 m). Statistics collected during the years 1999 to 2002 show, however, that the median value (50th percentile) of man is, as earlier, 1.75 m, so it is perfectly reasonable to adopt this as a norm for a typical visitor to a sports stadium. With respect to regional and social/ethnic composition, the random sample used in the above-mentioned study was selected in such a way as to be representative for the relevant age and occupational populations.

According to DIN 33402-2 of 2005, the secular trend of growth acceleration came to an end in Germany several years ago.

To characterize this variability, the values are stated by the DIN 33402-2 as so-called 'percentiles'. This is a numerical value which positions the individual under examination in relation to comparable individuals.

Entered alongside the median (or 50th) percentile are the 5th and the 95th percentiles. Individuals belonging to the smallest 5th percentile of the largest 5th percentile (i.e. of body height) are not incorporated into this norm. This delimitation, which incorporates 90% of the population, has been adopted for practical reasons, namely that the dispersal of sizes among the remaining groups, i.e. those lying at the extremes, is disproportionately large.

When the body size of a stadium visitor is expressed as a percentile, this means that his/her body size is expressed in relation to those of the remaining visitors. A visitor with a body size lying at the 50th percentile (median value) is larger than 50% of visitors. If the 50th percentile for men is 1,750 mm, this means that 50% of all men have body sizes measuring at most 1.750 mm. If the 95th percentile lies at 1.855 mm, this means that 45% of all stadium visitors have body measurements lying between 1.75 m and 1.85 m.

The German Federal Office of Statistics

Every year, they German Federal Office of Statistics publishes the results of a micro-census survey. In the survey, the dimensions of the population are collected according to age group. Recorded are body size and weight, which are used to calculate a 'body mass index'. The 'BMI' is calculated from body weight [kg] divided by the square of body volume [m²]. The BMI unit is therefore expressed as kg/m². An individual with body size of 175 cm and a body weight of 75 kg has a BMI of $75 \text{ kg}/(1.75 \text{ m})^2 = 24.5$.

As of April 2004, the gender-neutral average value for the size of the human body has been set at 1.74 m. For the sake of simplicity, the system value given by the DIN 33402-2 for a standard individual is 1.75 m for men and women alike (young/old), that is to say a neutral stadium svisitor.

The definition of EN/DIN 33402-2 for the dimensions of the human body according to statistical data collected by the German Federal Office of Statistics (2004)

5th percentile	5%	are smaller
50th percentile	50%	are smaller/larger
95th percentile	5%	are larger

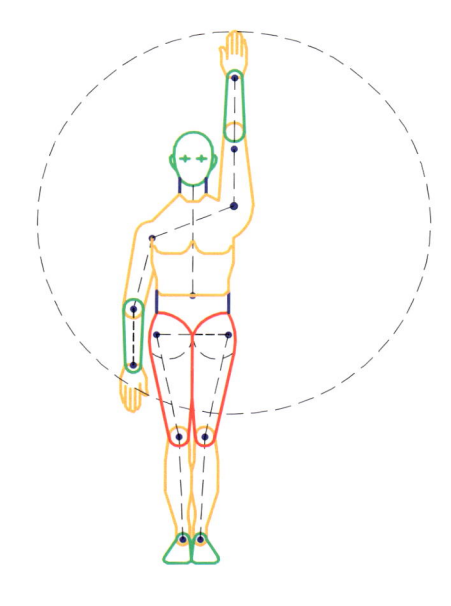

255 Frontal view, Modulor 'EN'
EN/DIN 33402-2 on bodily dimensions
arm extension heights
one arm: 2,160 mm (with 1.75 m),
2,260 mm (with 1.83 m)

page 128:
256 Table K: Overview of the essential bodily dimensions of a stadium visitor (Modulor 'EN')

Table 3.1.2 Body height [mm]

Ages	Men			Women		
	5%	50%	95%	5%	50%	95%
18–65	1650	1750	1855	1535	1625	1720
18–25	1685	1790	1910	1560	1660	1760
26–40	1665	1765	1870	1545	1635	1725
41–60	1630	1735	1835	1525	1615	1705
61–65	1605	1710	1805	1510	1595	1685

Table 4.1.3 Eyes height, standing [mm]

Ages	Men			Women		
	5%	50%	95%	5%	50%	95%
18–65	1530	1630	1735	1430	1515	1605
18–25	1565	1665	1785	1450	1550	1645
26–40	1545	1640	1750	1440	1525	1615
41–60	1510	1620	1720	1420	1505	1590
61–65	1490	1595	1690	1385	1480	1570

Table 23.2.2 Eyes height, seated [mm]

Ages	Men			Women		
	5%	50%	95%	5%	50%	95%
18–65	740	795	855	705	755	805
18–25	760	810	870	720	770	820
26–40	745	805	860	710	760	810
41–60	730	785	850	700	750	800
61–65	710	770	830	685	735	795

Table 66.5.4 Head height [mm]

Ages	Men			Women		
	5%	50%	95%	5%	50%	95%
18–65	210	220	235	190	210	235
18–25	215	225	240	195	215	240
26–40	215	220	240	190	210	235
41–60	210	220	235	185	210	235
61–65	210	220	235	185	205	230

Chapter 13

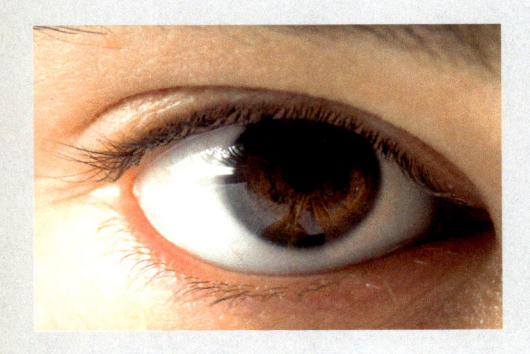

257 **Picture of the human eye**

Physiology of viewing
An evaluation of visual acuity, spatial perception and visual angle zones based on an examination of the human eye

Physiology of the eye

In order to perceive objects in our surroundings, an image must be projected onto the light-sensitive layer of the eye; a principle comparable to that of a camera. The optic apparatus of the eye contains a lens system, comprised of cornea, iris and lens, which focuses all rays of light emitted from one specific spot onto a corresponding spot in the retina and thus projects onto it a reversed image of the surrounding world. This process is called 'refraction' and is subject to the laws of physical optics.

Accommodation

signifies the capability of the eye to adapt its optic system to objects near and far in a flexible manner so that a sharp image emerges. The lens is suspended from the so-called ciliary body. When inactive, it is tightened by these zonular fibres and thus remains flat. When the ciliary muscle contracts, the zonular fibres are relaxed, and the lens returns back to its natural, more globular shape. This raises the refractivity of the eye, which enables us to see sharp at a short distance. Short-term changes in light intensity can be neutralized by a change in pupil size. Photo receptors react to long-term changes of light conditions by adapting to a medium light intensity ('adaptation').

The retina is located at the posterior surface of the eyeball where millions of individual photo-sensitive cells are arranged in a hemisphere-shaped layer. Photo-chemical degradation of the substance contained in a receptor triggers an electrical impulse, a 'neuronal excitation'.

Thus, visual perceptions are detached from the optic mechanism of mere transmission and processing of images inside the brain. At an angle of 5° temporal of the optic axis a sightline reaches the centre of the retina, also called 'fovea centralis', the central area of the 'macula' (also termed 'yellow spot'), or the point where sharp viewing takes place.

Retinal rods and cones

are two types of photo receptors. Visual perception of humans is differentiated between 'peripheral' viewing (retinal rods: light/dark) and 'foveal' viewing, that is, sharp viewing in the centre of the retina (retinal cones: colour perception). The macula region is made up of millions of ganglia (cones) responsible for colour vision.

The rest of the retina is covered by rods which produce only blurred black-and-white images, but function also under subdued light conditions. The image recorded by the sensory epithelium of the retina is further processed in the visual system of the brain.

The fovea centralis has a convergence of 1:1, that is, every receptor has a corresponding ganglium in the brain. These receptors are interconnected 1:1, resulting in the highest resolution, or highest visual acuity in this area. The so-called 'blind spot' on the retina is a zone without receptors since this is where the pathway to the brain (visual cortex) sets in.

Physiology of visual acuity (resolution)

Visual acuity denotes the capability to discern the details of an object.

The *minimum angle of resolution* (MAR) refers to the ability of perceiving two points of an object with a certain distance 'd' to one another as separate items. This value is independent of distance and relates to the resolution capacity, the minimal viewing angle in angular minutes. Visual acuity of the human eye is limited by three factors:[1]
1) the optic image of the eye
2) the diffusion of light in the retina
3) the neuronal processing in the retina

The term visus denotes the reciprocal value of the angular resolution measured in minute of arcs: $visus = 1/\alpha$ (in clinical terminology angular resolution (visual acuity) and visus are equally used, although they are differently defined.)

Full visual acuity is achieved at a visus of 1.0 or 100 percent.

$$\text{Calculation of visus value} = \frac{\text{actual distance}}{\text{target distance}}$$

An example: If a patient can discern a sign, which is normally recognized at 15 m distance, from a distance of 3 metres, the visus is measured at 3/15 or 0.2.[2]

Minimum separabile

As smallest dimension, the human eye can discern an object corresponding roughly to an angular minute of the visual field. Rendered to a distance measure, this means a human eye is capable of recognizing an object the size of 0.09 cm at a distance of 3.0 m.[3]

The regular minimum angle of resolution for adults is one angular minute, a value of visual acuity 1.0 or 100%.

Children have a higher resolution (1.28 to 1.5 visus). The density and distribution of receptors on the retina would theoretically enable a minimum resolution of 200%. For most activities of daily life a visual acuity of 0.5 to 0.6 is sufficient. A visus of 50% corresponds to a defective vision of 0.5 dpt; a person with a grade of 1.5 dpt short-sightedness (myopia) views at merely 12%. When determining refraction and visus, the size of optotypes is chosen so they can be discerned by a healthy eye from a distance of 5 to 6 m (Anglo-American measure = 20 feet). Based on the physiognomy of the optic components of the eye, visual acuity drops dramatically toward the periphery; at 10° eccentricity, a visual acuity of 10 – 20% of the foveal value remains.

258 **Light ray pathway of an optic image**
259 **Accommodation of the eye**
260 **Structure and measurements of an eyeball**
according to Prof Dr med M. Reim
(Department of Ophthalmology, RWTH Aachen)

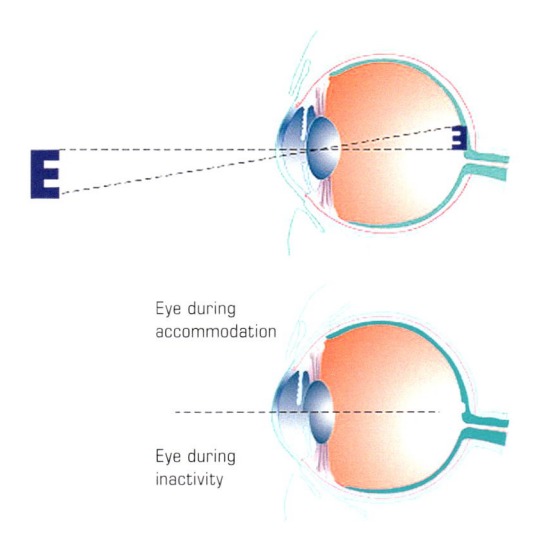

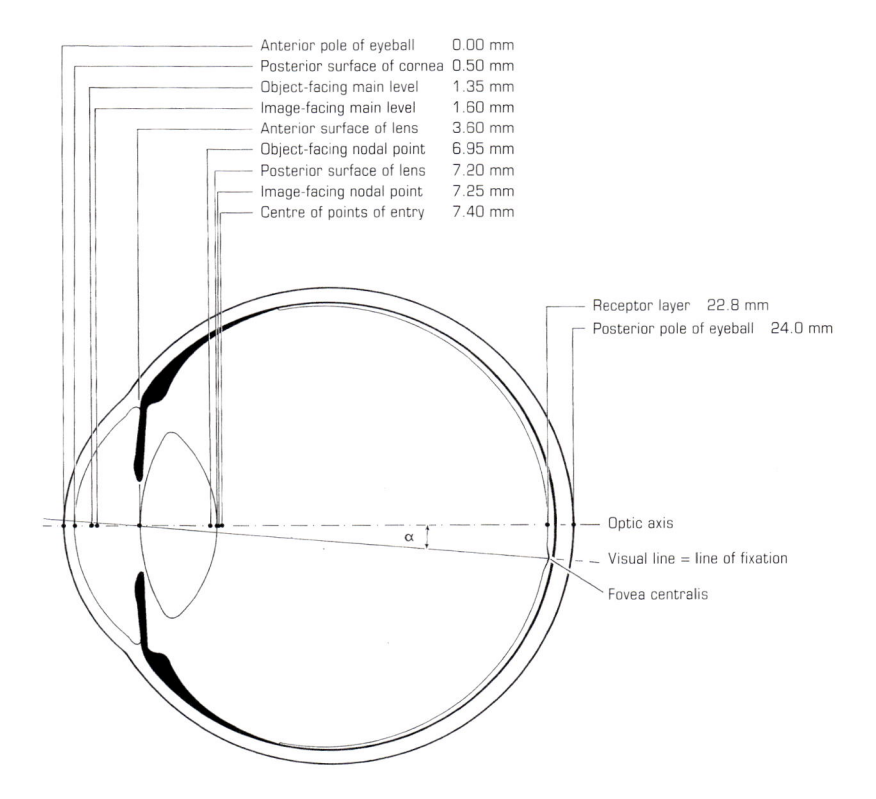

Eye during accommodation

Eye during inactivity

Anterior pole of eyeball	0.00 mm
Posterior surface of cornea	0.50 mm
Object-facing main level	1.35 mm
Image-facing main level	1.60 mm
Anterior surface of lens	3.60 mm
Object-facing nodal point	6.95 mm
Posterior surface of lens	7.20 mm
Image-facing nodal point	7.25 mm
Centre of points of entry	7.40 mm
Receptor layer	22.8 mm
Posterior pole of eyeball	24.0 mm

Optic axis
Visual line = line of fixation
Fovea centralis

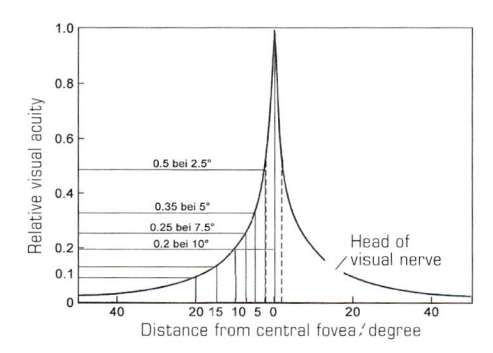

261 Image resolution of the retina
262 Landolt-Ring with an opening value of one angular minute

Visual acuity – visual angle

Visual angle of most sharp viewing Fovea centralis, minute of arc = 1.75 m reference measure / 200 m	0.5°
Visual angle of sharp viewing Simplified value as combination of natural eye movements ± 30° for scanning a visual field (motility)	60°
Depth of field perception Overlap of the nasal field of view, maximum horizontal visual angle ± 60°	120°

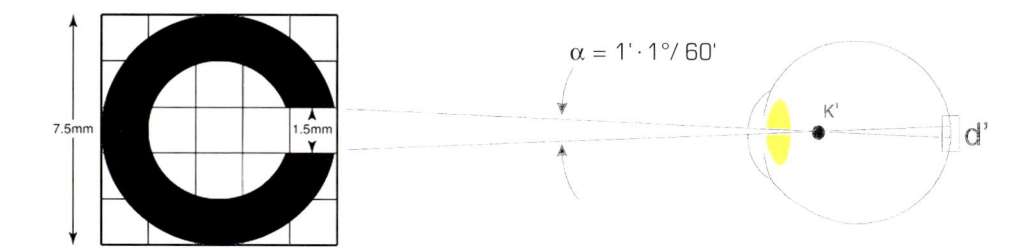

$\alpha = 1' \cdot 1° / 60'$

Angular minute
A minute (*minute of arc* or MOA) represents a subdivision of the unit of degree for the size of even angles. One degree corresponds to the 360th part of a full circle.

$360°$ = 1 full circle
$1°$ = 60 minutes of arc = $60'$
$1'$ = 60 seconds of arc = $60''$
$1°/60'$ = 1 MOA = $0,017°$
= α

Note: The angular units of minute and second are not to be confused with the value of 'rectascension' in astronomy (hours, minutes and seconds) where 24 h = 360°.

The diagram to the left shows the relationship between the opening angle and the visual acuity without eye movements. From this, the following figures are deducted for the area of sharp viewing (visus):

one angular minute 1/60 = $0.017°$ = 1.00
(approx. values) = $0.5°$ = 0.80
= $3.0°$ = 0.50
= $10.0°$ = 0.20

Adaptation luminance
In order to perceive an object, it has to be adequate in size or display sufficient contrast. By means of punctual visual acuity, we can determine if an object is still within the range of perception or not. A point cannot be discerned when illumination is so low that no receptor on the retina is stimulated.

This is why stars, although always present, cannot be seen during the day because contrast is too low. Resolution makes the details of an object visible. It is based mainly on the capability to discern the difference in luminance separating points or lines from each other.

'The luminance of objects produces different intensities of illumination on the retina. Spacious objects illuminate several receptors on the retina simultaneously. Punctiform objects smaller than 1' [minute] stimulate only individual receptors. For punctiform objects, instead of illuminance, luminous flux becomes relevant. ... The eye adapts to the luminance present within the visual field. ... Visual acuity diminishes with age: a 75-year old would generally possess only 50% of the visual acuity of an 18-year old.'[4]

As luminance in the direct surroundings diminishes, according to B. Wördenweber from the University of Paderborn, so does the speed of perception and the differential sensitivity. Glare also reduces the viewing performance of the eye and is caused by diffusing light which superimposes the projection of an object on the retina. Visual acuity is reduced at low illumination and at the periphery of the visual field.

An object which the eye fixates is always focused in such a way that its image is projected exactly onto the fovea centralis. Only in the central fovea of the retina (fovea centralis) is the human eye capable of seeing sharp properly; it is also the area of the highest resolution of colour perception.

In the peripheral retina we can also find retinal cones, and, to a limited extent, colour perception is possible.

Motility (eye movements)
To view something precisely, we have to direct our gaze directly onto the object so that it is projected onto the centre of the retina. At the periphery of our field of vision we can see objects only blurred and not in colour.

By constant minute eye movements, our optic apparatus ensures that different sections of the surroundings are projected sharply. Adaptation

Physiology of spatial viewing

of the eyes onto the fixated object is done with swift movements, the so-called 'saccades', a synonym for peering movements of the eye. Binocular viewing is an interaction of motility (muscular movement of the eyes) and receptive perception.[5]

Thus, through constant movements of the eye apparatus many small sharp images are combined into one entire image. Finally, the brain combines the different information into an image that appears to us uniform in sharpness and colour.

The effective binocular field of vision (driving regulations) is screened by the involuntary eye movements without head movements:

25° upwards
30° to the right/left
40° downwards

At 100% eye movement, the eye scans a field of 2 x 30° = 60°; sharp vision hence takes place at 60°.

Spatial viewing or focal depth is the product of processing two equal images. The term depth of field depth relates to the area that is fixated (the so-called Panum space); all else should be termed depth perception. The image of the environment on the retina is like all images two-dimensional, or rather a two-dimensional representation of a three-dimensional space. At the same time, we clearly discern distances, depth and other spatial effects. Stereoscopic viewing means that an object is viewed from two non-concurrent perspectives. Physically, we understand as spatial, binocular viewing the stereoscopic or transversely disparate viewing whose depth perception is a result only of the different transversely disparate projection of the points of an object onto the retina.

The eye muscles are arranged in such a way so that the left and right eye always view the same object: binocular viewing is the prerequisite for making the viewed object appear three-dimensional. Partial images produced by the eye are combined in the brain to form a spatial picture (medical term 'fusion') and by means of sensors processed thus as to form a binocular impression. In this manner, the different partial images are projected on top of each other, so that an impression of spatial depth and the three-dimensional shape of an object are produced.

Sight or visual angle

Since each eye records an image, the central projection of both eyes is 2 x 60° = 120°.

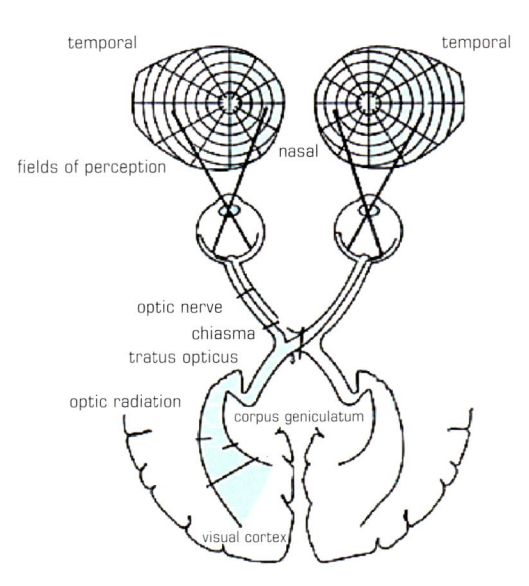

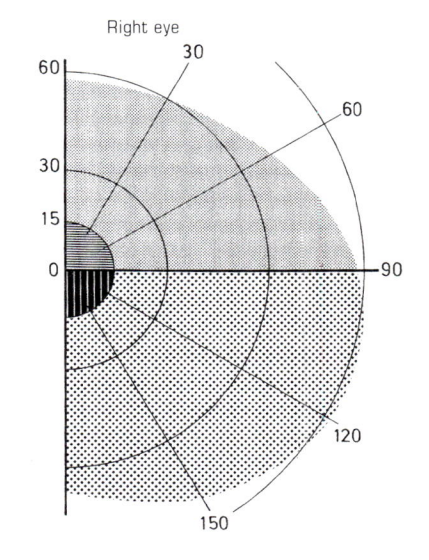

263 **Projection of field of perception into the optic centre of the cerebral cortex** (cf. Duke-Elder 1961)
264 **Projection diagram of the eye (right)**

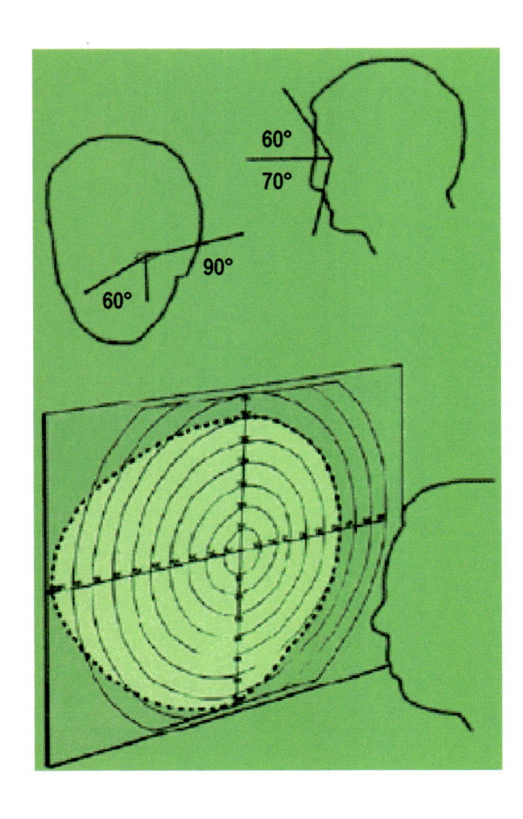

265 Normal field of perception of a regular spectator

Human field of perception

In 1974, specific body postures were classified to the OWAS method (Ovako Working Posture Analysis System).

Five categories were determined for movements of the head: postures from an angle of 0 – 30° were perceived as least straining, above this they were deemed as disagreeable.

Head and eye movements can be categorized as follows, with reference to classification of comfort for seating/standing accommodations:

Up to 90° 'full' rotation of the head: x 2 = 180°
– Shoulder view – unsuited for a permanent viewing direction.
– A turn of the body seems inevitable to maintain comfort.
– Very common for short-term viewing situations (match activity around the corner flags) or in exceptional cases.

Up to 60° 'half' rotation of the head: x 2 = 120°
A normal head movement without twisting of the shoulder – 'comfortable'.

Up to 30° 'slight' rotation of the head: x 2 = 60°
A slight head movement which produces the same view as through eye movements (without head movements).

The driving licence requirements for visual acuity determine for a normal visual field a horizontal diameter of 120° and for the central visual field a diameter of up to 30°. For transportation of passengers, a visual field of 140° horizontal has to be verified.

The image resolution capability of the retina, or visual acuity, diminishes rapidly toward the periphery without eye movement. Therefore, we may distinguish the terms *field of view* (or visual field) and *field of perception*: the first term defines the scope of an area which a person can see sharp without eye movements.

Each eye has a horizontal visual field of roughly 150° (90° + 60°). The visual fields of the right and the left eye overlap in the middle by around 120°. Only within this overlap the perception of depth is feasible.[6]

Normal visual field of an eye:
Upward   up to approx. 60°
Downward up to approx. 70°
Outward   up to approx. 90°
Inward   up to approx. 60°

Toward the temples (outward) we also use the term 'temporal' and toward the nose (inward) the term 'nasal' visual field.

As field of perception we understand the area that a person can overview with immobilized head and body but at maximum eye movements. This field is much larger than the visual field. For binocular viewing, the field of perception is around 240°.

Note: As a rule, the field of perception and the zone of attentive and sharp viewing should not be confused.

Visual perception

No other of the human senses requires a similarly complex brain performance than seeing. The 'reality' that our eyes record, is the result of a highly intricate process of interpretation. Visual perception (lat. *percipere*) defines the entirety of processes of perception (or sensation). Contrasts are intensified, blur is balanced and peripheral sensory impressions are filtered. Equally, movements of eye, the body or the outside world are compensated by the brain, so that a stationary image is created.[7]

All these functions serve to improve the quality of an image. Unconscious to the viewer, it is analyzed by the viewing apparatus and thus contributes to the understanding of the observed scenery. Consequently, many more influential factors determine the quality of viewing, some of which are:
a) perception of patterns
b) 'filling-in'
c) form and colour consistency
d) visual control of motor activity

With the help of pattern perception, familiar object are perceived and processed more quickly. From personal experience, familiar patterns are derived, which obviously involves a considerable degree of interpretation.

Perception and recognition of objects is largely reliant on the information content of an image. A lack of information may make an object hard to see or even unrecognizable. A classic example is the drawing on page 133 by W. E. Hill from 1915.

'Filling-in' is another method to complement images with the help of the brain. Objects are completed when mere physical optical perception is impaired. With the help of photo receptors, the eyes are capable of absorbing around 10 million bits of information, with each receptor

LIVE-effect

processing a maximum of 60 bits per second. As an example of visual control of motor activity and the eye's capability of perceiving movement, F. Grehn cites the header goal:

'In the penalty area, a striker shoots for the goal. The ball has a velocity of 75.8 km/h and requires a time of 0.76 s to hit the goal post, from which it ricochets at 42.4 km/h. The player who is still in a forward motion, heads the ball after 0.44 s from around 5.0 m distance toward the goal, hitting the net at 52 km/h after further 0.36 s.

The striker has just 440 ms time for an active decision. Of those, he requires already 110 ms for visual perception of the returning ball, 50 ms for anticipation and another 160 ms for the head-to-ball movement.'[8]

Pure visual acuity is not the sole decisive factor for human perception; neither is the opening angle of a minute of arc. Spectators, in particular, are confronted with a multitude of impressions.

Apart from viewing the action of a match or other event, spectators also take in other impressions on the periphery; for example the activities of other spectators seated or standing directly in front or in other parts of the stand, scoreboards or video screens with stationary or moving images, pyrotechnic smoke in the air, or even rain illuminated by floodlight.

The scenario of a sports and event venue is totally different from the intense atmosphere of a theatre. Distractions by parallel perceptions such as advertising, spectators or video screens are part of normal event activity.

All five human senses contribute to the understanding of our environment:

Sight – Hearing – Smell – Taste – Touch

For many, 'seeing' would presumably be the most important sense of orientation in the world since around 80% of sensory impressions are perceived through the eyes.

All information is further processed by the brain, supporting, or rather influencing 'pure' seeing; the physical optics of the eye. In spite of the currently very high image quality of television, which will be delivering crystal clear images with HDTV technology, football enthusiasts are still going to the stadium and, in the author's opinion, will keep on doing so in the future, as proven by the experiences of the FIFA World Cup 2006. (HDTV – a global digital television standard which contains essential improvements to image quality, in wide screen format with very sharp contours, rich colours and immense depth of field.)

Note: As a consequence of increased ticket demand for World Cup 1998, FIFA raised minimum stadium capacity by 10,000 seats.

The excitement of a game can be appreciated worldwide in front of a TV set, but the atmosphere of a filled stadium stand is still incomparable. Acoustic perception of the simultaneous excitement, and disappointment, of thousands of spectators, in combination with the real sensory impressions of smell and taste, make up the experience of a stadium visit for sports enthusiasts.

This holistic perception also influences the capability of each individual to follow the sporting or event activity subjectively and visually.

266 **'Old Woman / Young Woman'**
Drawing by W. E. Hill, 1915

13 Physiology of viewing

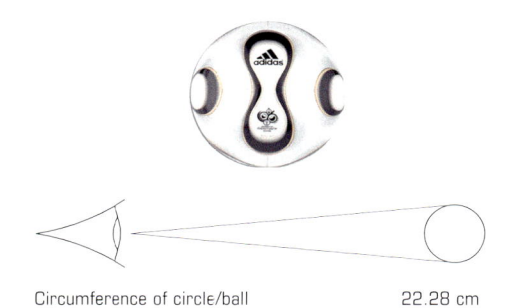

Circumference of circle/ball 22.28 cm

267 **Final of World Cup 1954, Switzerland**
268 **'Team Spirit', official football for FIFA World Cup 2006™**
(here as smallest dimension of perception regarding viewing distance)

269 **Table B1 from EN/DIN 13200-1: Recommended and admissible viewing distance**

Table B.1: Recommended and admissible viewing distance

Group	Indoors (in m)		Outdoors (in m)	
	D (recommended value)	D (maximum value)	D (recommended value)	D (maximum value)
A	110	130	190	230
B	85	110	150	190
C	60	80	70	100

Note: Distance between recommended focal point *P* and the eye point of a spectator.

Viewing distance

The ball
According to DFB, a football is according to the rules 'if it is spherical, made from leather or another suitable material with a diameter of circumference of 68 cm minimum and a maximum of 70 cm. At the beginning of a match it weighs between 410 and 450 gram and has a pressure of 0.6 to 1.1 atmospheres, corresponding to 600 to 1100 g/cm² at sea level.'[9]

If the viewing distance is related to the perception of a football as criterion for visual acuity of an angular minute, a football would still be discernible at 100% minimum angular resolution at a distance of around 770 m. This measure does in no way correlate to the specification in EN/DIN 13200-1. Consequently, the ball itself cannot be deemed as decisive factor for determining the maximum viewing distance.

The player
Calculating the angular values, specified in the framework of EN/DIN 13200-1, with the reference value of 'player' instead of 'ball' (= 1,75 m), then the following measures can be stated for the viewing angle to be observed (rounded):
– recommended (150 m) circa 0.7°
– maximum (190 m) circa 0.5° (football)
 (230 m) circa 0.4° (athletics)

The eyes of a spectator can perceive a ball action on the pitch with sufficient viewing comfort if the seat which is furthest away from the opposite corner flag is at a distance of less than 190 m. In this case, we may assume that the spectator can see the ball in a sufficient size, as 0.053° ≫ 1 angular minute = 0.017°.

First conclusion
The area of sharp vision with one angular minute and a motility of a pair of eyes of up to 2 x 30° = 60° (right/left) of the visual field is not sufficient as comfort value for the construction of sports venues. Following the details stated in EN/DIN 13200, instead of the maximum seat distance, a binding value of 0.5° maximum visual acuity angle is proposed, which thus harmonizes the values for athletics and football from presently 190/230 m, or rather 0.44°/0.53°.

Note: There is no 'pure' athletics stadium; a combination of tracks surrounding a central sports field is always a 'multipurpose' stadium.

Maximum viewing distance
The maximum distance from which the sporting event can be followed adequately depends on the size of the activity area and the size and speed of the object being observed (see table B.1 and image 268). For the classification of sports and the criteria for the determination of groups refer to Annex C of DIN 13200-1. The value for the maximum distance is referred to as 'mixed practical value'. Assuming a defined distance value with a maximum viewing angle of 0.5°, then a new mutual maximum distance for athletics and football emerges, at a reference size of 1,75 m (standard person size).[10]

Check calculation:

$$\frac{\text{opposite side}}{\tan \alpha} = \text{adjacent side}$$

$$\frac{1.75 \text{ m 'player'}}{\tan 0{,}5°} = 200 \text{ m}$$

Note: This value was recommended already in 1993 by the IAKS for multi-purpose stadia as limit for large complexes, but has never been properly derived.[11]

Distance of recognition for evacuation signs

For the purpose of comparison, the values are derived from DIN 4844-1, and the term of distance factor is described. The relation between the largest possible distance from which a safety sign is still readable (visible in terms of form and colour) and the height of the safety sign is established by the relevant DIN code together with distance factor Z:

$$h = \frac{l}{Z}$$

Distance factor $Z = 1/\tan\alpha$
(h and l have the same dimensional unit)

Distance factor Z depends on the fitted height of the sign, the size of the relevant detail, the luminance of the sign and the contrast to the background.

DIN 4844-1:2005-05, Table NA.1

The preferred sizes of illuminated signage for safety, support and letter signs are summarized in this table. The smallest unit of perception is the character size with $Z = 300$ and rescue signage with $Z = 100$.

Example:
1.0 m distance – character size 4 mm
30.0 m distance – character size 100 mm

$$Z = \frac{1}{\tan\alpha} = \frac{1}{\frac{OS}{AS}} = \frac{1}{\frac{0.1\ m}{30\ m}} = 300$$

OS   opposite side
AS   adjacent side

Step 1: If you relate the distance factor $Z = 300$ to the viewing conditions inside a stadium, at a maximum viewing distance of 190 m, a reference height of 60 cm emerges = one shoulder/module width.

Step 2: Relating the above formula to a 'player' height of 1.75 m with a maximum of 1.90 m, a value of $Z = 115$ emerges.

The distance factor Z is the reciprocal value of the viewing angle tangent (see above), which has the following effects on the angle:
$Z = 115 \quad \alpha = 0.5°$
$Z = 300 \quad \alpha = 0.2°$
$Z = 500 \quad \alpha = 0.017°$
     $=$ one angular minute

For single characters on a scoreboard, the FIFA specification applies at a maximum reading distance $= H/500$.

Second conclusion

This means that the angular minute is set as design calculation for reading signs; the reference measure for maximum viewing distance is sufficient at a maximum angular resolution of 0.5°.

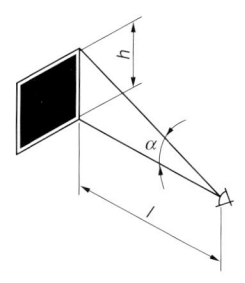

270 **Illustration (figure 21) from the DIN 4844-1 'Sicherheitszeichen' (safety signs) for determining the distance factor**

Viewing distance

Theatre	acting stage	24 m max.
	(maximum distance to actors)	
	opera	34 m max.
	(large gestures still visible)	
Football	recommended	150 m
	maximum	190 m
Athletics	recommended	190 m
	maximum	230 m (uncovered)
New	maximum	200 m at 0.5°

The maximum viewing distances of twelve World Cup stadia in comparison: [m]

01 Berlin	Football	207.50
	Athletics	239.00
02 Dortmund	Football	200.50
03 Frankfurt	Football	182.50
04 Gelsenkirchen	Football	187.50
05 Hamburg	Football	187.50
06 Hanover	Football	182.50
07 Kaiserslautern	Football	186.00
08 Cologne	Football	187.50
09 Leipzig	Football	167.50
10 Munich	Football	187.50
11 Nuremberg	Football	187.00
	Athletics	209.00
12 Stuttgart	Football	196.00
	Athletics	224.00

Distance factor – Z

$Z = 115$ maximum distance 'player'
$Z = 300$ maximum distance 'character size (safety signs)'
$Z = 500$ maximum distance 'scoreboard' (single characters)

Visual acuity / visus

Angular resolution
One angular minute $1/60 = 0.017° = 1.00$
(approx. values) $0.5° = 0.80$
                     $3° = 0.50$
                     $10° = 0.20$

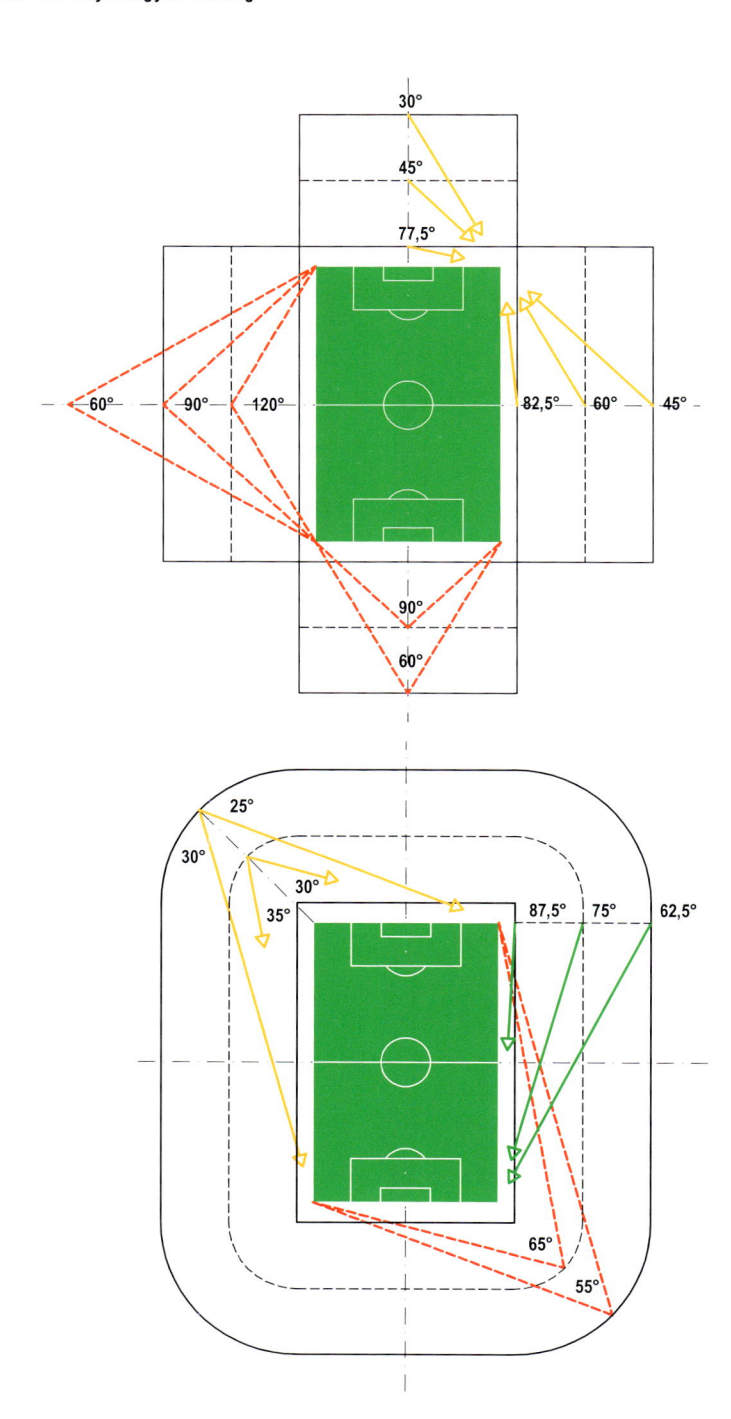

Visual angle zones

When planning a theatre building, the maximum dimensions of an auditorium are determined by viewing angles in reference to the viewing activity of the spectator. Similar notions of comfort may be derived for stadium construction.

Since activity at sports events is not reduced to a relatively small stage (for example an opera stage w = 26 m, d = 21 m, h = 27 m) but is spread out on a playing field almost five times this size (68 x 105 m), we may assume that a spectator's view has to 'wander' with the action on the pitch.

The combination of eye and head movements results in the following viewing angles:

< 30° 'good viewing': binocular effective field of perception without head movement, but with slight eye movement

30° – 60° 'comfortable viewing': supported by a slight head movement

< 120° 'depth of field/focus': for stereoscopic viewing, depth perception is possible only in the area of nasal overlapping of 2 x 60°.

Note: The planning guidelines of IAKS (1993) confirm the above assumption since a horizontal visual angle of 120° toward the scoreboard should not be exceeded.[12]

Centre-line method

Similar to lecture hall construction, a centre-line position of a seat is the starting point for stadium construction.[13] The minimum distance to the eye point at the long side is around 7.0 m and at the short side around 8.5 m.

Note: Safety distance, according to FIFA/UEFA, is at least 6.0/7.5 m, railing width around 20 cm, with 80 cm tread depth. Since the division of visual angle zones represents an attempt at 'qualitative' grading, we have omitted the reduction of distance between eye point/rear of tread.

Angle of perception of 120°

In order to see actions at the corner flags, a spectator should sit 30 m away on the long side of a 105-m long pitch. This distance matches the upper edge of the lower tier; minimum distance of stands plus average number of lower-tier rows of examined World Cup stadia.
(28 x tread depth = approx. 23.0 m + 7.0 m = 30 m)

Thus, the entire lower tier on the long side is located within the first *visual angle zone of 120°* between the touch line, or rather the outer athletics track, and the rear edge of the lower tier.

'From the corner of one's eye, all actions within this viewing angle can be perceived. Beyond, uncertainty sets in as 'something' seems to be evading the field of vision.

At full rotation of the head and shoulders, a field of perception of up to 260° is possible.'[14]

Most likely, an attentive spectator tends to follow the game and its actions directly with the eyes and with a turn of the head of 30° – 60° in the direction of the corner flag.

Note: The span of 30° – 60° arises from the individual movements of a spectator, subjectively combining eye and head movements.

At 165°, the necessary viewing angle of a centre-line seat in the first spectator row in the direction of the corner flag is much higher than the 120° maximum angle of depth perception, i.e. the ability to visually discern actions on the pitch.

In this position spectators are forced to move their heads in order to view the corner flags: by turning the head 82.5° to the side, they can see the object stereoscopically in the viewing focus of both eyes.

This is either done by gazing almost vertically over the shoulder to the side ('shoulder view') and/or by rotating the shoulder toward the viewing direction.

So if there is no activity at the corner flag – which at this point serves only as visual boundary and reference point for a player's position – spectators can see the entire half of the pitch by moving their heads only slightly by a quarter-turn and thus shifting the flag into the periphery of the field of perception (82.5° – 60° = 22.5°).

Moving the shoulder results in a slightly twisted posture of the occupant on the ergonomically formed seat shell.

Since the combination of eye, head and shoulder movements also depends on the predilection of the individual spectator, a further differentiation of comfort is not implemented. Spectators seated in the first row of the short side of an average World Cup stadium are capable of viewing the entire width of the pitch without head or eye movements (120°). If the FIFA minimum distance is kept, only slight head movements have to be performed by spectators.

The width of the ground-plan drawing opposite refers in both cases (angular or rounded) in diameter to the optimum 90-m viewing circle. The length corresponds to the chosen width, as all height profiles (step front edges) should be defined as circulating. Geometrically, this results in viewing angle zones of 90°/60°, which means that the rear edge of the upper tier stand (long side) corresponds to the rear edge of the lower tier stand on the short side.

Parallel terrace front edges toward the pitch represent some of the more complex geometries. Due to their high viewing angle divergence to the point of convergence at the centre circle of the rectangle, they are more uncomfortable in terms of sightlines but should not be generally ruled out.

The outer flagpoles are defined once again as reference points, resulting in viewing angles of 65°/55°. In the corner area beyond the goal line, the ground-plan line should never run rectangular to the pitch since spectator rows located at the corners of a rectangular grandstand would be facing the adjacent row of the opposite stand. The resulting neutral head position of around 45° would be extremely uncomfortable and therefore unacceptable.

Table L Distance D	Distance of stand to halfway line (m)	Distance of stand to goal line (m)	Distance of stand to corner flag (m)
01 Berlin	24.20	42.00	15.75
02 Dortmund	5.65	6.40	=
03 Frankfurt	11.80	13.70	6.75
04 Gelsenkirchen	11.00	11.50	6.75
05 Hamburg	10.85	11.30	9.50
06 Hanover	14.05	16.90	11.50
07 Kaiserslautern	13.60	10.10	9.00
08 Cologne	7.20	8.15	=
09 Leipzig	14.70	13.90	7.50
10 Munich	8.00	8.00	7.50
11 Nuremberg	19.85	36.55	18.50
12 Stuttgart	17.75	38.70	18.00
Average number	**13.22**	**18.10**	**11.08**
Minimum	5.65	6.40	6.75

page 138:
271 **Viewing angle zones**
Sketch showing viewing angles (stadium size refers to long side: max. 90-m viewing circle; short side: same depth as long side)

272 **Sketch of viewing angles at the corners**
(example: geometry of a rounded rectangle)

273 **Table L: Distances to the 'first row'**
Reference points are touch line/goal line and front edge of stand profile (moat or fences not considered)

13 Physiology of viewing

Best viewing radius – maximum viewing distance

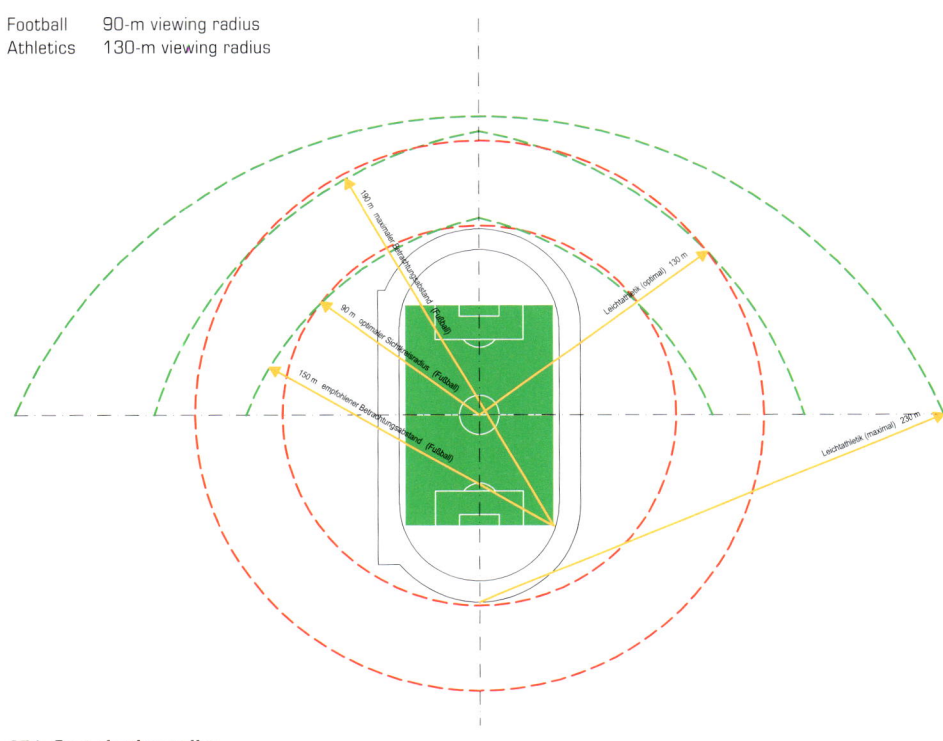

Football 90-m viewing radius
Athletics 130-m viewing radius

274 Best viewing radius
Drawing of the optimum viewing circle

Optimum viewing circle

Drawing a circle with a compass from each extreme point at the field with the recommended distance of 150 m would result in a *90-m circle* at the pitch centre.

This circle defines the spectator area with optimum viewing distances and is generally termed 'best viewing radius' or 'optimum viewing circle'.

It is also a useful planning parameter for examining the efficiency of spectator stands and a reference for the positioning of commercially important user groups, such as media or VIPs.

Applying the principle of optimal viewing distances to different stadium geometries illustrates what influence the ground-plan shape of a stadium has on the distribution of spectators and the corresponding viewing conditions toward the pitch.

Applied to athletics, the above compass method produces a *130-m circle* of optimum viewing distance, in approximation to the recommended viewing distance of 190 m.

Because athletics stadia are always planned as multipurpose venues that accomodate other sports, such as football, the subdivision of spectator areas generally extends the 90-m viewing circle.

All of the following system plans are geared in their dimensions to the 90-m radius of optimum viewing distance:

Example A shows four individual stands which are laid out parallel to the playing field and are limited to its length. The stand depth complies with the normal viewing distances. The system sketch clearly shows that within the 90-m circle the corner areas are 'underused'. There are no seats within a 'good' viewing distance zone.

Example B represents a continuous stadium bowl covering the corner areas but failing to exploit the 90-m circular shape at the long sides.

Example C This system sketch consistently complies in its ground-plan shape with the sightline circle. Thus, an ideal circular stadium is created, resulting in geometric consequences for the elevation and the profile of the stand's upper edge – an 'undulating upper tier edge'.

Example D shows the next geometric step towards an asymmetric stadium, e.g. through shifting the hub of spectators to the main stand.

Optimum viewing circle 141

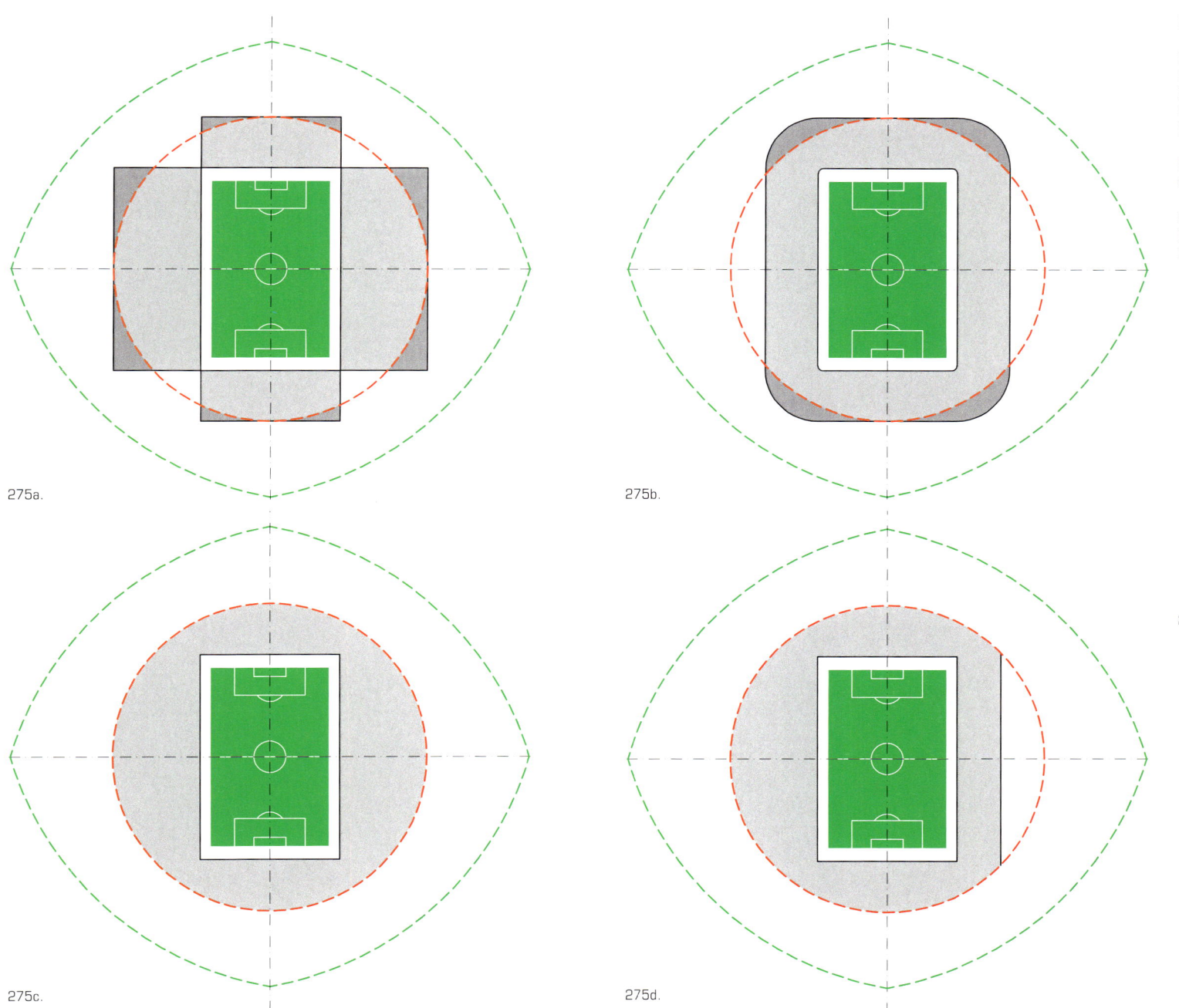

Field of perception / field of vision

Field of perception
Up — up to 60°
Down — up to 70°
Outward — up to 90°
Inward — up to 60°

Binocular 'effective' field of perception to the left/right 2 x 60° up to 120°

Eye movement
Unproblematic ± 30°
Maximum ± 60°

275a.

275b.

275c.

275d.

275 **Layout typologies auditorium**
(representation of relation viewing circle / grandstand shape)
a. Four single stands
b. Continuous bowl
c. 'Optimum' geometry
d. Spectator hub on main stand

Chapter 14

276 **Maracana Stadium, Rio de Janeiro, Brazil**
Capacity 77,750 persons

277 **Estadio do Brago, Euro 2004 Portugal**

Layout types and undulation
Defining geometrical families and explaining the geometrical context of ground plans and of inclination profiles for grandstands

Layout types

Stadia and arenas can be planned in virtually all formal variations and are in principle only constrained by the limits of design creativity.

Nonetheless, the grandstands of a modern sports facility must be fundamentally subordinated to functional performance, which is determined in the first place by the athletic or cultural event in question. Every type of sport has its own practical necessities, minimum requirements and dimensions, which tend to be broadly compatible. This is the basis for planning spectator facilities as multi-purpose arenas or 'convertible' stadia.

In Europe, two formal types are distinguished, each conditioned by function: football or athletics geometries with very different size requirements for the stadium interior. Nine principles can be derived from these formal types, which can in many cases be combined with one another. In such cases we speak of *hybrid forms*.

Asymmetrical derivations, as exemplified by the group of 'horseshoe' types, can be generated by any basic form and hence represent no independent layout group.

All layout types are geometric transformations of extreme forms of rectangles or circles.

According to the rules of sightline convergence, this results ideally in a purely circular shape for the body of the stadium, one in which all spectators are situated according to the same geometric conditions in relation to the centre spot.

Since the central playing field conforms to a clear rectangular form, the respective distances from the focal or viewing points of a sightline are altered accordingly.

An ideal transformation of rectangle to circle is only attainable by means of three measures. In this context, adequate spectator capacity (= number of stair rows) is a prerequisite for such geometrical reshaping.

Layouts *with* equal-height terrace edges

a) Overlapping/undercutting: When a rectangular playing field is transformed, e.g. via the oval shape typical for athletics utilization, into the ideal form of a pure circular geometry, projections or overhanging of the upper tier at the short sides are the result for the transitional areas, mostly at the height offset between different tiers.

b) Variations in seating row depths: As a result of differing seating row depths in the transitional areas leading from the corners to the axes of the stands, the body of the stadium is 'expanded'. The ensuing elevation is a wavelike form with continuously changing inclinations of the stands. In terms of sightlines, this can have an optimizing effect. If prefabricated construction is planned, however, this can lead to a disproportionate increase in the number of different stair elements.

Layouts *without* equal-height terrace edges

c) Undulating upper tier: In cases of stadium layouts based on non-affined geometries (to the stair profile), additional rows can be accommodated in the area of the central axis. Highly stable stepping profiles enforce a wavelike inclination profile of the front or rear edge of the stands.

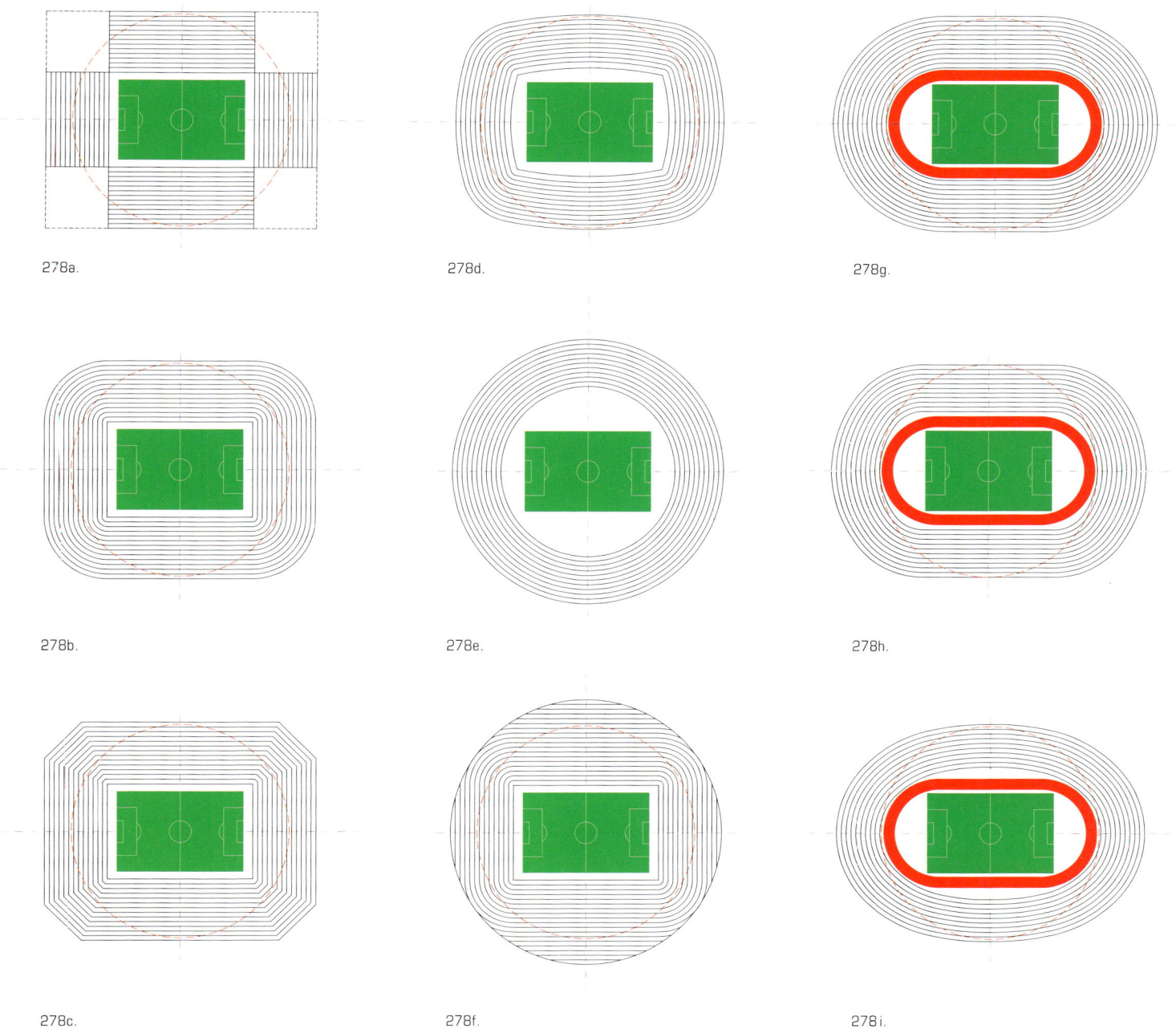

Super ellipse

A 'super ellipse' is a geometrical figure (curve) with elements of an ellipse and a rectangle (or a circle and a square).
The concept goes back to the Danish scientist Piet Hain (1905 – 1996). He used the brand name SUPERELLIPSE® ($n = 2.5$) in architecture, urban planning and furniture design. His 'Super Egg' is a three-dimensional super ellipsoid rotational body. Different from a regular ellipsoid body, it stands upright on a planar surface.

278a.

278d.

278g.

278b.

278e.

278h.

278c.

278f.

278i.

278 Layout types (auditoria)
a. Geometrically open rectangular geometry
b. Rounded polygonal geometry
c. Octagonal geometry
d. Radial geometry (super ellipse)
e. Ideal circular geometry
f. Undulating geometry
g. Semicircular geometry
h. Basket arch geometry
i. Oval geometry

Baseline procedures

In planning theatre buildings, the sight angle zone is marked out only for the central axis of the stage. Maximum spatial depth, therefore, is calculated in relation to the width of the proscenium. The centreline procedures described above are not adequate for assessing the sightline qualities of a stadium or arena. Given the large expanse of a sporting or event surface, a second point of analysis should be deployed for the ground plan line in continuation of the goal line. If a spectator is seated in a stand that runs parallel to the touch line at a minimum distance (6 + 1 = 7 m), then he must rotate his head from a neutral position – that is to say, from a straight-ahead position equivalent to the 'geometric viewing direction' – by approximately 86° in order to see the distant corner flag. This results in an uncomfortable shoulder view which can be optimized only by rotating the upper body.

To be sure, this condition is facilitated with greater distance and the minimization of the viewing angle. For this reason, when choosing the geometrical plan of the stands, the steppings at the corner areas should be arranged on a curved ground-plan line rather than parallel to the edge of the playing field.[1]

At a neutral head position, a viewing angle of 120° in the baseline position can be attained only from a distance of 42.5 m; an increase of 50% compared to the average position of the last row of the lower stand, with an approximate distance of 30 m to the touch line. Of all 12 World Cup stadia with between 16 to 33 rows in the upper tier, with an average of 25 rows, the relevant seating position is located at mid-level on the upper tier. Only at this distance and height, then, can a spectator survey the entire playing field without having to move his head. To view the neighbouring corner flag, again, a half rotation of the head by 60° becomes necessary.

Active viewing

The above-mentioned standard visual angle zones in stadia of 120°/90°/60° clearly show that for technical reasons related to sightlines, geometrically rounded plans are preferable. The group of parallel geometries (playing field / front edge of seating row) unavoidably result in sightline differentials between the 'true' sightline, i.e. the spectator's direction of view, and the predominant alignment of the seat of more than 30°. In six of 12 stadia, executive seats (excluding hospitality boxes) are located mostly in the lower tiers.[2] Assuming the data for eye and head movements registered above are relevant for the best-view seats of the main stand, this means that:

Note: Spectators are able to follow the action taking place on the playing field only by means of 'active viewing'. A comfortable and static position, one comparable with the relatively stable and fixed scenery of a theatre performance, is unattainable.

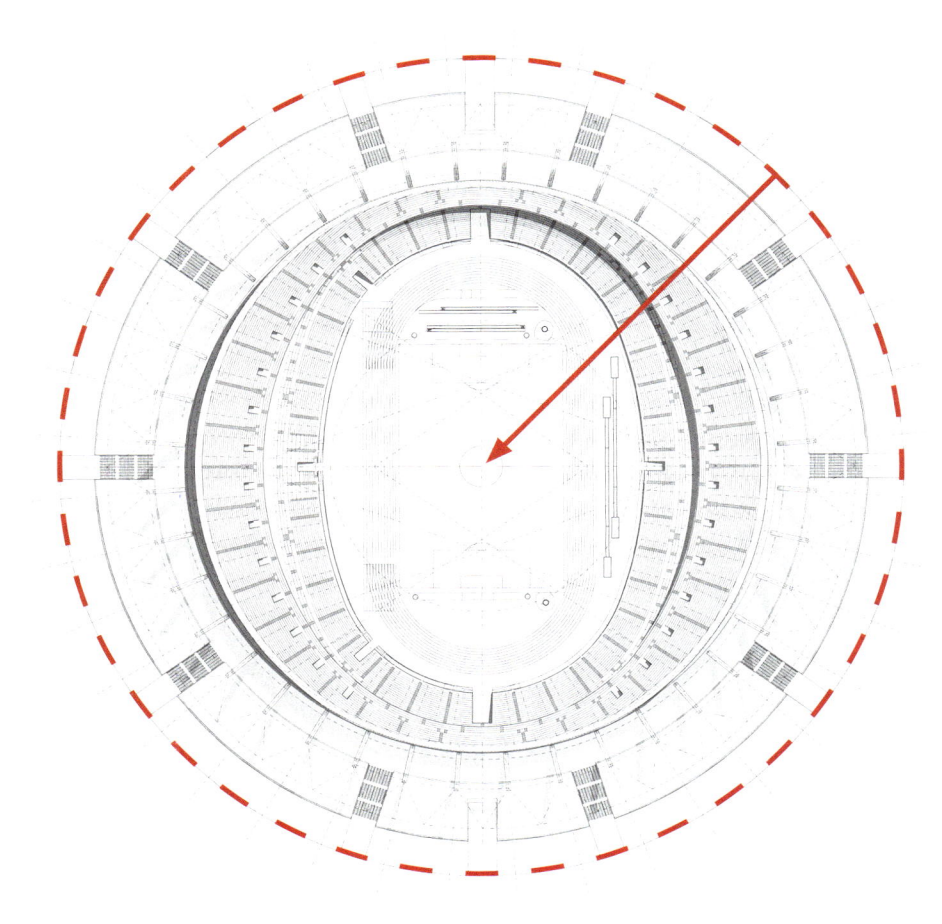

Head and body positions

0°–30°	with a minimum of strain
>30°	uncomfortable, inducing strain over long periods

Up to 90° full head rotation   x 2 = 180°
Shoulder view – unsuitable for sustained viewing direction, but quite common for short-term viewing (match activity at the corner flag)

Up to 60° half head rotation   x 2 = 120°
Normal head movements without rotation of shoulders – comfortable

Up to 30° slight head rotation   x 2 = 60°
or eye movements

279 **Ideal circular form**
Central convergence point of Foshan Stadium, China (planning by gmp architects)

page 145:
280 **Oblique aerial perspective Lotus Stadium**
Foshan, China (computer animation)

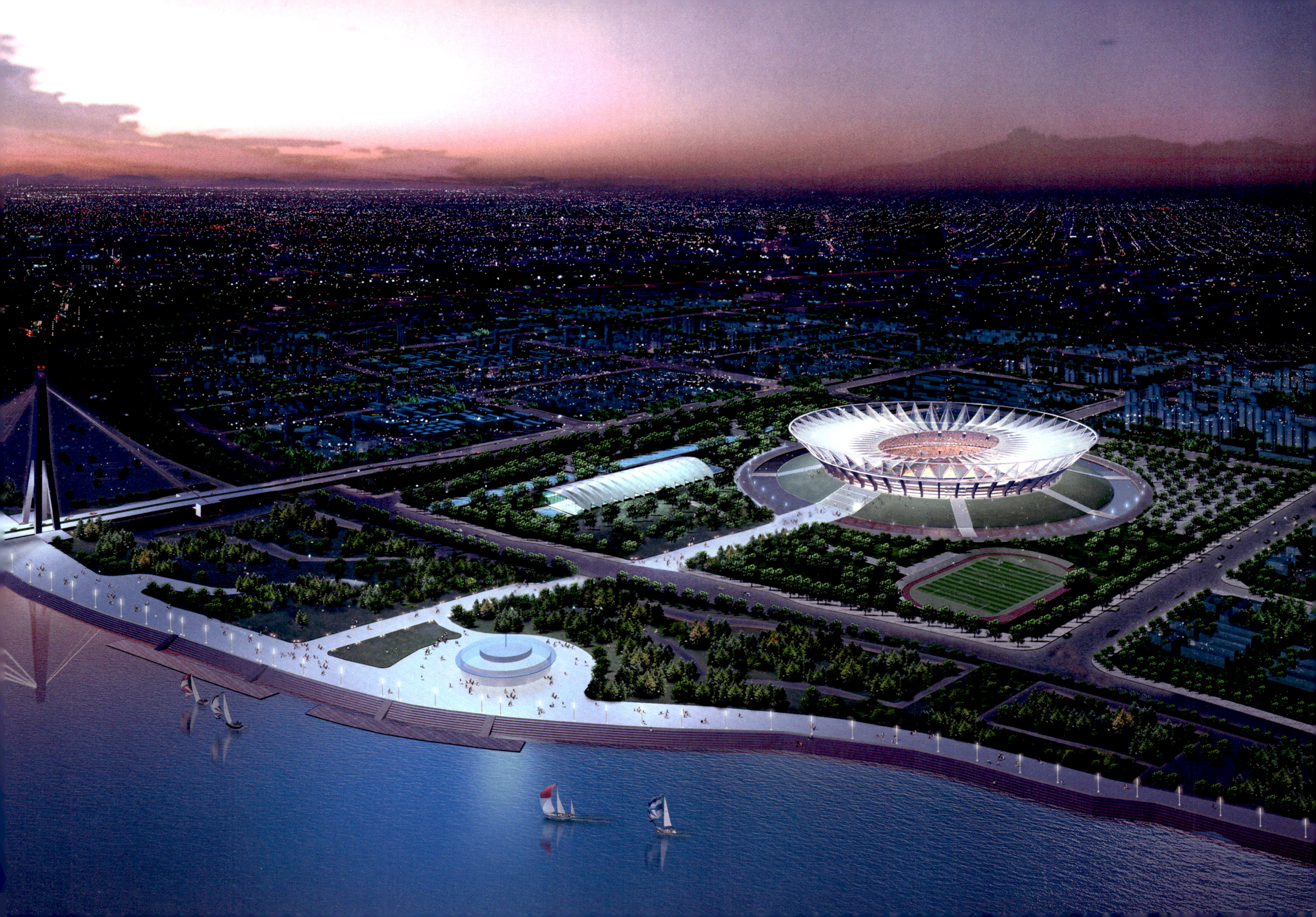

14 Layout types and undulation

Theatre undulation

Undulation

'Undulation' [French: *ondulation*] waving hair, produced by a heated curling iron'.³

In 1872, Frenchman Marcel Grateau developed the process of *ondulation* which was used right up until the 1960s.

281 **Upper tier / main stands**
Stand-alone long-side stand Gottlieb Daimler Stadium, Stuttgart
282 **Ideal circular form**
283 **Undulating long stand**
With open short side in the upper tier Zentralstadion, Leipzig

One of the two forms of the above-mentioned undulation we distinguish is represented by the curved edge of the stands, for example the upper tier. This geometrical phenomenon always emerges when the structural terrace edge is not an affine image of the general stepping geometry. The rounder or more curved the external geometry of the stadium body is to be, the more marked the effect in the undulating upper tier.

Genesis

In transforming the geometrical starting point, the rectangular playing field, into the rounded form of the stadium, the stands must mediate between the two different geometries by means of distances and layout.

The gap between a rectangular and the circle which circumscribes it is larger along the central axis than in the corners of the rectangle. Every ground-plan line of a grandstand structure also represents a height offset for each seat or standing step, which means that the stands along the central axis rise more steeply than those in the corner areas.

An example for an ideal circular geometry is the competition entry for a new stadium in Siena. Here, the idea of the design reconciles a self-contained circular shape with the pre-existing topography of the hilly Tuscan landscape. The geometrical profile of the undulating grandstand corresponds to the elevational profile of the surrounding terrain, which enables its harmonious integration into the landscape.⁴

A pedestrian perspective of the competition entry for Müngersdorfer Stadion in Cologne by the architects Schultz + Partner, Hanover, also shows this elevational profile quite clearly. As a rule, the principle of undulation is applied when the total form of the stadium body does not correspond to the chosen inclination of the stands.

Since every row ends with a different plan length at a different height, the result is a staggered grandstand terminus.

For compensation purposes, the circulation system can be organized so that the tiers of the stands are essentially filled from top or rear to bottom (example: World Cup Stadium, Leipzig). Alternatively, the necessary gangway can be accommodated along the rear of each block.⁵ In this transitional zone, the geometrical elevation can then be integrated.

The designer's goal of creating an ideal circular structure emerges either from formal or urban planning considerations, or else is generated from decisions concerning a suitable roof construction.

Ideally, tensile/pressure ring constructions which are laid out according to the spoke wheel principle follow a purely circular form, for in that case, all horizontal forces can be interconnected and the roof construction can be absorbed as a vertical load. Moreover, this type of load-bearing frame allows widely projecting support-free canopy roofing of the type recommended for large-capacity spectator facilities.

As early as 1954 Hans Gussman noted in the context of theatre construction that seating rows at the side tiers which are located close to the stage must be especially elevated. This means respectively a decreasing or inclined height profile for a given seating row.

Theatre undulation 147

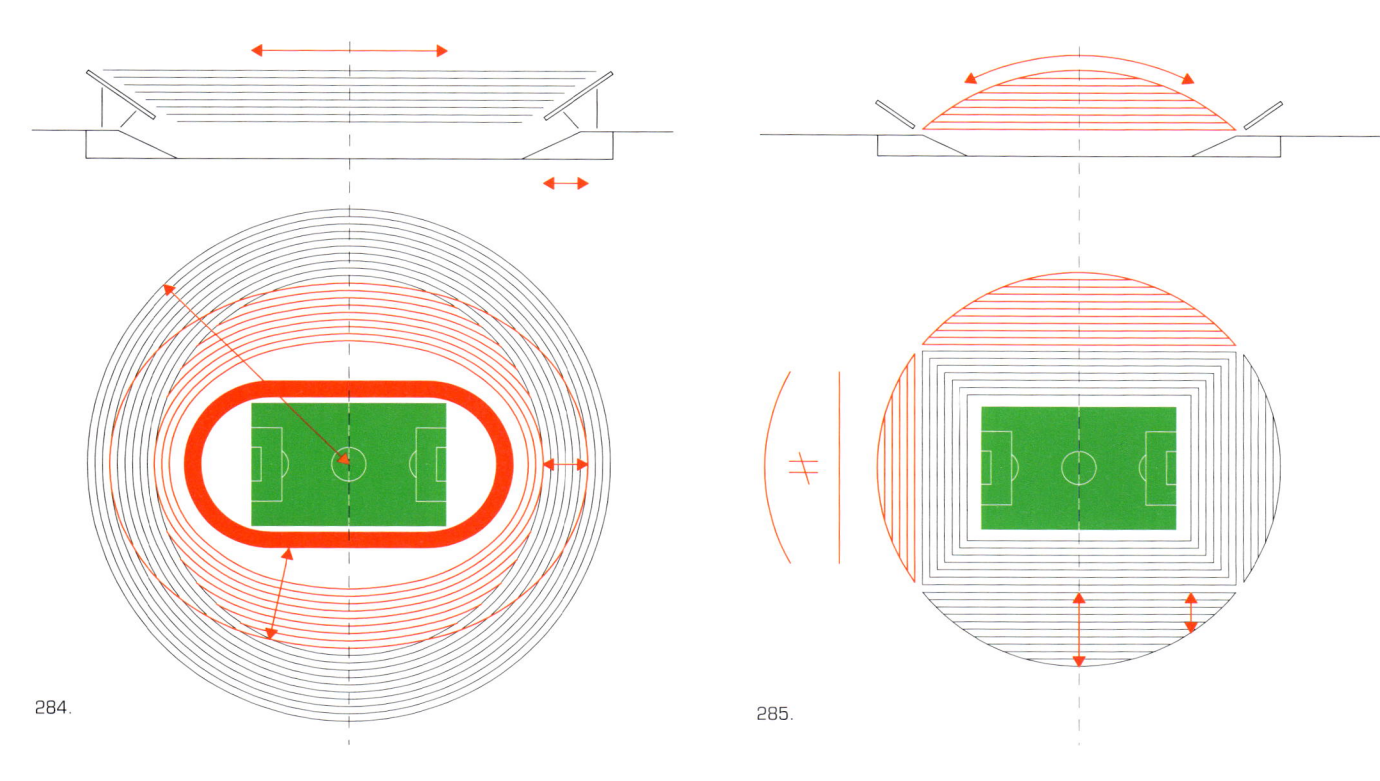

284 **Circle of convergence (sketch)**
a) lower tier in oval geometry for accommodating athletics activities
b) upper tier as ideal circle with straight inclination without undulation
Note: No circle without undulation or geometrical overlapping; centre-spot convergence only with ideal circular geometry.

285 **Undulation (sketch)**
a) lower tier as parallel geometry to the playing field (for example, for pure football use)
b) upper tier as continuation of the lower-tier geometry; inclination profile with undulation
Note: Undulation of the rear edge of a tribune is produced when the edge of the stand shows no affine conformity with the stepping profile.

bottom:
286 **Inclination profile of undulating upper tier**
Competition design by architects Schulitz, Hanover, for the World Cup Stadium Cologne, 2002

284.

285.

$a = 20$ cm $h = 1$ cm (5%)

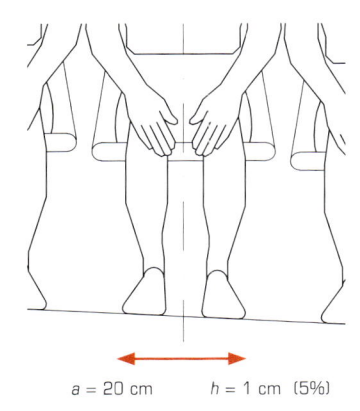

287 **Distance between feet of a seated individual – detail of drawing**
(seating grid 50 cm = 20 cm distance between feet differential elevation of feet = 1 cm at 5%)

288 **Model photo, design for a landscape stadium in Siena 2004**
(by Modellbau Kren, Aachen)

page 149:
289 **Drawing, detail of site plan for a circular ideal stadium with centre-spot convergence**

'Genuine' undulation

Investigations concerning the inclination of sightline profiles have shown that the gradient of the stands rises continuously the closer it lies to the point of focus. If the height of individual risers does not respond to the conditions from the initial rise and spacing, then the so-called 'C' value as a parameter for viewing quality is continuously reduced.

If, due to the expansion of its basic geometric form, a stand lies in a corner area, and hence closer in to the action than those aligned with the central axis, then the 'C' value is reduced, and the necessary steepness of the spectator stands increases.

This circumstance is generally referred to as the 'corner conflict', since generally, in the planning of stadia/arenas, designers assume a stable vertical development of the front edge of the steppings.

On the basis of the above-mentioned relationship between sightline quality and the inclination of the spectator stands, a genuine undulation (that is to say, an inclined vertical development that follows the seating/standing rows in a lengthwise direction) may be considered in exceptional cases. This means an adaptation of the sightline quality C by means of a modification of the vertical profile of the stands toward the critical stadium corner zones. Thus, spectators sit on a laterally inclined level.

Assuming that the corner zones, by virtue of their reduced distance from the playing field, are marked by a steeper vertical development (with the same depth of rise B), at a playing field length of 105 m this results in a height differential of 2.5 m to the start of the corner. The example of a typical box-tier offset of 2.5 m illustrates how the entire height of a box tier is used up for a 'genuine' undulation.

EN/DIN 13200-1 limits the recommended rise for a spectator ramp to 10% (6°). This gradient must surely be regarded as too high for the lateral inclination of a step. The resulting gradient will vary for each individual row depending upon the height differential between the axis and the corner. Assuming a block located along the halfway line has a depth of 30 rows, then a height difference of 2.5 m will be necessary at the corners; assuming the ground plan specifies an axial distance of the stand of approximately 10 m toward the touch line in the direction of the corner flag, which brings it 2.5 m closer to the playing field. In this case, the values range from 0.1% in the first row rising to 4.7% toward the corner areas, up to the maximum block size of 30 rows.

The longitudinal inclination of the rises of the stands should not exceed 5%.

The gradient of the steppings must be mirrored physically in the grandstand profile and may not project the successive upper tier. Remaining as a result is only the area of the differential in elevation between two tiers (the box tier). A stair row continuously ascending up to $h = 2.5$ m, however, compromises the usability of the hospitality suites and of other subsidiary functions found there.

Note: Employing the measure known as 'genuine undulation' should be considered only in exceptional cases, after weighing up alternative solutions for the hospitality viewing areas.

"Armonia di paesaggio e geometria" 1/2

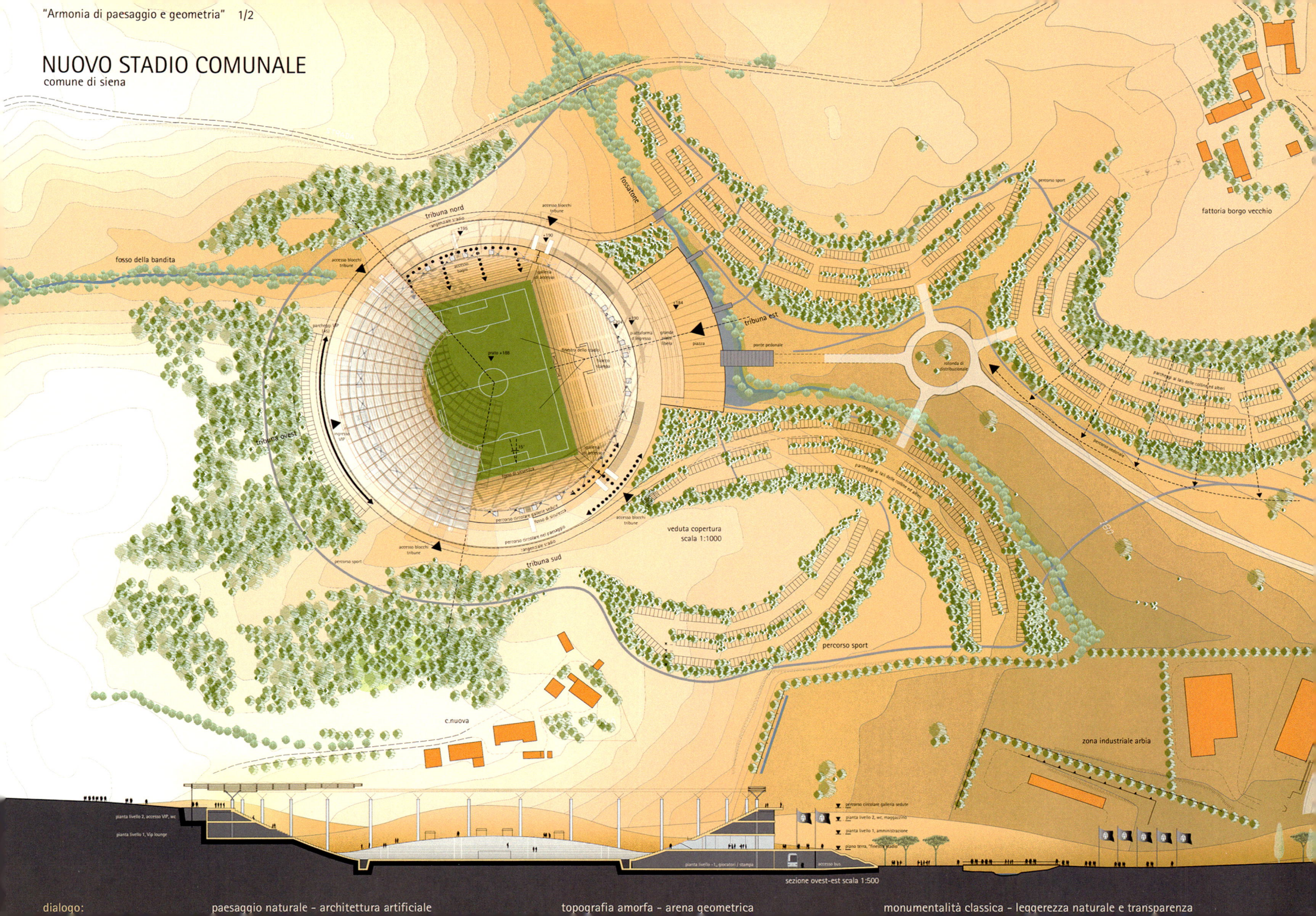

NUOVO STADIO COMUNALE
comune di siena

dialogo: paesaggio naturale - architettura artificiale   topografia amorfa - arena geometrica   monumentalità classica - leggerezza naturale e transparenza

Chapter 15

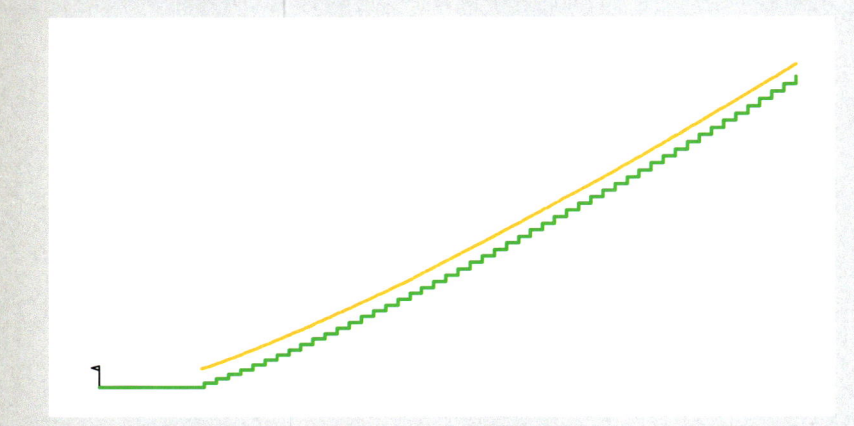

290 Eye-point curve and steppings
(diagram showing touch line)
yellow: eye-point height (sightline construction)
green: stand profile (derived)

Principles of sightline calculation
A methodical approach to determine optimized sightline profiles for places of congregation

Guidelines for sightline profiles

There are two options for sightline determination:
– Option 1: New calculation of a grandstand geometry
– Option 2: Evaluation of an existing grandstand geometry

According to the 'centre-line method', the elevation section in both cases lies on the axes of the long sides and the short ends.

If the front edge runs parallel to the pitch, the centre-line position is an affine image of the entire long/short side. In case the ground-plan line of a stair row does not run parallel to the pitch perimeter and the plan is based on a rounded or expanded geometry, the position of the spectator row with the shortest distance to the point of focus should be checked in a second step. This pertains in particular to the corner areas of a grandstand.

When evaluating an existing grandstand or plan, the main elevations along the terrace axes should be clarified first, since these are representative for the majority of spectators.

As a rule, the steepest viewing angle is the most relevant for planning purposes and the elevational section runs parallel to the geometric sightline, i.e. vertical to the front edge of a stair row or at the shortest distance to the point of focus.

The foregoing point of focus is always located on the extreme points of a playing field and is to be geared to the prevalent sports and event types as determined by the utilization profile of a stadium/arena. As a rule, in European multipurpose stadia this would be football or athletics.

a) For football, touch lines and goal lines as well as corner flags are relevant.
b) For athletics, it would be the outer running track. However, the distance to the field is so large that sightline comfort remains unchecked. More problematic seem to be the many fittings in the activity area, such as advertising banners, as they may obstruct viewing.
c) For ice hockey fields, the point of focus is not clearly determined. DIN 18035 merely refers to barrier heights. Thus, the upper edge of the barrier should be assumed as point of focus.

Note: Determination of a sightline follows the principle of converting standing into seating places, but is generally based on seating accommodation. Sightline profiles for standing places are merely examined, and if need be, 'C' value reduction is offset by the higher eye point (starting).

The dependency of the first eye point from the position/height of the advertising hoarding forbids the planning of standing/seating place profiles.

To calculate a grandstand profile the following guidelines apply:

The further away a spectator is located from the point of focus, the shallower the slope of the stand.

The higher the starting point of a sightline, the steeper the slope of the stand.

The higher the 'C' value of sightline elevation, the steeper (and faster) the stand ascends.

General terminology

D Distance
is the horizontal distance from point of focus P to the 'first' spectator's eye point.

X Total distance
is the horizontal distance from point of focus P to a random spectator's eye point. It represents the x-coordinate of an eye-point elevation curve or sightline profile.

A Eye-point height
is the vertical distance from point of focus P to the 'first' spectator's eye point.

Y Eye-point height
is the vertical total distance of the respective eye point to point of focus P (e.g. at the pitch). It represents the y-coordinate of an eye-point elevation curve or sightline profile.

B Stair row depth
is the depth of a specific seating/standing step of a row in a stand.

G Gangway width
is clear width as distance between the front edge of a seat and the step/tread front edge.

C 'C' value
is the so-called sightline or eye-point elevation. It characterizes the sightline comfort of a terrace and is the quality indicator of modern spectator complexes.

P Point of focus
is the point of view or sight reference point, depending on the sports (e.g. the byline for football).

R_H Riser height
This value represents the actual result of the calculation formula for sightlines and determines the 'new' height of the successive seat/standing row.

S_A Sightline distance
is the 'new' eye-point height as distance of the old to the new eye-point height and is calculated with $(S - C = S_A)$.

S_H Seat height
Fitting height of the seat front edge above the tread area of a stair row, 40 cm.

S_T Seat depth
Seat dimensions from the seat front edge to the back rest (incl.).

α Sightline angle
Vertical visual angel of the sightline or sightline to the point of focus.

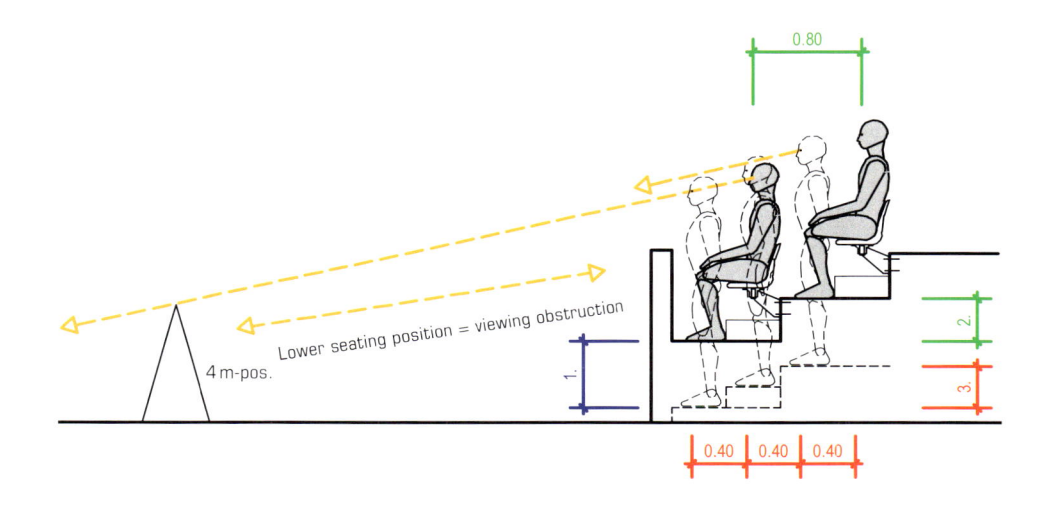

Sightlines and safety

Grandstand profiles of modern sports and event venues, dissertation RWTH Aachen (2006)

Calculation formula for a 'single step'

$$R_H = \frac{(A + C)}{D} \cdot B + C$$

Calculation formula for sightline elevation 'C' value

$$C = \frac{R_H D - AB}{(D + B)}$$

Calculation formula for

$$f(X_0) = Y = A + R_{H(X)}$$

291 **Illustration of standing/seating place conflict of a sightline elevation curve**

Design of seating vs. standing places

1) Starting points of standing/seating places are geometrically incompatible.
   Eye-point height (seated) 1.25 to 1.65 m = − 40 cm
   Advertising hoarding as definite sight obstruction
2) Gradient variation for standing or seating places (the higher, the steeper)
3) Primary calculation of sightline is never based on standing places since, after conversion, the first seating position cannot view over a standard advertising hoarding.

Calculation of single riser height

First, the formula for calculating a sightline/eye-point elevation curve is geometrically developed. The first approach is based on the mathematical conversion of a drawn design which is based on the geometrical laws of triangles.

Four planning parameters are decisive for sightline calculating. They need to be determined at the beginning of design.

A   Eye-point height – the vertical distance from the point of focus to the 'first' spectator eye point.

B   Tread depth – the depth of a determined seating/standing step/tread of a terrace row.

C   'C' value – the sightline or eye-point elevation (9/12 cm).

D   Distance – horizontal distance from point of focus to the first spectator eye point.

Note: The relatively complex relationships which form the basis of the foregoing parameters will be elaborated in a later chapter.

Trigonometry

An initially random eye point in a starting distance D to a point of focus P – for example the touch line, which in any case should serve as point of focus for spectators:

Together with the first sightline they form at angle α triangle no. 1.

Step 1:   Triangle no. 1

$$\frac{A}{D} = \tan \alpha_1$$

In a second step, the parameters tread width B and sightline elevation C need to be integrated. A spectator's sightline must run over the head of the person sitting in front. The corresponding value to be determined is C.

In relation to the stair riser, the rectangular triangle no. 2 results from the new sightline and runs from the new eye point over the sightline point; the 'C' value vertically measured from the eyes of the person in front.

Step 2:   Triangle no. 2

$$\frac{(A+C)}{D} = \tan \alpha_2$$

The new sightline angle of triangle no. 2 is determined with this calculation, depending on the sightline quality C.

From the new sightline angle α_2 and the known tread width B, the opposite side S_A may be determined (triangle no. 3).

Step 3:   Triangle no. 3

$$\frac{OS}{AS} = \frac{S_A}{B} = \tan \alpha_2 \qquad \begin{array}{l} OS \quad \text{opposite side} \\ AS \quad \text{adjacent side} \end{array}$$

$$S_A = \tan \alpha_2 \cdot B$$

$$S_A = \frac{(A+C)}{D} \cdot B$$

Through addition of S_A – the sightline distance (i.e. vertical distance between the eye point of the person in front and the new eye-point height) and the determined 'C' value – follows the opposite side of triangle no. 4.

Step 4:   Triangle no. 4

$$R_H = \frac{(A+C)}{D} \cdot B + C$$

$$\frac{OS}{AS} = \frac{R_H}{B} = \tan \alpha_3 \qquad \begin{array}{l} OS \quad \text{opposite side} \\ AS \quad \text{adjacent side} \end{array}$$

The angle α_3 corresponds not only to the inclined distance of the pairs of eyes of persons seated in front and behind, but also to the terrace gradient at the calculated position. Triangle no. 4 corresponds therefore to the inclination of a seating/standing terrace, depending on the four initial parameters:

A   'First' eye-point height
B   Tread depth/width
C   Sightline elevation
D   'First' eye point distance

Sightline elevation ['C' value]

After conversion of the foregoing formula and in the framework of the existing grandstand geometry, the 'C' value of the individual seating row may be determined with the help of the derived calculation formula.

Origin:
$$R_H = \frac{(A+C)}{D} \cdot B + C$$

Conversion:
$$-C = \frac{(A+C)}{D} \cdot B - R_H$$
$$-C = \frac{(AB+CB)}{D} - R_H$$
$$-C = \frac{AB}{D} + \frac{CB}{D} - R_H$$
$$-C - \frac{CB}{D} = \frac{AB}{D} - R_H$$
$$-DC - CB = AB - (R_H \cdot D)$$
$$C(-D-B) = AB - R_H D$$
$$C = \frac{AB - R_H D}{(-D-B)}$$
$$C = \frac{-AB + R_H D}{(D+B)}$$

Result:
$$C = \frac{R_H D - AB}{(D+B)}$$

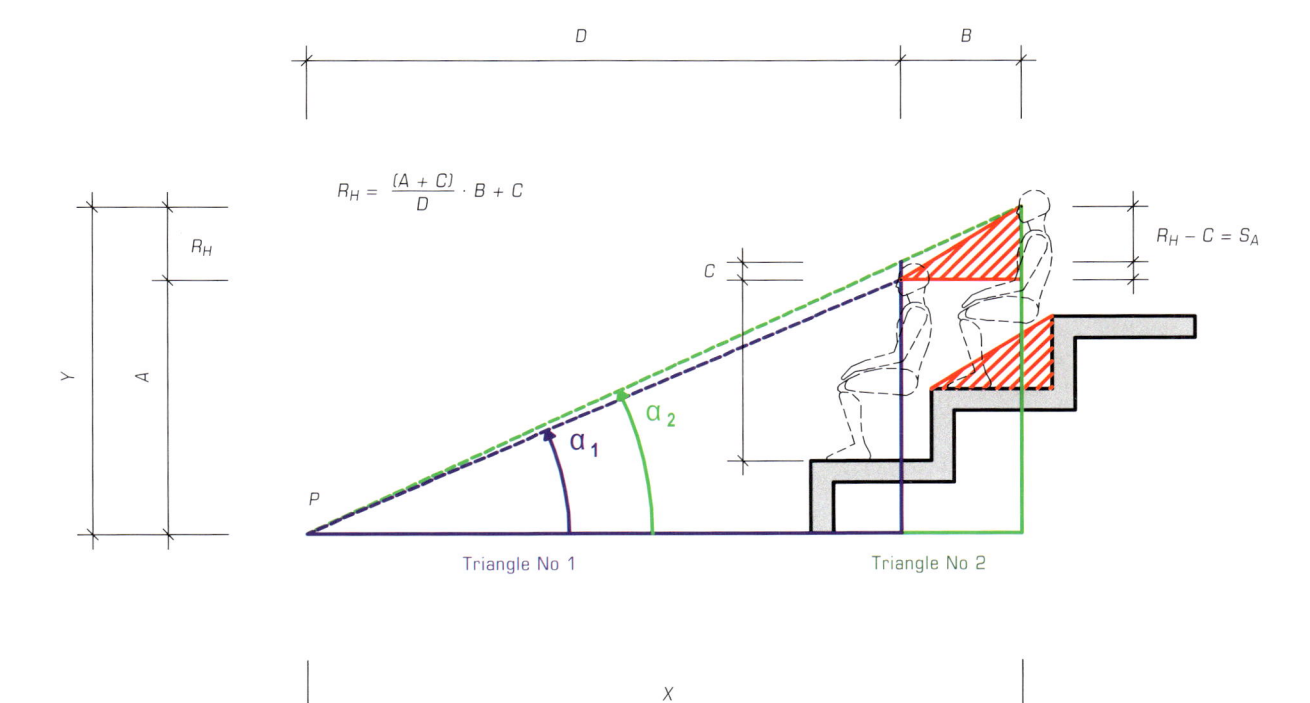

292 Illustration of the method of sightline determination
Eye-point elevation curve for the sightline profile of modern sports and event venues

15 Principles of sightline calculation

The term 'sightline'

Five-step method

In order to illustrate and compare the quality standard of sightline elevation between different plannings for stadia, the following 'five-step method' is defined:

Step 1:
Define point of focus P
(depending on the discipline, see table C.2, appendix C of EN/DIN 13200-1)

Step 2:
Type of pitch securing and determining parameters
1) first eye-point height   A
2) planned tread width   B
3) choosing sightline elevation   C
4) distance to the 'first row'   D

Step 3:
Construct grandstand geometry and position outfits (box tier, circulation system)

Step 4:
Extreme points (corners) with 'C' value calculation formula

Step 5:
'Polygonal transformation'

Note: Due to conversion, geometrical sightline calculation in standing areas is carried out equivalent to a seating place situation.

Calculating a sightline seems to be one of the most important planning objectives for a grandstand. In the first part of DIN 13200-1, the criteria for the spatial layout of spectator places are stated.

Good viewing of the sports activities must be guaranteed from all positions on the spectator stand (DIN 18035). Likewise, the relevant associations for football DFB/FIFA/UEFA, or IAAF for athletics, demand good and clear viewing conditions of the entire field for all spectator seats.

EN/DIN 13200-1 (3.10) defines the term 'sightline' as follows:
'Line which connects a spectator's centre of the eye to a point of focus in the activity area without obstruction.'

The position of focal point P as reference point is essential. EN/DIN 13200-1 (Annex C) promotes a classification of sports and 'criteria for the determination of groups'.

This event-specific definition of the point of focus takes into account diverse utilizations. The type of sport is determined according to the speed of activity and the size of the used sports equipment (table C.1):
– Speed of activity
   A = slow   B = medium   C = fast
– Size of the sports equipment
   A = large   B = medium-size   C = small

Athletics events are allocated to group A, football or American football to group B and ice hockey to group C.

These details are important since the relevant objects to be viewed are very different in size, from a small ball/puck to match participants. As a rule, the point of focus P to be evaluated lies at the extreme positions of a playing field such as the corner flags, the touch lines/goal lines, the running tracks or hoardings.

Information on the respective height of the point of focus is given in table C.2:

For ball sports such as *football*, etc. $h = 0$ cm, with the touch line as decisive reference point.

For *athletics*, the outside line of the outer track with $h = 50$ cm (alternatively, the centre of the outer track lane).

And for *ice hockey* the height of the opaque hoarding is to be counted (around 1.0 to 1.10 m).

Stand elevation

293 **Sightline parameters**
a. Eye-point height = A
   curves: initial height
b. Tread / seat width = B
   curves: tread depth
c. Quality of sightline elevation = C
   curves: 'C' value
d. Distance to point of focus = D
   curves: initial distance

293a.
Erst-Höhe 1,60 m (DFB)
Erst-Höhe 1,00 m
Erst-Höhe 0,00 m

293b.
Stufentiefe 0,80 m
Stufentiefe 0,90 m
Stufentiefe 1,00 m

293c.
"C"-Wert 15 cm
"C"-Wert 12 cm
"C"-Wert 9 cm
"C"-Wert 6 cm

293d.
Erst-Distanz 6,0 m
Erst-Distanz 7,5 m
Erst-Distanz 21,0 m
Erst-Distanz 38,0 m

A Eye point height
The higher the starting point of a sightline, the steeper the slope of the stand.

B Tread width
The deeper the tread width, the shallower the slope of the stand.

'C' value
The higher the 'C' value of sightline elevation, the steeper (and the faster) the stand ascends.

D Distance
The further away the spectator is positioned from the point of focus, the shallower the slope of the stand.

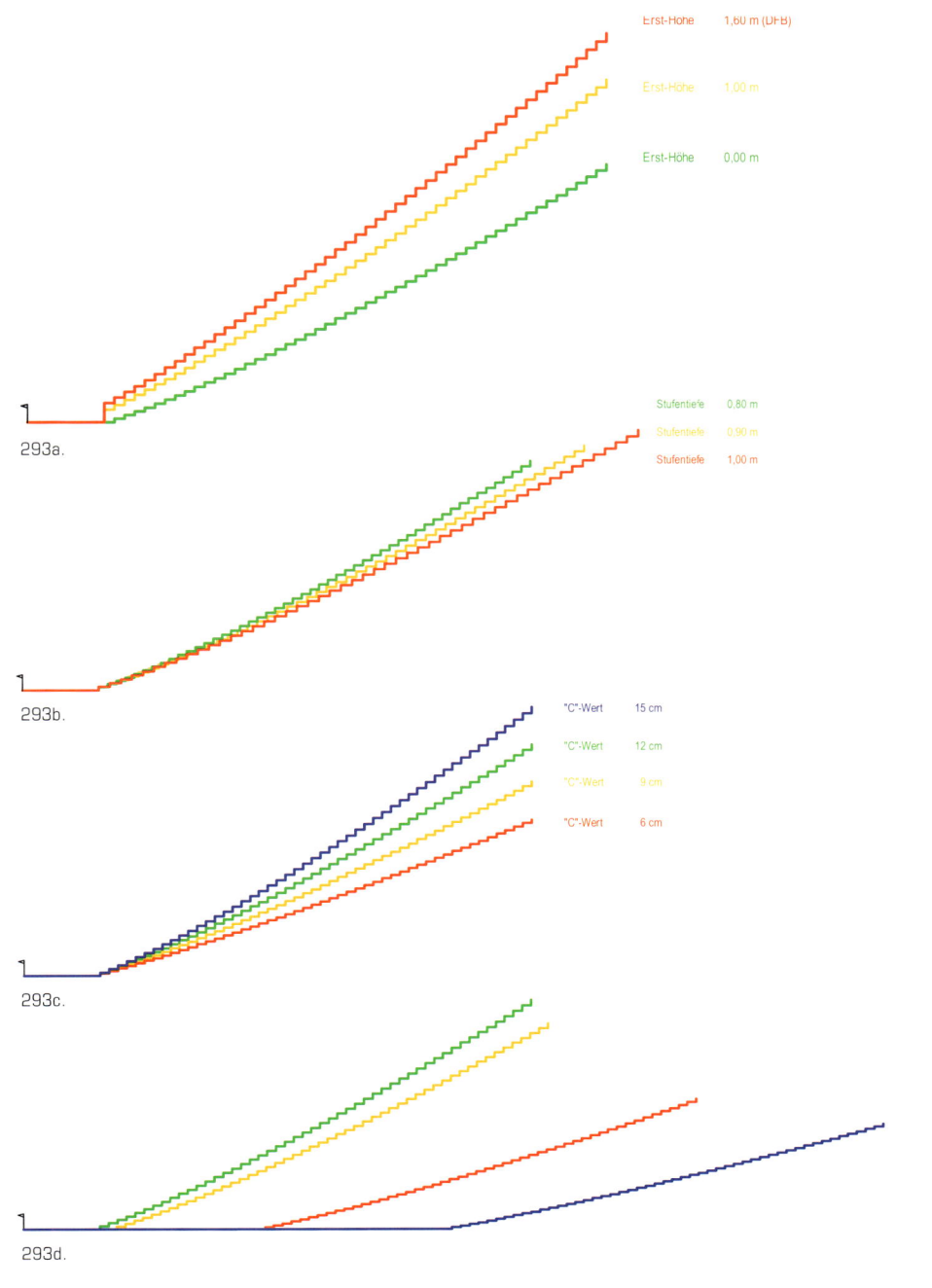

15 Principles of sightline calculation

A Eye point height

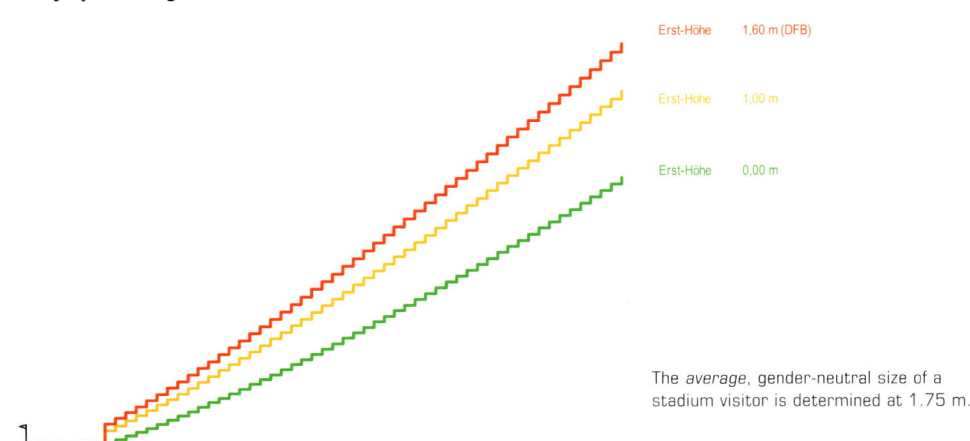

Erst-Höhe 1,60 m (DFB)
Erst-Höhe 1,00 m
Erst-Höhe 0,00 m

The *average*, gender-neutral size of a stadium visitor is determined at 1.75 m.

(293 a) Inclination of a stand in relation to eye-point height curves: initial height

294 **Eye-point height**
a. Regular single-seat position
b. Executive single-seat position
c. Standing accommodation

Eye-point height in relation to seating / standing place A_S

Eye-point height in relation to seating / standing place A_S is the vertical distance of a spectator's eye point measured from the tread of the respective seating / standing place.

Distance to tread (seated)	1.20 m
Distance to seat base	0.80 m
Height of the seat base	0.40 m
Distance to tread (seated) new	1.25 m
Distance to tread (standing) new	1.65 m
Eye-point distance (ground plan) to	
Rear edge of steppings	0.20 m
Front edge of steppings	0.15 m

Note: The reference value of 1.65 m was stated already in 1993 by IAKS, but the new DIN 13200 establishes a value of 1.60 m. A graphical translation of human anthropometrics into the idealized Modulor 'EN' as standard spectator results in a detail measure of 1.64 m, but for calculation purposes a standard measure of 1.65 m should be assumed.

Seating positions

After the definition of eye-point height A_S, this chapter will examine how different seating positions and postures affect eye-point positions and the resulting 'quantitative viewing conditions'.[1]

Although EN / DIN 13200-1 determines eye-point height at 1.20 m (seated), it fails to set a ground plan distance for the eye point in relation to the rear of the seating / standing row.

Stadium Atlas defines four different seating postures of a spectator (simplified figures):
– Scenario 1 'relaxed / normal'
20 cm distance at 1.25 m eye-point height
– Scenario 2 'tense'
55 cm distance at 1.15 m eye-point height
– Scenario 3 'demotivated'
05 cm distance at 1.15 m eye-point height
– Scenario 4 'attentive'
30 cm distance at 1.25 cm eye-point height

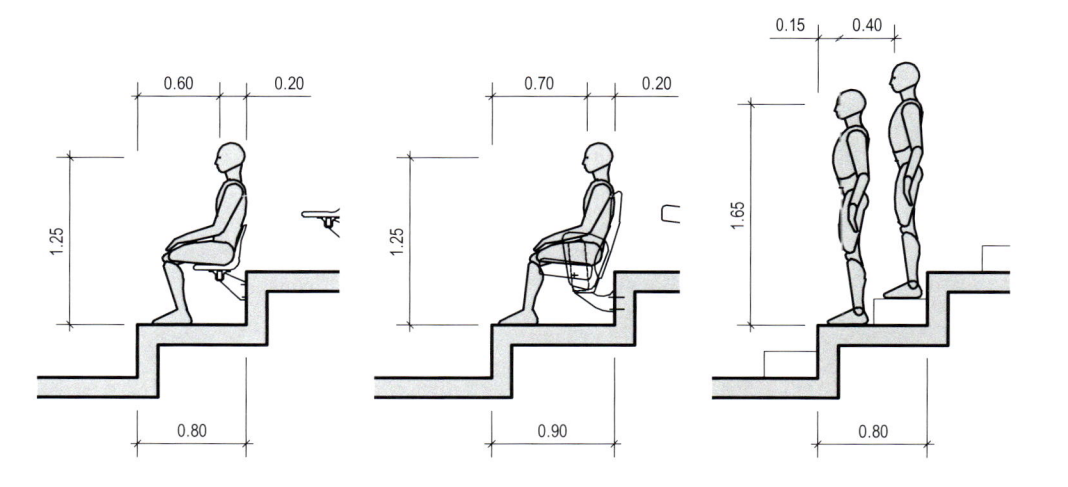

In a moment of heightened attention many spectators might jump to their feet, for example to better view a corner kick. This scenario cannot be taken into account when calculating a sightline since standing or seated positions are principally incompatible with regard to sightlines. As a typical reaction, most spectators tend to rise simultaneously, which in turn reduces viewing obstructions.

Conclusion

The more closely modern grandstands are moved towards the pitch, the more difficult it is to maintain a 'clean' sightline profile. Generally, for drawing or calculating purposes, the definition of eye-point height A_S and its ground plan position have no major effect.

The consequences become mathematically relevant only for the first seating row on first determining parameter A/D. Thus, the sightline profile is only marginally influenced since simultaneous seating postures of spectators are assumed. Since seating positions of individual spectators may vary and possible sight obstructions for others cannot be considered in planning, these issues are commonly settled amongst spectators themselves.

Alternative

An alternative approach to evaluate the viewing quality of a grandstand would be a statistical examination. Based on the details stated in *Stadium Atlas*, the visual possibilities would be examined, without considering the factors of simultaneous seating postures and individual spectator sizes within one block.

The survey could then be compared for various elevational geometries, and the outcome would result in a qualification of optimization measures (elevating the maximum gradient, alteration of 'C' value).

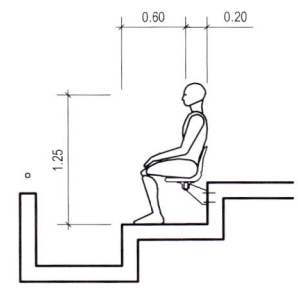

295 **Analysis of seating positions**
296 **Comparative illustrations**
of various seating positions and their vertical viewing angles

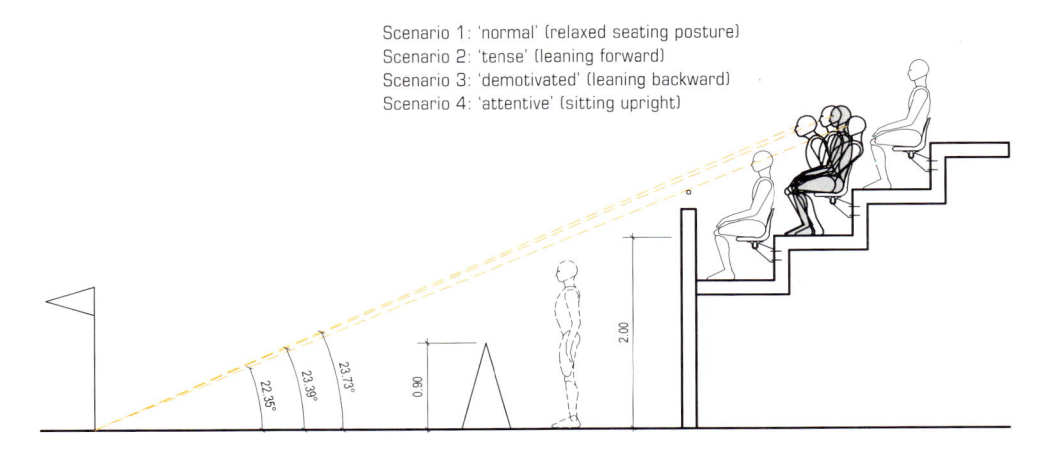

295a. 'normal' (relaxed seating posture)
295b. 'tense' (leaning forward)
295c. 'demotivated' (leaning backward)
295d. 'attentive' (sitting upright)

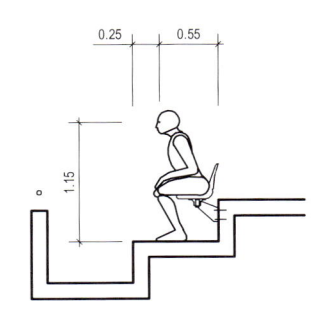

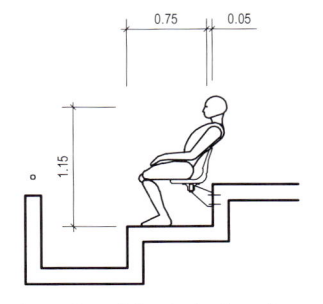

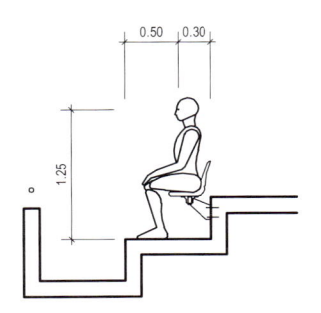

Scenario 1: 'normal' (relaxed seating posture)
Scenario 2: 'tense' (leaning forward)
Scenario 3: 'demotivated' (leaning backward)
Scenario 4: 'attentive' (sitting upright)

15 Principles of sightline calculation

B Tread width

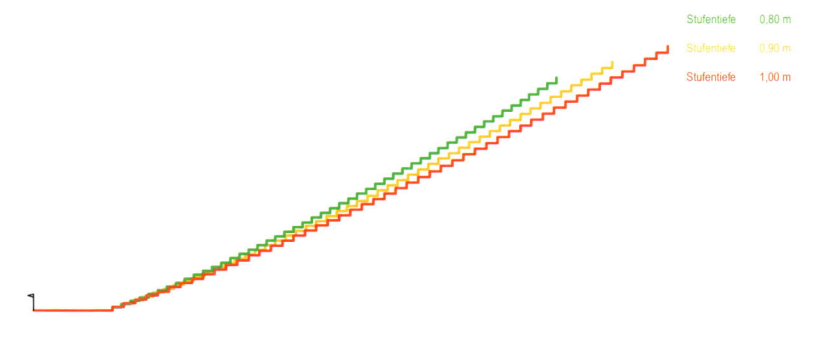

(293 b) Inclination of a stand in relation to tread depth
curves: tread depth

297 **Tip-up seat, Allianz Arena, Munich**
(Seat design, HdM Architects, Basle CH)

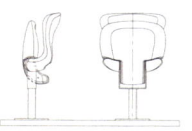

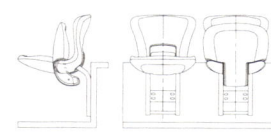

Allianz Arena, München
Insgesamt 61.500 Klappsitze FCB-M mit feuer-
verzinkter Stufenbefestigung oder Sockelfuß.

Höhe: 760 Achsmaß: 500
Breite: 450 Eigentiefe: 265/500
Abmessungen in mm.

Seating row depth and width

Tread depth B determines, at a given distance of the first row to the pitch, the entire slope of the tier. The lower the tread depth, the steeper the tier gradient (at constant sightline elevation).

Whereas formerly a clear width G of at least 45 cm was assumed, the new regulation on places of assembly has reduced this value to 40 cm: tread depth in addition to the minimum depth of the seat (40 cm) thus results in a total depth/width of a seating/standing tread of 80 cm.[2]

Seating comfort is affected if these measures are reduced.[3] If seating depth is increased for comfort purposes, this has a qualitative effect on the steepness of the stand. In this case, the maximum gradient admissible by law would be reached earlier as the stand ascends. In order to not supersede this limit, the sightline parameter of 'C' value (sightline elevation) would have to be reduced.

Planners may generally assume the following grids for seating places:
Regular seat: 50 x 80 cm
Executive seat: 60 x 90 cm

Comfort
An increase in tread depth to 1.05 m for higher comfort in executive zones was realized for World Cup 2006 in Germany.

Clearway
The obligatory escape route width G for the seatway of a stair row can be achieved simply by installing tip-up seats. The additional costs for the tip-up mechanism should be well considered.

Further types of seating place grids will be examined in separate sections on 'media/wheelchair places', here only summarized:
Wheelchair: 90 x 150 cm
Media/Press: 50 x 150 cm
Commentators: 160 x 180 cm

Sightline elevation 'C'

'Each row should rise as much over the previous row that the sightline from the eye of the listener to the face of the speaker is directed clear over the top of the head of all persons sitting in front.' (C. Gellinek, 1934)[4]

'The term sightline denotes the imaginary line from the eye of the spectator to a reference point on the stage.' (G. Graubner, 1968)[5]

'Sightline elevation', or the so-called 'C' value, is the vertical distance between the sightlines of two successively seated spectators. The distance is respectively measured above the eye of the person seated in front.

Thus, 'C' value is the measure of the desired viewing comfort which is generally determined by stadium operators as quality standard for spectator facilities. It is also established by FIFA / UEFA / IAAF and EN / DIN 13200-1 as minimum standard.

'C' value
Optimum / recommended: 12 cm
Acceptable / admissible: 9 cm
In exceptional cases: 6 cm

In fact, in earlier plannings for the construction of sports stadia by IAKS (1993) a 'C' value of 15 cm was stated as optimal, in consideration of spectators wearing hats. Today, of course, this is quite rare, although baseball caps are widespread, but their effect on sightlines is negligible.

Further notes from IAKS mention values derived from English football, which are similar with 12/9 cm, but expressly point out that for large holding capacities a 'C' value of up to 6 cm is possible.

Staggered seat layout

In 1954, Hans Gussmann describes a method of theatre construction which sets the layout of seats 'at gap', which means that each visitor may view over the head of the person sitting two rows in front.

This type of seat layout is quite common for theatres since sightline elevation may be halved from 12/2 to 6 cm, which results in a significantly reduced slope for the auditorium.

Theatre and stadium construction can only be marginally compared in this respect. According to the definition of sports groups in EN / DIN 13200-1, the activity speed of ball sports is considered medium to fast and activities take place on a large playing field.

A staggered seat layout, where spectators view between the heads of the two persons sitting in front, implies a field of vision of merely 35° (distance between rows = 80 cm and seating grid = 50 cm).

The head of the person sitting in front comes into view already at 17.5°. A staggered layout would be pointless for large stadia since eye and head movements between 30° and 120° are no exception.

The closer the stand is to the point of focus, the steeper the stand will be.

Planners usually have to deal with a maximum slope, as stated by local building regulations. Because of legal limitations to the gangway slope of 3 x 19 cm = 57 cm maximum, the upper part of the stand has to accommodate a reduction in sightline comfort to meet these regulations.[6] If a stand reaches its maximum gradient before the necessary holding capacity is achieved, planners are forced to lower the quality of the sightline.

Makrolon Effect [ˈmækɹəˌlɒn ɪˈfɛkt] = **thinking laterally, realizing visions effectively**

High-status national and international projects are built on and with Makrolon®. Indeed, anyone invited by architects to turn their vision into a reality has to be equally visionary in all they plan and do. Take for example the construction of the RheinEnergie stadium in Cologne, which required the know-how of our product designers right from the planning phase. The result is that today's crowds of up to 51,000 fans are sheltered under 15,414 m² of Makrolon® sheets. Tested against fire, storms and UV radiation, they can even withstand the ecstatic frenzy of football World Cup crowds.
You can find out more about the Makrolon effect at **www.makrolon-effekt.com**. And to arrange a meeting to discover how it can be tailored to your specific requirements, call **+49 6151 1303134**.

makrolon®
the high-tech material

15 Principles of sightline calculation

'C' value   eye-point elevation

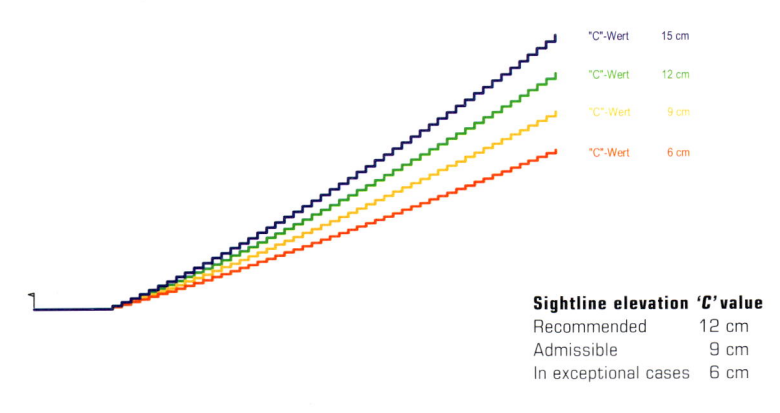

"C"-Wert   15 cm
"C"-Wert   12 cm
"C"-Wert   9 cm
"C"-Wert   6 cm

Sightline elevation 'C' value
Recommended   12 cm
Admissible   9 cm
In exceptional cases 6 cm

Sightline tolerance
Maximum of admissible
deviation   ± 5 mm

If this tolerance is observed,
the guidelines for 'C' value
are met.

(293 c) Inclination of a stand
in relation to eye-point elevation
curves: 'C' value

298 **Seating place grid (drawing)**
Comparison of seat layouts
Theatre, staggered vs. stadium, non-staggered

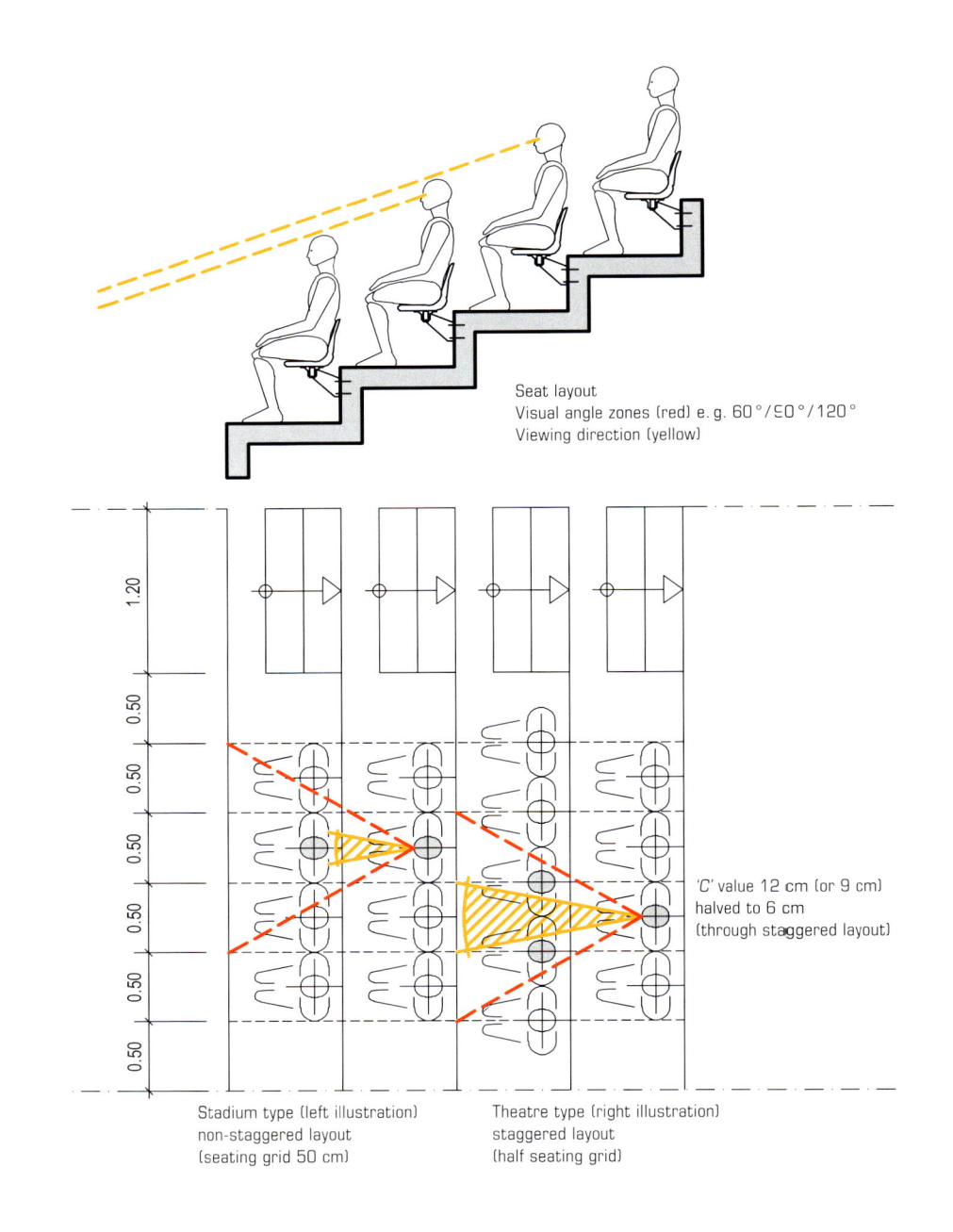

Seat layout
Visual angle zones (red) e.g. 60°/90°/120°
Viewing direction (yellow)

'C' value 12 cm (or 9 cm)
halved to 6 cm
(through staggered layout)

Stadium type (left illustration)
non-staggered layout
(seating grid 50 cm)

Theatre type (right illustration)
staggered layout
(half seating grid)

Distance of the 'first row' D

As demonstrated by the calculation formula for designing a sightline profile, height and distance to the point of focus are crucial. This applies in particular to the first standing/seating row since the starting point of the sightline affects the further profile and steepness of the entire stand. Its location depends on four different planning parameters:

a) The closer the stand is located to the activity area, the better and 'denser' the atmosphere would be.
b) The higher the terrace base point is positioned, the steeper the rise of the stand, which thus reaches its maximum slope earlier.
c) The further away the first row of a stand is located from the activity area, the shallower the terrace slope will be, resulting in better viewing conditions.
d) The further away the stands are located from the activity area, the larger and therefore more uneconomical the complex will become, at equal capacity.

The entire grandstand geometry is determined by the height and the distance of the 'first row' – a decision influenced by a number of factors.

Primarily, security-related aspects tend to dominate. The enclosure of the activity area or interior area serves to separate players/athletes and spectators. The necessity of such measures and their structural realization depends on the sports and its respective security objectives.

Wherever spectators run the risk of being injured by flying sports equipment, nets or transparent walls (e.g. ice hockey perimeter) have to be fitted. For the protection of players and referees in football, aggressive or emotionalized fans should be withheld from entering the stadium interior unauthorized and from causing injury to participants by hurling objects.

In combination stadia the stands are inevitably at a higher distance from the field than in dedicated football stadia.

Consequently, the respective stands have a lower inclination of the tiers at comparable sightline elevation ('C' value), with reduced viewing comfort for spectators of football tournaments.

Athletics competition complex type A:
Distance – short side: 21.00 m or
Distance – long side: 38.00 m
(Standard running track radius: 36.50 m)

FIFA-standard football field:
Distance – short side: 7.50 m
Distance – long side: 6.00 m

D Distance of 'first row'

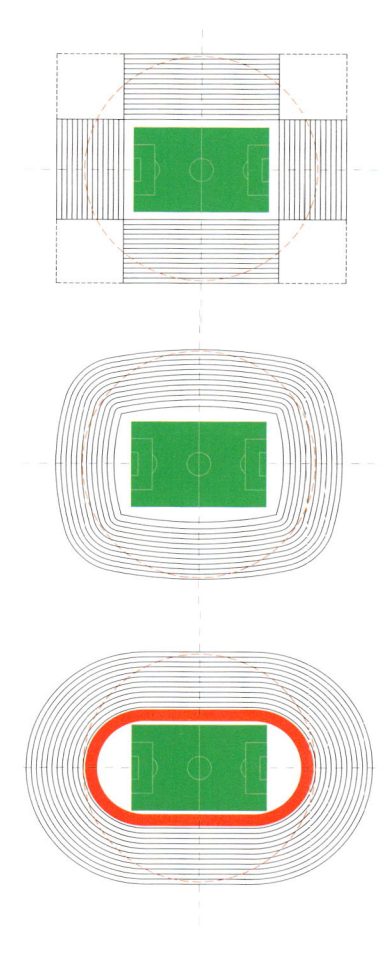

299 **Three variants for a stadium ground plan**
  a. Pure rectangular geometry
  b. Expanded rectangular geometry
  c. Semicircle geometry (athletics multi-purpose stadium)

Chapter 16

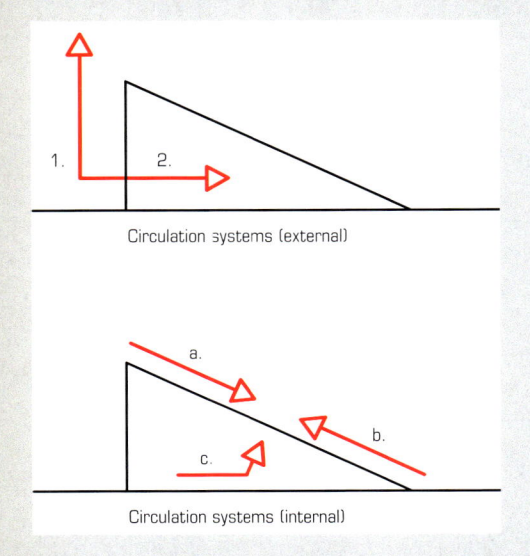

300 Circulation systems – external/internal
1 vertical access (external)
2 within/under terrace

a tier access from above
b tier access from below
c central vomitory access

Inclination and 'polygonal transformation'
The optimization of viewing conditions by means of maximum slope and 'polygonal transformation' of an eye-point curve

General introduction

A grandstand is a stepped structure where spectators may sit or stand. The inclination of a terrace riser is clearly limited by MVStättV 2005.

The maximum gradient can be determined by the relation of the highest possible number of goings and the maximum riser height within a radial gangway (3 x 19 cm = 57 for 80 cm). The inclination of a stand is related to the determined 'C' value of sightline quality.

In particular for the upper tiers, optimization and new definition of the maximum slope of a stand seems to be important for improving viewing quality. For standing areas, EN/DIN 13200-1 allows a maximum riser height of 20 cm, rather than 19 cm. In the following section this assumption is to be applied to the upper tiers. Security aspects related to a higher maximum slope are to be considered and to be compensated by supplementary measures; the objective being to determine and verify a sufficient 'C' value (9 cm) for sightline elevation of increasing holding capacities, also in the upper tiers.

In order to reach their seats, spectators access the viewing areas via vomitories within the inclined grandstand (vomitories or open access joints at the upper edge of the lower tier).

From there, spectators are guided in horizontal direction to a gangway which leads them in vertical direction to their respective seating/standing row. Once again in horizontal direction, they finally reach their numbered individual seats. In order to move within the stand, visitors have to manage heights.

Radial gangway versus stairs

Radial gangways are emergency evacuation routes, which are to be kept clear.[1] Colour signage of escape routes in large venues serves visibility and complies with relevant regulations.

Although radial gangways are stepped, they are not considered as general stairs in buildings, as defined by DIN 18065, but rather count as an independent type of vertical circulation, limited for use on spectator stands.

Explanation

1) DIN 18065 requires a landing after a maximum of 18 steps.[2] Most World Cup stadia have between 25 to 28 rows per tier and an average value of 55 cm riser height (upper tier). For each seating row there are three stair goings, which means that after 6 rows a landing would have to be provided. This scenario was not realized in any of the examined World Cup stadia.

2) Tolerances between two riser heights are limited to a maximum of 5 mm deviation from nominal value to actual size. Consequently, the rise/run ratio of obligatory stairs within a flight may not be altered. The relevant DIN provides no clear definition.

For reasons of sightline elevation, modern grandstands generally have a parabolic slope and therefore cannot be exclusively assigned to the group of general stairs.

Handrail as safety measure

The relevant regulation demands handrails for both sides, grip-secure with covered ends.[3]

No explicit request is made for further measures to secure gangways.[4] Rises below a height difference of 50 cm are within the scope of recommended rules for increment, comfort and safety, as stated in DIN 18065.

Higher inclinations up to a maximum of 57 cm have shown to be quite steep for users of such viewing accommodations.

Above a riser height of 50 cm a barrier should be installed, which then needs to be continued up to the vomitory and should not be restricted to riser heights above 50 cm.

Handrails can be mounted on both sides along the seating depth at the side of the gangway (gangway stirrup) or as central handrail (mid-gangway), also serving as visual block boundary.

The rail must be subdivided (1–3 treads) in order to allow changing to the other side of the gangway, in case a spectator wants to evade people walking in the opposite direction or has by mistake chosen the wrong side of the gangway.

A central handrail ensures a longer guiding of the hand. The useful width on both sides of a central handrail results in a widening of the entire gangway. Regulations stipulate a minimum width of 1.20 m.[5]

When divided by two, the resulting measure of 60 cm is sufficient for a one-person module width, but not to allow transports of injured persons on a stretcher, etc.[6]

A clear width of 90 cm would be adequate for emergency evacuation routes at venues for less than 200 spectators (compare to World Cup stadium Hamburg, with a 2.0-m total width at the upper tier).

Instead of a 1.20-m wide gangway, designated for 600 people, a central handrail (with 2 x 0.90 = 1.80 m) would result in a 50% increase of gangway width without a subsequent increase in capacity. This would mean losing one row of seats for a block size of 30 rows, i.e. 30 spectators less per gangway. Planners should consider this for each individual case.

Handrail height

According to the German regulation on spectator facilities, handrail height would have to be at least 1.10 m (comp. to *Guide to Safety*: hand rail height 0.90–1.0 m).[7]

Such elements have proved to restrict viewing severely, whereby a spectator stand fails to fulfil its foremost objective.

Examinations and inspections by representatives of building regulation authorities have shown that for gangways with maximum gradient a handrail height of 1.0 m is secure and sufficient.

'Green Guide' (B028) to Safety at Sports Grounds

Contrary to the presented proposal, the Scottish Office (item 7.5) stipulates for the design of stairs and for the optimization of admissible maximum gradient:

Goings	280 mm	minimum
	305 mm	recommended
Risers	150 mm	minimum
	180 mm	maximum
	190 mm	(existing)
	75–180 mm	(standing place)
Headroom	200 mm	minimum
	240 mm	recommended
Gradient of stand	34°	seating area (3 x 18 cm)
	25°	standing area maximum

The Scottish Office further points out (item 11.9) that riser heights might be increased in larger grandstands for reasons of sightline optimization, safe circulation provided.

301 **Upper tier, gangway stirrup**
World Cup Stadium Nuremberg
302 **Upper tier gangway, central handrail**
World Cup Stadium Hamburg
303 **Upper tier, gangway stirrup**
World Cup Stadium Cologne

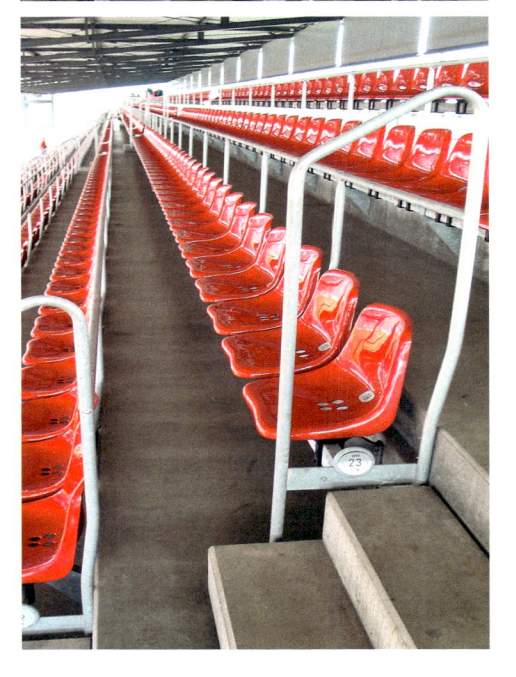

Request for deviation

In spite of clear legal guidelines and recommendations, during final planning the need to deviate from material requirements of the relevant building codes or other provisions arises, as well as the need to determine alternative measures.[8]

Planning rises of up to 21 cm

As stated in the fire protection survey for Cologne Stadium: 'In the lower tier [south], two risers (double risers) should be provided for gangways between two seating rows, with the riser height of two successive treads varying by around $\Delta h \leq 4$ mm. For a tread depth of 40 cm respectively, this results in rises of up to 21 cm, which means that the admissible value of 19 cm (VStättVO § 10, section 8) is exceeded in these areas.

Due to the comparatively large tread depth, the author has no reservations with regard to fire protection. In addition, the provision of triple steps within the standing areas of lower tier north and lower tier south would increase the risk of accident.'[9]

Note: The floors of the passageway between seating rows and the floors of rows providing standing accommodation are arranged level with the adjacent treads of the gangway.

304 **Drawing of gangway stirrup**
Spectator seated or walking
305 **Excerpt of table 1**
DIN 18065 on stairs in buildings

Commentary on table 1
(*1) includes maisonettes in buildings with more than 2 apartments.
(*2) but not < 14 cm; for determination of rise s/a see section 7.
(*3) but not > 37 cm; for determination of rise s/a see section 7.
(*4) For stairs with goings (a) below 26 cm, undercutting has to amount to a total measure of 26 cm (a + u).
(*5) For stairs with tread surface (a) below 24 cm, undercutting has to amount to a total measure of 26 cm (a + u).

Table 1: DIN 18065	Size limit	Finished size		Measures in cm
Type of building	Type of stairs	Useful flight width min.	Riser height s max. (*2)	Going depth a min. (*3)
Residential buildings with not more than two apartments (*1)	Stairs leading to common rooms	80	20	23 (*4)
	Basement stairs, not leading to common rooms	80	21	21 (*5)
	Ground stairs, not leading to common rooms	50	21	21 (*5)
Other buildings	Other buildings, stairs required by building legislation	100	19	26
All buildings	Stairs (additional) not required by building legislation	50	21	21

Maximum gradient of stands

The basic prerequisite for calculating a sightline profile lies in the stepped/inclined height profile of a stand.

Even though a radial gangway belongs to a separate group, DIN 18065 for 'building stairs' may be consulted for evaluating rises. Item 7 states the following principles:

The rises may be planned and checked with the help of the rules of increment, comfort and security. Riser height is denoted by (the value of) s and the going length by a. The median length of a human's stride is stated with 59 to 65 cm. An ideal rise of 17/29 meets all regulations.

Rule for stair increment
$2\,s + a = 59$ to 65 cm
Rule for comfortable passage
$a - s = 12$ cm
Rule for safe passage
$a + s = 46$ cm

A German expert commission on construction supervision comments in its explanation on the relevant regulation regarding seating and radial gangways:

'The regulation of section 8 […] is necessary, since only radial gangways with an uninterrupted succession of at least three goings between two levels are covered by DIN 18065 (stairs).

A special regulation is required since there are radial gangways with only one or two risers each between seating levels, and since radial gangways also function as escape routes.

Dimensions are adapted to the values of DIN 18065 […] This shall be the foundation for all further considerations.'[10]

Objective
The size of a grandstand is limited by the maximum gangway inclination.[11] The objective of the following calculation is to establish a connection to sightline calculation, in order to better exploit the sightline quality of a grandstand.

Restrictions for radial gangways
1) MVStättV 2005 § 10 section 8 limits the admissible rise-to-run ratio of a radial gangway to at least 10 cm and a maximum of 19 cm rake at a minimal tread depth of 26 cm.
2) European Standard EN/DIN 13200-1 recommends an inclination of gangways with a maximum riser of 20 cm and a recommended tread of 25 cm.
3) The angle of inclination of a radial gangway of 35° should not be exceeded (EN/DIN 13200-1, item 5.1.6)

306 **Radial gangway stirrup**
as fitted in upper tier of World Cup Stadium Cologne
307 **Maximum rise (at present)**
3 x 19 cm = 57 cm
Obligatory installation from a limit value of riser height > 50 cm

16 Inclination and 'polygonal transformation'

Safety compensation (20 cm)

Example 1
Classic linear rise 1 : 2 (see planning basis – Vitruvius) and principle for standing accommodation; two standing place treads amount to one seat tread of 80 cm.
Application of maximum inclination according to EN/DIN 13200.
2 risers, 20 cm/40 cm = 40/80 cm
Gradient of stand = 26.57°

Example 2
Maximum rise (MVStättV) with recommended tread width of 80 cm.
3 risers, 19 cm/26.6 cm = 57/80 cm
Gradient of stand = 35.47°

Example 3 (Proposal for optimization)
Proposal for adapting maximum rake upper tier equivalent to standing accommodation
3 x 20 cm riser = 60/80 cm
Gradient of stand = 36.87° = 37°

Example 4
A 'group' of four steps (= 4 x minimum tread 26 cm). Checking the inclination of this type of upper-tier terrace is as follows:
4 x 19 cm (risers) = 76/104 cm
Gradient of stand = 36.16°

308   Inclination gradients (system drawings)
  a. Example 1: 40/80 cm = 26.57°
  b. Example 2: 57/80 cm = 35.47°
  c. Example 3: 60/80 cm = 36.87°
  d. Example 4: 76/104 cm = 36.16°

The challenges for modern sports stadia lie in the reconciliation of high spectator capacities with equally high viewing comfort.

Examination of various sightline profiles has clearly shown that limiting the inclination of grandstands in return also limits their capacity or affects viewing conditions. The foregoing calculations for inclination have proven that exceeding the values for increment, comfort and safety seems inevitable, also with regard to the relevant regulations and usability (example: tread depth 80 cm, 'C' value). In order to maintain circulation safety on the terraces, the following preconditions are formulated, and appropriate structural measures are proposed.

Safety barrier
For riser heights above 50 cm safety barriers are still required for seats, projecting over the next row by at least 65 cm.

Gangway stirrup
At maximum gradient of 37° gangways are starting to become very steep. To balance this effect gangway stirrups are required for these gangways. With regard to utilization, installation of an interrupted handrail is recommended for riser heights of 50 cm and upwards, and becomes obligatory for 57 cm and more.

Note: In order to have a continuous line of handrail, gangway stirrups need to be installed also in gangway sections which are below the limit value of (R_H < 50 cm), i.e. they should be mounted down to the vomitory along the entire gangway length of the upper tier.

Capacity limitation
The problem for the rise of a stand lies in circulation safety and in the exertion or fatigue caused by using a gangway. Hence, the number of people sitting in a part of the stand with riser heights between 57 cm and a maximum of 60 cm should be reduced: from 600/200 (in the open/covered) to 200/60 persons per 1.20 m gangway width. In the open, a maximum of only 5 rows > 57 cm per riser height would be allowed.

Check calculation: 2 x 20 seats for each side of the gangway = 40 persons x 5 rows = max. 200 persons.

Proposal:
New definition maximum gradient
Inclination of stand max. 37° (at 80 cm tread depth)
Gangway rise 3 x 20 cm (maximum, equivalent to EN/DIN 13200-1)

Additional measures:
1) 65 cm safety barrier for R_H > 50 cm [12]
2) Gangway stirrups along the entire upper tier, if R_H > 57 cm to 60 cm is exploited
3) Reduction of capacity from R_H > 57 cm to 60 cm for the relevant rows from 600 to 200 persons per 1.20 m

Limitation of rows (maximum inclination)

In case the stand of a spectator facility reaching its maximum inclination without meeting the desired capacity, an adjustment is usually undertaken at the expense of the projected viewing quality 'C'.

Test geometries to examine the relationship between form and capacity have shown: on a four-sided lower tier stand (without vomitory) with a total capacity of 25,000 people on average 835 spectators can be accommodated in one seating row.

A price quotation may illustrate the economic relevance of such a deviation. As stated on the German ticket website *www.bundesligakarten.de*, a seating ticket for the game of FC Bayern Munich vs. FC Cologne on the 1 April 2006 in Allianz Arena Munich (in the category of seat at the curved corner / straight) costs from € 89 to € 109 →, on average € 99 → x 850 spectators ≈ € 85,000.

Note: Ticket prices in football stadia of the Bundesliga are on average between € 10 and € 50, but for large events such as the FIFA World Cup 2006™ in Germany prices many times that amount, from around € 100 and up to € 750, are to be expected.

These figures are not absolute, but they do show the critical relation between seating rows / capacity and the profitability of a sports venue. Thus remains the general necessity of a highly exact elevational curve as basis for efficient grandstand planning.

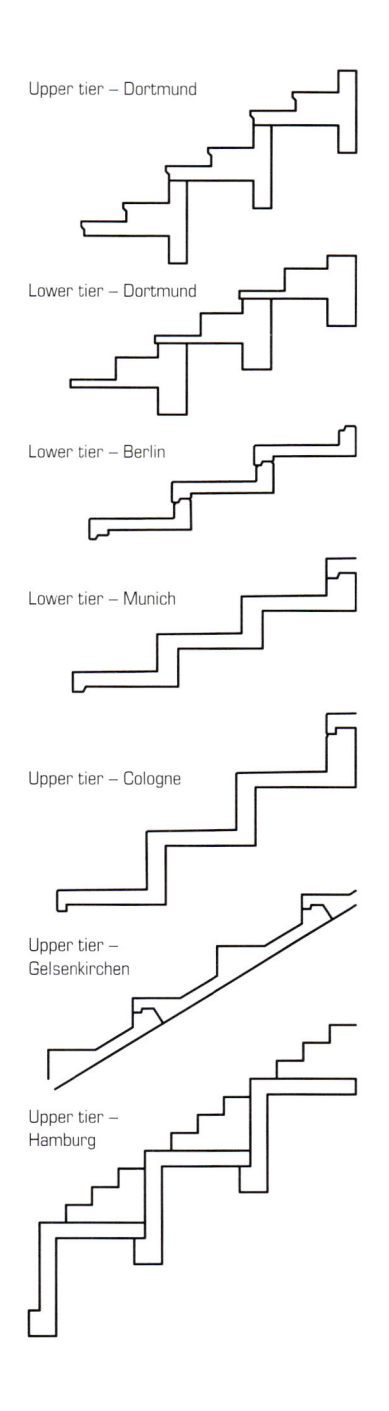

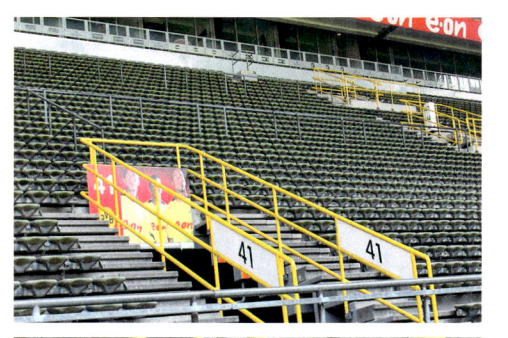

Upper tier – Dortmund

Lower tier – Dortmund

Lower tier – Berlin

Lower tier – Munich

Upper tier – Cologne

Upper tier – Gelsenkirchen

Upper tier – Hamburg

309 **Factory-built modules (drawing)**
Seven exemplary cross sections of steppings from twelve German World Cup stadia

310 **Vomitory situation, east stand**
World Cup Stadium Dortmund (2006)

311 **Prefabricated elements, east stand (1974)**
World Cup Stadium Dortmund (2006)

312 **Block steps, south stand**
World Cup Stadium Cologne

16 Inclination and 'polygonal transformation'

Polygonal transformation (part 1)

Example: 'Triple steps'
a) The ascending curve of sightline construction is subdivided into 'linear' parts:
Example Cologne, factory-built terrace member with three seating steps
b) Linear adjustment of riser heights at mid-level.
c) Seat height compensates polygonal divergence.

Polygonal transformation (part 2)

Example: 'Triple steps'
a) During polygonal transformation, inflexion points emerge between two linear rises.
b) Riser height may not exceed a tolerance of 5 mm.
c) A value of > 5 mm can be compensated for the last two gangway steps with 5 mm respectively.
d) I.e. at the inflexion point max. 3 x 5 mm.

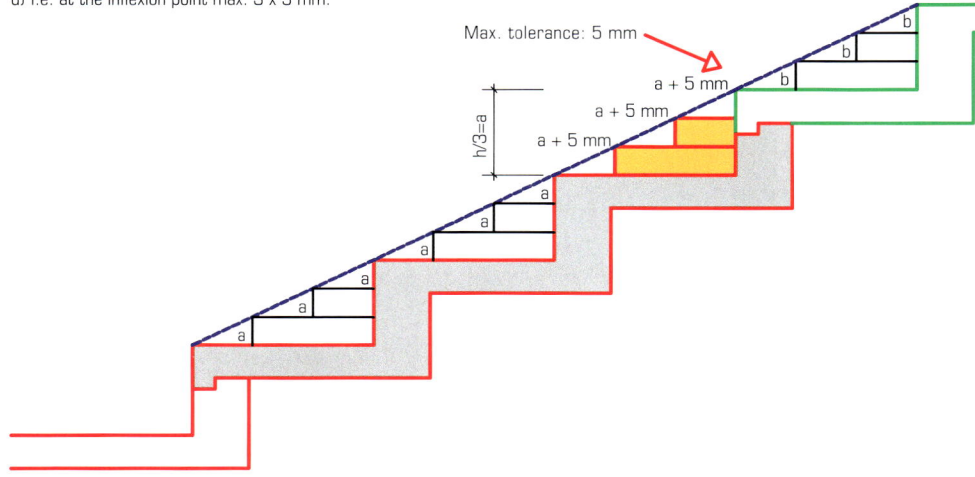

'Polygonal transformation'

Applying the calculation formula for the determination of a sightline results in a parabolic inclination curve for the terrace elevation.

This would result in a continuous modification of the built structure. The casting of reinforced concrete parts would become increasingly uneconomical.

Stadia of the third generation (World Cup 2006) mostly have ascending terrace profiles; except for the World Cup Stadium Hamburg and the four 'evolving' stadium complexes which have been continuously modernized since the 1970s.

Since theatres, cinemas and sports venues have started to be fitted with sloping auditoria in order to achieve a satisfying viewing quality, all efforts have been concentrated on achieving an economical construction by producing factory-built modules with the same rise. The constructional aim is therefore to generate a 'polygonal' terrace profile below the normally parabolic eye-point curve. Already in 1934, Dr. Christian Gellinek writes about lecture halls in university architecture:

'For an even construction and easy assembly of the seating area, a straight-line rise, only broken once or twice, would be more advantageous than a curved slope. At the same time, the subsequent ease of construction should not affect day-to-day utilization.'

Therefore, during the stages of final planning, when preparing the prefabricated parts, the number of different-sized steps has to be reduced to a minimum.

A solution would be the transformation of the elevation line into a traverse, the 'polygonal transformation'.

At the distance of eye-point height A_S, the terrace structure generally lies *below* the eye-point curve. Due to the bent geometric quality of a polygon, compared to a curve function, continuously changing differences between the curve and the polygon are created.

Planners may determine this height difference by means of a simple calculation programme depending on the parameter 'number of linear group of steps'.

These can be so called 'triple' steps, three risers in one prefab part or several single steps / 'triples' successively.

In any case, the change in gradient / inclination takes place at the deflection point (the 'bend'). This is of no consequence to the single seating / standing row, but in this context the affine image of the gangway in the same terrace profile is realized as 'stairway'.

Note: Although the radial gangway represents a special type of stairway, at the same time, it always retains its vital function as escape route.

313 **'Polygonal transformation'**
Part 1 – adjustment of seat height
(according to *Stadium Atlas*)
314 **'Polygonal transformation'**
Part 2 – adjustment of block steps
(according to *Stadium Atlas*)

Continuous difference in rise

The obligatory gangways for a stepped escape route are to be kept within tolerance at the 'bend' of a terrace profile. Therefore, the following proposal, as stated in DIN 18065, seems appropriate:

'The actual measure of riser s and going a within a (built) stairway may deviate from the nominal dimension by not more than 0.5 cm.'[13]

Although this is a construction tolerance rather than a planning tolerance, it seems reasonable to restrict the continuous difference in rise to 0.5 cm.

Explanation

'Concerning fire protection, we hold no reservation with regard to foregoing differences in rise [World Cup Stadium Cologne] because the divergence is distributed on a multitude of risers, and is therefore hardly noticeable on walking. (Note: Riser height between two successive goings varies by in total $\Delta h \leq 4$ mm).' (Fire protection survey, Stadium Cologne)

A superpositioning of the parabolic elevational curve of a sightline with the actual polygonal profile of the prefabricated elements must be compensated, in order to retain the sightline profile.

This elevational adjustment is done through the positioning heights of the seating accommodations.

Elevation of eye-point height

The differences in height of the back of the knee (fossa poplitea) allow adaptation and adjustment for all types of seats. (When determining the height of a seat, its type, covering and/or upholstering should to be considered.) EN/DIN 13200-1 limits height difference between riser and tread to 45 cm, which means that a compensation of up to 5 cm is feasible for a recommended regular height of 40 cm.[14]

An eye-point height of 80 cm above the tread is retained, and the correct eye point is thus located at 40 cm + 80 cm = 1.20 m + 50 mm, so at 1.25 m maximum. Tolerance is compensated only *after* the general sightline has been determined.
1) The 'eye-point curve' is determined.
2) Below this, at distance A_S, the terrace profile is set.
3) A 'polygonal transformation' is carried out by retaining the inflexion points on the curve and by 'bending upwards' the linear inclination.

For this reason, the height difference of A_S is always below 1.20 m, i.e. the 'softest' structural element, namely the seating shell as single seat or placed on a traverse, compensates the height. Depending on the measure of 'polygonal transformation', it will be installed at a lower position by x cm. Thus, the eye point remains within the theoretical sightline elevation curve.

Consequently, if the seat is always installed in a lower position, the recommended seating height of 40 cm is generally undercut.

Taking into account the ergonomic figures of the table opposite, then the basis for a sightline method would be better defined with a height of 45 cm.[15]

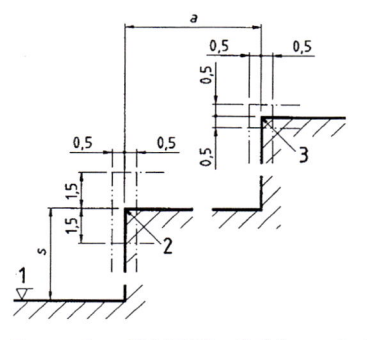

315 **Excerpt from DIN 18065 – 'Building stairs'**
Tolerances of step front edges (picture 8)
Measure in centimetres

316 **Excerpt of table from DIN 13200-3**
Seat height

317 **Excerpt of table from DIN 13200-3**
Distance between rear of buttocks and back of the knee

318 **Comparison of a linear inclination**
with a parabolic elevation curve

1 Surface stair landing
2 Nominal location of step front edge, first step
3 Nominal location of step front edge
s Riser (nominal dimension)
a Going (nominal dimension)

Seat height	Men		Women	
Share in %	5%	95%	5%	95%
Seat height	394 mm	490 mm	356 mm	445 mm

Gluteal-popliteal distance *)	Men		Women	
Share in %	5%	95%	5%	95%
Gluteal-popliteal distance	439 mm	549 mm	432 mm	533 mm

*) Gluteal-popliteal distance = distance between rear of buttocks and back of the knee

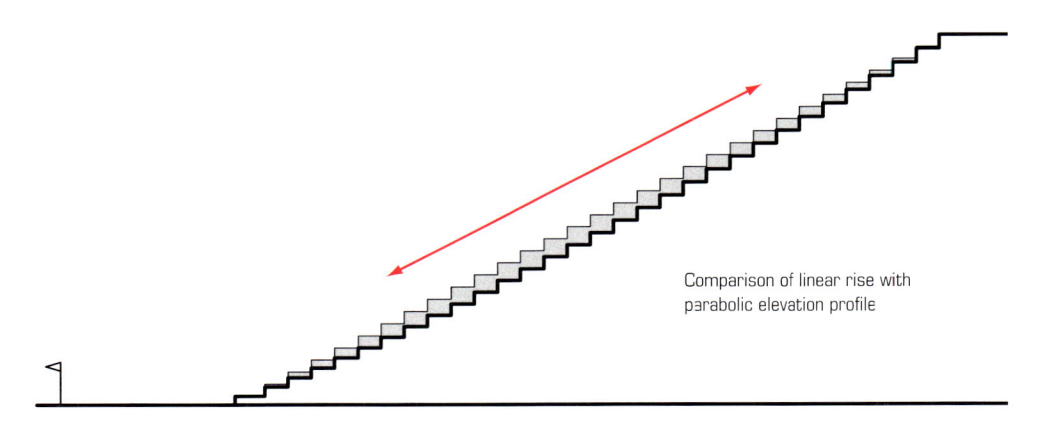

Comparison of linear rise with parabolic elevation profile

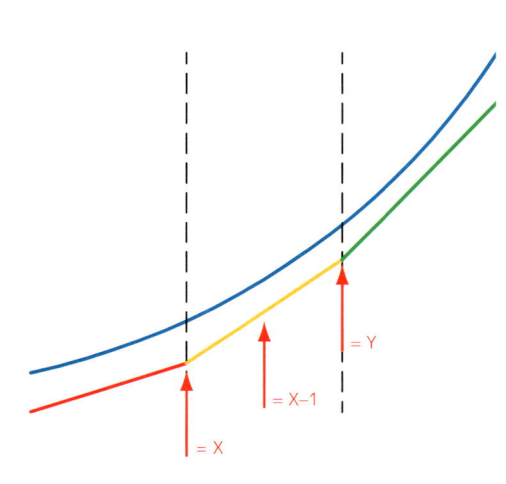

319 Sketch of the changing height differences between polygonal profile and eye-point curve

Polygonal Transformation

Maximum tolerances 0.5 cm
(within gangway)
Maximum height compensation 5.0 cm
(single-seat height)

Tolerance for rise compensation

With polygonal transformation, the 'bent' line (red/yellow/green) runs below the eye-point curve (blue). As a polygon, it displays successively alternating linear rises.

Should the maximum rise of a gangway step remain at 19 cm, then the number of single steps within a seating/standing row is determined as follows:

If $R_H/2 > 19$ cm, then double step, otherwise $R_H/3$ = triple step (gangway)

In the calculation table for single riser heights, a gap emerges at the inflexion point (alternation between the linear rises) whose tolerance may not exceed 5 mm.

The above sketch (polygonal transformation, part 2) illustrates the option to increase the maximum tolerance at the inflexion point to 3 x 5 mm, as the last two triple steps with a maximum of 5 mm each may be integrated, also within the tolerance of step-to-step rise.

As a rule, gangways consist of factory-built block steps which are mounted on top of the regular seating/standing accommodations.

The tolerance in height decreases continuously as the number of rows increases. Therefore, the critical value for compensation lies around the first two inflexion points.

Negative inflexion point

The numerical value resulting from the subtraction of the next higher rise may have a negative sign, if seat riser height allows a double gangway step based on the condition of < 2 x 19 cm. This gap cannot be compensated through the block steps. Therefore, the steps within a gangway should always be realized as 'doubles' or 'triples' and rises should be reduced to a maximum of 10 cm.

Prefabricated parts / block steps

Usually, the structural grid of prefab steps ranges from 5 to 10 m. The further apart the indented beams (stepped supports for seating/standing prefab risers), the more rigid the concrete elements must be.

'Bon Jovi Test' at Frankenstadion, 1998

Prof. Hans Jürgen Niemann of the Aerodynamics Team in the department of Civil Engineering at Ruhr University Bochum writes on the topic of 'oscillations caused by humans' at rock concerts:

'The resonance level which might potentially endanger a building needs to be determined by experiment. Resonance happens when the applied frequency comes close to natural frequency, which is when human-induced vibrations of the stand per second approach the natural frequency of the stand per second.'[16]

As a measure of susceptibility to spectator-induced vibrations, a natural frequency of below 6 Hz was stipulated for World Cup Stadium Frankfurt. The static height of the indented beam and the rigidity of the prefab stand risers are increased inasmuch as the measure is observed for all imposed load scenarios. (No complaints have been submitted during present use.)

In the wake of multi-purpose utilization of most spectator facilities for concerts, etc., this aspect has taken over an important role. With span widths of 10 m for the main structural grid in World Cup Stadium Cologne, the most economical solution would be the subdivision into triple steps. Three steps form one large factory-built module which is sufficiently rigid and reduces the total number of prefabricated structural components within the stand.

Note: General planning of a modern stadium has to consider which ancillary utilizations are to be accommodated underneath the terraces. A grandstand with high-quality utilizations placed underneath is considered as a general roof structure with all the subsequent requirements of fire protection (F 90 and smoke-proof) and waterproofing, etc.

Parameter studies

After creating a suitable eye-point or sightline elevation curve, during a second revision a 'polygonal transformation' can be carried out for economical reasons.

In cooperation with structural planners, building contractors (e.g. the supplier of prefabricated parts) and the architect, a decision must be taken on load capacity and positioning of the support beams (intervals of indented beams = step bearings), logistics and material demands.

Long spans demand an increased single-step rigidity; short spans enable smaller pre-cast elements but demand more supports in the ancillary areas, or prefabricated parts which comprise several seating steps. In present construction practice, triple steps have emerged as the most economical solution. They have the added advantage of reducing the number of joints by two thirds. This solution is gaining popularity, since viewing accommodations need to meet the quality standards of general roof structures. As a rule, block and gangway steps are only placed on top of impervious and fully adequate terrace structures and then fixed in position. With the help of an Excel table (cf. 'five-step method' as set out in the dissertation on sightlines and safety by this author, RWTH Aachen 2006), the stand is gradually 'polygonalized'; at the same time examining the effects on the necessary height compensation of seating and the height offset (tolerance) in the walking line of a gangway. The following recommendations are given for an economical and practicable prefabricated 'polygon':

1) Either the 'C' value deviations are accepted as sight quality standard, automatically producing by linear transformation 1 x 2 x 3 x 'triples', etc., or the seating accommodation must compensate this 'wave-like' deviation in height.

2) Necessary compensation measures decrease continuously as the number of rows increases, tending toward zero, i.e. the steeper the rake of a stand, the more obsolete the need for compensation in height becomes.

3) A further principle may be derived from the aforementioned geometrical rule: the more shallow the inclination of a stand, the fewer similar riser heights can be set as linear rise, or transformed as polygon.

Parameter studies have shown a clear trend. In the upper tier, groups of prefab parts with identical rises may be combined without problems.

The steeper a stand, the lower the discontinuous 'C' value deviation will be. If this trend is compensated in height in terms of the seat fixings, the numerical value for sightline elevation as stated in the invitation to tender can be observed.

Note: A parametrical examination seems practicable in any case, in order to produce an optimized and precise sightline profile.

320 **Prefabricated terrace structure**
South stand of RheinEnergieStadion, Cologne (during construction 2002)

Chapter 17

Calculation example
Determination of eye-point position:
  0.15 m Standard railing
+ 0.80 m Tread width
− 0.20 m Eye-point elevation
= 0.75 m Distance to front edge of barrier
1.25 m Eye-point height, seated

Corner area:
– may be freely selected.
– depends on stadium shape.
– example:
  7.50 m FIFA minimum distance
+ 0.75 m
= 8.25 m

Parameter studies
Geometrical relations between the sightline parameters eye-point height, tread width, 'C' value and distance of the 'first row'

Calculation of a grandstand profile

After creation of the sightline curve, the following method is separated from the planning of the terrace structure.

After the fundamental parameters $A\ B\ C\ D$ have been stipulated, the circulation system (lateral and radial gangways) and the different utilizations of a spectator facility (media outfits etc.) determine which special elements will be incorporated into the stand.

It is therefore strongly advised to set the continuous sightline profile (eye-point curve) as geometrical foundation.

The parameter of distance D is generally related to the ground-plan distance of the eye point and not to the actual structural front edge of the stand's base point, in order to guarantee flexibility up to final planning, e.g. with regard to the decision on a particular railing system (Vario fence). Its thickness is only relevant at the beginning of the calculation for the first row. Reliant on the draft, it is structurally variable and should be selected so as to avoid falling short of the minimum distance to the playing field. Above the minimum values, all values $A-D$ may be selected freely.

Lower tier

After determining the desired capacity, the number of tiers is set as starting point in accordance with the pre-dimensioning study.

The general distance D of the first row is determined by the stadium layout set in the draft and by the utilization of the interior area. The desired viewing quality is reflected by the 'C' value, and seating comfort depends on tread depth B.

Following clarification on the relevant safety concepts, eye-point height A, which determines the stand's gradient, is established. The lower tier is a product of the foregoing factors.

Upper tier

Modern stadia and arenas of the third generation have shown that height offsets are generally used to accommodate high-quality hospitality suites.

After the access joint is set in the draft, the height profile of the upper tier can be determined in the second spreadsheet. Separating the two parts of the table makes sense since it has emerged that the 'C' value normally set for the lower tier cannot be observed in the upper tier.

In the wake of polygonal transformation a geometrical phenomenon emerges: with increasing steepness the possible linear length (number of rows with the same riser height) increases.

Exceeding maximum values

If the desired capacity is not sufficient when limit values have been reached, the following procedure has proved practicable:
1) Lowering of initial height through alternative security measures.
2) Increasing distance (possibly by rounding/expanding the stand geometry while controlling the decreasing 'C' value in the corners)
3) Reduction of offset for hospitality boxes, or reduction of sightline quality C (as in upper tier for example).

Parameter: height and distance

Lower tier

The starting height is the most relevant planning factor for sightline calculation. Below a railing height of 2.50 m it has to be coordinated with alternative security measures for the first spectator row. Example:

A = 2.85 m: railing height 2.50
B = 80 cm: tread width
C = 12 cm: sightline quality 'C'
D = 6.75 m: minimum distance 6.0 m

The development of numerical values for the single riser height clearly shows: elevating the first row at a minimum distance of 6.0 m (touch line) in combination with a 12-cm sightline quality results in the admissible maximum inclination of 3 x 19 cm = 57 cm being reached already in row 13.

An increase of the maximum riser height of 3 x 20 cm = 60 cm would enable the accommodation of 50% more rows in the same quality (19 rows). In several steps, distance D is slowly increased by 75 cm until the lower tier has acquired the desired capacity. In this case, up to a theoretical maximum size of a lower tier block of (at present) 30 rows at a continuous 'C' value of 12 cm:

– 6.75 m: corresponds to 6.0 m minimum to touch line
– 7.50 m: (intermediate step)
– 8.25 m: corresponds to 7.0 m minimum to touch line
– 9.00 m: (intermediate step)
– 9.75 m: (intermediate step) at 2.50 m barrier height

The maximum rises are reached in the following rows:

at 6.75 m → 13 rows (57 cm) / 19 rows (60 cm)
at 7.50 m → 22 rows (57 cm) / 30 rows (60 cm)
at 8.25 m → 32 rows (57 cm) / 44 rows (60 cm)

At an ideal distance of 9.75 m, and when raising the first row to a barrier height of 2.50 m, the lower tier ascends sufficiently shallow so as to reach the limit height of 50 cm only in row 28, from where a 65-cm seat safety barrier becomes obligatory. (The last two rows, up to a maximum of 30 rows, could serve as the two end rows of the block in front of the box tier.)

'Rounded' stand geometry

When the grandstand is not an affine image of the playing field, i.e. terrace front edge and touch line are not running parallel, then, according to the 'centreline method', the stand axis with 9.75 m eye-point distance lies in ideal distance, at 12-cm 'C' value. In the empty column of the table to the right the so-called 'corner conflict' becomes apparent, i.e. when the corner area of a rounded stand geometry lies at a minimum distance of 6.0 m to the touch line (or 7.50 m to the goal line). Clearly, at a minimum distance, the critical value of 6 cm sightline elevation is undercut in the first ten rows, and for the entire lower tier no admissible minimum 'C' value of 9 cm can be attained. If the curvature of a stand takes into account the goal line at the corner area with a 7.50 m security distance, then 'C' values are improved.

Note: For rounded geometries, the corner area should be provided with a minimum value of 9 cm. Based on geometric rules, we may assume that 'C' values are increasing successively towards the respective centre of the stand.

Interpolation of the maximum block size

In this example, at an optimized distance of 9.75 m, a lower tier rise of 53.2 cm is reached only in row 40. Here, a maximum block size of 1,200 persons with 40 rows could be interpolated.

Minimum capacities of modern sports and event venues for international football tournaments are at around 40,000 seats. Depending on the ground plan type, a two-tier layout becomes inevitable in excess of 20,000 to 25,000 spectators.

Note: If the upper tier is projected by a horizontal tier with a vertical height offset, this has a severe effect on the necessary inclination. In spite of an 'optimal' lower tier, a detailed examination of the upper tier height profile seems therefore paramount for the next stage, in order to determine how many rows can be accommodated and at which sightline quality.

First Row

	X-value	Y-value
Option 1: Elevation – 2.00 m	variable	2.85 m
Option 2: Moat – 1.80 / 0.95 m	variable	1.90 m
Option 3: Fence – 2.20 m	variable	1.55 m
Option 4: Wembley – 1.80 / 0.65 m	8.25 m	1.90 m
Option 5: Double fence – 2.20 m	8.25 m	1.90 m
Option 6: Presence of stewards	8.25 m	1.90 m
Option 7: Lowering 5%		
35 cm / 7.5 m	8.25 m	1.90 m
50 cm / 8.5 m	9.25 m	2.30 m

Important guide note!
Parameters A B C D refer exclusively to the eye point for sightline construction and, only in a second step, are transformed into terrace parameters. Structure and circulation systems remain flexible in design.

The higher the starting point of a sightline, the steeper the rise of the stand.

The deeper the tread width, the shallower the rise of the stand.

The higher the 'C value' of sightline elevation, the steeper the stand ascends.

The further away the spectator is positioned from the point of focus, the shallower the rise of the stand.

Second step: Upper tier

When a railing height of 2.50 m is employed as a security measure, the parameters for the upper tier have to be changed, since the recommended sightline elevation by FIFA/UEFA of 12 cm cannot be realized for the upper tier.

A = 2.85 m: barrier height 2.50 m
B = 80 cm: tread depth
C = 9 cm: sightline quality 'C'
D = 10.75 m: starting distance eye point

If the distance of the first row is merely increased by one metre from 9.75 m to 10.75 m, and is planned with a reduced 'C' value of 9 cm, an upper tier can be realized up to row 29, with a maximum rise of 19 cm.

This mean value corresponds roughly to the distance of the long side to be found in all twelve World Cup stadia with a rounded or expanded stand geometry.

'C' value (corners)

When the determined riser height at the central axis of a stand is applied to the corner areas, and the distance to the corner flag diminishes geometrically, the 'C' value decreases. This effect is also known as 'corner conflict'.

Parameter: Tread width

Research has shown that in upper tier stands which are offset by a standard box tier and where the stand's base point is elevated to a barrier height of 2.50 m, this can result in stair riser heights of over 50 cm. In this case, protective measures (barriers) behind each stair row are obligatory.

The structure is either located within the seat back or is provided by a stand-alone 65-cm stirrup. If for economic reasons, at a standard tread width of 80 cm, no tip-up seats are to be employed, then less than 40 cm remain as minimum clear width for planning purposes since the protective structure has to be deduced.

By increasing tread width from 80 cm to, for example, 85 cm, the maximum stepping height (3 x 19 cm = 57 cm) is reached earlier, that is, already in row 10. The foregoing width of 85 cm is unproblematic if a maximum limit of 3 x 20 cm = 60 cm is permitted. Otherwise, the maximum inclination of 57 cm should be observed by gradually decreasing the 'C' value. This measure results in a noticeable deterioration of the sightline quality of a grandstand below the 9-cm limit stipulated by FIFA, unless client and planners resort to the more costly alternative of installing tip-up seats.

Note: This example clearly illustrates the sheer complexity of the relationship between sightline quality and grandstand geometry.

Adjustment of 'C' value

The larger or more multiple the polygon to be transformed is chosen, the more necessary a height adjustment of the seating accommodation becomes, in order to retain the eye-point curve on the altered terrace structure and to maintain sight quality as planned.

Single steps
(without 'polygonal transformation')

The rise corresponds exactly to the curve profile since every row has a new height. 'C' value remains unchanged, and the viewing quality remains consistent, as requested. The height offset within the walking line of a gangway starts below the limit of 5 mm and decreases continuously.

Double steps
(smallest 'polygon transformation')

The necessary height compensation of seating starts at around half a centimetre (−0.42 cm) and decreases steadily. 'C' value oscillates alternating by around the same measure up and down. From a structural viewpoint, the height levelling of the structural element 'seat' seams more practical than for the less adaptable element of the stair riser. But in the framework of practical tolerances, the alteration of seating height is far below the present building tolerances, cf. DIN 18203-1.

Note: For this reason, height compensation for seating accommodation can be omitted, as long as a possible deviation from the planned viewing quality 'C' at ± 5 mm is laid down in the contract.

The numerical value for the necessary height compensation within the lower-tier seating remains noticeable since it is significantly higher and falls outside of the continuous profile. This gap in the profile is created at stair risers < 38 cm, as the gangway steps change from a double- to a triple-step system. In this case, stair risers (for example rows 1 to 5) are also to be realized as 'triples' (10 cm minimum for each gangway step).

Triple steps (normal polygon transformation): Many of the newly built venues feature so-called 'triples', i.e. three stair risers to one prefabricated part. Although the eye-point curve rises parabolically (for example from 23.58° to 27.70°), the height offset in the gangway walking line is declining. Up to row 50, it drops to insignificant 2.0 mm (< 5 mm). Height compensation for the seating accommodation remains appropriate. If a tolerance of 5 mm is applied to 'C' value, height compensation is rendered obsolete.

Element comprising six steps: These are normally realized as double 'triple' steps, since prefabricated parts comprising more than three steps are logistically impractical.

The fewer different prefabricated elements, the more economical the employment of prefabricated parts.

For the prefabricated structure as a whole, it is quite possible to set different heights within a 'triple' prefab component. For circulating stadium geometries, i.e. equal-height seating rows, one prefab part theoretically exists as many times as the circulation length / number around the stadium. Simplification can only be attained when two successive seating-row rings can be realized in the same rise / run ratio. A triple step would be such a type of 'ring'; simplification can be realized with 'double' use, amounting in calculation to a group of six, nine or twelve steps. If six equal seating steps are to be formed, then these are averaged between the first and the sixth riser. The 'C' value oscillates three times respectively above and below the planned sight quality (for exam-

Table 1: Tolerances on dimension on length and width

Construction members	Tolerances on dimension in mm at nominal dimensions in m							
	up to 1.5	over 1.5 up to 3	over 3 up to 6	over 6 up to 10	over 10 up to 15	over 15 up to 22	over 22 up to 30	over 30
Length of bar-type construction members (e.g. columns, trusses, girders)	± 6	± 8	± 10	± 12	± 14	± 16	± 18	± 20
Length and width of roof plates and wall panels	± 8	± 8	± 10	± 12	± 16	± 20	± 20	± 20
Length of prestressed building parts	–	–	–	± 16	± 16	± 20	± 25	± 30
Length and width of façade panels	± 5	± 6	± 8	± 10	–	–	–	–

Table 2: Tolerances on (cross-) sectional dimensions

Construction members	Tolerances on dimension in mm at nominal dimensions in m					
	up to 0.15	over 0.15 up to 0.3	over 0.3 up to 0.6	over 0.6 up to 1.0	over 1.0 up to 1.5	over 1.5
Thickness of roof plates	± 6	± 8	± 10	–	–	–
Thickness of wall and façade panels	± 5	± 6	± 8	–	–	–
Cross-sectional dimensions of bar-type construction members (e.g. columns, girders, trusses, ribs)	± 6	± 6	± 8	± 12	± 16	± 20

Table 3: Angular tolerances

Construction members	Angular tolerances in mm at lengths in m					
	up to 0.4	over 0.4 up to 1.0	over 1.0 up to 1.5	over 1.5 up to 3.0	over 3.0 up to 6.0	over 6.0
Unfinished wall panels and roof plates	8	8	8	8	10	12
Finished wall panels and roof plates	5	5	5	6	8	10
Cross-sectional dimensions of bar-type construction members (e.g. columns, girders, trusses, ribs)	4	6	8	–	–	–

321 **Table 1: Tolerance on dimension and angles**
(excerpt from DIN 18203-1)

176   17 Parameter studies   Sightline summary

ple 12 cm in the lower tier). Should this occur to such an extent as to exceed 'C' value tolerance of 5 mm, the polygonal transformation is adjusted downwards. Otherwise, height compensation becomes necessary for the seating accommodation in the lower part of the terrace, which would be around 3 centimetres between rows 3 to 5.

Critical upper tier

The terrace gradient becomes increasingly more steep (ca. 33.5 – 36°) in the upper tiers. For single steps with different riser heights, 'C' value remains the same, and no height compensation of the seating accommodation is necessary. The continuous change in gangway gradients remains negligible. The values for height compensation within the seating accommodation correspond to the respective height offset in the gangway's walking line when using triple steps. It is approximately identical to the height offset in the walking line of a gangway. With three adjusted risers, the 'C' value alternates by the level of height compensation once up and down. Based on the low differences of around ± 1–2 mm, a triple step in the upper tier area would be an effective means of polygonal transformation. As opposed to 'triples', the figures for 'C' value deviation between walking-line height offset and seating height compensation already start diverging when elements comprising six steps are employed.

Only for the third generation of modern stadium new builds and conversions has the concept of sightline attained a particular significance. Compared to the sports utilizations of previous stadia, we now mostly find multi-purpose stadia with integrated athletics tracks.

The resulting distances of the first seating rows to the point of focus, touch lines and bylines of a football pitch generally create comfortable sightline conditions. These stadia easily reach 'C' values of 15 to 20 cm in the lower tier and up to 27 cm in the upper tier. Omitting the above mentioned athletic tracks in the plannings for the new World Cup stadia has moved the body of the stand very close to the pitch. Based on the geometrical condition 'the closer, the steeper' described earlier, sightline determination gains an exceptional significance.

'C' value has become a quality standard for the viewing areas in modern sports and event venues.

Beginning with the construction of Stade de France in St. Denis, Paris (World Cup 1998), the experiences made during the last decade have lead sports associations FIFA / UEFA to agree on the following:
C = 12 cm: 'recommendable'
C = 9 cm: 'admissible'
C = 6 cm: 'unfavourable'

322 **Under-terrace view of steppings**
South Stand of RheinEnergieStadion, Cologne

323 **Drawing on the method of sightline construction (repeat)**

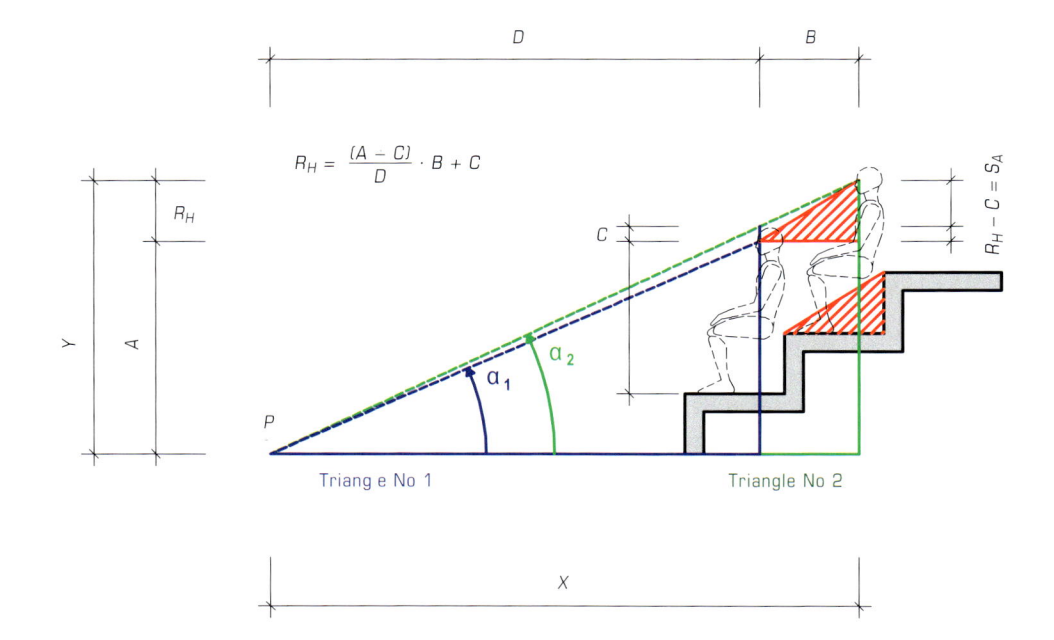

$$R_H = \frac{(A - C)}{D} \cdot B + C$$

Five-step method

In order to illustrate and compare the quality standard of sightline elevation between different plannings for stadia, the following 'five-step method' is defined:

Step 1: Define point of focus P (depending on the discipline, see table C.2, Appendix C of DIN EN 13200-1)

Step 2: Clarify type of securing for playing field and determine basic values (parameters)
1) first eye-point height: A
2) planned tread width: B
3) choosing sightline elevation: C
4) distance to the 'first row': D

Step 3: Construct grandstand geometry and position outfits (box tier, circulation system)

Step 4: Check extreme points (corners) with 'C' value calculation formula

Step 5: Carry out 'polygonal transformation'

Note: Based on the capability for conversion, the geometrical sightline calculation for standing accommodation in always carried out equivalent to a seating-type situation.

Chapter 18

324 Evening sun on the east stand
Olympic Stadium Berlin

Playing fields
A summary of the major sports types and relevant dimensions as a means to determine the minimum distance of a grandstand

Facilities for athletics competitions

DIN 18035 states in item 5.2 the requirements and disciplines for athletics facilities. Three categories from $A-C$ are distinguished accordingly (item 5.2).

Competition facility type A
Features an eight-lane circulating track with short distance track and large playing field located in the central area; discus and / or hammer throw, high-jump and javelin in the southern segment; pole vault, shot put, discus and javelin facilities as well as water ditch for the hurdle race in the northern segment; long-jump and triple-jump at the eastern straight, beyond the circulating track.

Competition facility type B
Features a six-lane circulating track with six- to eight-lane short distance track and large playing field located in the central area; discus and/or hammer throw, high-jump and javelin in the southern segment; facilities for pole vault, shot put, discus, javelin, long-jump and triple-jump as well as water ditch for the hurdle race in the northern segment; long-jump and triple-jump may also be accommodated, as in type A, beyond of the circulating track.

Competition facility type C
Features a four-lane circulating track with four- to six-lane short distance track and large playing field located in the central area; high-jump and javelin facilities; pole vault, discus, long-jump and triple-jump facilities as well as shot put facility in the northern segment.

Orientation

As stated in DIN 18035-1:2003: 'Large and smaller playing fields are aligned in their longitudinal axis in *north-south direction*. Deviations from NNW/SSE to NNE/SSW are possible, the first direction being generally preferable.

For difficult terrain or for unfavourable site layouts, a different orientation is admissible in order to avoid major ground movements and redundant ancillary areas.

For tennis courts, the north-south direction is favoured. High-jump and javelin facilities are to be so arranged as to avoid athletes being blinded by the sun.

For all other running, throwing and jumping facilities, orientation should correspond to the main times of use to avoid glare.'[1]

Principal utilizations

In the northern hemisphere the main stand is generally located on the west side of the stadium since daytime events normally take place in the afternoon (football, etc.). Otherwise, the course of the competition of individual disciplines is adjusted accordingly (all day for athletics).

Aligning the field in north/south direction ensures that goalkeepers and players are not blinded by the sun low down on the horizon.

The main stand with VIP and press areas is hence located on the west side so that the late afternoon sun is at the back of the spectators.

This exposition has been taken over as guideline by nearly all sports associations.

Note: A sport venue of the third generation (as pure football geometry) with its very high terraces and the structural density of the stadium interior (without athletics tracks) is normally faced with the problem of insufficient sunlight falling onto the natural grass turf to guarantee adequate regeneration or drying.

Athletics geometries display fewer problems with the turf due to the relatively large opening of the stadium's interior roof.

Glare from the sun low down on the western horizon may still be expected.

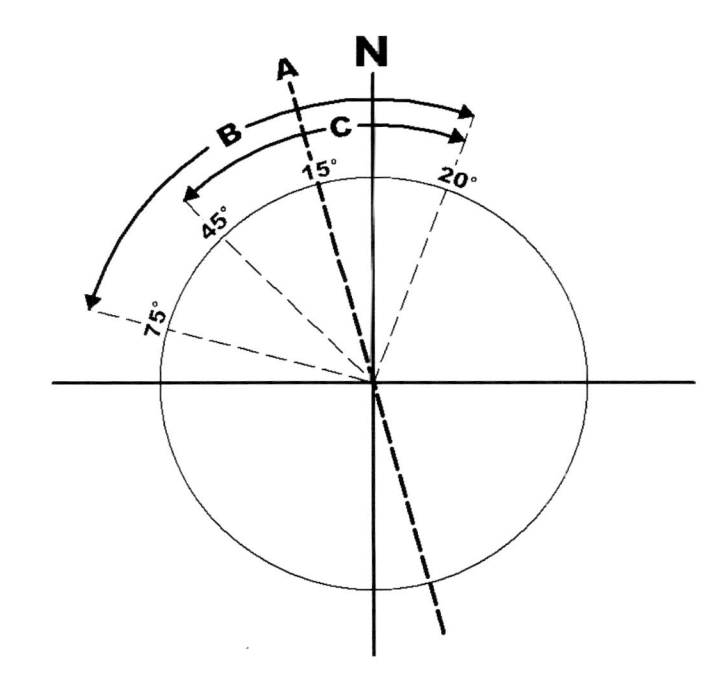

325 **Stadium exposition toward the sun**
As in *Stadia – A Design and Development Guide* (3rd edition)

18 Playing fields

Table 1.5.3 – Requirements of the construction categories	Construction category I	Construction category II	Construction category III	Construction category IV	Construction category V
400 m Standard Track as described under section 2.2 with 8 lanes and 8 straight lanes for 100 m and 110 m hurdles	1	1	1	–	–
400 m Standard Track as line 2, but with 6 lanes and 6 straight lanes	–	–	–	1	–
400 m Standard Track as line 2, but with 4 lanes and 6 straight lanes	–	–	–	–	1
Water ump for the steeplechase	1	1	1	–	–
Facility for long- and triple jump, with landing area at each end	2	2	1	2	–
Facility for long- and triple jump as line 6, runways in the same direction	–	–	–	1	1
Facility for high jump	2	2	1	2	1
Facility for pole vault with provision for landing area at each end	2	2	2	2	1
Facility for pole vault with runways in the same direction	–	–	–	1	–
Facility for discus- and hammer throw combined	1	1	1	–	1
Facility for discus throw	1	1	–	1 a)	–
Facility for javelin throw	2 b)	2 b)	2 b)	1	1
Facility for shot put	2	2	2	2	1
Provision of ancillary rooms (Chapter 4)	*	*	*	*	*
Full facilities for spectators	*	*	*	*	*
Warm-up area, comprising 400 m Standard Track with min. 4 lanes and min. 6 straight lanes (similar surface to the competitian track); throwing field for discus, hammer, javelin; 2 facilities for shot put	*	–	–	–	–
Warm-up area, comprising preferably a 200 m oval track each with 4 lanes and 4 straight lanes (similar surface to the competition track); throwing field for discus, hammer, javelin; facility for shot put	–	*	–	–	–
Warm-up area, comprising park or playing field in the vicinity for warm- up, preferably with Standard Track with 4 lanes and min. 6 straight lanes; throwing field for discus, hammer, javelin; facility for shot put	–	–	*	–	–
Warm-up area adjacent to the stadium: park or playing held	–	–	–	*	–
Ancillary rooms eg. for conditioning and physio-therapy, adequate space for athletes resting beween events, with area of min. [m^2]	250	200	150	200	–

* Essential Requirements
a) For large events a second facility outside the stadium, but in the same throwing direction is desirable
b) One at each end of the area

326 Table 1.5.3: Excerpt from IAAF – Track and Field Manual

Athletics arena
Type A (IAAF standard)
Scale 1:1,000

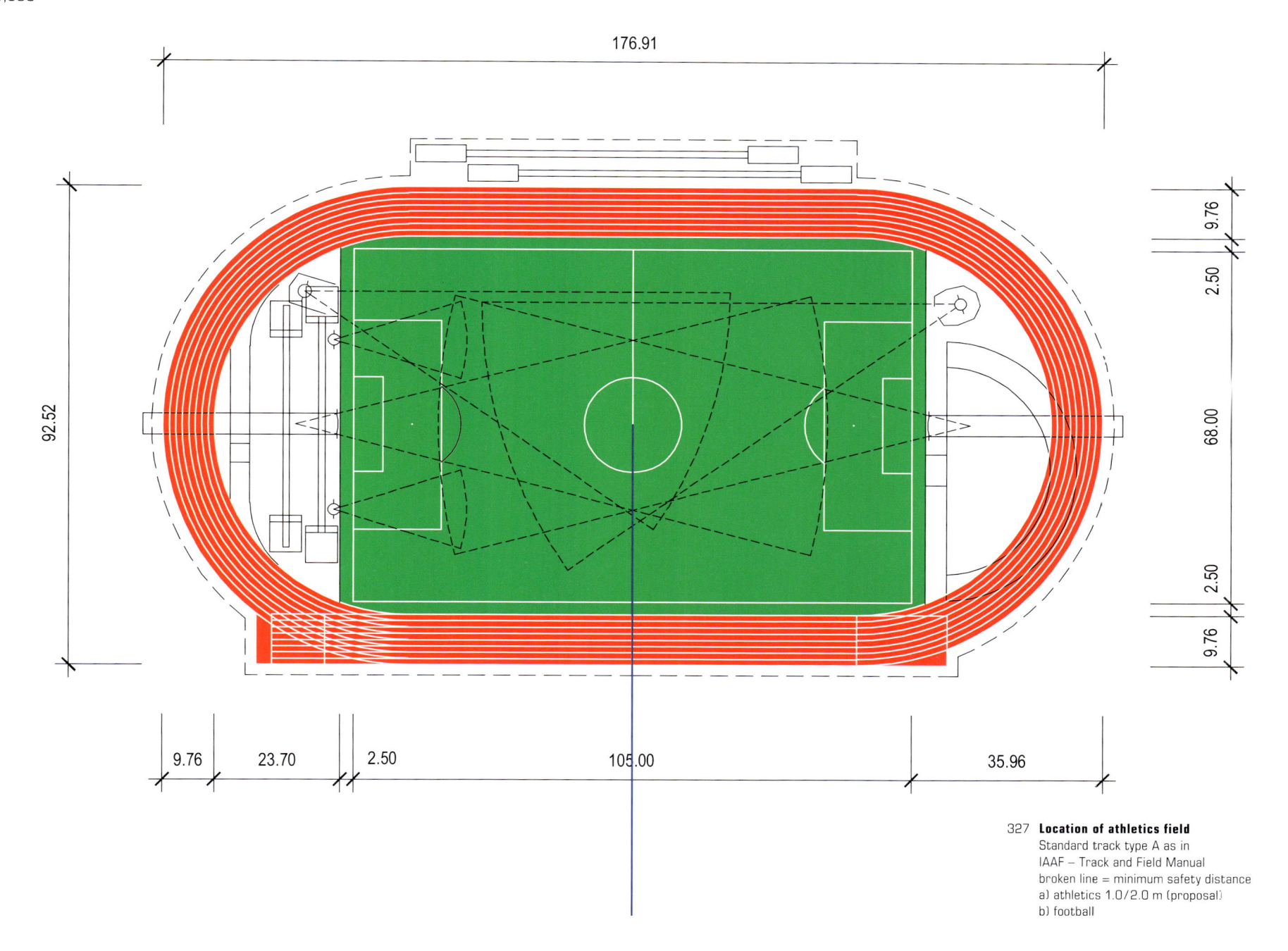

327 Location of athletics field
Standard track type A as in
IAAF – Track and Field Manual
broken line = minimum safety distance
a) athletics 1.0 / 2.0 m (proposal)
b) football

18 Playing fields

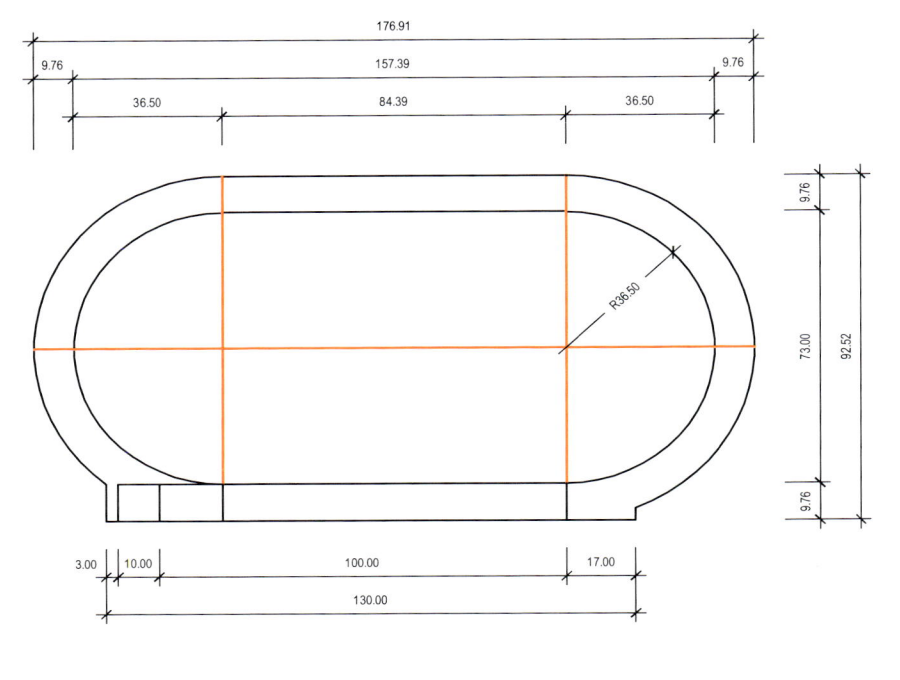

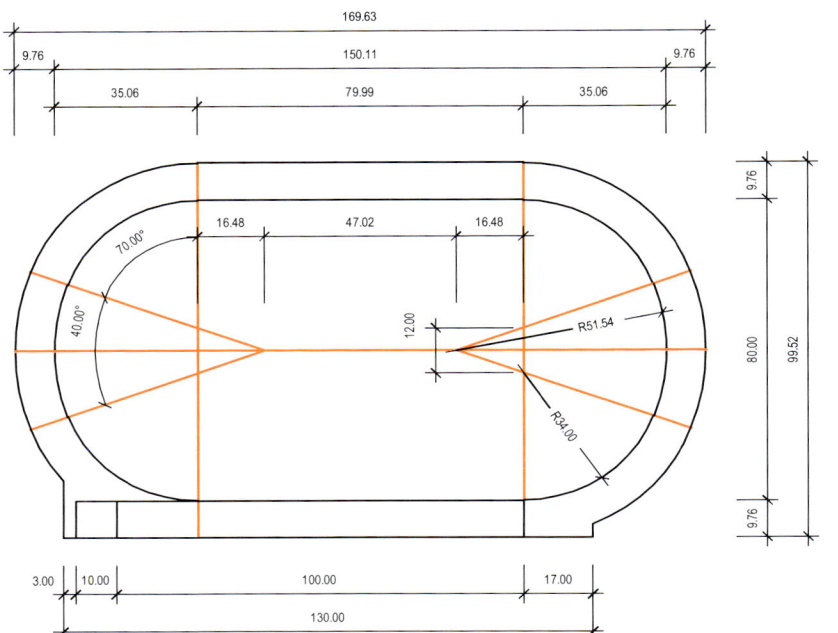

Form type 01 (IAAF)
IAAF standard 400-m running track

Geometry
Semicircle
Long sides
2 x each = 84.39 m
Curve radius
Radius = 36.50 m, 2 x
Running tracks
4 – 8 tracks
1.22 m wide each = 4 ft
Width = 9.76 total
Length = 400 m total
Running track
Safety distance (acc. to IAAF) = 28 cm lateral, outer track
Proposal by *Stadium Atlas*
At least 1.25 m circulating

Sprint/running track
(anticlockwise direction)
Length = 100 m
Starting zone = – 10 m
Run-out = + 17 m
Sprint track
Total = 130 m
Hurdling
Length = 110 m
Run-up = – 3 m
Orientation
in NS direction; in exceptional cases (acc. to IAAF) = in NNW or NNE direction

328 **System drawing, layout**
IAAF Standard 400-m track

Form type 02 (IAAF)
IAAF Double Bend Track 40/70

Geometry
Basket arch 40/70°
Long sides
2 x each = 79.995 m
Curve radius
Radius A = 51.543 m, 2 x
Distance + 16.485 m
Radius B = 34.000 m, 4 x
Running tracks
4 – 8 tracks
1.22 m wide each = 4 ft
Width = 9.76 total
Length = 400 m total
Running track
Safety distance (acc. to IAAF) = 28 cm lateral, outer track

Proposal by *Stadium Atlas*
At least 1.25 m circulating
Sprint/running track
(anticlockwise direction)
Length = 100 m
Starting zone = – 10 m
Run-out = + 17 m
Sprint track
Total = 130 m
Hurdling
Length = 110 m
Run-up = – 3 m
Orientation
in NS direction; in exceptional cases (acc. to IAAF) = in NNW or NNE direction

329 **System drawing, layout**
IAAF Double Bend Track 40/70

Form type 03 (IAAF)
IAAF Double Bend Track 60

Geometry
Basket arch 60°
Long sides
2 x each = 98.52 m
Curve radius
Radius A = 48.00 m, 2 x
Distance + 20.78 m
Radius B = 24.00 m, 4 x
Running tracks
4–8 tracks
1.22 m wide each = 4 ft
Width = 9.76 total
Length = 400 m total
Running track
Safety distance (acc. to IAAF) = 28 cm lateral, outer track

Proposal by *Stadium Atlas*
At least 1.25 m circulating
Sprint/running track
(anticlockwise direction)
Length = 100 m
Starting zone = –10 m
Run-out = +17 m
Sprint track
Total = 130 m
Hurdling
Length = 110 m
Run-up = –3 m
Orientation
in NS direction; in exceptional cases (acc. to IAAF)
= in NNW or NNE direction

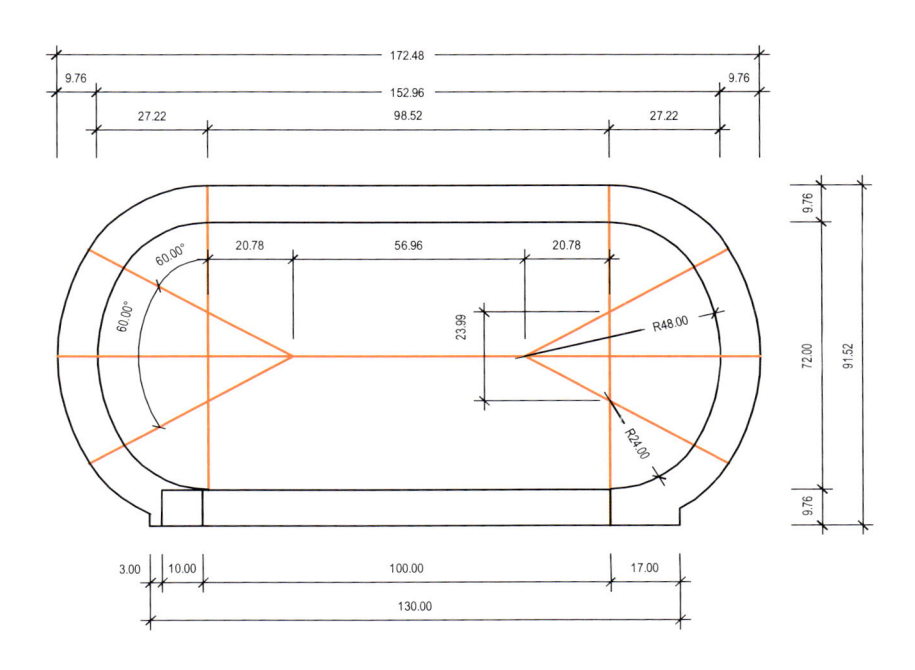

330 **System drawing, layout**
IAAF Double Bend Track 60

Form type 04 (IAAF)
IAAF Double Bend Track 74/53

Geometry
Basket arch 74/53°
Long sides
2 x each = 97,265 m
Curve radius
Radius A = 40,022 m, 2 x
Distance + 10,333 m
Radius B = 27,082 m, 4 x
Running tracks
4–8 tracks
1.22 m wide each = 4 ft
Width = 9.76 total
Length = 400 m total
Running track
Safety distance (acc. to IAAF) = 28 cm lateral, outer track

Proposal by *Stadium Atlas*
At least 1.25 m circulating
Sprint/running track
(anticlockwise direction)
Length = 100 m
Starting zone = –10 m
Run-out = +17 m
Sprint track
Total = 130 m
Hurdling
Length = 110 m
Run-up = –3 m
Orientation
in NS direction; in exceptional cases (acc. to IAAF)
= in NNW or NNE direction

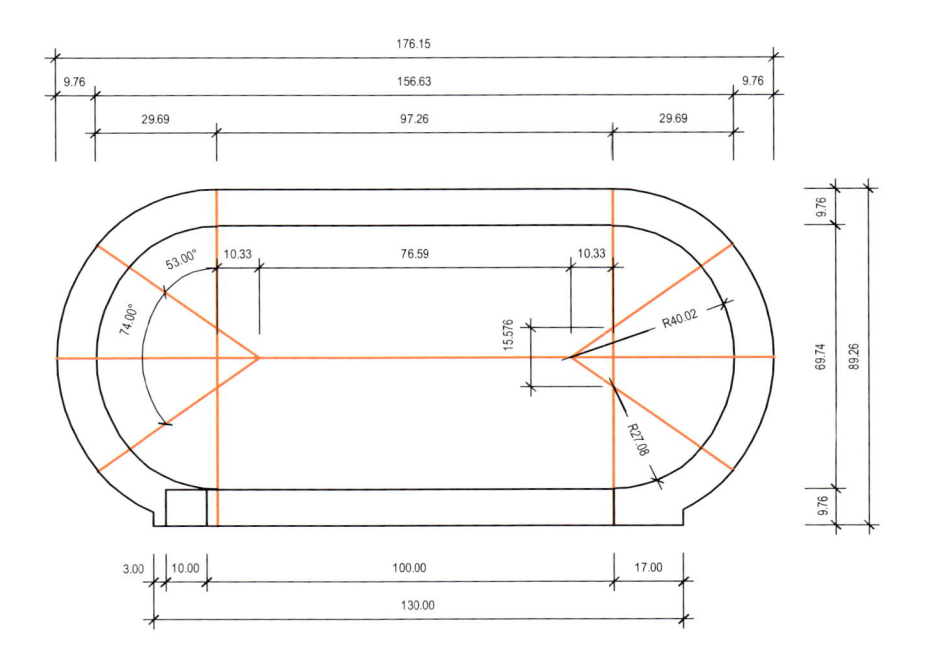

331 **System drawing, layout**
IAAF Double Bend Track 74/53

18 Playing fields

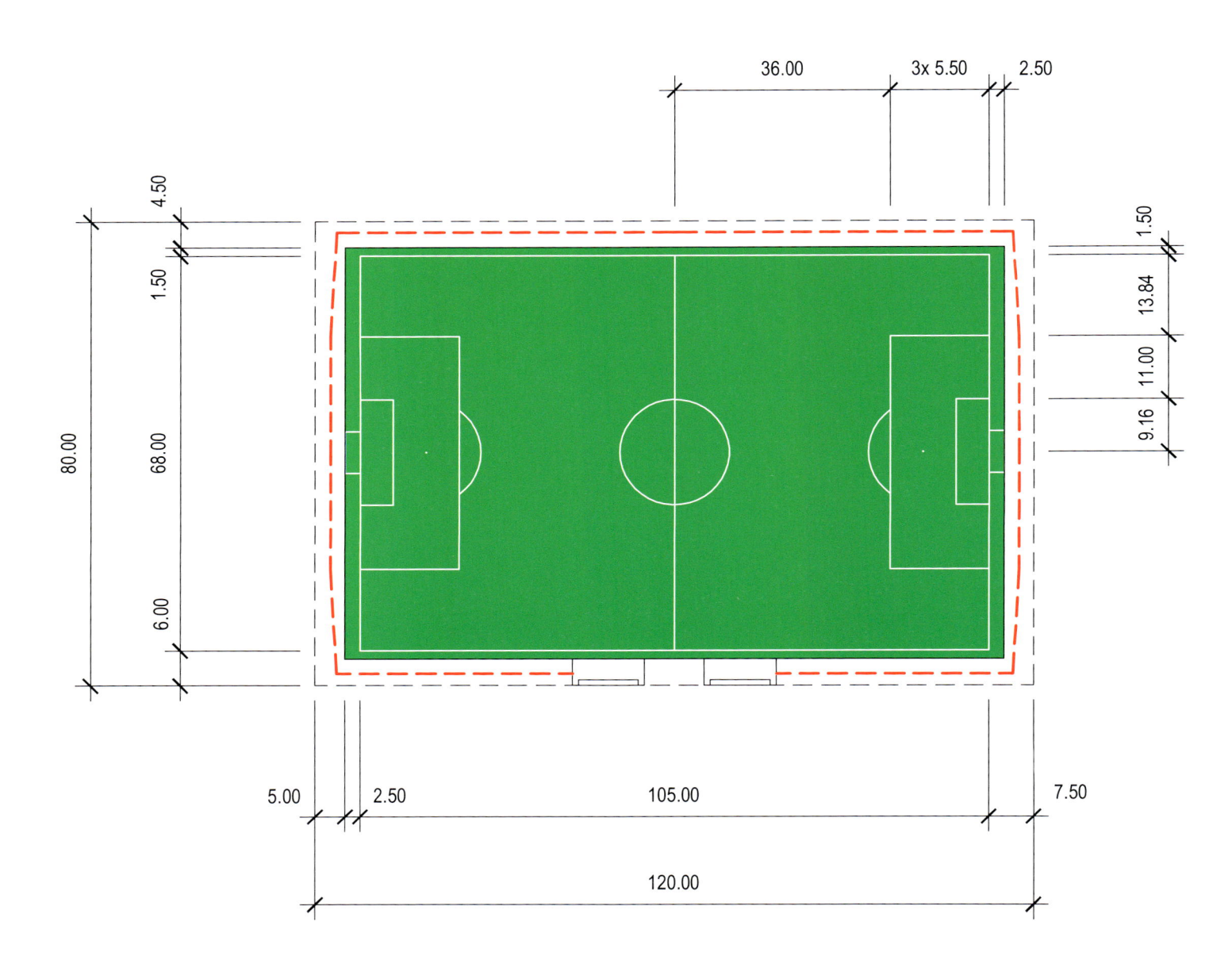

332 Field plan for football
FIFA/UEFA standard size: 105 x 68 m

Football playing field
(FIFA/UEFA standard) 68 x 105 m
Scale 1:1,000

Geometry
Rectangle
Pitch size
Long sides = 105 m
Short sides = 68 m
Turf side strip
Long sides = 1.5 m min.
Short sides = 2.5 m min.
Safety distance
Long sides = 6.0 m
Short sides = 7.5 m
Stadium interior
Minimum width = 80 m
Minimum length = 120 m

DIMENSIONS
Goal dimensions
$w = 7.32$ m
$h = 2.44$ m
Goal mouth
$w = 18.32$ m, (Goal + 2 x 5.50 m)
$d = 5.50$ m
Penalty area
$w = 16.50$ m, 2 x
$d = 16.50$ m
Penalty arch
$r = 9.15$ m
Penalty spot
$a = 2 \times 5.50$ m
Centre circle
$r = 9.15$ m
Corner flag
$r = 1.0$ m
$h = 1.5$ m
Players' bench
20 players = 10 m
Single seat
50 cm each
Managers' zone
$w = 10$ m + 2 x 1.0 m from touch line
$a = 1.0$ m distance
Bench
$a = 5.0$ m
Advertising hoardings (FIFA 1995)
Long side = 4.0 m
Goal area = 5.0 m
Corner area = 3.0 m
Vertical position
90° to the horizontal
Contact
www.fifa.com/de

Baseball playing field

(MLB standard) 335 x 400 yards

Scale 1:1,000

Geometry
Diamond

System dimensions
1 inch = 2.54 cm
1 foot = 30.48 cm
1 yard = 91.44 cm
1 yd = 3 ft = 36 in

Field dimensions
Foul Lines = 335 ft = 102.10 m (90 – 120 m)
Outfield = 400 ft = 121.90 m
Infield = 90 ft = 27.43 m
The three bases and the home plate are located in the corners of the diamond geometry.
Grass line (radius) = 95 ft = 28.96 m beyond which lies the outfield.

Stadium interior
Foul line + 18.30 m
Minimum width = 120.40 m

Safety Distance
Fair territory = 15 ft = 4.57 m
Foul territory = 10 ft = 3.05 m

Contact
http://mlb.mlb.com

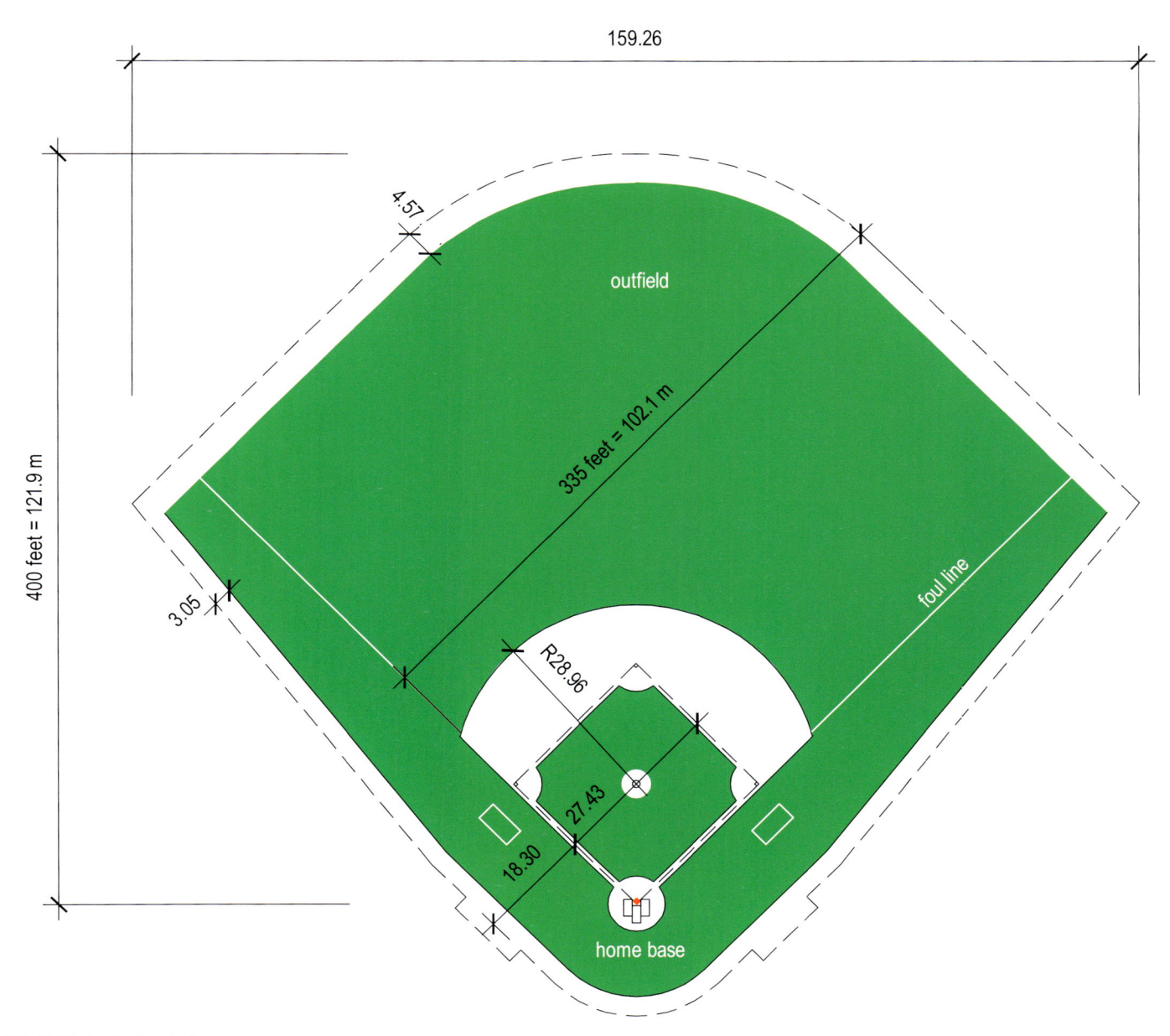

333 **Field plan for baseball**
MLB standard size: 120 x 160 m

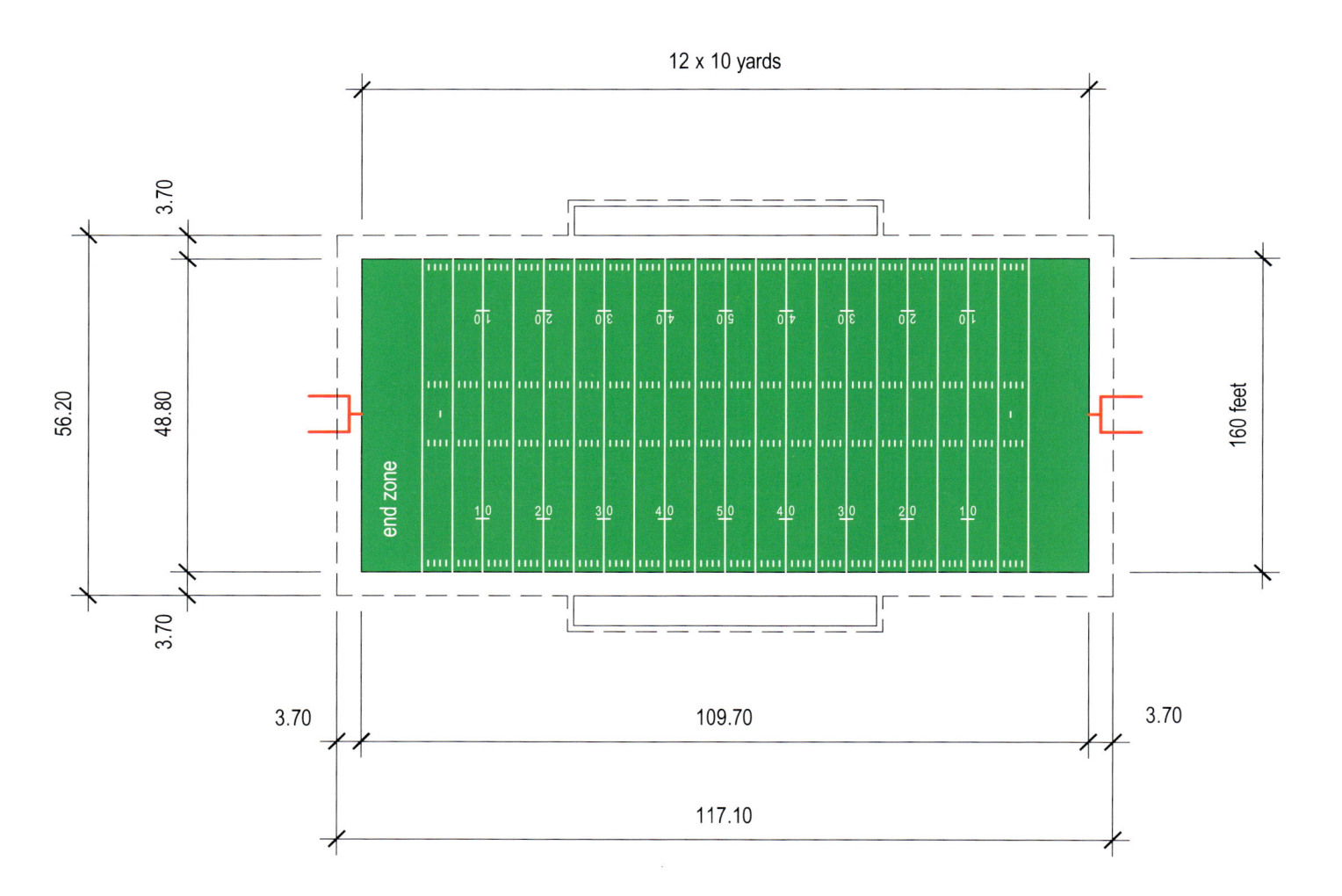

American football playing field

(NFL Europe standard) 53 x 120 yards
Scale 1:1,000

Geometry
Rectangle
System dimensions
1 inch = 2.54 cm
1 foot = 30.48 cm
1 yard = 91.44 cm
1 yd = 3 ft = 36 in
Field dimensions
Long sides = 120 yd = 109.70 m
Short sides = 53 yd = 48.80 m
Safety dimensions (safety margins)
Long sides = 3 yd = 1.8 – 3,7 m
Short sides = 3 yd = 1.8 – 3.7 m
Stadium interior
Minimum width = 67.20 m
Minimum length = 117.10 m
Goal measures (goal post)
Width = 18.5 ft = 5.6 m
Length = 20.0 ft = 6.1 m
Starting with 10.0 ft = 3.0 m
End zone
Depth d = 10 yd
Field markings d = 5 and 10 yd
With one-yard marks and inbound lines

PLAYERS' ZONE
Team/coach's box
w = 50 yards
d = 10 yards

Contact
www.nfleurope.com

334 Field plan for American football
NFL Europe standard size: 117 x 56 m

Rugby playing field

(IRB standard) 70 x 100 m
Scale 1:1,000

Geometry
Rectangle
Field dimensions
Width = not over 70 m
Length = not over 100 m + in-goal area 2 x 22 m
Safety dimensions
Circulating = 3.5 – 5.0 m
Stadium interior
Minimum width = 80 m
Minimum length = 154 m

Line markings
22-m line = distance from goal line
Broken line
10-m line = distance to halfway line
Touch line
5-m distance to touch line
Irregular line
a) 1-m length, 6 x
5 m distance to goal line, lateral 5/15 m, 2 x in front of goal post
b) 1-m length, 5 x
15 m distance to touch line, in halfway, 10-m and 22-m line

GOAL MEASURES
Goal posts
w = 5.6 m
h = 3.4 m (min. + 3.0 m)
Flags
h = 1.20 m
8 x in the corners of the in-goal area
6 x 2-m distance to touch line

Contact
www.rugby-verband.de

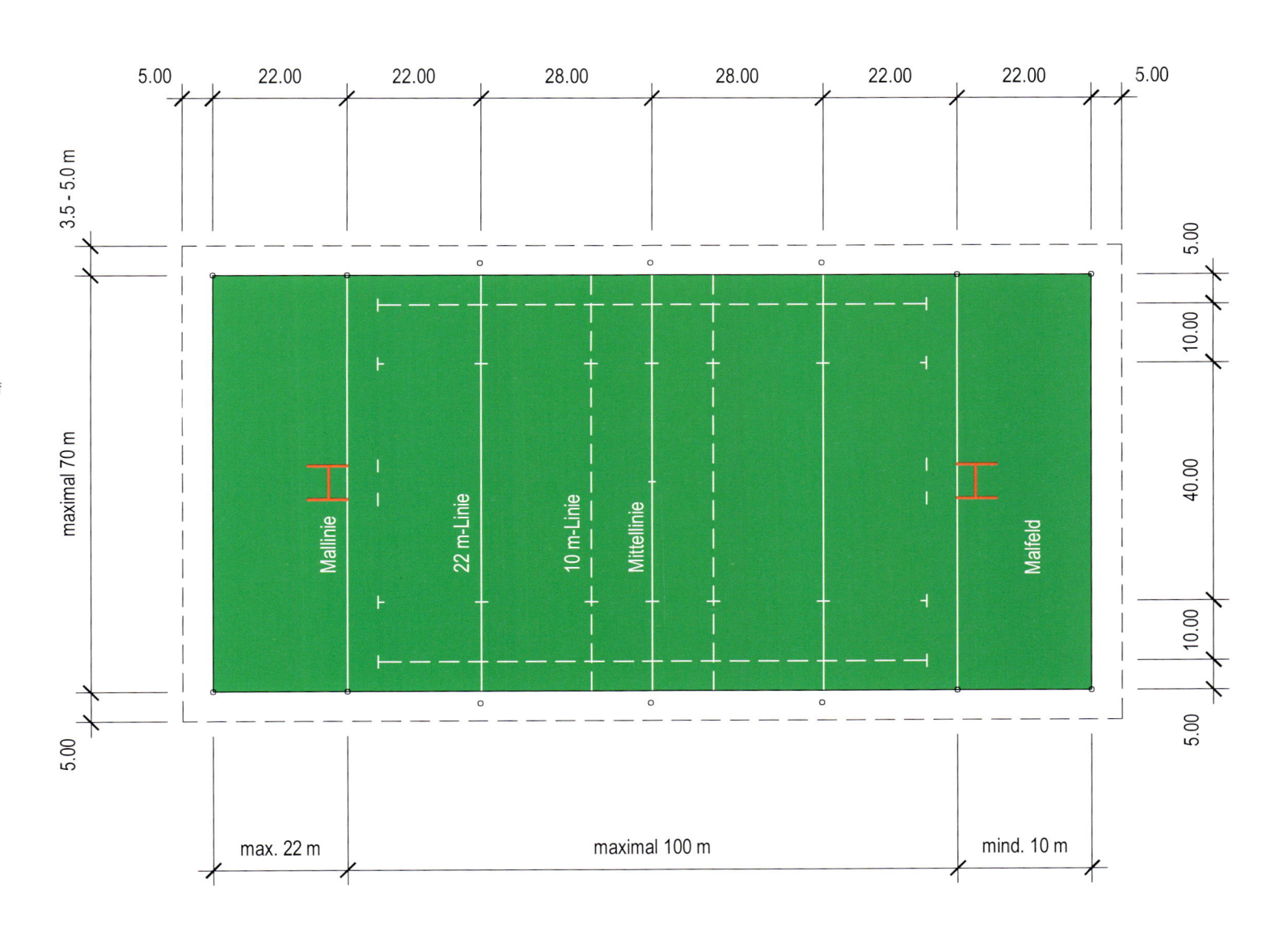

Orientation 187

335 **Field plan for rugby**
IRB standard size: 70 x 100 m

18 Playing fields

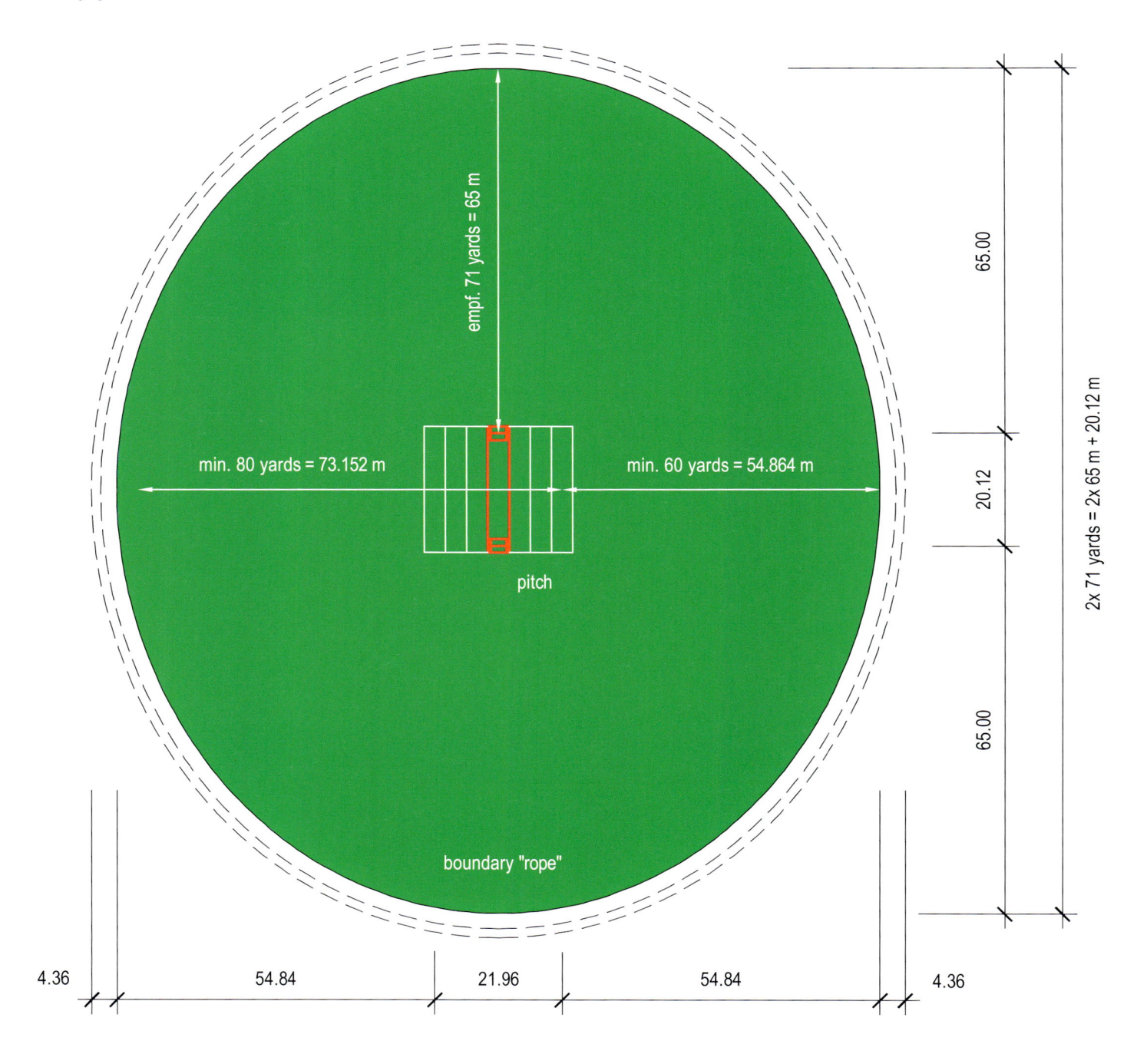

Cricket playing field
(ICC standard) 65-m radius
Scale 1:1,000

Geometry
Oval/circle

System dimensions
1 inch = 2.54 cm
1 foot = 30.48 cm
1 yard = 91.44 cm
1 yd = 3 ft = 36 in

Field dimensions
Recommended = 65.00 m = 71 yd

Safety dimensions
Circulating = 4.34 m = 4.75 yd
Boundary 'rope' = 2.74 m = 3.00 yd
Moat/netting = 1.60 m = 1.75 yd

Stadium interior
Minimum width = 140 yd = 128 m (without securing)
Minimum length = 142 yd = 129.85 m

Goal measures
Pitch/cricket table = seven tracks
Single length = 20.12 m + 2 x 1.22 = 74 ft = 22.56 m
Single width = 12 ft = 3.65 m
Bowling crease = 8 ft = 2.64 m
Wicket distance = 22 yd = 20.12 m

Wickets
Width = 28 ft = 71.0 cm
Length = 9 ft = 22.8 cm

Contact
www.icc-cricket.com

Deutscher Cricket Bund – DCB (German Cricket Federation)
Form and size of a cricket field are not mandatory: an oval with a diameter of 100–140 m enclosed by a white line, a rope, a fence or small flags.

336 Field plan for cricket
ICC standard size: 130 x 150 m

Orientation 189

Australian Football playing field
(AFL standard) 145 x 165 m
Scale 1:1,000

Geometry
Oval
System dimensions
1 inch = 2.54 cm
1 foot = 30.48 cm
1 yard = 91.44 cm
1 yd = 3 ft = 36 in
Field dimensions
Width = 120 – 170 yd
Boundary line = 110 – 155 m
Length = 150 – 200 yd = 135 – 185 m
AFL Germany standard
145 x 165 m playing field
Safety dimensions (Surrounding gap)
Circulating = 3.0 m
Goal measures
Goal post h = 2 x 15.0 m
Behind post h = 2 x 9.0 m
Goal distance = 6.4 m each
Field markings
Centre circle = 9.10 ft = 3.0 m (diameter)
Square (centre) = 50 yd = 45.72 m (Lateral length)
Goal area = 6.4 m (width) = 9.0 m (distance)
Contact
http://afl.com.au

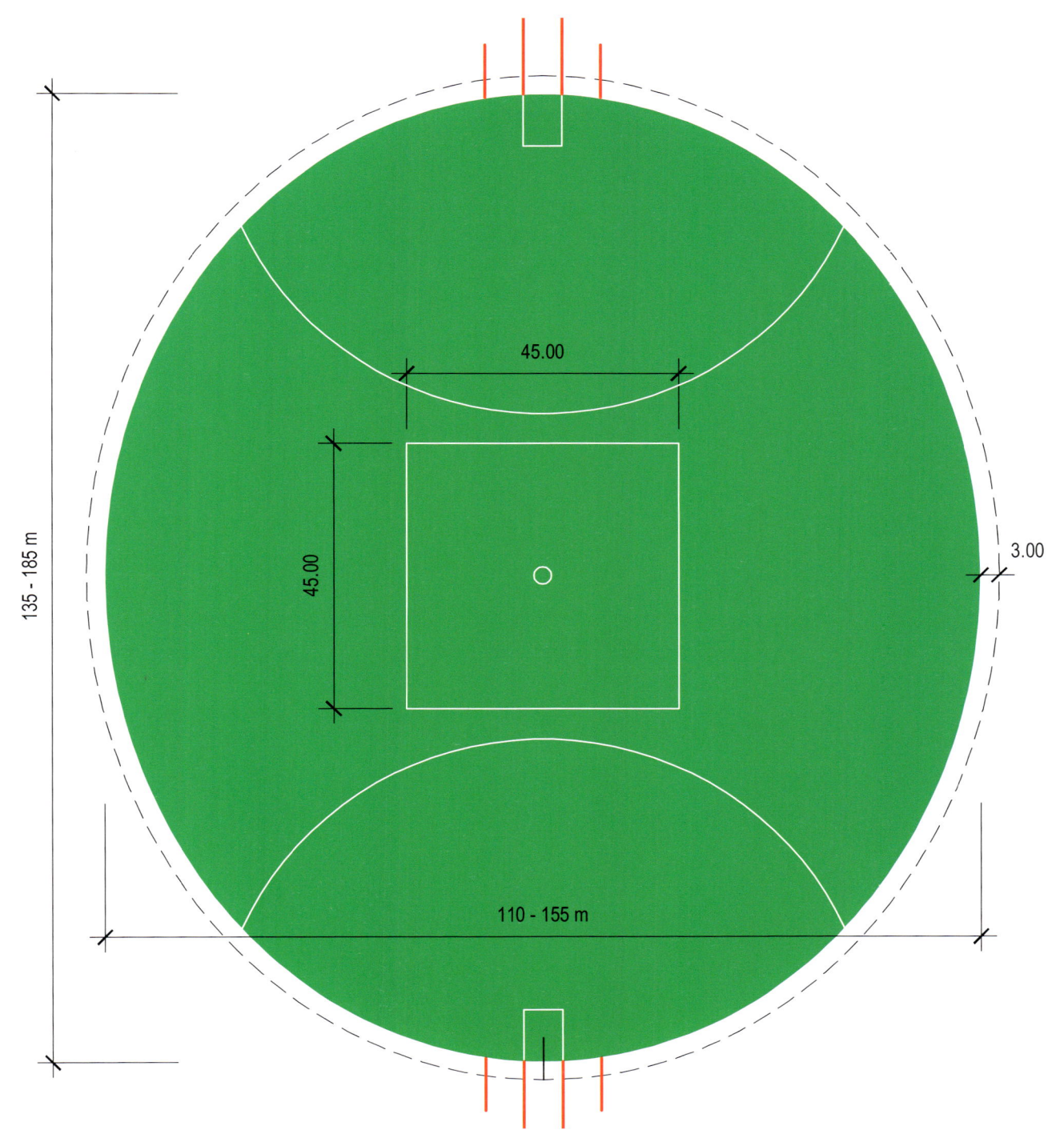

337 **Field plan for Australian football**
 AFL standard size: 145 x 165 m

18 Playing fields

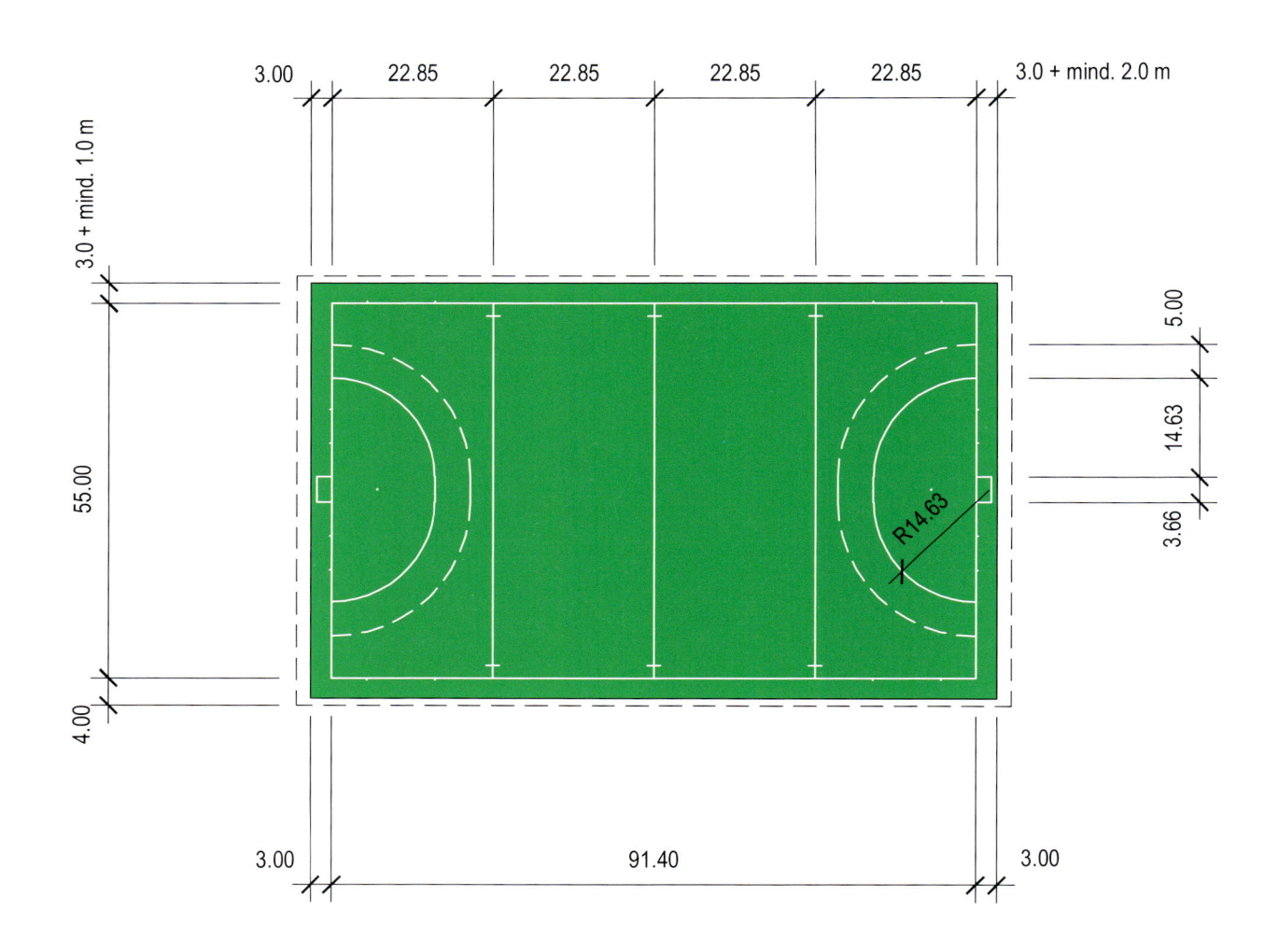

Hockey playing field
(FIH standard) 60 x 100 yards
Scale 1:1,000

Geometry
Rectangle
System dimensions
1 inch = 2.54 cm
1 foot = 30.48 cm
1 yard = 91.44 cm
1 yd = 3 ft = 36 in
Field dimensions
Long sides = 100 yd = 91.4 m
Short sides = 60 yd = 55.0 m
Turf side strip
Long sides = 3.0 m
Short sides = 3.0 m
Safety dimensions
Long side = minimum 4.0 m from short-side touch line
Short side = 5.0 m min.
Stadium interior
Minimum width = 63.00 m
Minimum length = 101.40 m
Goal measures
w = 4 yd = 3.65 m
h = 7 ft = 2.14 m
Penalty shot (7-m kick)
Distance d = 7 yd = 6.40 m
Shooting circle
Radius r = 16 yd = 14.63 m

338 Field plan for hockey
FIH standard size: 60 x 100 yd

Playing fields for indoor sports

(standard dimensions)
Scale 1:1,000

Ice hockey
(IHF standard)

Playing field dimensions
Width = 26 – 30 m
IIHF standard = 29 – 30 m
Length = 56 – 61 m
IIHF standard = 60 – 61 m
Corner radius = 7 – 8.5 m

Field markings
Centre circle radius = 4.50 m
End-zone circle radius = 4.50 m

Goal measures
Goal height = 1.22 m
Goal width = 1.83 m
Perimeter height = 1.17 – 1.22 m

Safety glass
End zone = 1.60 – 2.00 m
Long sides = 0.80 – 1.60 m

Contact
www.iihf.com

339 **Field plan for ice hockey**
IIHF standard size: 30 x 60 m

Indoor hockey
(FIH standard)

Playing field dimensions
Width = 18 – 22 m
Length = 36 – 44 m

Goal measures
Goal width = 3.0 m
Goal length = 2.0 m

Penalty spot
Distance = 7.0 m
Penalty area circle (radius) = 9.0 m (goal post)

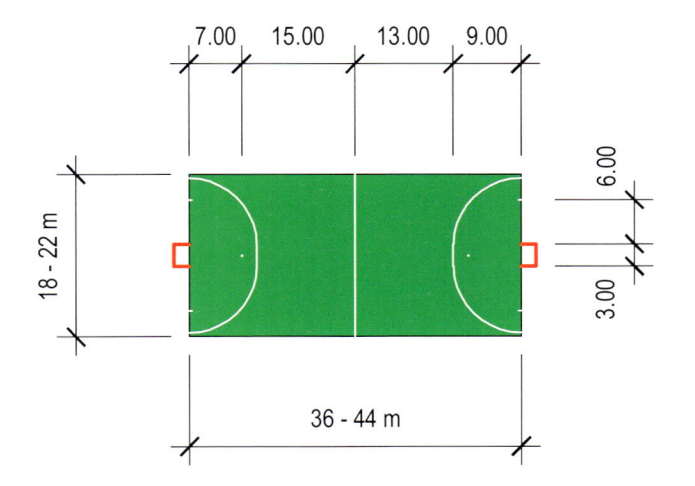

340 **Field plan for indoor hockey**
FIH standard size: 22 x 44 m

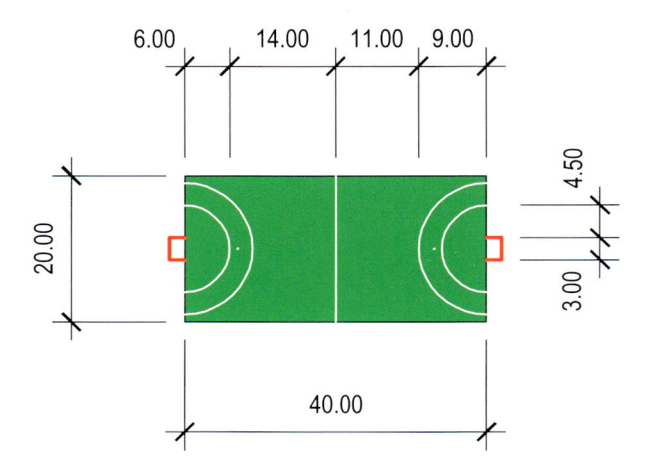

18 Playing fields

341 Field plan for handball
IHF standard size: 20 x 40 m

Handball
(IHF standard)

Playing field dimensions
Width = 20 m
Length = 40 m
Goal measures
Goal depth = 3.0 m
Goal height = 2.0 m
Field markings
Penalty kick distance = 7.0 m
4-m line = 4.0 m
9-m line radius = 9.0 m (Field axis)
Goal-area line = 6.0 m (Field axis)
Contact
www.ihf.info

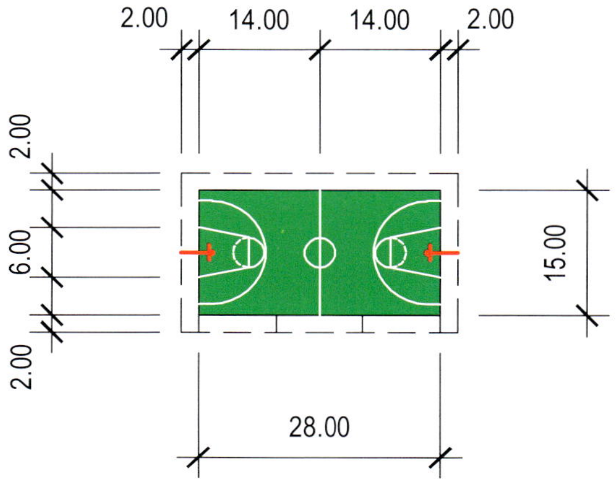

342 Field plan for basketball
FIBA standard size: 15 x 28 m

Basketball
(FIBA standard)

Playing field dimensions
Width = 15 m
Length = 28 m
Safety dimensions
Circulating = min. 2.0 m
Field markings
Free-throw line = 5.80 m (baseline width 6.0 m)
Free-throw circle radius = 1.80 m
Centre-circle radius = 1.80 m
3-point area = 6.25 m (from centre of basket)
Minimum ceiling height = 7.0 m
Basket/board
Top edge of basket = 3.05 m
Board (h x w) = 1.05 x 1.80 m
Distance = 1.20 m (from baseline)
Team zone
Distance = 5.0 m (from centre line)
Contact
www.bbsr.de/regel

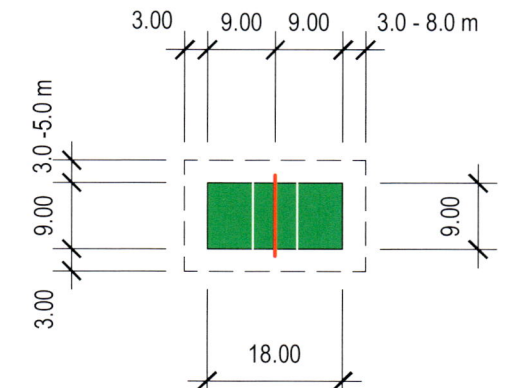

343 Field plan for volleyball
FIVB standard size: 9 x 18 m

Volleyball
(FIVB standard)

Playing field dimensions
Width = 9 m
Length = 18 m
Activity area
Service zone = + 3.0 – 8.0 m (short side)
Attack zone = + 3.0 – 8.0 m (long side)
Field markings
Attack line (behind net) = 3.0 m
Top edge of net
Men 2.43 m
Women = 2.24m
Mixed = 2.35 m
Contact
www.fivb.ch

Orientation 193

Tennis
(DTB standard)

Playing field dimensions
Width/single = 27 ft = 8.23 m
Width/doubles = 36 ft = 10.97 m
Length = 78 ft = 23.77 m
One side = 39 ft = 11.89 m
Safety dimensions
Long side = 10 –12 ft = 3.04 – 3.66 m
Short side = 18 – 21 ft = 5.50 – 6.40 m
Field markings
Service line = 21 ft = 6.40 m (parallel to net)
Net
Height (centre) = 3 ft = 91.4 cm
Protrusion = 3 ft = 91.4 cm
Height (outside) = 3,5 ft = 1.07 m
Contact
www.dtb-tennis.de

Badminton
(IBF standard)

Playing field dimensions
Width/single = 17 ft = 5.18 m
Width/doubles = 20 ft = 6.10 m
Length = 44 ft = 13.40 m
One side = 22 ft = 6.70 m
Field markings
Front service line = 1.98 m (parallel to net)
Rear service line
Double = + 3.96 m
Single = + 0.76 m
Net
Height (centre) = 1.55 m
Protrusion = 0 m
Minimum ceiling height = 9.0 m
Note
The IBF Event organisation Manual 11.2 of Sept 1993 (Section 11 – Seating) states as follows: 'There may be more than one arrangement, eg one at an early stage, a second at the semi-finals, and a third at the finals. Seating on the same level should never be more than two rows deep.'
Contact
www.internationalbadminton.org

Boxing
(WBA standard)

Dimensions
Box ring size = 16 – 24 ft
Standard = 20 ft = 6.1 x 6.1 m
Platform height = 1.0 m
Size = around 7.50 – 8.0 m (square)
Ropes 4 x 46, 76, 107 and 137 cm (distances from the ground)
Contact
www.boxen.com

Gymnastic
(FIG standard)

Playing field dimensions (example)
Width = 23.0 m
Length = 47.5 m
(individual dimensions in consultation with organizer)
Contact
www.fig-gymnastics.com

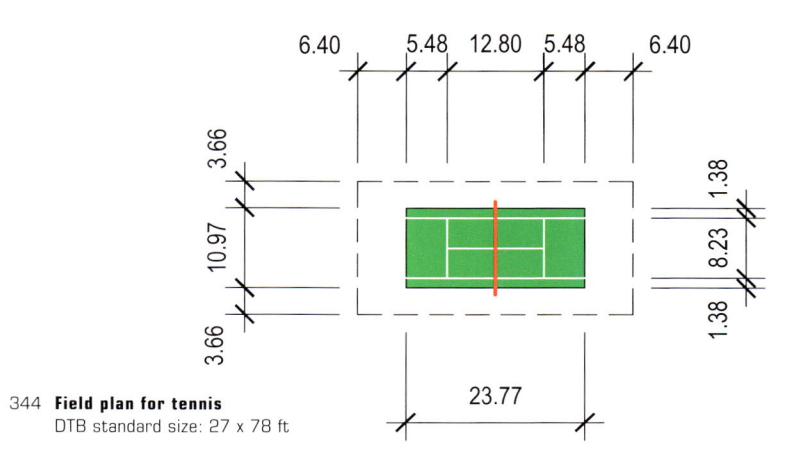

344 **Field plan for tennis**
DTB standard size: 27 x 78 ft

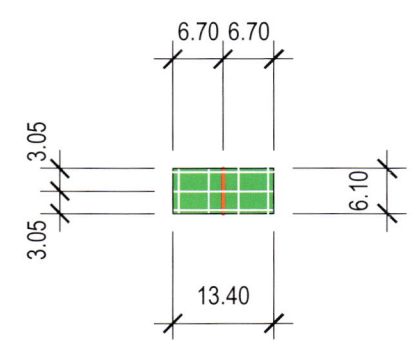

345 **Field plan for badminton**
IBF standard size: 17 x 44 ft

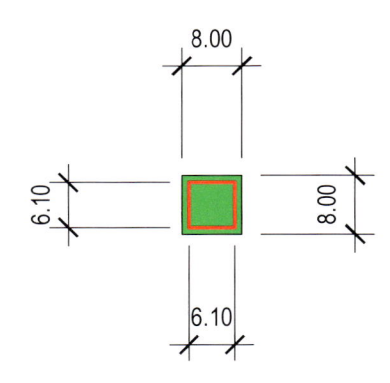

346 **Plan of a box ring**
WBA standard size: 8 x 8m

18 Playing fields

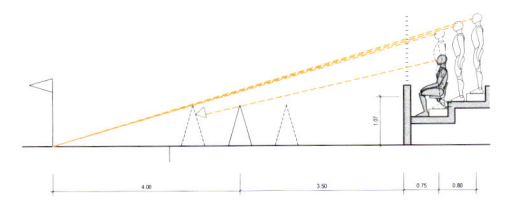

350 **Sight obstruction by advertising hoarding**
at the short end of the stadium between 3 to 5 m

5-m position:	no obstruction, as terrace front edge is at least 7.5 m away
3-m position:	problematic
4-m position:	no obstruction, but according to FIFA only on the long side

Note: Starting point 'first row' – Riser height is generally determined by the main stand. If the short side has to be adjusted at 3/5 m (7.5 m), a height offset has to be realized in the corner area.

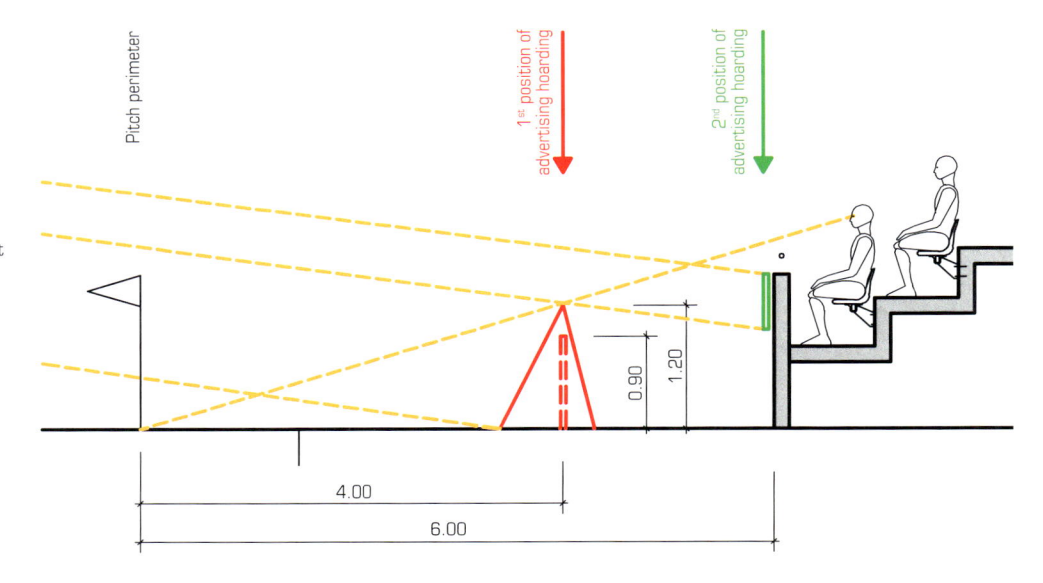

351 **Position 1 and 2 for advertising hoarding**
Double or 'split' hoardings

| Position 1: | standard FIFA advertising position (see above 3 m/4 m/5 m-distance) |
| Position 2: | depending on the camera position on the opposite side |

Note: During Bundesliga increasingly higher hoardings are set up. Instead of 90 cm, a new height of 1.20 m emerges as standard.

Attention: The sightline profile should reflect this development for new plans. Otherwise sight obstructions in the first rows toward the touch line should be expected!

347 **Advertising perimeter board as 'A'-board**
348 **Second (rear) element of the XXL-board**
349 **Visualization of XXL-board**
(APA Co.)

Minimum angle of view at the advertising hoarding

The first but unavoidable sight obstructions for a grandstand geometry are the perimeter boards. Their position is laid down by the sports associations, as by FIFA for the Football World Cup 2006:

'Approval is given for mounting perimeter boards (90 to 100 cm high) around the playing field in one single continuous row. FIFA marketing partners are entitled to install advertising hoardings in a vertical position, i.e. at 90° to the horizontal. For television viewing purposes, hoardings may also be installed above the ground up to a certain height determined by FIFA.'[2]

Individual height provisions may be admissible for some large sports events, but they remain highly questionable when it comes to clarifying general initial parameters for sightline calculation since the height/position of a hoarding represents the most important regulator for the profile of a stand.

In any case, the first sightline *should/must* run above the advertising hoarding, resulting in a general elevation of the first row by 30 cm, at a position of 4 m. Depending on the positioning of a perimeter board, the first sightline angle is:

Long side: $h/a = 0.90 \text{ m}/4.0 \text{ m} = \text{arc tan } 12.7°$
Short side: $h/a = 0.90 \text{ m}/5.0 \text{ m} = \text{arc tan } 10.2°$
$h/a = 0.90 \text{ m}/3.0 \text{ m} = \text{arc tan } 16.7°$

Note: The sketch on the left illustrates the problem of 3-m hoardings in the corners. They cause viewing problems after the conversion of standing places at the short ends into seats.

The further away the standard hoarding is positioned, the shallower the first vertical viewing angle, or the lower the terrace base point (at least 2.0 m intermediate space between perimeter board and terraces):

– at 4.0 m = 12.7° → upper edge of finished floor h = 30 cm
– at 4.7 m = 11.3° → upper edge of finished floor h = 20 cm
– at 5.0 m = 10.2° → upper edge of finished floor h = 15 cm

System variants for perimeter board advertising

In modern sports and event venues, perimeter boards are the most prominent form of sponsoring and advertising. From the lettered plate fixed onto a fence to the most up-to-date and costly technology such as LED-boards, there are numerous different perimeter board systems available.

Due to constant optimization and development, a broad spectrum of systems has emerged, suited to the conditions on site as well as the respective financial means.

Standard perimeter boards at the box tier (lower edge, upper tier)
Permanently installed advertising space: plates fixed to railings, fences or especially manufactured support structures.

In modern sports venues this type of advertising is placed as 'second row' before/behind the railing perimeter. The classic advertising perimeter board in front of the first-row barrier of a spectator tier should be positioned at a height of at least 90 to 105 cm.

Perimeter board (rotating board, LED-boards)
A-boards: advertising display inclined on both sides with individually variable angles.
L-boards: vertical advertising, almost reflection-free, usable for advertising also at the rear side.
90°-boards: combination of vertical and inclined advertising space, universal use, or as combi-boards with specially developed folding mechanism allowing individual adaptation to on-site conditions (sunlight, rain, floodlight).

XXL or Power Pack boards
One option to enlarge board advertising without constricting the field of vision of spectators is the XXL or Power Pack board (APA Co.). 'Two banners are positioned one behind the other, giving the impression of a single large area without constricting the view for fans and spectators.'[3] Exact measuring and calculating produces this effect, which can be appreciated by television audiences only from the vantage point of the leading camera.

Static radial perimeter boards® (APA Co.)
Compared to regular perimeter boards and at the same height, radial boards offer more advertising space with a curvature of around 40%. They appear larger and more dominant than regular boards. The curvature of the radial perimeter board® is adapted to the relevant sports venue. Camera angle and distance and the distance of spectators to the board are individually calculated. The visible advertising surface is significantly increased – an added advantage for sponsors, without impeding spectators.

Additional advertising measures
3D-carpets or Roll-up: By means of a specially calculated distortion of a sponsor's logo, an impression of a vertically standing, three-dimensional object is created on the television screen.

Similar to the XXL-board, optimum viewing is restricted to television images from the leading camera viewpoint.

Perimeter board advertising,
circulating the pitch in a single continuous row, may not produce reflections detriment to television cameras and may not represent a structural obstacle for evacuation in case of emergency. (FIFA)

Vertical position 90° to the horizontal axis

FIFA advertising boards (1995) Goal line 5.0 m
Touch line 4.0 m
Corner area 3.0 m

352 **'Roll-up'-board at truck entrance**
Mönchengladbach Stadium
353 **'Roto'-board**
Dortmund Stadium
354 **Advertising boards at the running track**
at ISTAF Athletics Berlin

Chapter 19

355 **Moat as field safety measure**
Müngersdorfer Stadion, Cologne (old)

356 **'Geral' (portug.) as field safety measure**
Stadium Maracana, Rio de Janeiro, Brasil

357 **Volksparkstadion Hamburg**
Final of German Championship 1928
HSV vs. Hertha BSC

Securing of the playing field
The role of the 'first row' and securing of the stadium interior toward the grandstand

Significance of the 'first row'

Safety measures applied to the 'first row' significantly affect the sightlines of the viewing accommodations ascending behind it. The following principles should be considered when deciding on the planning parameters:

The higher the starting point of a sightline profile, the steeper the rise of the stand.

The closer the base point lies to the point of focus, the steeper the spectator's view needs to be slanted downward.

The steeper the view, the sooner a stand will reach its maximum gradient.

Note: The options of securing the pitch are to be coordinated between planners, stadium operator and security staff very closely in the build-up of construction. The requirements stated in the invitation to tender should also be weighed up against the geometric consequences for the calculation of the sightline.

Securing measures for the playing field

Option 1: Elevating the first seating row

Option 2: Insurmountable fences, permanent or dismountable

Option 3: Moat difficult to overcome or combination of fence and moat

Option 4: Presence of police / security staff

Option 5: Double-fence system, two successive fence lines (only in exceptional cases)

Option 6: 'Wembley' system, horizontal, flexible intercepting fences (only in exceptional cases)

Option 7: Lowering of the service ring (only in exceptional cases)

Principles of pitch securing

The general safety objective of such measures is documented clearly in the FIFA security guidelines. Synonymously, DFB guidelines on the improvement of security at Bundesliga matches state:

The interior area is to be secured against unauthorized access.[1]

Players and referees are to be protected against spectator incursion of the playing field. The Technical Recommendations by FIFA for the construction and modernization of stadia state that a separation of spectator and field area should ideally not be realized through fencing. EN/DIN 13200-3 cites this advice by UEFA: 'Safety requirement allowing, a civilized and pleasant atmosphere should be created in a stadium without the help of fences and partitionings'.[2]

The UEFA 'Stadia List' deems the setting up of fences undesirable. On top of that, all security measures are to be approved by local authorities.

The German regulations state that the interior space and auditorium have to be fitted with a barrier at least 2.20 m high (in sports stadia with more than 10,000 spectator places).[3] In comparison, another clause explicitly states that in front of seating places no barriers are necessary; consultation with the authorities responsible for public safety provided, as part of a coordinated safety concept.

Note: The DFB guidelines for the improvement of safety (§9) strongly demand that no elements (trash bins, floor elements, etc.) in spectator areas should be removable for use as missiles.

British stadia

Every country seems to have a singular approach to stadium construction and different measures to secure the playing field are favoured. The above mentioned principle to hinder unauthorized persons from entering the stadium central area is equally presupposed by all national football associations.

Pure football stadia are often compared to the examples of British stadium geometries, regarding their proximity to the pitch. The live experience of being closely involved in the action on the field is the main attraction for many football fans worldwide. This not only entails a spatial immediacy in height and distance but, above all, an emotional experience. The geometric-mathematical relationships of the four sightline parameters clearly show that such proximity ensures moderate gradients and quality sightlines for the terraces.

The English term 'hooligan' has become synonymous worldwide for violent football fans. At the same time, phenomenon is not a justification for setting up the 'first row'. British stadia, in particular, are distinguished by their reduced securing of the stadium enclosure, the pitch perimeter barriers.

358 **Pitch perimeter securing at Anfield Road Stadium**
FC Liverpool
359 **Pitch perimeter securing at Stade de Suisse**
Berne
360 **Pitch perimeter securing at Estadio de Guimares**
EURO 2004 Portugal

19 Securing of the playing field

Option 1

364 **Elevation – 2.50 m**
The railing height of 2,50 m results in a first sight angle around 10° steeper as possible.

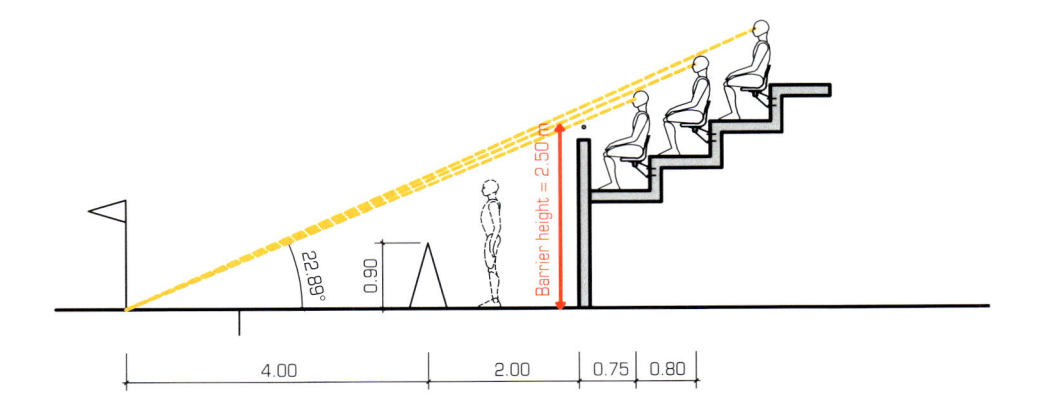

Option 2

365 **Fencing – 2,20 m**

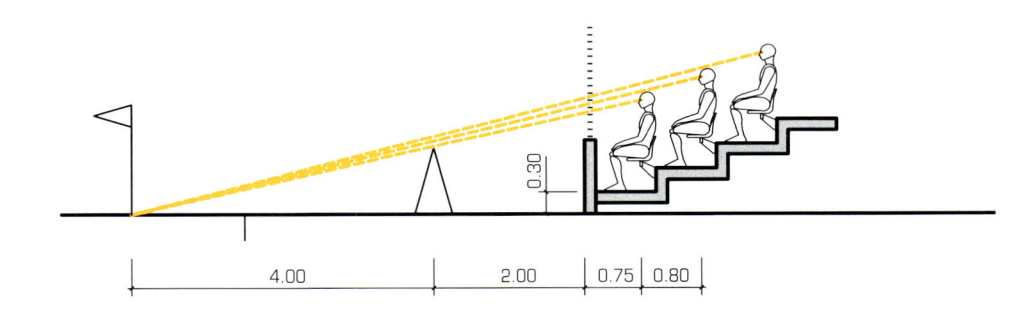

Option 3

366 **Moat – 1,80 m**

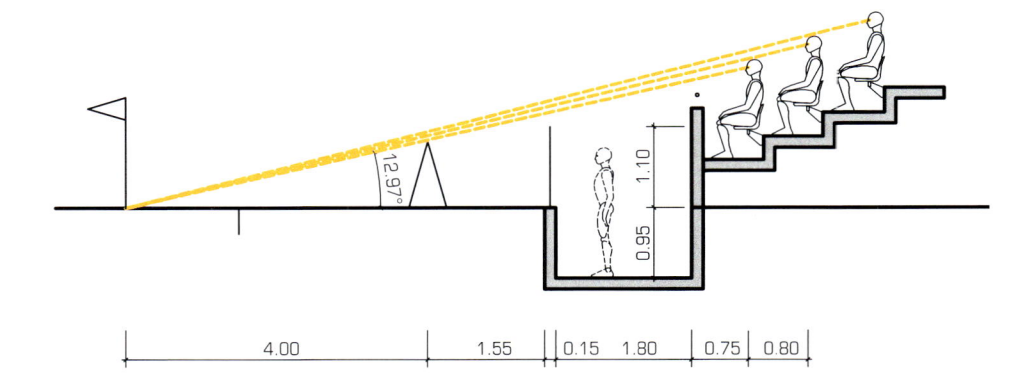

361 **Elevation by 4.45 m as securing measure**
World Cup Stadium Leipzig
362 **Fencing as securing measure**
Westfalenstadion Dortmund
363 **Moat/turf as securing measure**
World Cup Stadium Gelsenkirchen

Principles of pitch securing   199

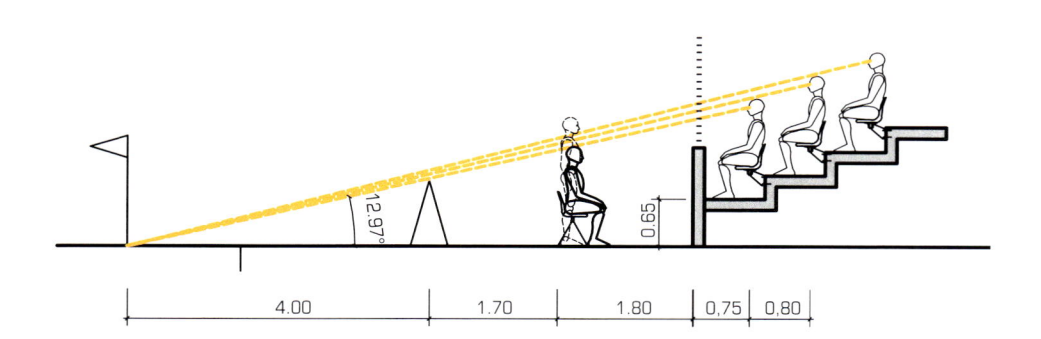

Option 4

367 **Stewards – 7.50 m**

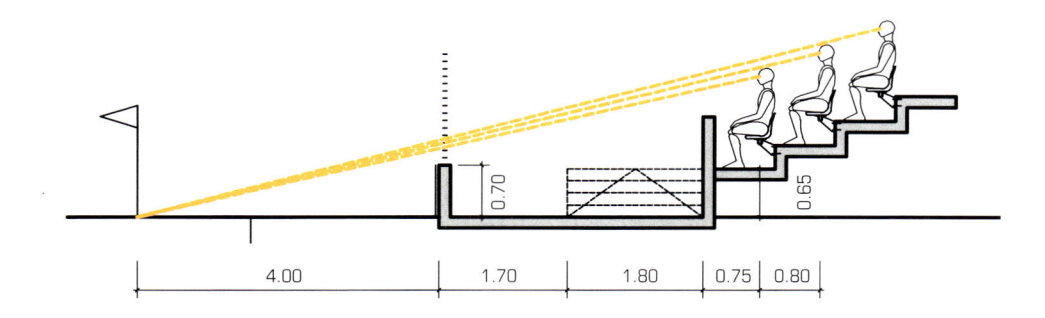

Option 5

368 **Double fence – 7.50 m**
(in exceptional cases)

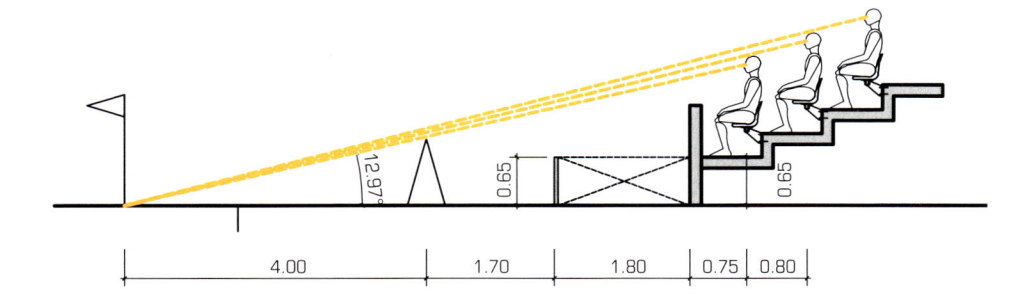

Option 6

369 **'Wembley' – 1.80 m**
(in exceptional cases)

370 **'Vario Fence'**
RheinEnergieStadion, Cologne
371 **'Fence'** (right: building site fencing)
World Cup Stadium Stuttgart
372 **'SecuFence'**
ConFedCup 2005, Hanover

Option 1: Elevation of the 'first row'

The Technical Recommendations of FIFA state as first option for pitch securing the elevation of the first row:

'... *up to a level which makes it impossible or at least improbable for spectators to invade the pitch.*'

A height of 2.50 m is strictly recommended in the FIFA Technical Recommendations for World Cup 2006. The reference point for this measure is not further defined. DFB safety guidelines consider raising the first seating row to at least 2.0 m above pitch level to be sufficient for securing the central area (former DFB reference height = seating tread, compare to update).[4]

By excluding athletics tracks, the grandstand structure has moved very close to the pitch and the terraces with their required capacities have reached the limits of admissible steepness.

Definition of the measure 'raising the first row' becomes clearer if barrier height is requested to be 2.50 m.

    2.50 m: upper barrier edge above field level
− 0.90 m: barrier height
= 1.60 m: height above field

+ 0.45 m: seat area above upper edge
            of finished floor
= 2.05 m: 'new' seat height

Option 2: Installing a fence system

International football association tournaments may only be hosted in all-seater stadia. For games of the German Football League DFL standing places may also be offered, resulting in two entirely different scenarios:

Taking into account safety personnel experience, standing accommodations can only be secured by fences and moats, as large masses of people (acc. to MVStättV 2005 § 27 a maximum of 2,500 people per block) may be liable to run to the pitch. DFB favours fencing as first option of securing a pitch.

This type of fencing (metal structure, safety glass, etc.) shall be at least 2.20 m high. The mesh size shall prevent a mounting and climbing of the fence. According to EN/DIN 33402-2 for body measures, the width of a foot is set at 110 mm, plus footwear. The model 'Anticlimb' by fabricator HERAS has a grid of 35 x 150 mm. FIFA classifies these systems as anti-climb, opaque partitionings or fences (permanent or dismountable).

Note: UEFA Technical Recommendations (5.7) for EURO 2008 prohibit fencing between spectators and the pitch.

According to the *Guide to Safety at Sports Grounds* (10.16), 'the use of pitch perimeter fences is not recommended under any circumstances' (p. 99). Further, a reduction in capacity is recommended, if fencing restricts viewing. Fencing height should not exceed 2.20 m.

Update

New formulation of § 7 Section 1 of the DFB guidelines for improving security at Bundesliga matches *[Richtlinien zur Verbesserung der Sicherheit bei Bundesligaspielen]* according to 'official announcements' of 31 May 2007:

The central area is to be separated from the viewing accommodations either by fencing which is 2.20-m high (metal construction, laminated safety glass, etc.); by a moat which is difficult to cross; by a combination of fence and moat or by elevating the first spectator row by at least 2.0 m above pitch level.

Prior consent of the stadium owner and local security bodies provided, the enclosure of the interior area in front of seating accommodation may also be ascertained by other suitable means.

---

373 **'Vario Fence'**
RheinEnergieStadion, 1. FC Köln
374 **Fence / Elevation**
World Cup Stadium Kaiserslautern
375 **Stewards**
World Cup Stadium Cologne

Option 3: Moats

The FIFA safety guidelines (§ 5) state the setting up of moats which are difficult to overcome or sufficiently deep and wide. At the same time, FIFA indicates in its Technical Recommendations that, due to the resulting space requirements, this would make sense only for spacious stadia.

A bridge solution for the moat must be structurally provided to allow emergency use of the pitch as relief space. The dimensions of the moat systems diverge since they are partly open to traffic or are needed as service or supply routes. After consultation with the head of safety at the LOC of World Cup 2006 in Frankfurt, the following minimum dimensions for a moat could be proposed, as in the example of the Olympic Stadium in Berlin:

Width 1.80 m: corresponding to the minimum width for relief gates (acc. to DFB 1.80 m)

Height 2.50 m: height of the terrace railing in the moat (upper edge finished floor of the first row at sightline angle 12.97° above the advertising banner)

Depth 0.95 m: base point of the moat below pitch level

Note: In the famous football stadium Estadio Centenario in Montevideo, capital of Uruguay and host of the World Cup 1930, a combination of moat/water ditch and elevation of the first row was realized to secure the pitch.

Option 4: Presence of security staff

According to § 5 of FIFA safety guidelines and in coordination with the stadium owner and local security bodies, the central area can also be secured by other means such as increased presence of stewards.

For international games hosted by UEFA the presence of police and security staff at the pitch zone or in its proximity would be the first option (TR 3.5).

The disadvantage for regular matches would be relatively high personnel costs. On top of this, stewards could possibly obstruct the field of view of spectators seated in the first rows. To avoid sight obstructions through advertising hoardings, etc., UEFA points out that tickets for first two rows should be not be sold to the general public and should instead be given to certain groups such as stewards or voluntary workers, etc.

Note: Due to the high loss of seat revenue, this proposal surely appears to be only an exceptional measure for particular FIFA/UEFA tournaments. Generally, viewing restrictions should be strictly avoided in advance, during the planning stage.

The example of the FIFA World Cup 2006 illustrates the situational deployment of stewards. Depending on primary security (options 1–3), persons are positioned between advertising banner and first row. Depending on the space available, stewards are standing in two rows, alternating and with a distance of 10 m/5 m/2.50 m to each other.

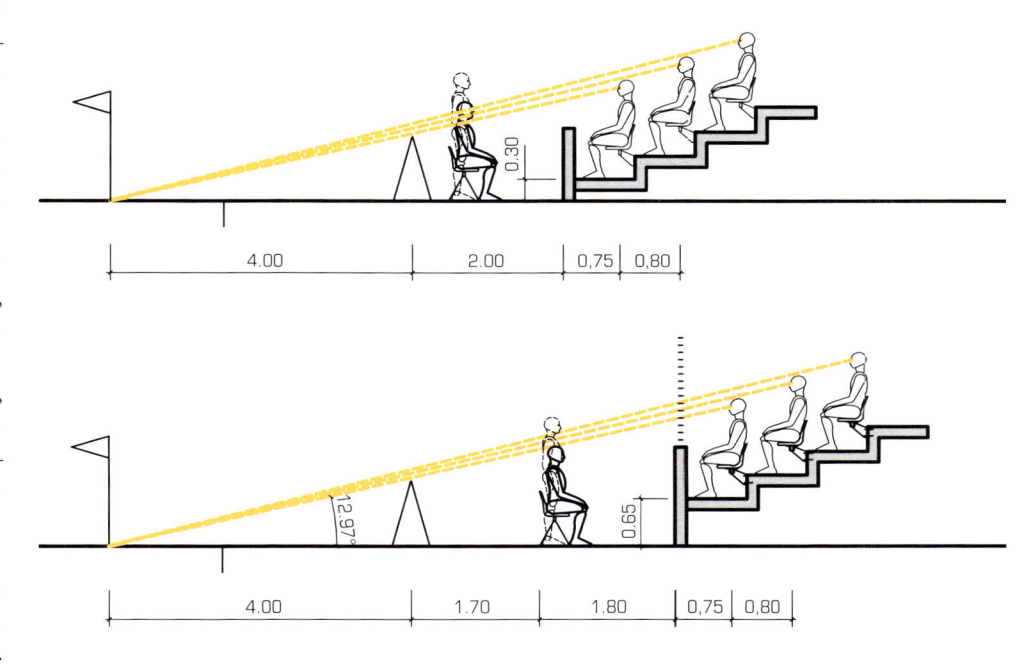

View restriction caused by advertising banner
Stadium, long side, 4-m position

376 **6.0-m minimum distance**
Minimum for possible starting height with viewing angle of around 13° and only 30 cm height difference to pitch. Stewards, seated and standing, within field of vision.

377 **7.5-m minimum distance (sketch above)**
Minimum for possible starting height with viewing angle of around 13° but with 65 cm height difference to pitch. Seated Stewards represent no sight restriction (first row).

378 **Differences in base heights for a 'first row'**
 a. World Cup Stadium Berlin
  (System: moat + 0.60 m)
 b. World Cup Stadium Leipzig
  (System: elevation + 4.45 m)

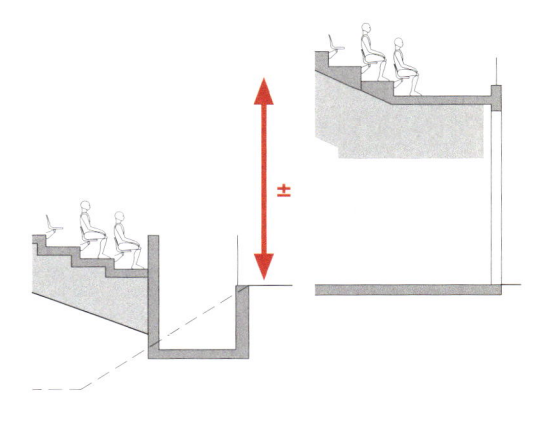

Mobile security measures

As multi-purpose utilization of sports and event venues becomes more and more important, the parties responsible need to respond very flexibly to the issue of security requirements for specific events. Spectators at an athletics competition are differently agitated compared to those attending a football match. Here, security demands may multiply when a local derby or decisive game takes place.

Whenever a fence would be unfeasible, unpractical or would disturb regular operations, mobile fences or simple warning tape are employed as temporary barriers for:
– certain zones and buildings
– definition of passageways
– organization of events
– partitioning of parking spaces
– internal 'sensitive' zones
– loading and unloading areas
– storage space and construction work

The mentioned systems have to be transported, unloaded, installed and dismantled, requiring more personnel and resulting in higher costs. Frequently, they are also visually unappealing. Yet, ever more importance is attributed to the overall appearance of sports arenas and large event venues, and the demands of guests and spectators increase steadily. Therefore, modern stadia should employ flexible systems allowing easy and convenient management of large crowds. Most important, though, remains the effective enclosure of a particular area. Increasingly, mobile fence systems are employed in order to maintain responsiveness of security systems during events. Their area of use ranges from the outer safety ring to the sector divisions in the stands and circulation areas. Innovative fence and barrier systems seek to meet these new demands.

Example 1: Mobile permanent fence system
Bayer Leverkusen
(Epping Metallbau, Bocholt)
Towards the pitch, vertically sliding fence units are installed at the terrace base point. These come in unit sizes around 5.0 m wide and are moveable up or down via threaded rods by ca. 1.50 m (adjustment to minimum height toward auditorium: 2.20 m). Manual operation by cordless screwdriver or by electronic remote control. The cladding may also serve advertising purposes, as employed in World Cup Stadium Cologne.

Example 2: Insertion systems
Another option to adjust fence heights to demand would be all variants of slip-on and portable fence units for railing sockets, as employed in World Cup Stadium Frankfurt.

Example 3: Folding or sliding fence systems
These are mainly used in circulation areas, as adequate storage space is available mostly in niches and secondary rooms. Wickets often have a sleeper profile and are therefore unsuited as escape routes.

In order to retain these separate areas as flight options, such gates have to be part of a coordinated security concept and are to be manned at all times.

379 **Securing of sector by glass partitions**
World Cup Stadium Gelsenkirchen
380 **Securing of sector through inserted elements**
Allianz Arena, Munich
381 **Securing of sector through rolling fence**

Example 4: Rolling fence system

Model 'FenceBox' (SecuFence AG)

Stored in a box, the vertically rolled-up fence (length up to 50 m, height up to 250 cm) is pulled out of a side opening and inserted into a special contrivance at the end. This device can be fixed to the next box or to existing fittings such as columns, pillars, etc. According to the manufacturer, 50 m of fence can be easily erected in less than 10 minutes.

A guide rope functions as 'curtain rail'. Depending on the fence length, the upper guide rope is supported by vertical posts. When erected or dismantled, the material slides along the rope and is also firmly fixed to the base point. The system thus becomes very stable, making it very hard to climb over or crawl under.

Example 5: Buffer blocks

Model 'Portugal 2004'

Downward, in direction to the pitch, three seating rows are covered by a tearproof plastic sheet with a smooth surface (3 x 0.50 m = 1.50 m). By means of a steel cable, a tunnel is created on top of the seats with tarpaulin spanned and arrested above this structure. The lower steel cables are fixed with clips to the seats. As a result, pulling up the cover is very difficult and time-consuming.

Note: The system was already employed with good performance at large events such as EURO 2004 in Portugal. It can be adjusted to all present conditions without additional construction or foundation works. The issue of a second flight direction in the stand is to be critically reviewed and coordinated.

Example 6: Wembley fencing

Model 'SecuFence' (SecuFence AG)

This type of anti-climb outfit for sports complexes enables a 'fence-free' stadium. The system consists of a foldable frame structure with an especially built-in, extremely flexible rubber-cable system. Operated by one person and fixed to the existing concrete wall or fence, it needs around 20 cm of storage space when folded up. As a horizontal retaining fence, it tightens automatically and is fitted with an emergency function for instant refolding. Climbing over or crawling under the system is hindered significantly as movement becomes almost impossible due to the instability and flexibility of the cable structure.

382 **Zone securing by rolling fence**
Allianz Arena, Munich
383 **Sector securing by buffer blocks**
FC Liverpool, England
384 **Pitch securing by 'SecuFence'**
Test for World Cup Stadium Cologne (not realized)

Service ring with downward slope

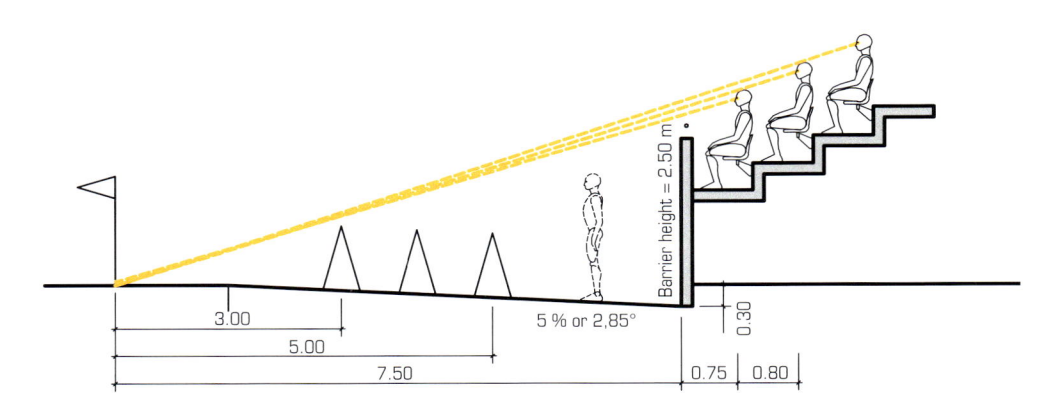

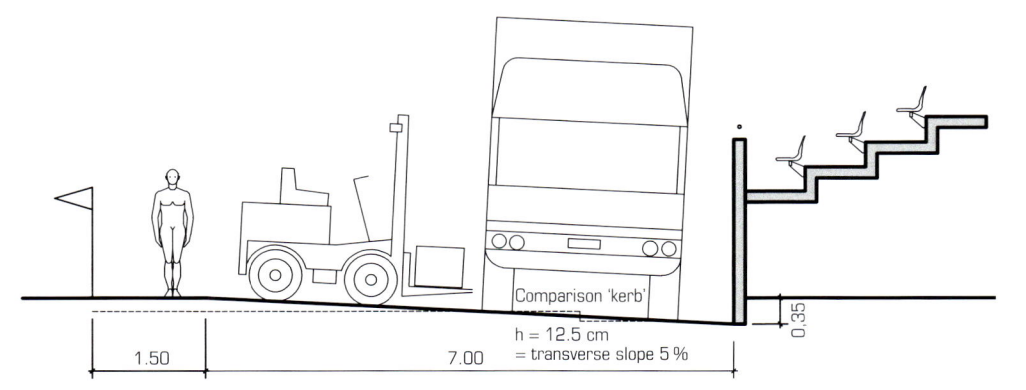

A ground-level service ring is the preferred option for most stadium bowls (maximum slope for drainage purposes: 2%). Examination of the rise of a sightline profile has revealed that a lower base point has significant added-value in terms of viewing quality.

Option 7 of pitch securing is represented by a lowering of the service ring. Alternatively, the field level may be raised against the stand. In both cases, the service ring is inclined and has a transverse slope.

During an event, the service ring is mostly used for the personal support of players (manager's bench) and by the press (photographers, etc.).

Outside of matches, the service ring is used predominantly by machines for turf care and by delivery traffic for erection and dismantling works for events. In case of emergency, large rescue vehicles must be able to drive into and around the stadium interior.

Normally, when erecting a show stage, an HGV ($l = 16.50$ m) must be able to stop in the service ring to unload. To allow the passing of another vehicle without riding over the turf area, the service ring should measure at least 2 x 3.0 m.

Requirements for the transverse slope

The limitation to a maximum transverse slope is a result of the limitation of use, such as the loading and unloading of an HGV by fork lift.

According to the general road an traffic regulations, 'the load is to be stored and secured so as to not slip, fall over, roll around, fall down or produce avoidable noise. […] Vehicle and load together may not exceed 2.55 m in width and 4.0 m in height'.[5] At present, there is no official limit to a maximum transverse slope for HGV loading scenarios.[6] As a rule, personal security is paramount.

For the comparable scenario of 'one wheel on the kerb' a slope of max. 5% applies for an HGV of 2.55 m width and a kerb of 12.5 cm height. Taken as planning base for a defined 5% transverse slope in the service ring, the following restriction applies, which needs to be coordinated in the overall planning with the client.

A transverse slope of 5% is permissible for standard cargo, provided it is neither exceedingly small nor top-heavy.

Note: Otherwise, corresponding measures of compensation must be carried out in each individual case. This applies also to fork-lift traffic during the whole loading and erection process.

Option 7

385 **Lowering by 5%**
(in exceptional cases)
Transverse slope 5% maximum
1) Restriction of use
2) Exercise caution with narrow cargo or top-heavy loads.

386 **Service ring (delivery)**
Transverse slope at service ring 5%
Minimum distance (FIFA) 6.0 m to touch line
                         7.5 m to goal line
Turf perimeter strip   1.5 m
Trafficable service ring
2 x 3.0 m = 2 x HGV
  b = 6.0 m   h = –30 cm
  b = 7.0 m   h = –35 cm
  b = 8.0 m   h = –40 cm
  b = 9.0 m   h = –45 cm
  b = 10.0 m h = –50 cm

Additional rows up to maximum rise through lowering of the service ring by 5%

30-cm lowering (7.50 m distance)
Maximum rise, lower tier + 12 rows

35-cm lowering (8.50 m distance)
Maximum rise, upper tier + 11 rows

40-cm lowering (9.50 m distance)
Maximum gradient upper tier + 24 rows

Rescue or relief gates

The relevant German regulation states that places for spectators in a stadium with a capacity above 10,000 persons are to be separated from the central area by a barrier.[7]

The requirements on field spaces and rescue gates are equally described in the national concept on sports and safety and other regulations for Bundesliga games and for the regional league.[8]

'For relief of the viewing accommodations in case of a panic among spectators, these fences must be fitted with rescue gates, which are assigned to the gangways of the stands. These gates cannot be opened by spectators themselves by means of panic locking devices. Rather, they can only be opened from the interior area or after authorization by the central command or the security service officer in case of danger. Further requirements serve to ensure the function of these gates.'[9]

Measurements and requirements:
1) barrier minimum height = 2.20 m
2) per gangway at least 1.80-m wide gates (one-wing)
3) central, remote-controlled or manual opening toward central area
4) equal-height transition

EN/DIN 13200-3 (5.2.2): Fence gates must be provided toward the central area in case of emergency.

DFB on rescue gates: no advertising banners in direction of flight, panic locking device and marked visually by colour, numbers or characters. Different from DFB, FIFA demands $w = 2.0$ m; optional measure in case of sufficient alternative escape routes.[10]

Retaining nets

According to DFB safety regulations, stadium fields without running tracks (pure football stadia) must have, at least around the penalty area, sufficiently high and close-meshed nets to avoid objects being thrown over or through them. In most cases, there are flag-pole structures to which the nets may be hoisted.

Definition

Set-up heights need to be coordinated with local safety authorities. At 7.5 – 10.0 m, they generally cover the lower tier. Depending on construction and location of the roof front edge, the net is partly taken down. Mesh size is sufficient at ca. 45 mm.

387 **Relief stairs**
Commerzbark Arena, Frankfurt
388 **Retaining nets** (suspended from roof edge)
Commerzbark Arena, Frankfurt
399 **Retaining nets** (set up in fence area)
Allianz Arena, Munich

19 Securing of the playing field

Securing against streakers vs. securing against panic

Generally, two security scenarios are to be considered:
1) The most important safety goal when planning securing measures is the personal security of spectators.
2) Event operations are to be maintained, i.e. avoiding disturbances and unwanted disruptions.

Venues of the World Cup 2006 are obligatory fence-free, at least within the transitional area to the stadium interior (FIFA TR 2006).

By introducing all-seater stadia, crowd control has become much more effective and a more civilized atmosphere now prevails. Disasters and accidents caused by violent supporters, as in 1985 in Heysel stadium, Brussels or in 1989 at Hillsborough Stadium, Sheffield, have diminished significantly. Nowadays, more problems are caused by individuals running onto the pitch – an issue that has become more acute due to increased media presence. While some are ambitioned visitors trying to pay tribute to their idols, personally and publicly, others are so-called streakers who like to 'exhibit' themselves to the public or TV audiences. In this case, security staff should be enabled to gain as much time as possible to hinder spectators from entering the stadium interior.

Panic response

For years now, many stadium operators and scientists active in the field of panic research have been looking for solutions with regard to the incompatibility of the mentioned security goals: the decision to provide for cases of panic, or to avoid disturbances.

Note: Theoretical models of human flight responses are incorporated into evacuation calculations. Analogies to flow behaviour (fluid mechanics) might suggest the provision of single-passage systems at escape gates. So far, no conclusive and practical proposal has been put forward.

A human being is equipped with a natural flight response. Flight directions which might constitute a danger to life and limb, like jumping out of a window, are mostly considered 'ultima ratio', and are decided upon intuitively in a matter of seconds. Large heights as escape options, such as the structural offset between upper and lower tier, are only resorted to in an absolute case of emergency. Fences are usually also avoided, unless all other flight options are unavailable or not perceived. Fence gates opening toward the stadium interior are not permitted as escape routes but are intended as relief gates or access for rescue and security staff. The interior space is not considered a secured area since all escape routes must lead into the open toward public circulation areas.[11] Recommended bridges over a stadium moat as continuation of a gangway only serve the purpose of relieving the accumulation of masses.

Access of spectators to the stadium interior can only take place as part of controlled emergency measures, which is why relief gates can only be opened from the field side. 'First-row' security measures only serve to avoid disturbances. They must be suitable for cases of panic, which is exactly where individual room for manoeuvre becomes limited for planners and operators / security staff. In this context, development of a sightline profile is paramount, as each measure requires distance and height. Whichever measure may be chosen in the end, the geometrical phenomenon of sightline construction remains: the closer and higher, the steeper.

390 **'SecuFence' pitch securing**
Test, World Cup Stadium Cologne (not realized)
391 **Sector plan (horizontal)**
Ernst-Happel-Stadion, Vienna
392 **Vario Fence (realized)**
RheinEnergieStadion, Cologne

The objective of adequate viewing conditions should always be integrated into the difficult process of decision-making regarding safety. Due to the low terrace base point, the horizontal intercepting fence represents an alternative in favour of the 'soft' security factors of disturbance prevention. Not applicable for standing areas, its use is also restricted in cases of panic. The model 'SecuFence' was tested during ConFedCup 2005 in Hanover and Stuttgart. The 'Wembley System' as its predecessor had already been introduced in the FIFA Technical Recommendations of 1995. As a rule, two measures may be implemented:

Visible net

The terrace base point is relatively low at 65 cm height. At the same time, the system requires a ground-plan area of approx. 1.80 m width so that the minimum distance to the pitch perimeter is 7.50 m, in order to create an intermediate space of at least 1.70 m between advertising hoarding and retaining fence. In case of a mass panic, the intentional 'tripping' of the streaker caused by the system could result in obstructions and injuries because the low difference in height does not suggest danger to the people fleeing and the flexibility of the fence is not deemed problematic. Therefore, the model should be fitted with an emergency retraction mechanism. As this system is highly reliant on the operational skills of security staff, though, it remains disputed among the approving authorities.

Net in stand-by position

Approximately 80 cm of net are folded inside a vertical U-shaped rail and may be released at several points. Stewards sitting at the pitch perimeter facing the audience overlook a relatively large area and can activate the system in case of spectator incursion. The retaining net springs up in fractions of a second, holding back the streaker. Once seized by security staff, the system is then retracted and instantly operational again.

Relief gates or staircases should only be opened from the stadium interior (DFB) and should therefore be resorted to only as an alternative for evacuation. Other solutions are required for the first and second escape route. The example shows how far decisions on safety measures – panic or disturbance prevention – are connected to the geometry of the 'first row'. Here, decision-makers are presented with a truly difficult choice.

STADIUM DISASTERS

1946 Bolton, England
Over 33 Bolton supporters were killed after a barrier gives way to massive pressure caused by exceeding the admissible total capacity of around 65,000 by more than 20,000 people.

1964 Lima, Peru
More than 300 people were killed and over 500 football fans injured, after riots, caused by unpopular referee decisions, broke out amongst spectators.

1971 Glasgow, Scotland
66 Ranger fans were killed during an outbreak of panic at Ibrox Stadium. After the final whistle, people on the stairways were pushing toward the exit. A wooden barrier gave way after some people had tried to climb over it.

1982 Moscow, USSR
During an international football game, shortly before the final whistle, police led a large group of spectators over a narrow, iced-over way out of the stadium. After a late goal, people pushed back into the stadium and 340 people were killed.

1985 Bradford, England
People were fleeing from the timber stand, due to a fire caused by a cigarette. 56 people died in the flames and over 200 were injured in the ensuing panic.

1985 Brussels, Belgium
Fighting broke out between rivalling fan groups on the terraces of Heysel Stadium in Brussels. A sector partition failed as English fans were trying to get into the block for the Italian fans. More than 400 people were injured and 39 fans were trampled to death.

1989 Sheffield, England
A panic ensued in a completely overcrowded block at Hillsborough Stadium, as masses of people were pushing forward from the top, in order to gain a better viewing position. 96 spectators were crushed to death at the separation fence in front of the first row.

2001 Accra, Ghana
More than 120 spectators died during a mass panic, after police had shot tear gas into a block of rioting fans.

393 **Bradford, England, 1985**
Devastating fire in the stand
394 **Streaker Mark Roberts, 2004**
(UEFA Cup Final)
395 **Sheffield, England, 1989**
Mass panic caused by fighting spectators

Chapter 20

396 Coliseum, Rome
Isometric of the event technology

Adaptable stadia
Structural conversion of modern sports and event facilities in the light of multifunctional utilization

The adaptable coliseum

A capacity for conversion has always been among the key features of any place of assembly. Such adaptability guarantees suitability for accommodating the most diverse types of performances and competitive events.

The 'Ampitheatrum Flavium', erected in 70–80 AD by the Emperors Vespasian and Titus, probably represents the highpoint of monumental form and well-considered logistics in the history of the ancient Roman amphitheatre.

The structure in question, better known as the Roman Coliseum, is the prototype of all later arenas, including those found at Verona, Arles and Nimes. Its dimensions are 188 m in length, 155 m in width and 51 m in height. It is claimed to have had a capacity of 50,000 spectators, including 45,000 seats and 5,000 standing places.

Under the Emperor Domitian, the authentic arts of the stage were progressively displaced by spectacular and gruesome performances. 'Bread and circuses' came to serve as a tool of power politics, becoming an essential aspect of Roman social life, thereby promoting the construction of vast amphitheatres.

Adaptable stage sets and décor allowed the interior space to appear as a wooded landscape, or even as an aquatic environment. Events provided to the Roman populace free of charge included gladiatorial contests, and even 'genuine' sea battles carried out in the centre of an arena. Found below the central surface of the staging area, and designed to allow the setting up of diverse theatrical structures in the briefest possible time, were passageways, staircases and storerooms.

The entire interior space was set up as a false or raised floor in order to facilitate the release of wild beasts from cages or the lowering of theatrical structures through large openings by means of wooden lift mechanisms. The elevation of the first row for security reasons also made possible horizontal deliveries through arcade systems. The logistical system of the Roman amphitheatre continues to be detectable in modern theatre facilities.

Four stories rise above an elliptical ground plan. As in a theatre, the cavea is constructed in concentric rising rings around the arena. Below lies the curved circulation level of the tribunes. Rising at an angle of 37° (1:1.3), these were provided in Roman times with marble steps for sitting. (Compare with the maximum rise specified by the regulation on places of assembly 57:80 cm 1:1.4 = 35.5°.) In order to allow spectators to be seated as close to the events as possible, the 50 rows of seats were arranged very steeply in succession.

The recommendations of Vitruvius, who for acoustic reasons proposed a consistent rise of 1:2 for each tribune, were violated for the sake of bringing the public as close as possible to the events taking place in the arena. For safety reasons, a base measuring 360 m in height separated the spectator area from the arena and protected the public from the action of the combat (and from animal hunts) taking place within. The consistent gradient and the elevation of the spectator area led, however, to relatively poor viewing for the public. In their orientation, the combat areas of antiquity were set out along north-south or west-east axes.

During morning or afternoon events, most of the spectators would sit with their backs toward the sun to avoid being blinded.

The 'arena' was even adapted to Rome's climatic conditions through the provision of a closing roof based on a technology involving sailcloth and cord. Ring brackets set on the capstones held the masts of this solar sail.

The auditorium was subdivided by storeys into individual tiers, and spectators were distributed through them according to their social standing. Accordingly, the continuous base was intended for Roman senators and provided with comfortable seats and richly decorated balustrades. The loges in the transverse axis were reserved for the emperor and high tribunes. Above these were the 'maenianum primum', which were reserved for the nobility without the rank of senator.

The rows set above these (the 'maenianum secundum') were subdivided into three sectors:

The lowest sector (the 'immum') served prosperous citizens, while the topmost sector (the 'summum') was conceived for the most impoverished residents of Rome. Worse still were the places where women from the lowest social strata were accommodated. For them, there were standing places on a wooden construction on the uppermost level (the 'maenianum secundum'), which Titus had had constructed.

Sectional view

Assumed point of focus at 8.50 m (determined by sightline eye-point, first seating row above upper barrier edge)
Lower tier: 42 rows
Tread depth: 69 cm
Rise: 52 cm
Constant gradient: 37°
Sightline elevation: 13.6 – 4.6 cm

From row 13 upwards, the 'C' value of 9 cm is not attained, and from row 29 up, it even lies below the minimum value of 6 cm. The maximum distance of vision to the assumed focal point lies at:
Maximum distance: 49.30 m
Maximum distance (horizontal): 40.65 m
Maximum distance (vertical): 27.85 m

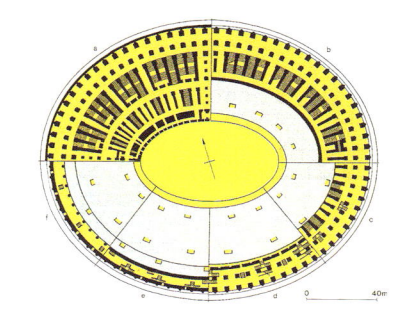

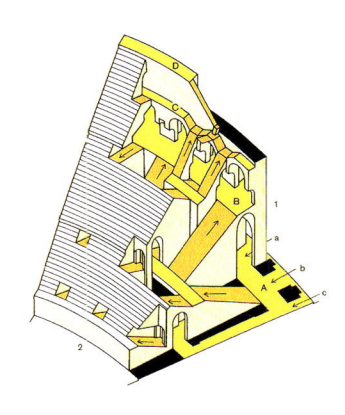

Coliseum, Rome
397 **Ground plan system and circulation system (isometric)**
398 **Sectional view with sightline profile**
399 **Photograph of the interior, today**

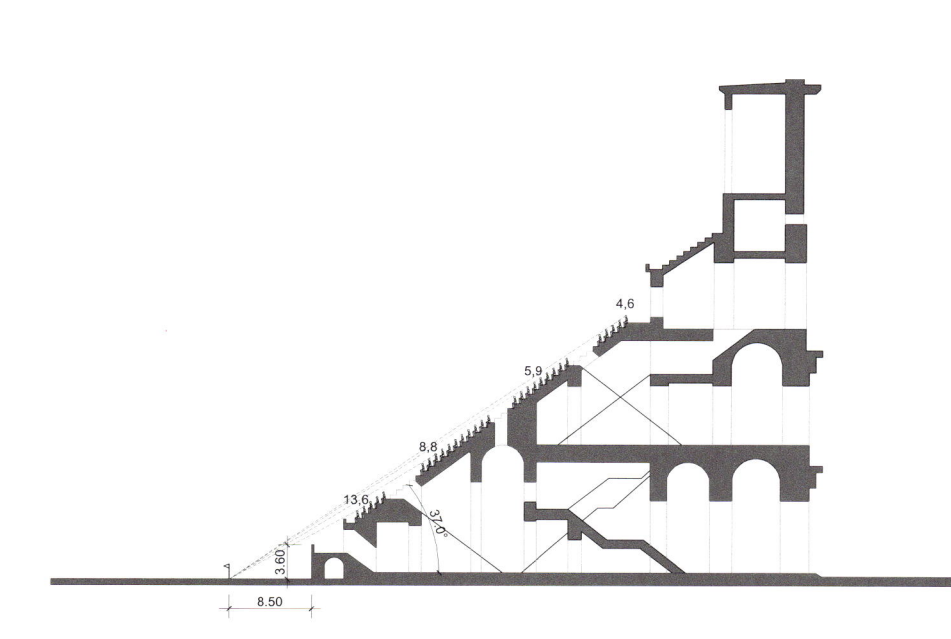

Multifunctional stadium interior

Multipurpose utilization is one of the fundamental features of a modern sports and event venue. Open-air concerts, for example, have been integrated into stadium plans for years. To be competitive on the event market, minimum capacity is set at 50,000 people. (The subsequent loss of back stage seats needs to be considered.)

RheinEnergieStadion in Cologne is based on a multipurpose principle, without featuring a particular structural flexibility. Event-related fittings are merely temporary.

Block definition

Barriers for standing areas in front of stages are to be set up as follows: a 'service aisle' at least 2.0 m wide is arranged directly in front of the stage.[1] If more than 5,000 people are standing in front of the stage, two more barriers must be erected with a lateral spacing of 5.0 m and at least 10.0 m toward the stage. Regarding maximum capacities, the German regulation valid in 2005 states: 'Standing areas enclosed by barriers are to be dimensioned so as to accommodate not more than 1,000 people.'[2]

A block of around 750 m² is created, assuming a minimum width of playing field + service ring = 80 m x 10.0 m clearance between the two barriers, taking into account the diminution to 5,0 m at the perimeter.

With an average number of two persons/m², around 1,500 people can be accommodated in these areas.[3] The arithmetical maximum of six persons/m² (e.g. cable car) could rise up to eight persons/m² in jostling crowds close to the stage, and is thus widely underestimated at two persons/m².

'Corresponding to the common practice of event organizers, a higher density is permissible only for standing places on stepped rows, e.g. in sports stadia, resulting in larger dimensions for escape routes. For interior spaces, accommodating four persons/m² is neither practicable nor justifiable from a security standpoint' [commentary by ARGEBAU, June 2005]. To fill the space between the first two crush barriers in front of the stage more adequately, the going approval practice at present is to admit 3 persons per square metre in the first two blocks in front of the stage, as long a the total capacity of the stadium interior does not exceed 2 persons/m².

Note: 'During the approval stage, operators of multi-purpose complexes or large halls should already have the respective multi-purpose seating variants authorized since later alterations to the seating plan can only be executed in the framework of the existing structural escape routes' [commentary ARGEBAU, 2005].

400 **Stage for Phil Collins concert, 2005**
RheinEnergieStadion, Cologne

page 211:
401 **Master plan for multipurpose seating, type 1**
(south stage, small stage 20 x 50 m)
402 **Master plan for multipurpose seating, type 2**
(south stage, large stage 30 x 60 m)
403 **Master plan for multipurpose seating, type 3**
(centre stage, with seating auditorium)
404 **Master plan for multipurpose seating, type 4**
(centre stage, with standing places for stadium interior)

Multifunctional stadium interior 211

Type 1

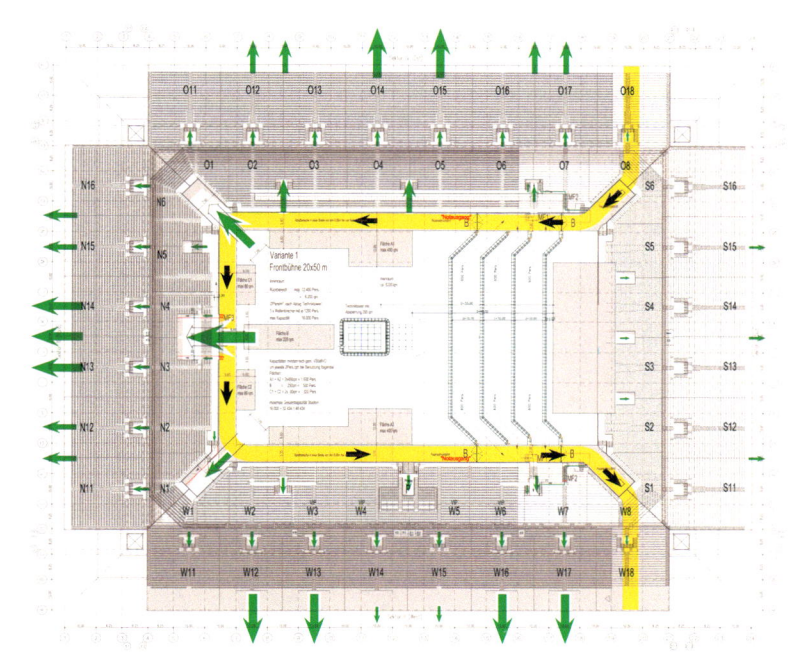

Type 2

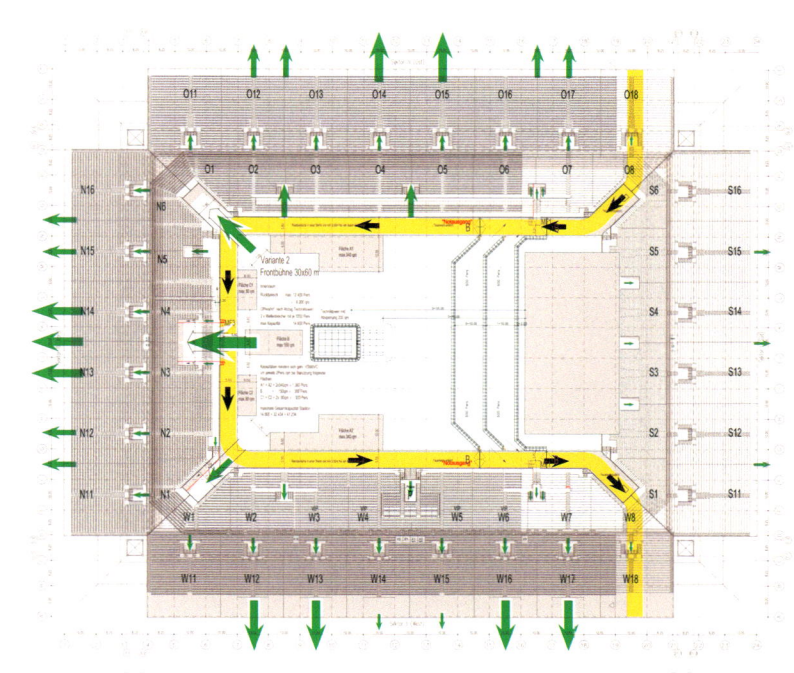

Type 3

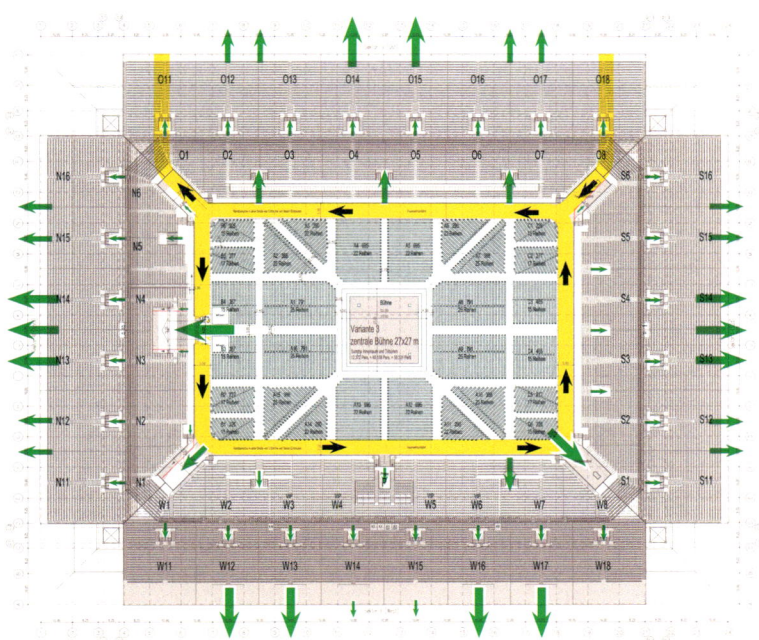

Type 4

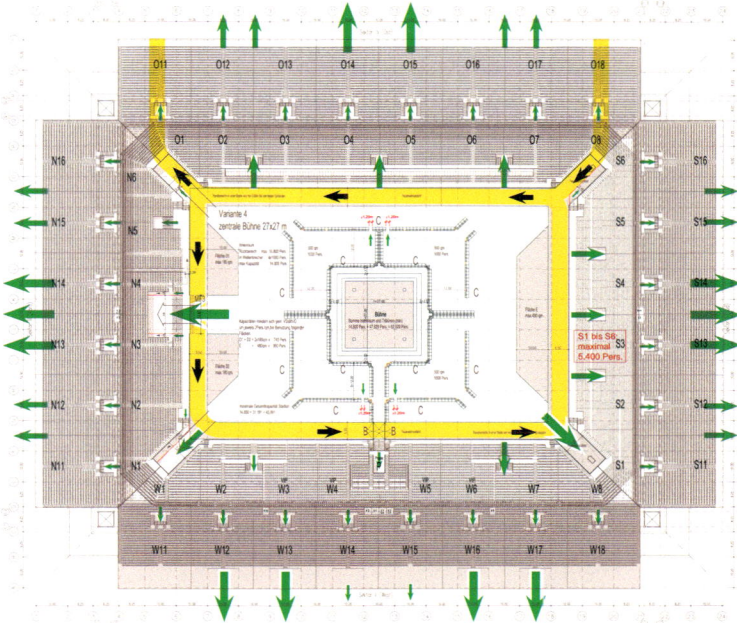

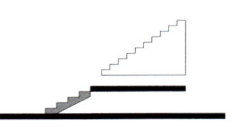

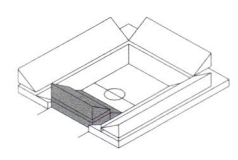

Multipurpose hall

Assembly places are described in the German MVStättV § 2 as structural complexes 'which are intended for the concurrent presence of a multitude of people at events of an educational, commercial, social, cultural, artistic or political nature; at sports or entertainment events as well as food and drinks facilities.' Assembly halls and foyers, lecture halls and studios are also included. The commercial operation of such an event venue nowadays requires a multifunctional utilization profile.

Structural demands may vary widely from event to event, and alterations can turn out to be very costly. Specialized stadia, whose utilization capacity is not a high as that of a multifunctional 'allrounder', should be evaluated in advance in terms of revenue.

Nowadays, the management of a large arena will endeavour to expand the standard utilization for sports and concerts. This trend can be clearly observed in the example of Veltins Arena in Gelsenkirchen. Hosting a broad range of events, it may be termed a multifunctional multipurpose arena. With its movable grass surface and closing roof, the complex guarantees instant re-use by providing a 'playable' central area and ensuring independence from weather conditions. Indoor-biathlon championships or motocross races are taking place outside the regular football matches of FC Schalke 04. The layout of the grandstands may be altered by setting up temporary spectator areas, responding to the reduced dimensions of handball events, for example. After moving out the turf tray, the lower tier, by folding down several rows, can also be pushed back and stored 15 m underneath the bridge-like structure of the south stand. Thus, the usable interior area is expanded and the erection of a stage facilitated. (The more a stage can be retracted, the more spectators may be accommodated in the auditorium and in the lateral upper-tier stands, which otherwise would disappear behind the stage front without viewing relation.)

Acoustic aptness of the closed hall provided, the utilization plan may be complemented by hosting concerts (e.g. the opera 'Aida').

To be profitable, a venue would have to host an event almost every day. The Schalke 04 grandstand is planned as a non-air-conditioned open-air stadium (minimum temperature around 0° C). Although not realized, installing heating elements into the roof structure would provide a convenience level comparable to that of a theatre, even during winter.

Note: The necessary exit width triples for assembly places without an open playing field (§ 7 MVStättV 2005). Therefore, envisaged spectator figures must be in accordance with the existing or planned evacuation widths. The stadium interior in Gelsenkirchen measures around 90 x 125 m, or 11,250 m². At 2 persons/m², this would amount to 22,500 persons: 200 x 1.20 m = 135 m. (Compare to stadium Cologne, with circa 40 m evacuation width.)

405 **Seating plan, Arena AufSchalke**
Opera 'Turandot' (9 July 2005)

406 **Opening celebration, Arena AufSchalke**
Concert of the group 'Pur' (13/14 August 2001)

407 **Stadium interior, Arena AufSchalke**
Covering of steel rails after moving out the turf surface

408 **Pictogram**
Movable lower tier for stage erection

page 213:

409 **Seating alternative for handball use**
Stadium interior, TBV Lemgo

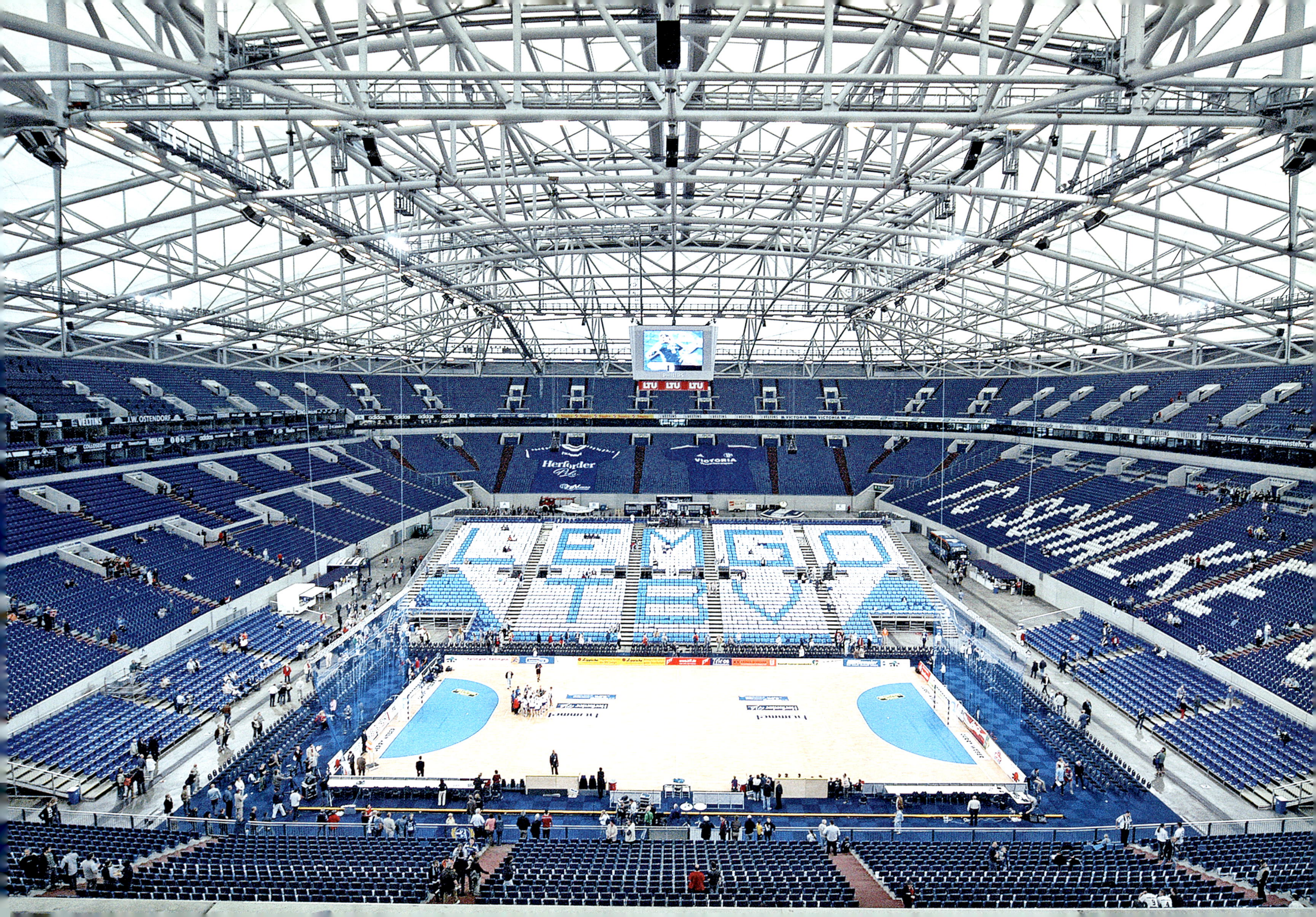
LEMGO
TBV
Herforder
Victoria
LTU LTU LTU

Movable grass pitch

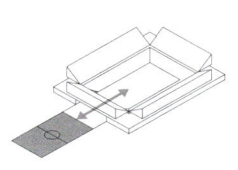

There are three approaches to create an optimal growth situation for a natural grass surface:
1) maximum opening of the stadium for sufficient airing and lighting of the playing field.
2) modular turf for gradual dismantling and exchange, at provision of two additional grass surfaces outside.
3) movable turf, as in Veltins Arena, Gelsenkirchen, which has a combination of closing roof and movable grass pitch.

The high investment and operating costs of a movable field have to be evaluated in the financing concept. This solution is only economically viable if the regenerative effect for the turf, when outside, can be combined with a simultaneous utilization of the vacated stadium interior.

Instantaneous re-usage of the interior space strengthens the utilization concept of a modern multi-purpose stadium. Consequently, the decision for a retractable roof covering the event space, in the light of weather-independent usage, seems to suggest itself.

The turf is placed on a 120 x 80 m reinforced concrete slab, forming a ca. 1.50 m high concrete trough. The structure weighs 11,400 t and can be pushed outside within five hours on coated steel rails by four hydraulic units, the so-called gripper jacks. Alternative transport mechanisms are provided by a roller system or air cushions.

In order to execute the moving process, similar to pulling out a drawer, one of the stands has to be built as a bridge structure. Veltins Arena has a steel bridge comprised of three trusses at 350 t each with a span of circa 85 cm. The structure is supported in three places to bear the loads of a filled south stand.

For moving the pitch, supports are folded up temporarily. The storage position of the turf is naturally at the south side, in order to avoid shadowing from the stadium body and to exploit a maximum of sunlight.

With the grass turf inside the stadium at football matches, the paved exterior area can be used as media or event parking space. Various events, from concerts and operas to motocross and biathlon competitions, are feasible within the stadium bowl, complemented by temporary accommodations. However, the relevant points of focus with reference to the regular terraces must be taken into account.

Note: A closing roof entails a change in building-type classification. A 'stadium' (uncovered playing area in the open) becomes an 'arena/hall' (roofed), resulting in a threefold exit width; instead of 1.20 m per 600 people only 200 people may be allowed.[4] An extension of the 30-m escape route by 5 m, per 2.5 m of smoke-free layer, is limited by the roof covering (exception: sufficient cross ventilation, as in Frankfurt).

Sapporo Dome in Japan is also a multi-purpose arena, completed in May 2001 for the FIFA World Cup 2002™ in Korea/Japan. In response to the local climate, the Sapporo Dome was planned as an all-weather stadium.

The Dome impresses with its extraordinary architecture and singular technical extras. The central element of the roofed complex is a movable grass pitch, which can be slided in or out, depending on demand and the weather situation.

Expert planners were expecting that outside of the arena the grass would grow under natural conditions and would be usable during the summertime.

The 'double dome' combines an open-air stadium with an inside arena. When a football match takes place inside the stadium, the pitch slides into the dome on an air-cushion and, once

410 **Aerial view, Arena AufSchalke**
During construction (2000)
411 **'Gripper Jack'**
Moving mechanism for the grass field
412 **Terrace bridge, south**
Movable lower tier (Gelsenkirchen)
413 **Pictogram**
Movable grass surface

in the centre, turns by 180°. Accordingly, over a width of 90 m, the seating rows of the lower tier stand can be folded back and then moved tangentially. The entire manoeuvre lasts around two hours.

Converted into a football stadium, the Sapporo Dome provides accommodation for 42,000 spectators, placing them as close to the pitch as possible on a one-tier stand with an average slope of 27°.

The roof is a single-layer, teflon-coated fibreglass membrane. The teflon material is fitted sufficiently tight to protect against the prevalent weather conditions. Its translucency admits only around 16% of natural sunlight and shields the stadium bowl against overheating.

Material resistance and service life is quoted with 30 years.

The interior fittings are highly flexible and adaptable to every type of event, such as indoor baseball. During summertime, an air-conditioning system provides pleasantly cool air; in winter, heated seats and a powerful heating system keep spectators warm.

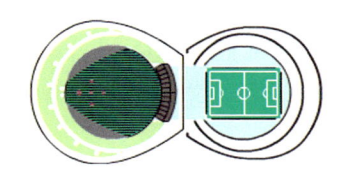

414 **Sapporo Dome, exterior**
Air-cushion grass field
Capacity: 42,100
415 **Ground plan system drawing**

Adaptable stadia

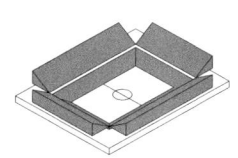

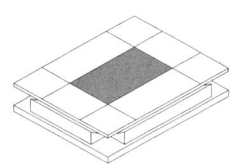

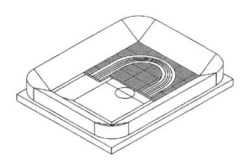

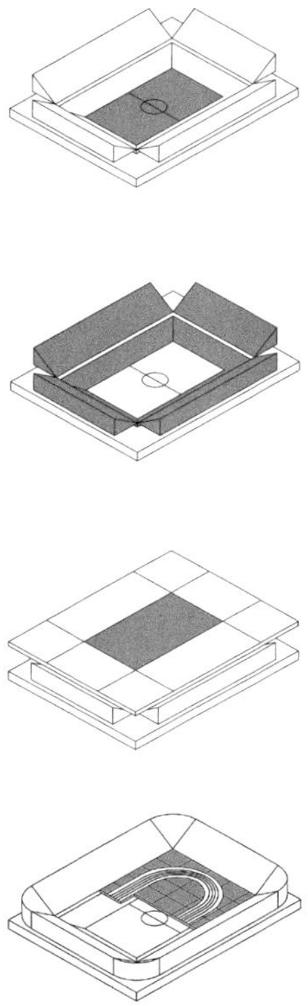

The adaptability of a stadium or arena has been a vital characteristic of this building type from its very beginnings.

The relevant terms 'multi-function' or 'multi-purpose' are used time and again in the context of building modern sports and event venues of the third generation. The concept of adaptability is distinguished by four different aspects:

Part A: duration of use
1 temporary
2 permanent

Part B: purpose
3 utilization
4 capacity
5 identity/image

Part C: structural part/type
6 playing area
7 stand
8 roof

Note: Frequently, different adaptive measures are combined for a particular event.

These temporary insertions or fittings are employed only once or in exceptional cases; for example additional commentators' and press positions fitted short-term for a large sports events, or the provision of additional cable routes or barriers. Other temporary structures are the 'tent cities' of so-called sponsor villages. The definition of sectors within the auditorium, in accordance with the stipulated safety concept, is only provisional. Since they can be repeatedly employed in each individual case, suchlike sector divisions can be permanent, depending on demand. Telescopic terraces or closable roofs are also considered permanent measures, even when used only sporadically. The question if a permanent installation of adaptable components makes sense can be settled for each individual case only, in the framework of the relevant business plan and operating concept.

Combination of measures

A draft from the faculty of Architecture at RWTH (Technical University) Aachen for a 'New National Stadium' in Warsaw vividly illustrates how different adaptations may be combined. The draft is based on modular terrace components, comparable to the stadium San Nicola in Bari, Italy (architect Renzo Piano).

The fundamental objective of a sports and multi-purpose stadium results in a classic oval geometry, which releases the athletics track by retracting a telescopic lower tier. This movable terrace reaches up to the lateral gangway, in order to ensure usage of the lower tier in front of the hospitality suites.

Lift turf

The height offset caused by a retracted lower-tier terrace affects viewing as spectators can only see the running track opposite. Hence, a lifting stadium interior is proposed, capable of adjusting viewing conditions to an athletics scenario.

Such an approach has been proposed repeatedly in architectural competitions. The general feasibility has been proven with the lift system by the planning group Schüssler-Plan for LTU-Arena in Düsseldorf.

The cost of a movable turf has to be weighed up economically against the advantage of the directly available interior space. Cost considerations have prevented a realization of this system in Europe so far.

The possible regeneration of the turf in the exterior surroundings (roof position) is offset by the ensuing operating, personnel and energy costs. Moreover, viewing from the stands is restricted by the lift support of a mobile grass roof.

If a grass field is elevated by several metres to balance viewing conditions, a natural grass turf would have to be covered for non-sports events in the stadium interior.

The turf can be damaged by non-professional execution or prolonged covering. Great athletics events such as the Olympic Games are not hosted as often as other sports events such as football championships.

On the occasion of the 2012 Summer Olympics in London, the new stadium in Wembley takes a different path: planners are intending to elevate the entire stadium interior.

A raised installation height uses the extended space geometrically created by the ascending terrace of a lower tier. The necessary height corresponds to the required surface for athletics usage. Such a measure would be considered a temporary, individual-case solution not intended for constantly changing utilization.

If a mobile grass pitch is decided upon, the utilization profile of the viewing areas is extended; provided that operation becomes independent of weather conditions by means of a closing roof.

416 Pictogram
Types of structural transformation:
Playing field, stand, roof, raised floor

Adaptable stadia 217

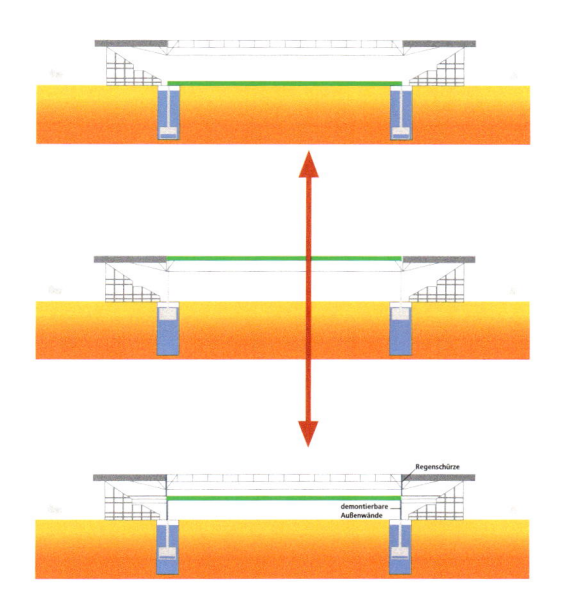

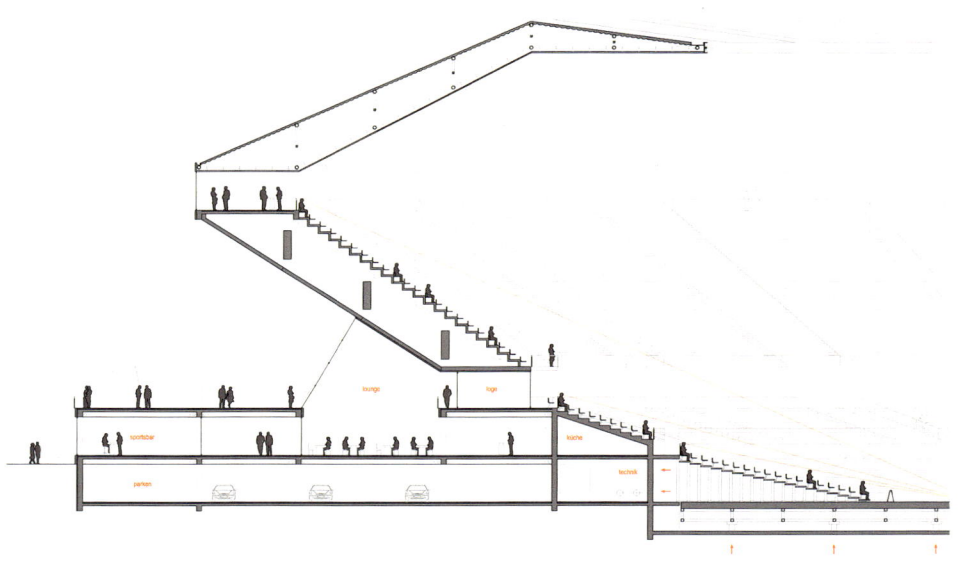

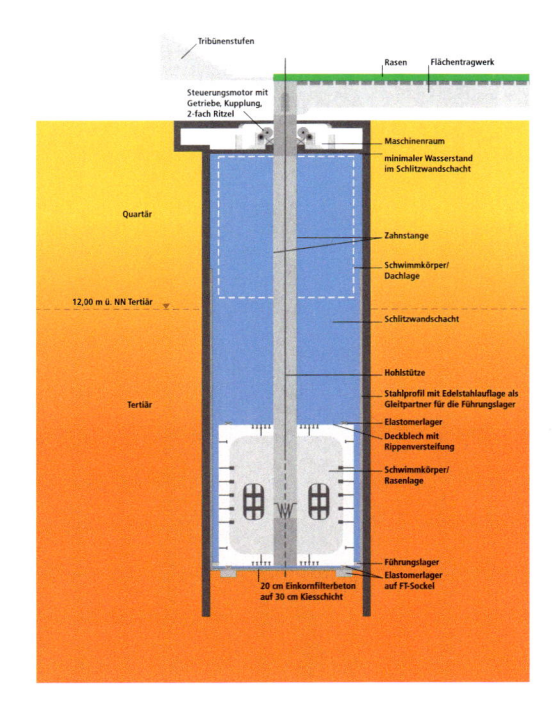

417–418 **Lift-slab roof concept**
by Schüssler-Plan Ingenieursgesellschaft mbH

419–420 **Design drawing**
Ground plan, sectional view and pictograms
RWTH Aachen, Kai Grosche, 2005
Supervisor: Stefan Nixdorf
(Chair for Urban Planning)

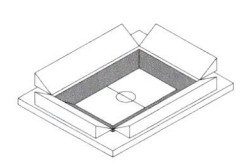

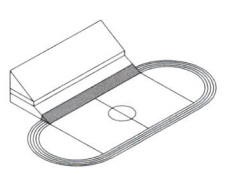

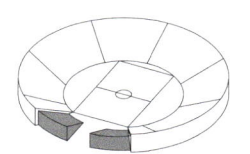

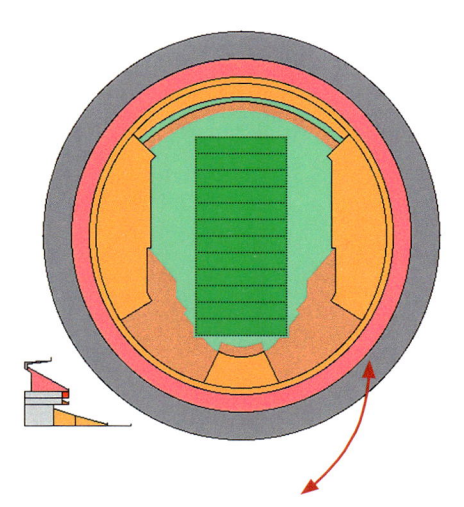

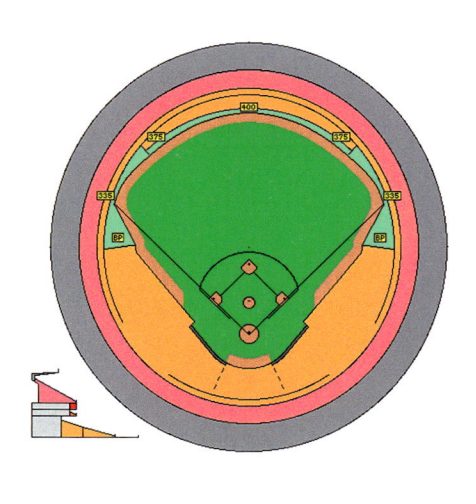

Retractable lower tier

The Sapporo Dome in Japan is an example for multiple usage of an arena interior, in this case another main discipline, namely baseball.

Unlike Veltins Arena in Gelsenkirchen, where the grass surface is retracted below a terrace bridge structure, in Sapporo, the telescopic eastern stands are pushed inward and, in a circular movement, are stored tangentially underneath the neighbouring stand.

The rather extraordinary geometry of a rotated and rounded lozenge can be matched by moving the two crescent-shaped elements of the lower tier in such a way as to create a baseball-ground shape or, turned by 90°, take the form of a football pitch.

Three Rivers Stadium in Pittsburgh, first erected in 1970 and renovated in 2001, has a capacity of around 48,000. With its movable lower tier the stadium interior can be adjusted for baseball or American football usage.

Grandstand at Stade de France

For Football World Cup 1998, Stade de France was built in Paris, St. Denis, as new national stadium, featuring three tiers.

A special system was employed to allow conversion of the 74,000 capacity facility from a football stadium into a venue capable of hosting athletics events. In three steps the lower tier can be moved outside: first, the rear part of the lower tier stand is moved via telescopic supports downwards. The mobile terraces are turned and then stored in the basement between the main structural supports. Thus, capacity is lowered by around 5,000 seats.

In a last push, the steel structure of the front stand is retracted by around 7.50 m, uncovering the circular athletics track underneath.

Telescopic stand

Another option to adjust the capacity of terraces is the classic telescopic stand. The new EN/DIN 13200-5 describes retractable stands as a row of platforms or steppings resting on movable structures or connected to the surface of the ground in such a way as to open the system in a forward movement and close it by moving backwards. Each row or level is connected to the next so as to make the entire system behave like one single scaffolding, covering the entire height from the front seats to the highest steps at the rear side.

Although telescopic terraces are normally fixed to the framework of a building, they can also be constructed in such a way as to be movable vertically and horizontally, mostly in schools and small sports halls up to very large halls with, for example, 10,000 fully upholstered single seats distributed over 20 rows in staggered layouts.

---

421 **Sectional view/isometric: Athletics ground**
New Wembley Stadium, London, HOK s+v+e (2004)

422 **Pictogram**
Adaptability of terraces

423 **Three Rivers Stadium**
 a. American Football
    Seating variant by Andrew G. Clem
 b. Baseball
    Seating variant by Andrew G. Clem

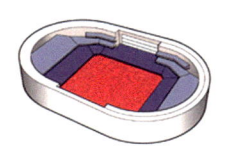

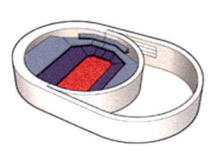

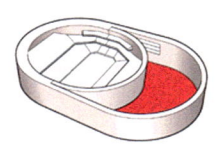

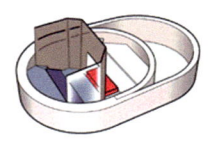

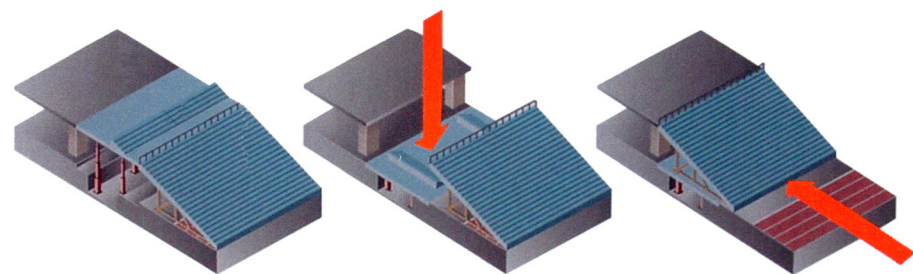

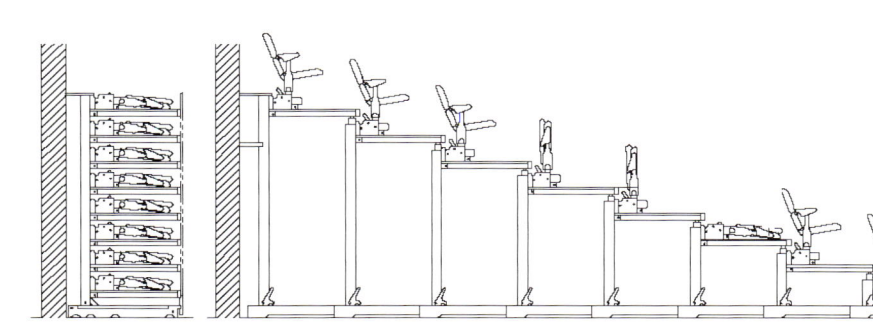

Retractable lower tier 219

left:
424 **Stade de France**
Isometric representation
425 **Picture of movable lower tier**
426 **Telescopic stand**
Drawing taken from EN/DIN 13200-5

427 **Pictogram**
Fitting variants of Saitama-Arena, Japan
a. Largest-possible volume
b. Small arena
c. Multi-purpose hall
d. Variants for stage set-up

right:
428 **Sapporo Dome, Japan**
a. – c. Sequence of movable telescopic stands

Convertible roofs

One of the most common structural options is the conversion of the roof, as it allows weather-independent usage of the venue.

Either the stadium interior is closed by a completely circulating roof, or the roof structure as a whole is movable and opens up the area above the spectator accommodation. Two general types are distinguished:
1) soft structures
2) rigid structures

The adaptability of convertible structural systems can be categorized according to the type and the direction of movement.

Soft structures are generally light membrane systems which are mounted as rope-spanned bearing systems and can be retracted or rolled back. Besides, there are also movable frameworks whose membranes can be moved, folded or turned by means of support structures. Some membrane fields with large spans, as used for stadium interior roofs, are supported pneumatically (cf. Toyota Stadium by Kisho Kurawa Architect & Associates with Ove Arup & Partners Japan Limited).

Rigid structures can be moved in a parallel, central, circular or peripheral direction.

This type of convertible frame has to be constructed as a lightweight structure since it not only bears the roof load and its own load but also has to absorb dynamic loads from the movement of the roof. The concept of rigid, movable roof slabs is commonly employed in the US and Japan. The roof cover frequently consists of metal sheeting and membranes.

Sometimes, these structures can be enormous in size, with spans of up to 150 m and domes of equivalent static height, as represented by one of the first 'very large-span' convertible roofs (120 m, ca. 3,000 t, 17,000 capacity) at Mellon-Arena, erected 1961 in Pittsburgh. Since the movement of the roof resulted in material deformation, the arena was only ever operated with a closed roof.

The Fukuoka Dome (built in 1993) in the north of Japan covers with its dome-shaped roof structure ($h = 84$ m) around 52,000 spectators. The three titanium-coated circular segments with a total span of ca. 210 m can be opened to almost two thirds.

Roof elements moving in horizontal direction (single and two-part) transform the stadium interior into a weather-independent event arena, as for example the Millennium Stadium (built in 1999) by HOK + LOBB Partnership and WS Atkins in Cardiff, England, with a holding capacity of 75,000. Four 90 m high pylons carry a 27,000 m² Kalzip metal roof. The interior roof (80 x 105 m) is moved shut from both sides via arched roof supports. (A modular turf system allows a partial turf exchange.)

The cover of the central area of Gerry Weber Stadium in Halle (Westphalia), Germany, consists of transparent roof slabs with ETFE cushions, which can be stored underneath the circulating roof at the short ends.

A very elegant solution from the year 2000 is represented by a bullfighting arena in Vista Alegre, Madrid (Ayuntamiento de Madrid).

Framework planners Schlaich, Bergermann und Partner from Stuttgart covered the circular 50-m diameter opening with a statically independent and vertically opening air cushion. The atmosphere of the hall interior is impressive. At the touch of a button, ropes are set in motion, which pull the entire cushion 10 m up, thereby creating, through additional light and fresh air, the typical 'open-air' feeling.

The fascinating idea of using helium-filled, freely suspended temporary 'pneus' to cover the

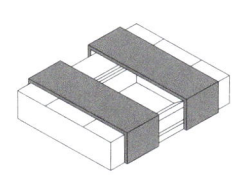

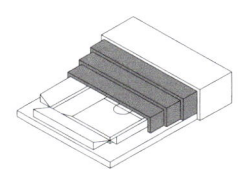

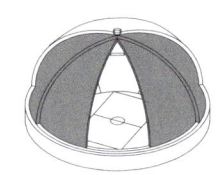

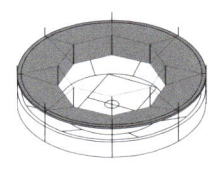

429 **Olympic Stadium, Beijing 2008**
Competition drawing gmp Architects, Berlin
430 **Movable roof, 'radial' concept**
Competition drawing Olympic Stadium, Beijing
431 **Movable roof, concept of 'folding' stand**
Competition drawing Olympic Stadium, Beijing
432 **Movable roof, 'photographic lens' concept**
Tennis stadium Qi Zhong, Shanghai, China
433 **Pictogram**
Structural adaptability of the main roof: movable, telescopic, radial, retractable roof

stadium interior has been occurring time and time again in architectural competitions. (To the knowledge of the author, no such project has yet been realized.)

Olympic Games, Beijing 2008

The rapid economical development and the ensuing construction activities have put China at the centre of global attention. Planning and construction of large projects are currently realized in incredibly short periods of time.

A global architectural competition was initiated for the Olympic Games to be hosted in Beijing in 2008. With a capacity of 100,000, a multifunctional sports and event complex with athletics usage will be setting new standards.

Within the newly erected Olympic Park stands the contribution by gmp Architects, Berlin: a 400-m circular tensegre dome, transforming the rectangle of the playing field above the athletics oval and a two-tier stand into an ideal circle.

The 140-m diameter opening in the transparently covered pneu-roof is supposed to open and close like a circular flower bud. This light and transparent roof structure is a perfectly ingenious innovation; at the same time, it functions as symbolic metaphor of a water flower in a park of undulating lakes. Among the various proposals for movable roof structures which were submitted for this competition, one design attracted particular attention. It has four mobile upper tier segments, which by folding down cover the lower tier, thus adapting capacity for future usage of the stadium.

Masters Cup, Shanghai 2005 – 2007

The Qi Zhong Tennis Centre in Shanghai will be Asia's largest tennis centre, when completed in 2006. The centre court is enclosed by a circular grandstand with 30 rows for 15,000 spectators. Like a photographic lens, it opens and closes by means of a complex sliding-roof structure.

The appearance of a stadium or arena is significantly determined by the structure of the roof. Form and geometry of a spectator complex determine the framework options and have to be coordinated at an early stage.

In a competition, adding significance and uniqueness to a design proposal is a recurrent challenge to architects and engineers. The progressive development of building materials and construction methods open up a seemingly inexhaustible creative potential.

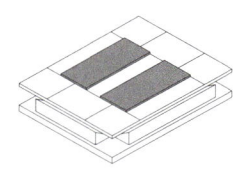

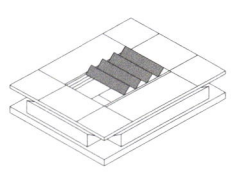

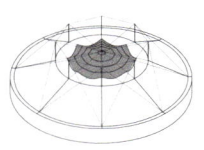

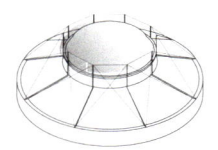

434 **Pictogram**
Structural adaptation scenarios for the main roof: Sliding, fold-up, retractable, lift roof

435 **Fold-up-roof concept 'lotus blossom'**
Competition drawing by gmp Architects

436 **Suspended-roof concept 'air cushion'**
Competition drawing for Olympic Stadium, Beijing

437 **Lift-roof concept 'air cushion'**
Bullfighting arena Vista Alegre, Madrid

438 **Sliding-roof concept 'air tube'**
Toyota Stadium, Japan

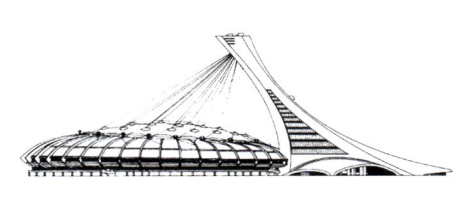

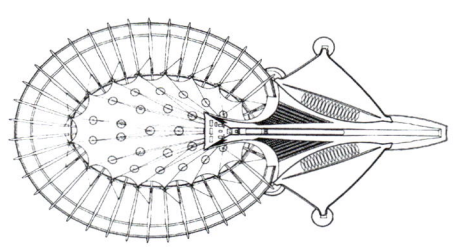

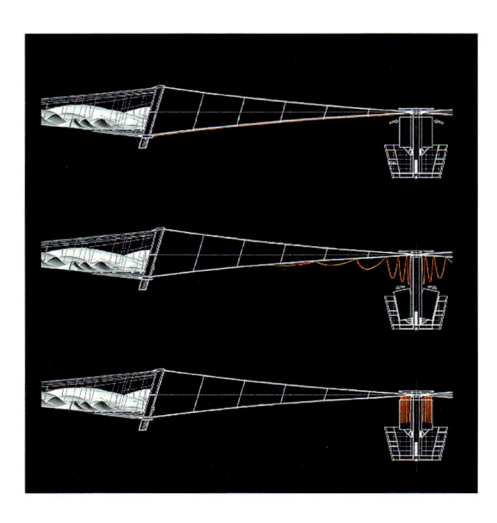

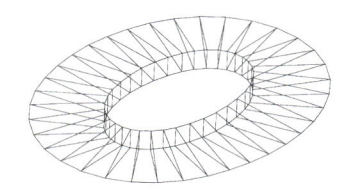

Retractable roof membranes

The gathering of a sail-type membrane is traditionally one of the oldest forms of adaptive framework; even in the Coliseum in Rome the arena interior was shadowed by a rope and sail mechanism, derived from shipping. During the 1960s, structural planner Frei Otto devoted himself to the development of convertible roof structures.

For the Olympic Games 1976 in Montreal, Canada, a multifunctional sports complex was to be erected that was 'to convince by its innovative and exclusive appearance', as stated in the invitation to tender. The roof construction of the opening part for the new Olympic Stadium in Montreal is reminiscent of the swimming pool covering of Boulevard Carnot, Paris, developed by Frei Otto 1968. At the institute for lightweight structures in Stuttgart, a 20,000 m² large, centrally retractable membrane roof was designed. The strong expressive form of the slanted central mast convinced the jury. Technical problems and the local climate forced operators in 1998 to exchange the wind-damaged membrane and keep it closed. Material and technological improvements as well as better climate conditions of FIFA World Cup 2006 in Frankfurt, promoted the development of a 'cabriolet roof'. Architects von Gerkan, Marg und Partner, Berlin, realized the project for the ConFedCup 2005 together with engineering firm Schlaich, Bergermann und Partner, Stuttgart, and the general contractor Max Bögl, Neumarkt.

The following text is a translated excerpt from the architects' report (Feb 2004):

'The interior roof spans the opening above the playing field formed by the hollow nub of the exterior roof. It consists of one hub or central node and 32 upper and 32 lower spoke cables which are connected to the upper, respectively lower, ends of the struts and are connected by suspended ropes to cable binders. The central node (approx. 20 t) and the spoke cables form a stable primary system whose load-bearing properties correspond to those of a spoked wheel. This primary structure with a total weight of 2,400 t, together with the curved lower radial cables, forms the dome-shaped frame for the enclosing membrane which is connected via rollers with the cables. To open the roof, the membrane, which as secondary support element transfers the load into the primary system, is gathered toward the centre. In the closed state, the membrane is tensioned, as is the cable support system. In this manner, loads can be effectively transferred without major deformation, and fluttering of the material caused by wind can also be prevented.

The same materials are used for the convertible interior roof: the membrane consists of PVC/PES.

Opening or closing the roof takes around 15 minutes. The folded interior roof is stored in the middle of the cable support system. To protect the membrane from adverse weather a cylindrical body is provided, which opens and closes like a blossom when the roof is moved. The storage position of the interior roof in the centre of the stadium is synergetically used as ideal location for the video matrix, offering optimum conditions in functional and scenic terms.'

439–440 **Retractable interior roof**
Commerzbank Arena, Frankfurt
Planning: gmp Architects with sbp Stuttgart

441 **a-b Mobile membrane-roof construction**
Olympic Stadium, Montreal (today closed)

442 **System drawing of interior roof**
Commerzbank Arena, Frankfurt
Planning: gmp Architects with sbp Stuttgart

443 **Spoked-wheel construction**
Structural system drawing

page 223:
444 **Stadium interior with video cube**
Commerzbank Arena, Frankfurt

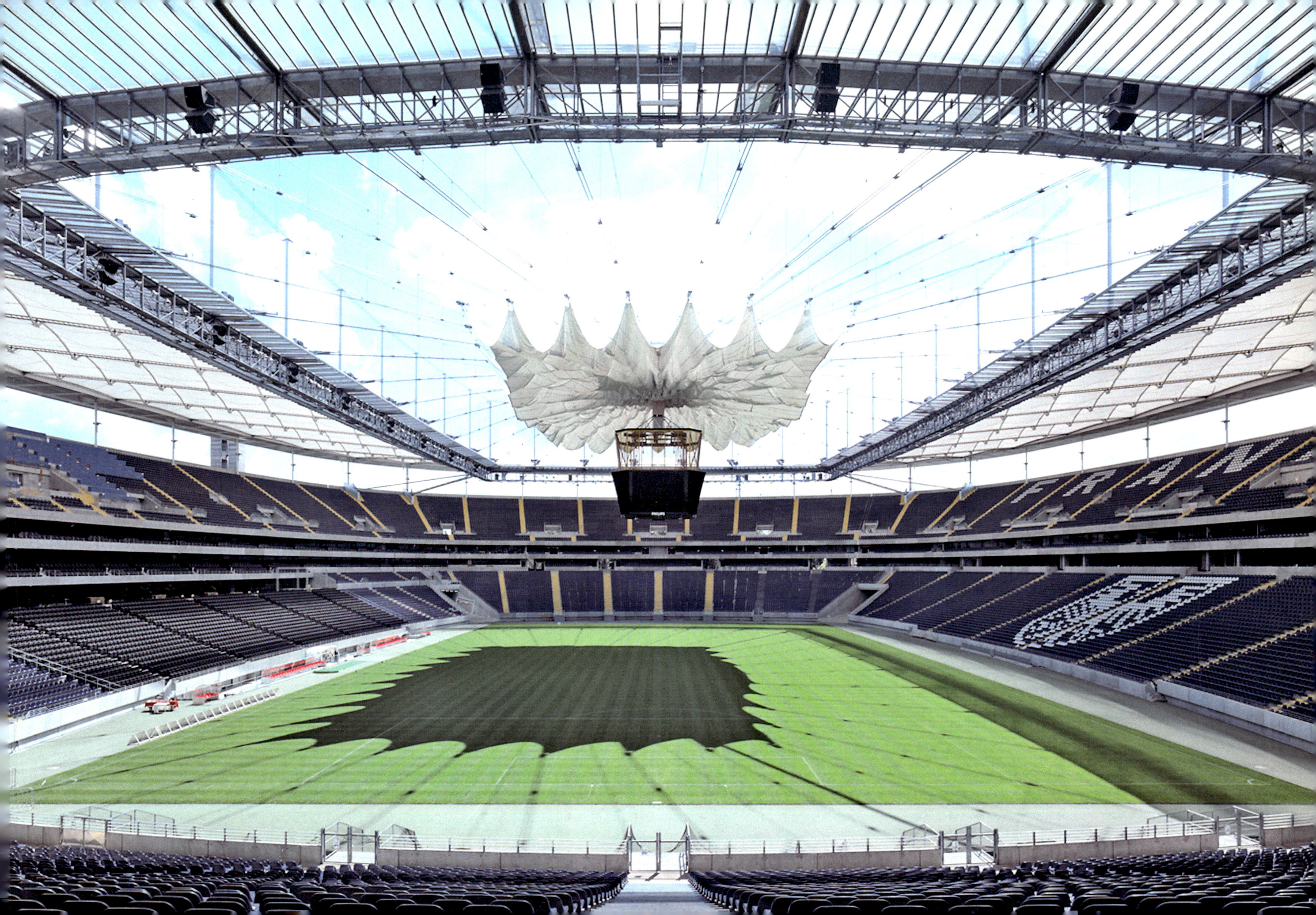

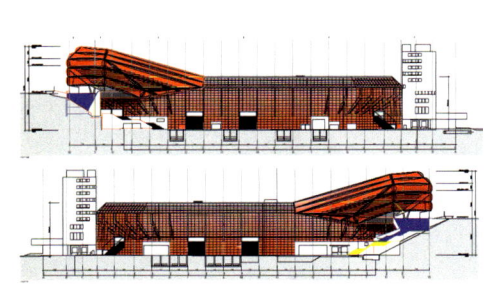

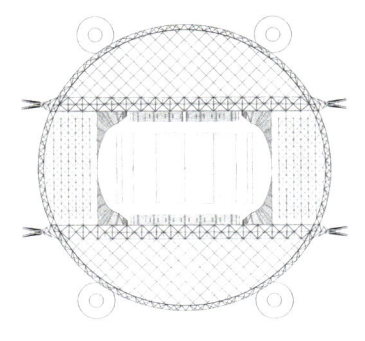

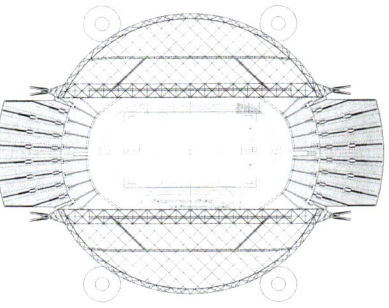

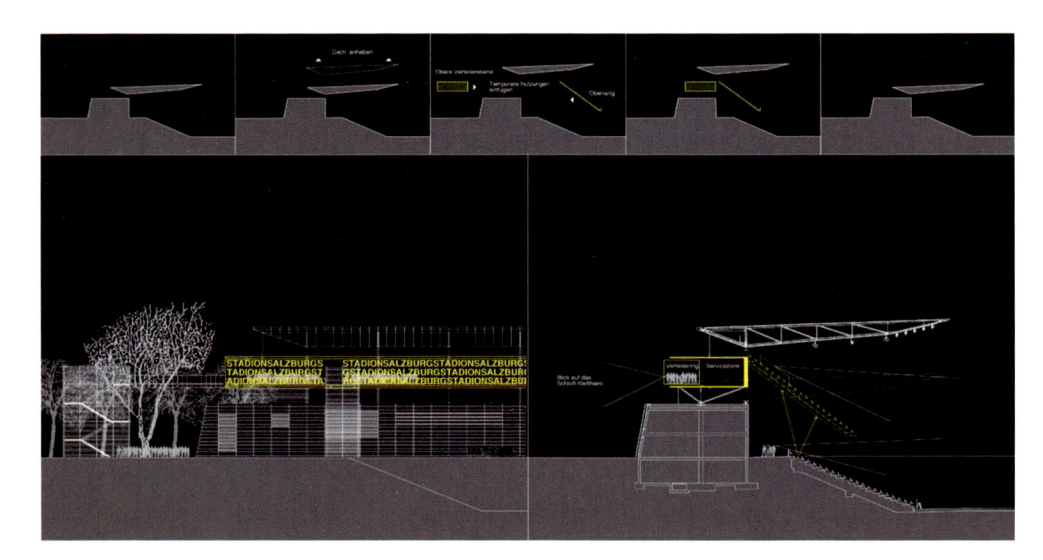

Adapting capacity

Upcoming major sports events are frequently the main motivation behind the modernization, renovation or new construction of event venues.

Since requirements for such events often by far exceed normal demand, planners and operators are keen to find suitable measures for adapting capacity. Underestimated operating costs of overdimensioned stadia or arenas in the follow-up of such large events might result in major financial difficulties for operators.

The various solutions available can only be decided upon as part of a refinanced follow-up concept. No clear recommendations can be made here, since the circumstances and constellations of operator / client / owner, as well as the intended concept of operation, can vary greatly from case to case. Generally, we may distinguish two concepts:
1) permanent capacity
2) temporary capacity

In the latter case, redundant seats are dismantled after the event. Architects HOK + LOBB suggested for the new build of Olympic Stadium Sydney two grandstands for 2 x 15,000 spectators, set up behind the lower tier of the short ends for the duration of the event. After dismantling, the missing roof structure was closed in a circle.

Lift-slab roof structures

In exceptional cases, roof structures are also planned in a vertically relocatable manner. Lift-slab roof structures are also suited for adapting capacity. Due to the short-term provision of additional terraces, such as the temporary set-up of an upper tier, the roof has to be moved upwards to release the necessary height.

This method is currently being analyzed in the run-up to the Football European Championship in Austria / Switzerland 2008. Here, a general capacity of around 12,000 spectators has to be upscaled to the short-term capacity of around 30,000 spectators demanded by UEFA.

To what extent these conversion measures (additions) to existing stadia for EM 2008 will be downscaled back to the original requirements was not uniformly decided upon by the end of 2005. Assembly and dismantling costs have to be offset against expected operating costs.

Note: An effect that should not be underestimated: hosting an event in a stadium which is too large might affect the atmosphere significantly.

Supplementary utilizations

Multifunctionality of a venue is frequently ensured by adding further units which are not directly connected to the actual sports utilization. Operated independently, they often have an indirect relation to the stadium/arena, or merely exploit the felt advantage of an attractive location.

Examples for such 'sports parks' can be found the world over. Frequently, investors are interested in exploiting the extensive requirements of potential visitors. Again and again, the 'family' is at the focus of such scenarios: visiting an event, using the catering facilities and shopping in a mall. Individual interests might be so varied that one member of the family would go to the fitness club, while the others go shopping, run errands or picks up the kids from daycare and later take them to the on-site cinema complex. And all of this with excellent parking facilities provided.

The sheer size of such a complex can often only be accommodated outside the urban centres. At the same time, it would be highly reliant on traffic infrastructure (cars, trucks and public transport). Should this development toward mega-complexes continue, urban-development solutions will have to be found.

In 2005, the practice of Dutch architect Wiel Arets in Maastricht developed a stadium complex with additional utilizations and adjacent buildings for Stadium Euroborg in Groningen.

This design transforms the stadium type into a singular urban element where living, working and leisure are interconnected. It thus questions classic concepts and views of a stadium, to be identified as a sports venue from its appearance.

Architectural practice gmp Architects, Berlin, in 2004 was commissioned with the planning and realization of a large multi-purpose sports complex with four stadia and an adjacent shopping mall in the United Arab Emirates: 'Dubai Sports City'. The approach to build four specialized stadia as core utilization of a new urban area is based on an urban development concept connecting all individual parts through a boulevard in the form of a ring-shaped mall.

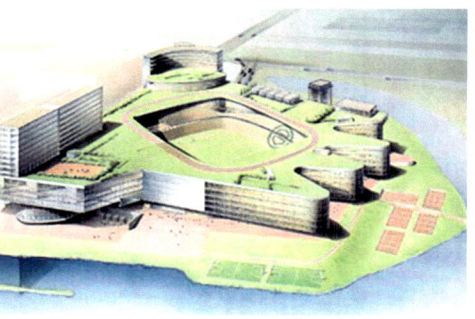

page 224:
445 **Temporary stands**
Olympic Stadium Sydney 2000 (HOK s+v+e)
446 **Ground plan drawing, Sydney**
Grandstand structure during the Olympic Games
447 **Ground plan drawing, Sydney**
Grandstand after dismantling of mobile terraces
448 **Planning of additions for EURO 2008**
St. Jakob Stadium, Basle
449 **Planning of additions for EURO 2008**
Salzburg Stadium
450 **Sliding roof structure (Kobe, Japan)**
after dismantling of temporary short-end terraces

451 **Sports and business centre**
Dubai Sports City (gmp Architects, Berlin)

452–453 **Multifunctional component for urban development**
Groningen (Arch. Wiel Arets, NL)

Chapter 21

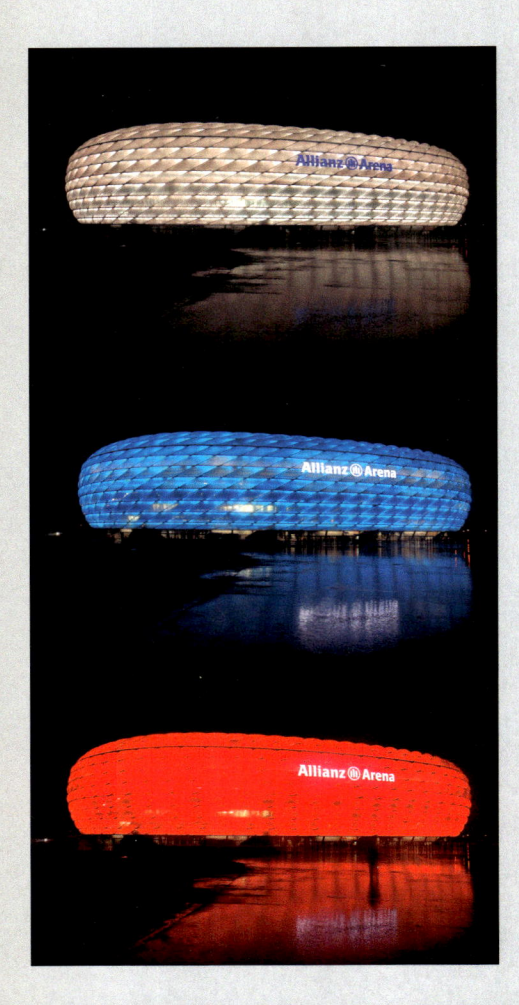

454 **Colour-light atmosphere of façade**
Allianz Arena, Munich
a. white: neutral
b. blue: TSV 1860 München
c. red: FC Bayern München

455 **Pictogram**
Adaptability of façades

Light and architecture
Light orchestration and media-compatible floodlighting for event venues

Allianz Arena, Munich

A stadium is transformed into a luminescent body. Positioned next to the motorway, it becomes a shining landmark and, together with the wind turbine and the Fröttmaninger Berg hillside, creates a new gateway to the city.

The new Allianz Arena is distinguished by the same special quality as the Olympic Stadium in Munich: two home clubs share the stadium, FC Bayern Munich and TSV Munich 1860.

Depending on which teams are playing, the variable illumination of the outer shell ensures identification by colour: white for neutral games, red for home matches of FC Bayern and blue for TSV 1860.

'The whole stadium breathes blue, red or white light. The luminous body welcomes approaching fans prior to the game with the club colours, heightening the excitement and anticipation. The atmosphere of the stadium pervades also areas and rooms that have no relation to the pitch. At the end of a football day the light accompanies fans on their way back home and reflects the feel of the events on the field' [from the official website of Allianz Arena, Munich].

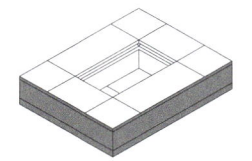

Façade illumination

2,750 lozenge-shaped cushions made of ETFE foil form a rounded overall shape of 66,500 m², at the same time roof and facade. It is currently deemed the largest membrane shell in the world.

1.056 cushions on a total surface of 25,500 m² are illuminated in white, blue or red.

Per cushion, four lights are positioned in pairs between two lens-shaped cushions (a total of 4,250 special lights); 25,000 Langfeld fluorescent lamps with a service life of 8,000 h each and a performance of around 1.47 MW.

A red, blue or transparent cover plate placed in each cushion enables the change of colour. An asymmetric parabolic mirror ensures even illumination of the cushions; light density on the shell with a maximum of 3000 cd/m² at full illumination. A gradual change of colour can be realized within two minutes; a faster execution is impeded by the potential risk of distracting drivers on the adjacent motorway.

The Allianz Arena logo placed at the north and south side of the arena on a 40-m steel substructure consists of twelve letters, which can be illuminated by 100,000 light-emitting diodes in blue and white.

Weighing 350 g/m², the ETFE foil (ethylene tetra fluoro ethylene) has a thickness of merely 0.2 mm, according to manufacturer Covertex. Depending on the number of foil layers and the type of print on the surface, ETFE foils have a light permeability of up to 95 %.

UV rays are not held back by the material, thus significantly sustaining the natural light conditions of the arena interior. The cushions are transparent at the roof area (south-western area for better lighting of the grass surface) and translucent (white) for the rest of the façade. Translucency can be reduced by 50 % through prints on the surfaces. The coefficient of heat

Comparison PTFE/PVC

transfer is circa 2.0 W/m², depending on the geometrical conditions and the number of foil layers.

The total energy value of a printed, three-layer cushion is around 0.77 (g-value). The 1,380 cushions vary in their overall size and form, ranging from 7.6 m² to 40.7 m² with an edge length of around 3 to 10 m. The largest span ranges from 1.9 m to around 4.6 m with a diagonal of 17 m maximum.

The cushions are inflated by ventilators at a continuous pressure of 350 Pascal (pressure can be increased up to 800 Pascal for snow up to 1.6 m).

The life cycle of these cushions, according to the manufacturer, is expected to be 25 years. They are non-inflammable and extremely resistant to heat (up to 300° C) and frost.

ETFE skins have very little mass with a basis weight of 0.15 to 0.35 kg/m², a strong advantage in case of fire. The fabric melts up and thus creates outlets for smoke and heat. At the highest point (around 51.5 m), 19 cushions can be opened and closed for heat and smoke extraction.

The special hydraulic cushions can bear a maximum load of 8 t, and they can resist a wind suction of 22 t. The foil is non-walkable.

High-quality synthetic foils, like PTFE/PVC and ETFE materials, make possible almost all types of textile architecture.

PTFE membranes are produced from a glass fabric to which a PTFE coating is applied. This coating protects the fabric from rain or UV-rays and provides the material with a first-rate anti-adhesive surface which makes the membranes dirt-repellent and easy to clean. Rainwater already cleans off most of the dirt. Depending on their usage and climate conditions, PTFE membranes have the longest service life of at least 30–40 years, according to Covertex. The tearing strength of the standard types ranges from 40–150 kN/m².

PVC-PES membranes are made from a polyester fabric (PES) to which a PVC coating is applied (comparable to truck tarpaulins).

This coat protects the fabric against rain or UV radiation. An additional sealing coat is applied to prevent plasticiser movement within the PVC coat. A high-quality protective PVDF varnish gives the surface also very good anti-adhesive qualities. As opposed to PTFE membranes, PVC materials have a shorter lifespan of 15 to 25 years, depending on their areas of use.

Due to their high tearing strength of 60–196 kN/m² and their extraordinary flexibility, PVC and PES membranes can be universally used.

Areas of use

With such properties, PVC-PES is exceptionally suitable for movable structures. It is also less sensitive in handling and thus easy to fabricate. However, usage of this material for retractable roofs appears less beneficial since the structural fibre may break.

PTFE glass fabric with its long life cycle is more frequently employed for permanent, mechanically tensioned, one-layer roof covers or façade structures. It can also be combined with other materials for pneumatic structures. Multiple-layer structures, thermally insulated and sound-absorbing, are also feasible. Due to their excellent mechanical properties, PTFE-glass membranes are very low-maintenance. Inspected every three to four years, the surface of the membrane is examined for mechanical damage and can frequently be repaired on-site, depending on the type and size of the damage.

Fire performance

Most PTFE-glass membranes fulfil the fire resistance grading 'incombustible A2' of DIN 4102 and are generally tested to French, British and American standards. With a basis weight of 0.8 to 1.55 kg/m² these membranes display very little mass. Depending on the thickness of the material, they have a light permeability of up to 13% (in exceptional cases more than 20%, with a diminished resistance to tearing).

All PVC-PES membranes generally fulfil the fire resistance grading 'hardly inflammable B1' of DIN 4102 and form non-burning droplets. Depending on the thickness of the material and the type of coating, PVC-PES membranes have a light permeability of up to 20%. Maintenance properties are comparable with those of PTFE membranes.

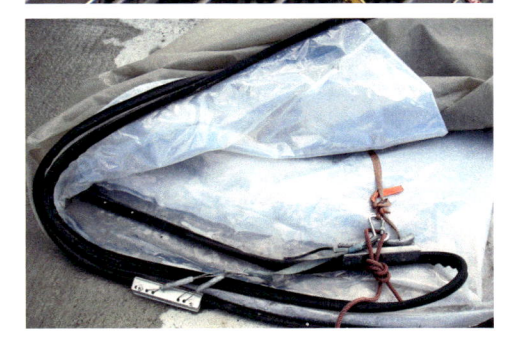

456 **Pneu-façade structure**
Allianz Arena, Munich
a. Façade section
b. Installation of air cushions
c. Detail image of an empty pneu (before installation)

Light and structure

The presence of a building is not extinguished at night. Particularly event venues seem to come to life primarily at the onset of darkness.

Light is the prerequisite for visual appreciation of buildings, roofs, supports, walls and fittings.

Transgressing mere necessity, light offers the opportunity to orchestrate architectural design also at night time. By means of specific illumination, a building can be 'bathed' in a completely different light.

Many bearing structures are displayed to their full advantage only through a suitable lighting concept, as in the roof structure of new Beijing Olympic Stadium or the illuminated façades of St. Jakob Stadium in Basle, both by Herzog de Meuron. The stadium structures of Cologne and Frankfurt, designed by gmp Architects, are accentuated by a silhouette of indirect illumination of the roof soffit, which makes them appear light and suspended. Jutting out like indicators into the urban silhouette, the four light pylons of the World Cup Stadium in Cologne strengthen its significance and recognition factor. The towers as necessary pylons for the suspension bridge not only serve static or technical purposes but also function as reference points in the Cologne Sports Park, with the main entrances to the complex located at their four base points. The perception and atmosphere of the stadium at dusk and during the evening hours is completely distinct from the daytime.

457 **Night-time illumination**
Beijing Olympic Stadium (HdM)
458 **Night-time illumination**
St. Jakob Stadium, Basle (HdM)
459 **Night-time illumination**
Commerzbank Arena, Frankfurt (gmp)
460 **Night-time illumination**
Olympic Stadium Berlin (gmp)
461 **Night-time illumination (south side)**
RheinEnergieStadion, Cologne (gmp)

Floodlight planning

From athletics competitions to international football tournaments: the capability of hosting events at any time of the day is a prerequisite for a sports stadium.

The uniform kick-off time for games of the UEFA Champions League is 20.45 CET, which means the entire stadium ground has to be illuminated sufficiently.

The requirements stated in DIN EN 12193 apply to illuminations.[1] For television, stricter requirements may apply, therefore the guidelines of the relevant sports associations are binding, as they have been agreed with TV broadcasters. The light orchestration of a stadium has a significant effect on the intended atmosphere. Following, only the minimum requirements to floodlighting are defined. The decision on which type of illumination is selected is determined by the type of sports and the requirements of spectators; athletes and players have to recognize small objects like balls, and spectators and media have to be able to follow the activities on the field.

Five grades of lighting equipment are defined, allocated to two types of sports complexes:
– Events broadcast on television:
Grade V   International
Grade IV National
– Events not broadcast on television:
Grade III National events
Grade II League and club events
Grade I   Training/leisure events

Kick-off times
(DFL season 2005/2006)

1st Football Bundesliga
Fri   20:30 h
Sat   15:30 h
Sun   17:30 h

2nd Football Bundesliga
Fri   19:00 h
Sat   13:00 h
Sun   15:00 h
Mon   20:15 h

462 **Night-time illumination (north side)**
RheinEnergieStadion, Cologne (gmp)

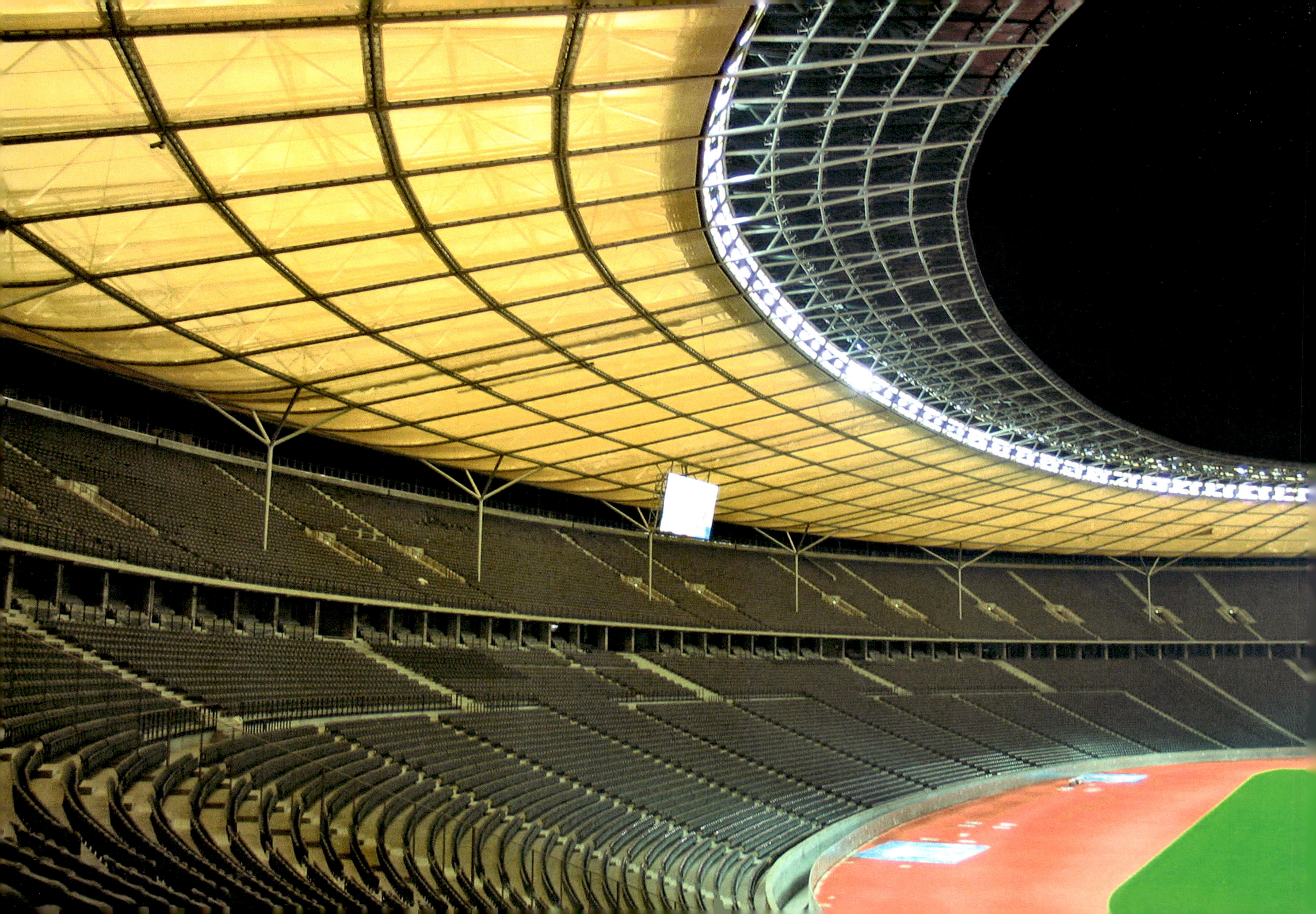

'Ring of Fire' – Berlin

The renovated floodlight and illumination system of Olympic Stadium Berlin represents a true progression of the possibilities to orchestrate events by architectural means. New standards have been set by this integral approach to floodlighting: the stadium light illuminates the stage and accentuates the playing field, similar to a theatre. The event itself takes the centre stage; all else becoming the backdrop. In contrast, the light concept 'ArenaVision' of Philips Lighting also comprises direct illumination of tiers; not as brightly lit as the turf but sufficiently enough to bring out choreographies, flags and the stadium atmosphere in general. In World Cup Stadium Gelsenkirchen the tiers were specifically floodlit to enhance its character as a spectator facility. These are two very different approaches to stadium lighting. The decision remains with planners and operators how to use light orchestration to support the event profile of a modern sports venue. A new development by Frans Sill GmbH (specialist light technology manufacturer) for a 'spotlight, specifically for stadium lighting' [title of patent application] allows direct illumination. Through improved reflector technology, the lighting in Olympic Stadium Berlin can be terminated at specific architectural lines (e.g. the terrace front edge), without the uncontrollable and undesired effect of diffused light caused by conventional stadium spotlights.

With its height and distance to the pitch, the stadium roof in Berlin enables a standard layout of spotlights in one continuous line (ring closure). This circulating 'Ring of Fire' eases the appearance of floodlight beams. Years of viewing habits have accustomed spectators to the four fields of light produced by four floodlight masts. In contrast, stadia of the 'third generation' are normally fitted with groups of spotlights aligned in one or two lines. Although blinding of spectators and participants is mostly avoided, the lights can still be identified as short catenaries at the roof. The Berlin approach is pointing the way whenever a circulating roof soffit is supposed to be 'toned down' by architectural means, in order to avoid visual incongruity of structure and luminaries. Height and distance of the ring of light at the transition of glass roof to membrane roof (position approx. 13.50 m behind the front edge of the glass roof) allow a circulating lighting ring: to the first touch line in a distance of circa 68 m, at an angle of 47° and at circa 35 m height (20° to the opposite touch line). The most striking feature is the creation of a continuous carpet of light on the playing field with 1,500 lux produced by the 155 double spotlights made of Halogen lamps (2000 W).

Previously, every spot would be illuminated from four sides, a mast layout with rotation-symmetric spotlights and additional light fixtures at the roof. Another type of floodlighting works with one or two 'rings'; a four-split shadow remains, only smaller. More spotlights would mean less shadowing. However, a television-compatible 'modelling' of the illuminated objects requires sufficient contrast. The Anglo-Saxon approach often employs the classic aspheric mirror which achieves a lower modelling and generally requires more energy. Light planning should be based on local daylight conditions: if the general weather situation produces little shadowing, this should be considered. The same goes for zenithal-light regions.

Through multiple overlaying of the broader cones of light produced by the circulating spotlights at World Cup Stadium Berlin, a uniform carpet of light is created.

The concept attempts to integrate the advantages of the foregoing methods: sufficient modelling at a highly uniform overall appearance. The architectural equanimity of the roof is continued by a smooth interior lighting of the roof body. 4,200 dimmable luminescent tubes produce an indirect light behind the mesh membrane, which as suspended ceiling is spanned between the bottom flanges of the roof structure. All technical-service fittings (PA system, ducting and maintenance, etc.) are integrated into the roof structure. During the day, the roof soffit thus appears very calm and refrains from deflecting attention. At night, the self-luminous roof appears very light and also serves to illuminate the stands. Various forms of light orchestration can be realized with different degrees of brightness.

page 230:
463 **Orchestration of stadium lights**
Olympic Stadium Berlin (gmp)

464 **Tree-like column and transparent roof**
Olympic Stadium Berlin

465 **Lower circulation ring**
Olympic Stadium Berlin

The development of floodlight technology

Over 50 years ago in Charleroi, Belgium, Philips installed a sensational type of floodlight. In those days, 14 masts with four lamps each produced 6–30 lux, depending on the respective position on the pitch. In comparison: the lighting of a regular training ground is recommended with 75 lux, and the FIFA request for World Cup 2006 was $E_v = 1,500$ lux for all areas covered by the main camera. Until the late 1960s, football stadia did not require excessively bright floodlights; the stadium at Bökelberg in Mönchengladbach, for example, was fitted with the first TV-compatible floodlighting with 440 lux. Colour television caught on in Germany with the Olympic Games in Munich and the Football World Cup 1974. In order to guarantee a realistic colour representation on TV, lighting had to be brighter, and aspects like colour temperature or colour display had to be taken into account. AufSchalke Arena, opened in 2001, has a lighting intensity not much higher than that of Rosario Central stadium (World Cup 1978, Argentina).

Energy consumption

In 1949, in Charleroi, energy demand was at 14 kW; not in the least comparable to that of modern stadia. According to manufacturers, RheinEnergieStadion in Cologne is lighted by 210 spotlights à 2,000 W, resulting in an energy demand of near to 450 kW. The old Müngersdorfer Stadion employed almost double as many spotlights with 800 kW. Still, according to TV producers, lighting quality is much better today.

Positions of floodlights

For decades, floodlight masts had been typical features of stadium complexes, frequently turning them into urban landmarks. Commonly, there were four masts, one in each corner, circa 10° behind the byline and 5° next to the touch line. The height of the lower edge of a floodlight field is stated at 25° horizontal to the pitch centre (inclination of spotlights at the mast head circa 15°). Although the playing area is sufficiently lighted, the cross-shaped four shadows which a player produces are mostly undesired. In this context, blinding through spotlights positioned at the roof area has to be taken into account with regard to television camera technology (16:9).

Floodlighting from the roof area

The progression of lighting technology and the problems related to unwanted light emissions have brought about a change in lighting layouts. With the loss of athletic tracks, terraces have been moved much closer to the pitch and thus the size of the circular roof opening has been reduced (circa 80 x 120 m).

From a geometrical point of view, this would complicate but not outrule the set-up of masts; detailed planning and the clear wish for prominent masts provided. Integrating floodlights into the stadium roof remains a possibility. Differences in brightness on the field should be minimized. Accordingly, the effects of shadowing caused by parts of the roof or video cubes should be particularly considered.

As a rule, one or two lines of luminaries are positioned at the roof; the first at the front edge of the stadium roof and the other at the rear. Depending on the height of the roof, sufficient lateral illumination has to be provided since it is vital for images produced by a television camera. In order to avoid blinding of the goalkeeper during a corner kick, an angle of 15° to the corner flag should be kept clear of lights. Fixing floodlights to the roof edge is problematic if terraces or their coverings are too low. When the angle of lighting the turf is too shallow, players and spectators are blinded; if spotlights are turned downwards, the opposite side remains in the dark. In cooperation with FIFA, Philips Co. has determined that the direction of the spotlights should never diverge more than 70° (mast position due to television glare) from the vertical. As a result, floodlights should be positioned at least 30 m above the playing field. The ideal transverse positioning of floodlights can be determined at 25° of the first row measured to the horizontal (in reference to the opposite touch line) and 50° of the second line of floodlights to the closest touch line.

466 **Spotlights at the roof edge**
Anfield Road, Liverpool, GB

467 **Cover**
Handbook for the illumination of football grounds by artificial light by Philips Co.

Illuminance E

Sports association guidelines

Horizontal illuminance E_h determines at which luminous flux (measured in lux = lumen/m²) horizontal surfaces such as the turf should be illuminated.

For television cameras, vertical illumination E_v is decisive, with regard to capturing football players or participants in competitions.

Level of vertical illuminance

(excerpt from DIN/EN 12193 section 5.3.2)
'Vertical illuminance is mainly reliant on the activity speed of the respective sports as well as the distances and angles of recording.

Sports may be divided into categories A, B and C, depending on the speed of activity and the dimensions of the objects to be recorded on camera.

CTV groups are listed in table 5.2. Knowledge of the maximum recording distances and of the CTV group for the respective sports allows the calculation of the corresponding value of vertical illuminance.

These values are not adequate for slow-motion recordings, which require higher levels of illuminance.'

Illuminance values are stated in the SI-unit lux (notation E, unit notation: lux).

Illuminance is the photometric equivalent to irradiance E (unit: W/m²) It is the quotient of luminous flux → per element of receiving area A_e. Hence, lighting intensity is a mere receiving value.

As property of a light source it is not reliant on the distance at which a spectator is located. It relates to the part of a luminous flux (unit: lumen) which is emitted into a certain direction. Light density or luminance is another quality (unit: candela).

DFB/DFL

In the implementation directions for the DFB Laws of the Game the following stipulations were laid down:

Floodlight systems must have a performance of at least 700 lux; existing floodlight systems in the regional leagues have to produce at least 400 lux. DFL/DFB request 800 lux.

Clubs or ground operators should have the lighting system thoroughly checked and cleaned at least twice a year by a specialist company, in order to avoid system failure or a fast repair of damages.²

FIFA/UEFA

The rules of the FIFA Football World Cup Germany 2006™ state that games may be hosted at daylight or under floodlight conditions. The regulations also stipulate that for games taking place in the evening the entire field has to be lighted with a minimum of 1,200 lux.

UEFA requirements range, depending on the relevant competition, from 800–1,400 lux.

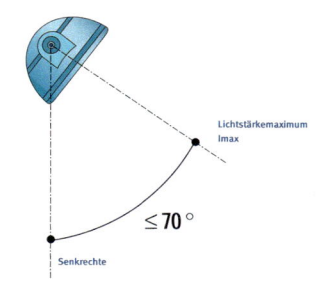

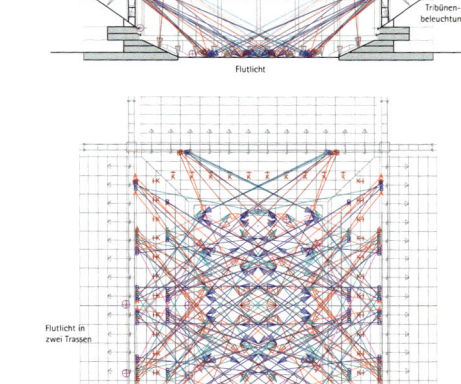

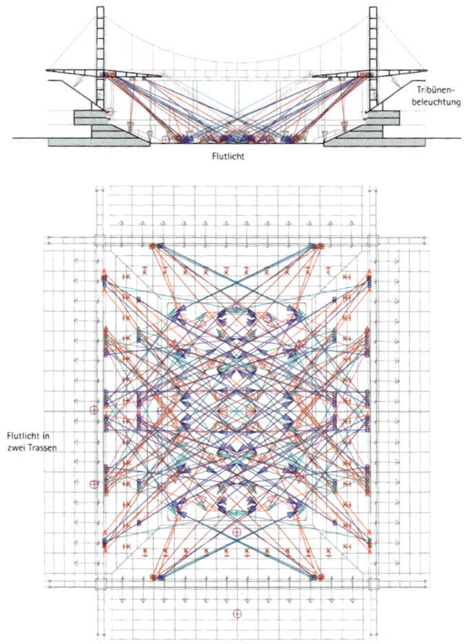

468 **Sketch of head of a spotlight**
70° maximum inclination

469 **Floodlight plan by Philips Co.**
Competition entry for Stadium Cologne (gmp)

470 **Typical floodlight mast**
Stadium of Dynamo Dresden

Vertical illuminance for floodlight systems

DFB
E_v = 700 lux   Regional league
E_v = 800 lux   Bundesliga

FIFA/UEFA/IAAF
E_v = 1,800 lux   for slow motion
E_v = 1,400 lux   for permanently fixed cameras
E_v = 1,000 lux   for movable cameras

Illuminance during lighting failures

FIFA/UEFA/IAAF
E_v = 1,000 lux   main camera

471 **Floodlight plan**
Letzigrund Stadium, Zurich, Switzerland
472 **Floodlight (roof-fixed)**
Stade de Suisse, Berne, Switzerland
473 **System drawing (glare)**
FIFA 1995
a. Floodlight positions
b. Sun and shadow

page 235:
474 **Illuminance**
Excerpt from DIN 12193 'Sportstättenbeleuchtung'
Maintenance value of vertical illuminance in relation to the recording distance (page 11)
475 **Table 4.3.D: Illuminance**
from the *Handbook for lighting of football grounds by artificial light* by Philips Co. in cooperation with FIFA (2002)

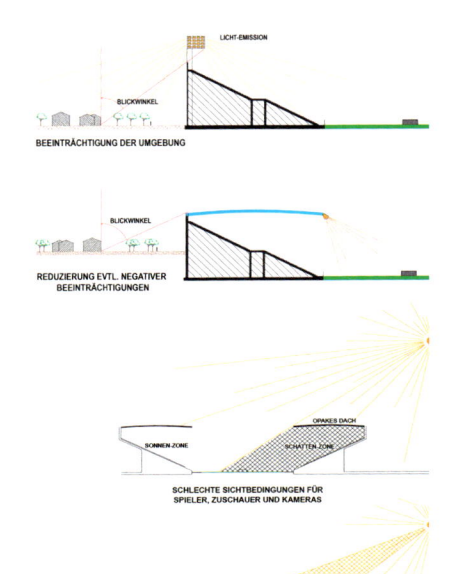

Diffused light
is 'light beamed onto vertical surfaces such as façades or penetrating through a window into a house or apartment, representing a nuisance to residents'.[3]

Glare
defines the brightness or intensity of a source of light which an observer sees when looking into the direction of where the light is emitted from; and light beamed upwards (light immissions), floodlight beamed into the sky as a so-called 'dome of light'.

FIFA Technical Recommendations 2006 raise these values to:
E_v = 1,500 lux   main camera
E_v = 1,000 lux   for other zones

IAFF (Track and Field Manual)
National and international competitions
E_v = 1,800 lux   slow-motion camera
E_v = 1,400 lux   mounted main camera
E_v = 1,000 lux   movable cameras and television emergency light

Generally, we may assume that light performance may have to be compensated due to dirt or the short-term failure of individual spotlights. Therefore, a reduction factor of 0.8 should be determined since the requested values reflect maintenance illumination.

Note: Consequently, the indicated numerical values as found in various publications, frequently vary. Example World Cup 2006: Due to relatively new floodlight systems, sometimes lower attenuation factors (0.85 x 1,500 lux = 1,275 lux → 1,200 lux minimum) were granted.

Glare rating (GR: a scale of 10–90 from non-perceptible to intolerable glare) should not exceed a value of 50, for both players and spectators; with a good floodlighting system clearly below 50. Therefore, the positioning and set-up of floodlights is particularly important.

The colour temperature T_k is measured in Kelvin and is an expression of the architectural appearance and the ambience that is produced by this light; red creating a warmer and blue a colder sensation. White balance for television is always important, and FIFA has stipulated in cooperation with broadcasters a value of 4,000 to 6,000 degrees Kelvin for interior areas.

The colour rendering index R_a defines the capability of a light source to authentically reproduce natural colours. FIFA demands at least R_a 65 (for training sessions), preferably higher or equal to R_a 90, i.e. colours are matching, which is vital for advertising exploitation. (A 15% loss in luminous flux is taken into account for the mentioned values.)

Light failure
The biggest potential problem for a television broadcast lies in a sudden power failure. Emergency power generators supply stand-by power to maintain safety lighting and to enable emergency announcements and, if necessary, guarantee a continuation of television broadcasting.

As stated in the relevant FIFA rules, an emergency power generator or a second independent power supply must be provided so that, in case of failure, illumination of the entire pitch is guaranteed at the above mentioned rate (1,000 lux). In comparison, UEFA demands a substitute floodlight system with 1,200 lux for EURO 2008.

In the event of a short-term power failure, a standard floodlight cannot be switched on again instantly as it has to cool down first. Therefore, a sufficient number of spotlights with a hot re-ignition mechanism has to be installed so as to guarantee the previous light intensity.

Another variant or combination would be continuous power supply through batteries, emergency diesel generators or a second power feed-in.

Safety facilities

In the event of a general power failure, the following facilities have to be provided with a safety power supply:[4]
1) safety lighting
2) automatic fire extinguishing systems, etc.
3) smoke extraction systems
4) fire alarm systems
5) warning systems

Complementing this, FIFA Technical Recommendations 2006 request a stand-by power supply for the following units:
6) sound/PA system for the entire stadium complex
7) floodlight system (2/3 of the required light intensity)
8) video monitoring system
9) computer infrastructure
10) press stand (media workplaces)

Safety lighting (escape routes)

During a possible failure of general lighting, visitors have to find their way easily to the relevant escape routes. All areas up to the public circulation levels have to be well-lit; the same goes for parking routes and spaces. DIN 13200-1 item 6.1 requests a lighting intensity of at least 10 lux. A seating and escape route plan must be placed well visible at the entrance area and may not exceed or alter capacity and approved planning.[5]

Loud-speaker facilities

As multi-purpose event venues, modern sports stadia require an excellent quality of sound for speech and music with individual sector alignment. Security staff suggest an either totally or selectively switchable PA system, e.g. for intermediate areas behind the goals, straight and back straight (esp. home and away fans) or the playing field.

UEFA states the following requirements: a modern and efficient system which produces clearly audible announcements during the game, inside and outside the stadium.

Intelligibility has to be maintained in spite of unfavourable conditions such as spectator noise, sudden rises in noise levels or loss of main power supply.

According to Technical Recommendations of FIFA/DFB, a police priority interrupt with dedicated microphone should be installed, and an automatic regulation of sound volume should be feasible for emergency announcements. Rooms for loud-speaker technology are to be located near the stadium control room. FIFA determines the sound pressure level with 105 dB (A), 10 dB (A) above the interference level.

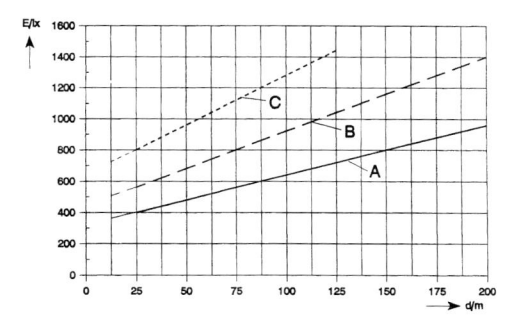

474. Maintenance value of vertical illuminance in relation to recording distance

475.

List of disciplines (alphabetic sequence)

Sports	Group for television and film recordings	
American Football	outdoors	B
Ice Hockey	indoors	C
	outdoors	C
Football	indoors (small pitch/hall)	B
	outdoors	B
Athletics	indoors (all disciplines)	A
	outdoors	A

The letters A, B, C refer to the relevant kind of sports for television broadcasts (CTV) and film recordings.

Values for illumination of non-broadcast events

Event or activity (Class)	Horizontal illumination	Uniformity gradient	Glare rating	Colour temperature	Colour rendering
Class	Ehcam average (Lux)	U2	GR	T_k	R_a
National events (Class III)	500	0.7	≤ 50	T_k > 4000 K	≥ 80
Training and club events (Class II)	200	0.6	≤ 50	T_k > 4000 K	≥ 65
Training and club events (Class I)	75	0.5	≤ 50	T_k > 2000 K	≥ 20

Values for illumination of broadcast events

Event or activity (Class)		Vertical illuminance			Horizontal illuminance		Colour features of lamps	
Class	Calculation according to direction	Ehcam average	Uniformity gradient		Ehcam average	Uniformity gradient		Colour temperature	Colour rendering
		Lux	U1	U2	Lux	U1	U2	T_k	R_a
International events (Class V)	Slow-motion cameras	1800	0.5	0.7	1500 to 3000	0.6	0.8	T_k > 5500 K	≥ 80 (pref. 90)
	Permanently mounted cameras	1400	0.5	0.7	1500 to 3000	0.6	0.8	T_k > 5500 K	≥ 80 (pref. 90)
	Mobile cameras (at pitch level)	1000	0.3	0.5	1500 to 3000	0.6	0.8	T_k > 5500 K	≥ 80 (pref. 90)
National events and safeguarding of television broadcast (Class IV)	Permanently mounted cameras	1000	0.4	0.6	1000 to 2000	0.6	0.8	T_k > 4000 K	≥ 80 (pref. 90)

Chapter 22

476 Artificial turf playing field
Stadium Salzburg (EURO 2008, Austria)

The installed playing surface should ...
... first and foremost ensure a risk-free and highly weather-independent exercise of the respective sports, according to the rules of the relevant sports association.

Natural versus artificial turf
Shedding light on a question of faith and the current technological status of developments for the playing field

Natural versus artificial turf

This issue has become almost a question of faith, largely influenced by the varying intentions of diverse interest groups.

Artificial turf surfaces will probably never achieve the countless positive features of natural grass in a very good state of maintenance at an ideal location in its best vegetation phase; even though they have come quite close. Many experts do agree now that artificial turf is preferable to any hard sports covering and even to low-quality natural grass surfaces.

As a rule, four different surface types are distinguished for sports grounds:[1]
a) hard sports areas: training or leisure football field with cinder covering
b) grass surfaces: competition or leisure sports ground with natural turf
c) synthetic surfaces: covering for running tracks and tennis ground with synthetic polymer layer
d) artificial turf surfaces: alternative synthetic turf covering for field sports.

Hard sports areas

are mainly found in leisure sports. This type of surface involves relatively small maintenance and is mainly used for the training grounds of small sports clubs or as substitute areas when the main grass pitch is unplayable. prevention

Frequently, heavy dust during the summer and sludge in spring, autumn and winter are the results of inadequate construction and maintenance, as hard sports pitches require irrigation (cf. DIN standard).

Synthetic surfaces

have replaced years ago the cinder tracks as standard for popular and competitive sports. Albeit higher investment costs, they involve less risk of injury and are considered to be more economical sports surfaces.

Artificial turf surfaces

have been substituting hard pitches for decades. The relevant stages of development are allocated to three generations.[2]

The *first generation* of artificial surfaces was employed in the late 1960s for American Football in the US. At the time, these were surfaces with an unfilled pile layer resting on thick and soft elastic layers.

This synthetic turf is increasingly approved for disciplines such as field hockey as its properties help ensure a more precise game and a more even ball roll.

On top of this, artificial turf shows a higher frequency of usage compared to natural grass. Nowadays, artificial turf has become the norm for field hockey the world over and is requested for international games by the hockey association.

In the mid-1980s another development from the Netherlands came on the market.

This *second generation* is distinguished by a different type of construction: a filler material is added to the layer of synthetic fibres, preferably quartz sand.

At relatively low cost, this 'fibre-armed' sand pitch can substitute hard sports areas. Frequently directly applied to the unbound base layer, the functional properties, however, fall far below the expectations of sports associations.

In 2000, the *third generation* was introduced to the market. Instead of a pure sand in-fill, the pile layer is formed by grass-like fibres with sand/rubber granules. First laid on an unbound base, the method soon progressed towards an elastic layer.

International football associations FIFA and UEFA have voiced their support for artificial grass turf of the third generation.

UEFA started a pilot project where artificial turf was laid in five European stadia. A first test field was commissioned at its headquarters in Nyon, Switzerland. Since then, FIFA and UEFA have been executing experiments with regard to sports-medicinal factors and ball behaviour.

In return for their cooperation, the clubs based at the selected venues in Almelo (Netherlands), Moscow (Russia), Dunfermline (Scotland), Orebro (Sweden) and Salzburg (Austria) received a subsidy by UEFA for costs incurred.

The German Football Association (DFB) in its latest Laws of the Game (2004/2005) will take over a provision stating that, in agreement with competition regulations, matches may be played on a natural or artificial surface.

DIN 18035 for sports grounds has also responded to the evolution of this type of covering.[3]

Third-generation technical set-up

Synthetic turf carpeting mainly consists of the basic fabric, also called the pile layer, usually less than 15 mm high.

The fibres of the artificial turf are made from special polyamide PA, or modified polypropylene or PP/PE block polymers and are tufted onto a backing material. This may consist of polyester or fibre-glass fabric onto which mainly straight or curled monofilaments or pre-fibrillated bands are applied.

In order to avoid skin-burn these surfaces are watered and often not as highly perforated as previously, in order to slow down water flow into the drainage layer of the textile covering (second layer).

Note: Such unfilled pile layers are nowadays mostly used for hockey.

The quartz-sand infilled pile layers of the second generation are similar but with a layer height of 30 mm and highly perforated (football or multi-purpose usage). Today's artificial turfs have fibre lengths of 35–60 mm and are either backfilled with EPDM granules, or, as intermediate solution, with coloured polyurethane-covered recycling granules. The use of black recycling granules should have become obsolete by now.[4]

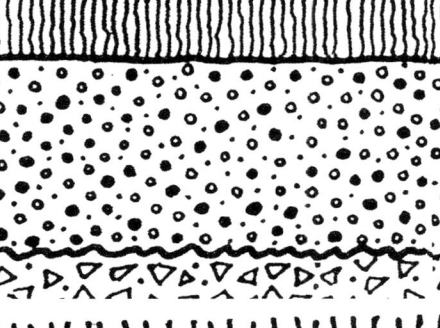

In appearance, this third-generation turf is the most similar to natural grass. The relevant materials as well as installation heights depend largely on the product and manufacturing process.

Layer 1: preparation of subsoil through earthworks
Layer 2: geo-textile or filtering layer
Layer 3: unbound supporting layer to secure bearing capacity
Layer 4: bitumen-bound supporting layer (a good solution from a technological vantage point, which is commonly not employed for cost reasons)
Layer 5: elastic layer or flexibly-bonded supporting layer
Layer 6: artificial turf with pile layer filled with quartz sand or rubber

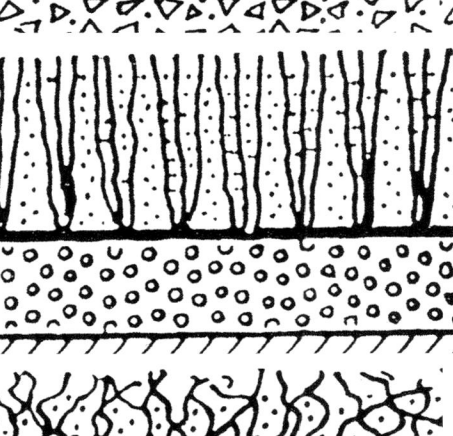

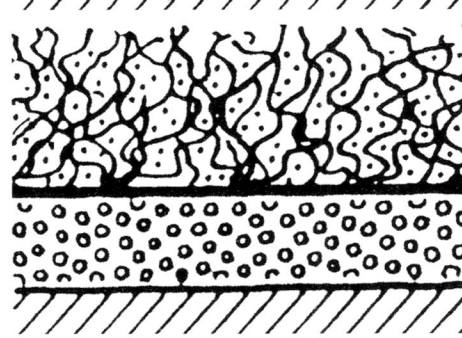

477 **Artificial turf layers**
 a. First generation: Non-filled pile layer
 b. Second generation: Quartz sand infill for pile layer
 c. Third generation: Higher pile layers
  (e.g. EPDM non-recycling granules)

22 Natural versus artificial turf

Friday, November 12, 2004

UEFA President/Director General Lars-Christer Olsson declares that, at the present state of affairs, the final round of the UEFA European Championship 2012 will be held on grass pitches.

478 **Laying of elastic layer**
Polytan Co. (artificial turf manufacturer)
479 **Filling of the pile layer**
Polytan Co.
480 **Cutting in lines**
Polytan Co.

The artificial turf fibre

The visible pile layer of the sports surface is formed by artificial turf fibres. As a rule, fibrillated fibres are distinguished from monofilament fibres.

Fibrillated fibres
are narrow foils with grid-type incisions, which improve the processing properties of the yarn and makes it softer. Continued wear tends to splice up the yarn rapidly so that form and structure of the surface are subject to changes, affecting the playing properties of the turf.

Monofilament fibres
are more costly to manufacture and consist of one extruded profile. They are much more resistant to wear and tear as they are non-splice. Abrasion occurs as loss of thickness and/or breaking of the fibre. With a special composition of the fibre this effect may be delayed.

All fibres can be curled which makes them less prone to migration on the surface (rubber granules). Curled fibres tend to lean over less, reducing maintenance costs. It has to be pointed out clearly that for athletes an increased rotational resistance has been observed, an effect which can be countered by choosing suitable shoes. The difference between natural grass and artificial turf is more marked for curled fibres than for upright ones.

Infill materials
Third-generation systems use sand first and foremost for weighting down and as filler material, normally a round-grain quartz sand of 0.3 to 1.0 mm. State-of-the-art and approved are coloured or pure EPDM granules.

Compatibility of fibres and granules is paramount. The rubber granules represent a key cost factor of the artificial turf system. Precise analysis of demand and an exact comparison of infill quantities for different systems are obligatory.

Laying artificial turf
The artificial turf is laid loosely, i.e. without bonding, onto the elastic backing layer. The individual 4 m wide lanes are rolled out lateral to the main direction of play and bonded at the joints. Fixing at the edge of the pitch is redundant due to the dead weight of the surface.

Continuous line markings in the laying direction (goal line, centre line) are pre-tufted at the factory. All other lines are cut into the synthetic turf (pasting in the coloured line bands) at the same overall quality.

Filling of the surface is done in two steps. First the quartz sand is evenly applied and brushed in, then a specified quantity of rubber granules is applied and worked in.

The quantity of quartz sand depends is on the relevant system and may range from 15–35 kg/m^2; the rubber granules from 4–8 kg/m^2.

'FIFA recommended 2 Star'

Irrigation of artificial turf surfaces

Synthetic turf of the third generation does not compellingly require watering – in contrast to unfilled artificial turf surfaces (fully synthetic surfaces) where moisture functions as a sliding film. The rubber granules in sand/rubber-filled systems provide sufficient mobile material, in case of athletes falling.

Note: Regarding the risk of abrasions and skin-burn, opinions vary significantly. Supporters of state-of-the-art artificial turf (FIFA Recommended 2 Star) argue that these properties are comparable to natural grass surfaces.

'Watering does make sense in order to lower temperatures on the artificial turf surface in extreme weather (high summer) through the cooling effect of evaporating water, to bond fine dust and to increase kinetic friction on the surface, which decreases fibre abrasion' [artificial turf manufacturer Polytan Co.].[5]

On its official website, the world football association FIFA supports installation and use of artificial turf in climate zones and stadia where the maintenance of natural grass fields is difficult:

'Many regions of the world suffer from extreme climatic conditions and as a result are often without adequate natural grass pitches. However, the trend towards building steep-sided stadia with roofs and terraces for additional spectator comfort creates an environment better suited to football turf. The newest generation of artificial surfaces combines the advantages of playing characteristics similar to natural turf, including player comfort and safety, with easy maintenance and extended usage.' (status 2005)

In the *FIFA Quality Concept Handbook of Test Methods and Requirements for Artificial Football Surfaces* from 2002, a quality concept 'FIFA recommended' was established for synthetic turf, in order to standardize the quality of artificial grass, to guarantee safety for players and to promote new developments in the field.

Over 80 artificial turf pitches with the quality standard 'FIFA recommended' were laid worldwide during the last three years alone. Mostly local or municipal properties, their advantage is to be playable almost without interruption. Many large football clubs like Ajax Amsterdam, Glasgow Rangers, Boca Juniors, Deportivo La Coruña and FC Porto have installed FIFA-standard artificial turf surfaces in their training facilities.

Note: Non-official reports by players have shown that these training grounds are not favoured or used only in exceptional cases.

Taking into account the opinions of players and physicians, the introduction of a 2 Star quality seal for 'FIFA recommended' has extended this demanding quality standard, which will further increase playing properties and safety.

Alternatively, the term 'International Artificial Turf Standard' may be used. Synthetic turf playing fields carrying this tag fulfil the same requirements as pitches with the one-star seal of approval, but do not entail a licence fee for use.

As first manufacturer, a German FIFA licensee acquired the seal of quality 'FIFA recommended' for its model LIGATURF 2040 ACS 65 and laid the first synthetic turf of this type at Borussia Park in Mönchengladbach.

In November 2004, the UEFA Executive Committee also decided that from the 2005/06 season, matches of the UEFA competitions can take place on artificial grass. The executive committee thus followed a recommendation by the UEFA commission for artificial turf. The playing surfaces are subject to UEFA approval.

481 **Certification symbols**
a. FIFA Recommended 2 Star
b. FIH Field Hockey World Association
c. IAAF Certification (Track and Field)
d. RAL seal of quality (artificial turf)

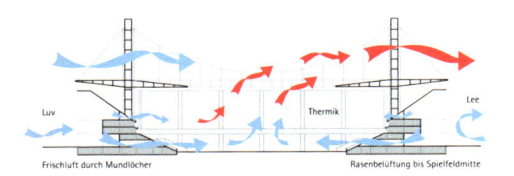

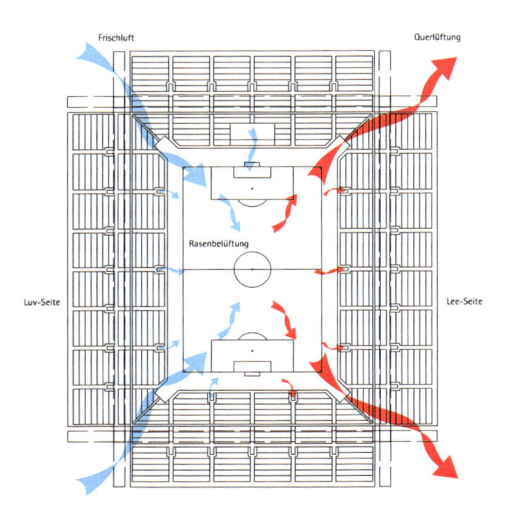

482 Pictogram
illustrating necessary turf aeration
(competition phase for World Cup Stadium Cologne,
gmp Architects, Aachen)

Natural turf

A desired proximity to the pitch produces large-capacity, steeply-sided stands. Hence, the planning of modern stadia today is accompanied by a system-immanent problem: the loss of athletics tracks and the resulting approach to the pitch perimeter produces a shadowing effect from the stand body onto the pitch.

For stadia complexes built in recent years, growth and quality of the turf have proven to be quite problematic issues. In particular during the growth phase in springtime, at an angle of just 37.5° solar altitude (50th degree of latitude) to 10° in the late afternoon, sunlight does not fall on all areas of the playing field.

Natural grass requires sunlight of a certain wave length for its photosynthesis, i.e. the assimilation of organic material from inorganic material through light energy. Photosynthesis is the metabolism that enables plants to grow and is therefore also vital for sports turfs. All green plants produce oxygen which they in turn consume at night in large quantities, in relation to their biomass. (250 m² of grass produces roughly as much oxygen as consumed by four persons.)

A typical problem is represented by the reduction of necessary light penetration through a cantilevered roof system.

Another issue is the lack of air circulation in the interior space of stadia. Such effects can be successfully countered by two measures.

1) sunlight: the roof area is altogether or in large parts transparent. Technical properties of the used materials should allow the passage of sufficient sunlight to ensure intense lighting of the turf area in its growth phase (around 400–600 Nm).

2) ventilation: ensuring that archways and passages close to the pitch are kept open consistently should ensure sufficient ventilation.

Caused by the chimney effect created by gateways and vomitories connected to the stadium exterior, fresh air can flow into the central area.

This effect is supported by subpressure caused by rising winds above the roof level.

In spite of its compact built and minimal distance to the pitch, RheinEnergieStadion in Cologne is a good example for adequate aeration and lighting based on structural concepts.

Cross ventilation through an open roof joint or openable corner façades in the lateral areas of the four light towers represented further control measures (not realized).

In this case, covering the split roof with a 30-m transparent Makrolon® sheet (PMMA double-web slabs) ensures a maximum exploitation of sunlight.

'GreenGoal' – FIFA WM 2006™

Designed as an 'ecological sports venue', AWD Arena in Hanover attempted to achieve optimum turf conditions by mere structural means. This was a response to an initiative of the LOC for World Cup 2006 together with FIFA, the German Ministry for the Environment and the Federal Foundation for the Environment [*Deutsche Bundesstiftung Umwelt*]. In March 2003, the project team 'GreenGoal' stipulated the following environmental framework and objectives for the FIFA Football World Championship 2006™ in Germany:

Water
Reduction of the present water requirements in stadia by 20%; meeting the remaining demand to 20% from rain, surface or well water.

Refuse
Overall waste consists of paper (32%), glass (16%), plastics/packaging (3%), biodegradables (14%) and general waste (35%).[6] Reduction of amount of waste in the stadium and its surroundings by 20%. Introduction of waste management and the use of package-free or package-reduced, reusable or returnable systems. Waste containers should be provided in all areas of the complex, in particular near kiosks (realization depending on waste disposal concept). As stated in §9 of the DFB Safety Guidelines, trash bins have to be installed so as to prevent misuse as missiles.

Electricity
Consumption (reference value for Bundesliga matches): lighting 30%, floodlighting 20%, kiosks 20%, kitchens 10%, scoreboard 3%, other 15%. Reduction of the current energy requirements in stadia by 20%. Meeting demand, wherever possible, with renewable sources of energy and introduction of energy management.

Mobility
Increasing the share of public transport up to a minimum of 50%. Reduction of greenhouse gases by 20%.

After geometric consideration of solar altitudes, the roof of AWD Arena was divided into a closed and a transparent zone. The secondary supports of the interior roof substructure were spanned with a 10,000 m² ETFE membrane. The single-layered, 2.5 mm thick foil is braced flat via cable loops every 1.10–1.40 m. A light transmission rate of 95% ensures the UV light necessary for growth of the turf. Entrance areas to the north and south have pitch-wide ventilation openings at approx. 6.0 m above the playing field (elevation supports); the broader the opening, the lower the nozzle effect with its unpleasant draught conditions for spectators.

In Allianz Arena Munich the entire roof and façade is enclosed with ETFE cushions. Responding to solar altitude during the growth period, the white cushions in the roof area S/W have been exchanged for transparent pneus. A suspended ceiling made of 60 m long membranes is retracted wave-like to the outside in order to allow sunlight penetration. During a match, however, the closed lower deck creates the desired atmosphere.

Principles of environmental protection

'Maintaining, improving and permanent securing of living conditions and foundations for humans as well as for wild animals and plants through'

01. construction methods saving resources and material, use of materials with low levels of harmful substances, recycling materials and building materials from renewable sources.

02. avoiding materials containing PVC, surface sealants, building refuse (separation of waste) and consideration of eco-balance of building materials and environmental management systems.

03. surrounding areas as grass surfaces or meadows, waterbound areas to create percolation areas.

04. avoiding disturbances to the neighbourhood.
Noise: clarification of protection needs in the neighbourhood, determination of noise levels to be expected and 'adequate noise protection facilities at justifiable means' (IAAF), e.g. lowering the stadium partly into the ground, or green noise-screen wall.
Light: avoiding light diffusion into the neighbourhood or glare through high luminance (traffic).

(Air pollution, odours and tremors do not generally emanate from sports complexes.)

483 **FIFA project logo**
Green Goal World Cup 2006

484 **ETFE membrane roof**
AWD Arena, Hanover

485 **Open circulation area N/S**
AWD Arena, Hanover

486 **Two-piece roof soffit**
AWD Arena, Hanover

22 Natural versus artificial turf

Turf exchange

Costs for a turf exchange

The price per square metre of rolled turf, including laying, at 8.00 Euro (2005) is much lower than only a few years ago. Surely, this is a response to the necessity of a turf exchange in modern stadia and arenas of the third generation.

487 **Preparation of turf sub-base**
RheinEnergieStadion, Cologne
488 **Laying of the rolled turf**
RheinEnergieStadion, Cologne
489 **Supplementary turf shadowing**
Camp Nou, Barcelona

In case a turf is damaged to such an extent self-regeneration becomes impossible, an exchange of the surface has to take place.

Amsterdam Arena, with its two-tier complex and a closable roof, is said to have had its turf exchanged in the last five to six years almost thirty times.

The high mechanical wear through ball sports and sometimes double use of the surface by a second home team, e.g. American Football NFL League Europe, frequently necessitates the replacement of large parts of the surface.

Installation of a natural grass surface with irrigation and drainage systems and the fitting of a grass heating system can be carried out in one or two weeks, depending on logistic efforts; the exchange of individual parts, according to size, is a matter of a few hours.

The upper grass covering layer is mechanically milled off by 5 cm and the new turf sod is laid as rolled turf.

Generally, three laying widths are distinguished: (details from Horst Schwab GmbH, Waidhofen, additions by Hendriks Co., PG Heythusen, NL)

Small roll: economical solution for small areas. A 40 cm wide roll is ideal for smaller surfaces. Private garden, lawns for sunbathing, etc. are thus easily covered.

With a surface of 40 x 250 cm the small roll can be easily laid by hand. A roll is around 1 m² and weighs around 15 kg. Delivery is on Euro pallets at 40 rolls each.

Standard roll: standard size for almost all areas of use. For larger surfaces, standard rolls (60 cm x 16 m) are laid by using a small motor hand-laying appliance. Delivery is on pallets or without.

Weight is, according to length, around 300 kg per roll; laying performance is 1,000 m² per day and machine.

Jumbo roll: for large areas. These large rolls are lanes with a new length of 30 m, which are laid by special machines. There are some significant advantages compared to the laying of smaller rolls:
– less joints within the laid surface
– better grow-on results and higher stability of the laid turf
– the turf can be played earlier
– lower costs
– laying of an entire football stadium (approx. 8,000 m²) within 2 days.

The playing field of Allianz Arena in Munich was laid with large rolls in April 2005. The 250 rolls (each 15 m long and 2.20 m wide) were delivered on 20 long-distance road trains from the cultivation area in the north of Munich to the arena. On the 8,000-m² playing field, a 30 mm thick sod of a newly developed 'power turf' was employed for the first time.

At present, also widths of 1.20 m (Hendricks Co.) or 1.05 m are common for prefabricated turf sod.

Factors supporting turf growth

Light

In the Netherlands, for some time now, research has been done on artificial growth light, as used in greenhouses, as a viable solution for the problem of turf shadowing. We may assume that this approach equally involves high investment and in particular high energy costs.

In World Cup Stadium Hamburg, the mobile turf solarium 'Mobile Lighting Rig' of Norwegian manufacturer Mobilt Drivhus AS is tested [Sept 2005].

132 UV lamps à 600 Watt are supposed to promote the growth and resistance of those parts of the turf which are hardly or not exposed to the sun.

According to the stadium management, the first results seem to be very promising. After only a few days, there was visible improvement regarding the rooting and density of the grass.

Mobilt Drivhus Co. has already tested the system in various stadia in Scandinavia and Great Britain (Manchester City Stadium). Here, they have proven to be a real alternative to synthetic turf.

The US patent of Mobilt Drivhus comprises a system for creating favourable growth conditions through a mobile lighting unit.

Air

The second problem lies in the demands on comfort in closed sports arenas, as this often prevents sufficient ventilation, which is necessary for drying the turf after watering or when morning dew has accumulated.

In many stadia, such as AOL Arena in Hamburg, mechanical drying systems are being employed against turf rot. These are large mobile blast engines which are set up at the pitch perimeter and produce air circulation in the stadium interior. Significantly, Leipzig Stadium lies within the historic circle of the former Zentralstadion. This embeddedness results in little air exchange close to the turf in spite of a circulating open joint above the lower tier and the stand-free short ends; at circa 15.0 m the distance to the turf surface is too large to guarantee its drying.

Maintenance of the turf has become a science of its own and the greenkeeper plays a very important role in stadium operations. The strong mechanical wear of a natural turf during football or other activities requires solutions for regeneration during play-free time.

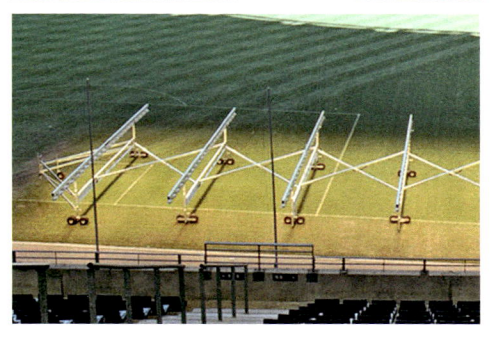

490 **Stadium interior (three-tier stadium)**
Amsterdam Arena with closable roof
491 **Additional mechanical turf ventilation**
Amsterdam Arena
492 **Additional turf lighting**
Allianz Arena, Munich

Chapter 23

The planning of significant roof structures
An overview of structural systems for fast orientation

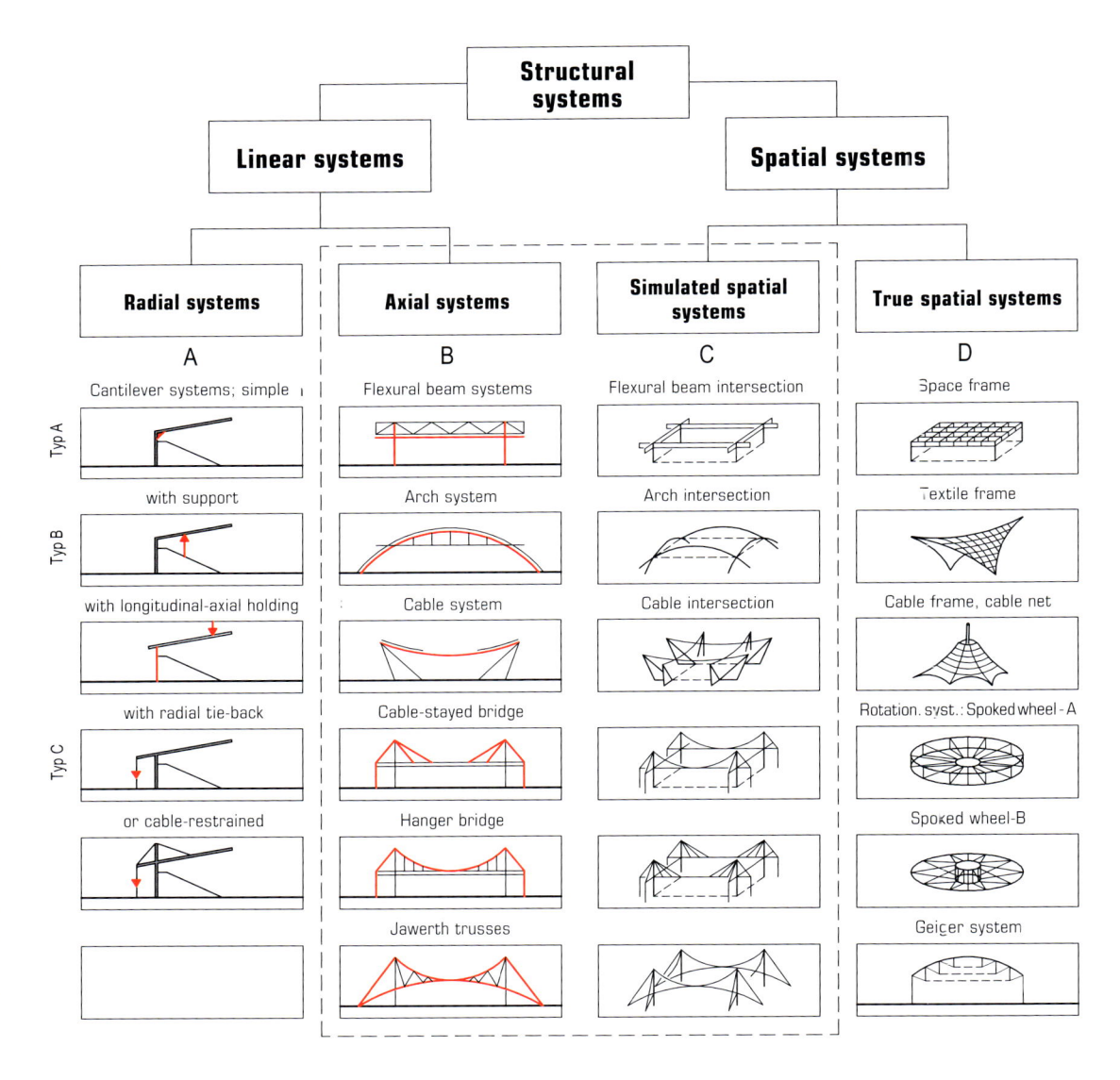

Structural systems

- **Linear systems**
  - **Radial systems** (A)
  - **Axial systems** (B)
- **Spatial systems**
  - **Simulated spatial systems** (C)
  - **True spatial systems** (D)

	A – Radial systems	B – Axial systems	C – Simulated spatial systems	D – True spatial systems
Typ A	Cantilever systems; simple with support	Flexural beam systems	Flexural beam intersection	Space frame
Typ B	with longitudinal-axial holding	Arch system	Arch intersection	Textile frame
	with radial tie-back	Cable system	Cable intersection	Cable frame, cable net
Typ C	or cable-restrained	Cable-stayed bridge		Rotation. syst.: Spoked wheel - A
		Hanger bridge		Spoked wheel - B
		Jawerth trusses		Geiger system

493 **Overview of structural systems**
(freely citated after Prof. Udo Peil)

Structural systematics

The load-bearing structure of a stadium or arena represents one of the most important design aspects since it has a significant influence on the overall appearance of the complex. Its market value and its potential for identification are essentially grounded in the structure itself.

A circulating roof unites the stadium interior and adjacent spectator stands, creating a particular atmosphere and a unique space for experience. Hence, the shape and framework of a sports complex should always be synchronized.

The presented classification of structural systems for stadia is based on considerations by Prof. Dr.-Ing. Udo Peil of Technical University Braunschweig, Institute for Steel Construction, and is differentiated with regard to cable frames.

First some structural elements and their static effects are explained. According to Führer/Ingendaaji/Stein (RWTH Aachen), four different types of load-bearing systems are distinguished in terms of their exterior geometry – the 'shape' or 'form'. The 'interior' geometry is defined by the load-bearing structure.

Zero dimension – punctual frame   (hinge)
First dimension – linear frame   (bar)
Second Dimension – surface frame   (plate)
Third Dimension – spatial frame   (body)

Two types emerge when distinguished after their stress effects:
Bending happens when forces act vertical on the structural element (beam or plate).
Normal force (axial force) happens when forces are flowing in the direction of the large structural dimensions (columns, walls etc.).
'Generally, we may assume that load transfer is much more preferable via normal force than through bending. Structural systems subjected to normal force, such as arches and cables, are reduced in mass, compared to systems stressed by bending, such as beams and plates.

At a higher dimension, structural systems tend to become more complex, and their calculation more difficult; their bearing behaviour, however, is improved since a large part of the force is transferred along the structural system dimension, and not vertically' [W. Führer et al., 1995].[1]

The formula $Vol./F \times L$ defines the effort necessary for load transfer over a certain distance; the required amount of material ($Vol.$) in relation to the force (F) and distance over which this force has to be transmitted (L).

The larger the span (L), the higher the strain: Altering the forces of tension/compression and bending can result in a tenfold increase in stress for bending-force structures compared to normal-force structures.

Hybrid structural systems

In 1967, the architect Heino Engel created the term 'hybrids' which denotes the property of load-bearing systems with different redirection of forces to be united into one single effective mechanism.

The underlying principle is the fundamental equivalence of original systems and their mutual interdependency.

Combination systems cannot be defined as an individual category of structural system as they neither exhibit a typical mechanism, nor do they develop a specific condition of forces and stresses or display characteristic structural features. Interlocking of structures is realized through:
Parallel joining = superposition or alignment
Successive joining = coupling
Cross joining = penetration

Hybrid systems are frequently employed in the construction of stadia and arenas, especially for grandstand roofing, usually by combining vector-active and section-active beams with cable webs into cable-supported or cable-restrained systems or with integrated cable bracing (tensegre structures). Large spans are often made possible only through the combination of specific material properties of load transfer.

Because modern stadium roof structures should be kept free of columns for viewing reasons, a structure widely cantilevering from the rear of the stand becomes almost inevitable.

An alternative to radial linear cantilevers is represented by spatial systems which manage to do without a bending moment in the inflexion point of a cantilevered structure, for example through ring connection (spoked-wheel system).

Families of structural systems

Since the 1960s, Heino Engel has been active in research on architectural load-bearing systems at the University of Minnesota, USA. He divides them into five groups of 'structure families'. The following definitions are taken from his book 'Structure Systems' from 1967. The predimensioning and the overview of feasible structural systems are intended to facilitate design decisions regarding the stadium structure:

Form-active structure systems
... are systems of flexible, non-rigid matter, in which the redirection of forces is effected by particular form design and characteristic form stabilization:
a) Cable structures: radial, biaxial or cable trusses
b) Tent structures: peak, undulating or indirect peak tents
c) Pneumatic structures: air-controlled indoor systems, air cushion systems, air-tube systems
d) Arch structures: linear arches, vaults, thrust lattices

Vector-active structure systems
... are systems of short, solid, straight lineal members (bars), in which the redirection of forces is effected by vector partition, i.e. multi-directional splitting of single forces (compressive or tensile bars):
a) Flat trusses: top chord trusses, bottom chord trusses, two-chord trusses, cambered trusses
b) Transmitted flat trusses: linear, folded, intersecting trusses
c) Curved trusses: cylindrical, saddle-shape, dome-shape, spherical trusses
d) Space trusses: flat space, folded space, curved space, linear space trusses

Section-active structure systems
... are systems rigid, solid, linear elements (including their compacted form as slab), in which the redirection of forces is effected by mobilization of sectional (inner) forces.
a) Beam structures: one-bay, continuous, pin-jointed, cantilever beams
b) Frame structures: one-bay, multipanel, storey frames
c) Beam grid structures: homogeneous, gradated, concentric grids
d) Slab structures: uniform, ribbed slabs, box frames, cantilever slabs

Surface-active structure systems
... are systems of flexible, but otherwise rigid planes (= resistant to compression, tension shear) in which the redirection of forces is effected by surface resistance and particular surface form:
a) Plate structures: one-bay, continuous, cantilever, intersecting plates
b) Folded plate structures: prismatic, pyramidal, intersecting, linear folded plates
c) Shell structures: cylindrical, dome, saddle, linear shells

Height-active structure systems
... are systems, in which the redirection of forces necessitated by height extension, i.e. collection and grounding of storey loads and wind loads, is effected by typical height-proof structures, highrises.
a) Bay-type highrises: framed, trussed, stabilized post-beam, shear wall bays
b) Casing highrises: framed, trussed, stabilized post-beam, shear wall bays
c) Core highrises: cantilever or indirect load cores, core combinations
d) Bridge highrises: girder, storey and multi-storey bridges

Note: Details and system sketches as shown in the overview opposite are taken from 'Structure Systems' by H. Engel in the 1997 revision. Reference values are presented for the primary components and spans, complemented by standard minimum/maximum figures. The overview does not claim completeness.

494 **Cable net structure**
Design: Frei Otto, Stuttgart

page 247:
495 **Families of structural systems**
freely adapted from Heino Engel
'Structure Systems', 1967

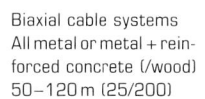

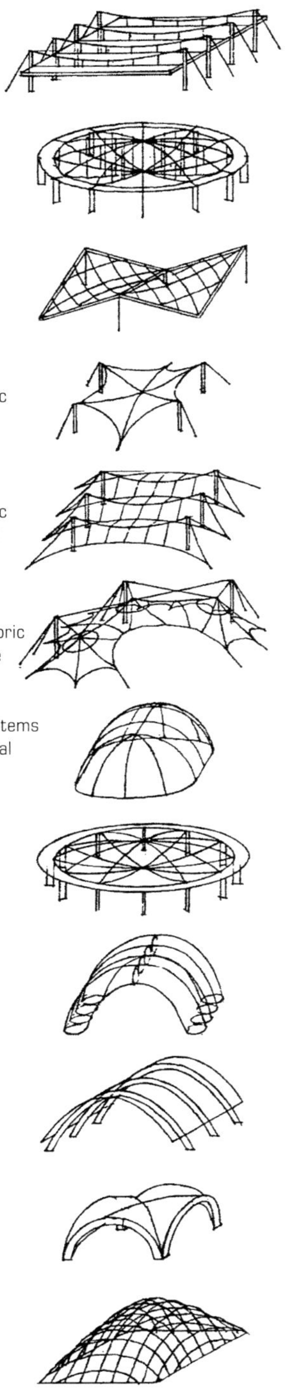

Parallel cable systems	All metal or metal + reinforced concrete 80–500 m (50/500)	Top chord trusses	Wood or metal (Steel) 15–30 m (8/40) 15–30 m (10/50)	One-bay beam	Wood 4–8 m (0/12) Steel 7–20 m (5/25) Reinf. conc. 4–10 m (0/15)	One-bay plates	Reinf. concrete or wood 10–40 m (8/50) 8–30 m (6/50)
Radial cable systems	All metal or metal + reinforced concrete 60–200 m (30/250)	Bottom chord trusses	Wood or metal (Steel) 20–50 m (10/60) 20–80 m (100)	Continuos beam	L-Wood 10–30 m (7/35) Steel 8–25 m (5/30) Reinf. conc. 10–25 m (7/30)	Continuous plates	Reinf. concrete or wood 15–50 m (10/60) 10–40 m (8/50)
Biaxial cable systems	All metal or metal + reinforced concrete (/wood) 50–120 m (25/200)	Two chord trusses	Wood or metal (Steel) 10–20 m (6/25) 10–25 m (10/35)	Pint-jointed beam	Wood 4–8 m (0/12) Steel 7–20 m (5/25) Reinf. conc. 4–8 m (0/12)	Hinged plates	Reinf. concrete or wood 8–20 m (5/25) 5–15 m (8/20)
Biaxial cable systems	Textile fabric or synthetic material + metal / + wood 10–25 m (5/40)	Cambered trusses	Wood or metal (Steel) 20–50 m (15/60) 25–100 m (15/120)	One-panel frame	Wood 15–40 m (10/50) Steel 15–60 m (10/80) Reinf. conc. 10–25 m (7/30)	Folded box frame	Reinf. concrete or wood 15–50 m (10/60) 10–40 m (8/50)
Undulating tents	Textile fabric or synthetic material + metal / + wood 30–70 m (20/100)	Folded trusses	Wood or metal (Steel) 12–25 m (8/30) 20–80 m (10/90)	Multi-panel frame	Wood 15–45 m (10/55) Steel 15–65 m (10/85) Reinf. conc. 10–28 m (8/35)	Folded barrel vault	Reinforced concrete or wood 25–150 m (20/200) 20–120 m (15/150)
Peak tents	Synth. mat. or textile fabric + metal / + reinf. concrete 30–80 m (20/150)	Intersecting trusses	Wood or metal (Steel) 15–35 m (8/45) 15–60 m (80)	Story frames	Wood 20–50 m (15/60) Steel 20–70 m (15/90) Reinf. conc. 15–30 m (10/40)	Pyramidal folding	Reinforced concrete or wood 25–80 m (20/100) 20–60 m (15/80)
Air controlled indoor systems	Synthetic material + metal 10–40 m (50) 90–220 m (70/300)	Cylindrical trusses	Wood or metal (Steel) 12–25 m (8/30) 20–80 m (10/90)	Homogeneous grids	Wood 12–25 m (10/30) Steel 12–25 m (10/30) Reinf. conc. 8–18 m (5/20)	Folded frame	Reinf. concrete or wood 20–70 m (10/90) 15–60 m (10/70)
Air cushion systems	Synth. material + metal / + wood / + reinf. concrete 20–70 m (120)	Dome-shape trusses	Wood or metal (Steel) 15–25 m (8/30) 20–80 m (10/90)	Gradated grids	Wood 15–30 m (10/35) Steel 15–30 m (10/35) Reinf. conc. 6–20 m (5/25)	Barrel shell	Reinforced concrete 20–60 m (10/75)
Air tube systems	Synthetic material 10–50 m (70)	Spherical trusses	Wood or metal (Steel) 40–160 m (20/200) 50–190 m (20/500)	Concentric grids	Wood 10–20 m (8/25) Reinf. conc. 8–15 m (5/18)	Dome shell	Reinforced concrete 40–150 m (20/200)
Linear arches	Reinforced concrete, laminated wood, metal 25–70 m (15/100)	Flat space trusses	Wood or metal (Steel) 15–60 m (8/80) 25–100 m (6/130)	Uniform slabs	Wood 0–5 m (6) Reinf. conc. 0–6 m (8)	Curved rotational shell	Reinforced concrete 25–70 m (15/90)
Vaults	Masonry 8–20 m (4/30)	Curved space trusses	Wood or metal (Steel) 15–60 m (8/80) 25–100 m (6/130)	Ribbed slabs	Reinf. conc. 7–15 m (5/20)	Saddle shell	Reinforced concrete or wood 25–60 m (15/70) 20–50 m (15/60)
Thrust lattices	Metal or wood 20–90 m (10/150)	Linear space trusses	Wood or metal (Steel) 20–50 m (15/70) 25–120 m (15/150)	Closed box frames	Reinf. conc. 4–8 m (3/12)	Shell-type frame	Reinforced concrete 25–80 m (20/100)

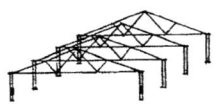

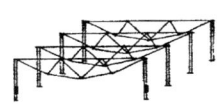

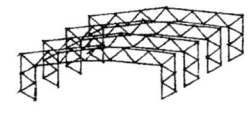

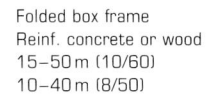

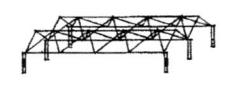

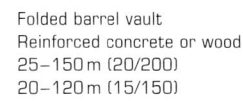

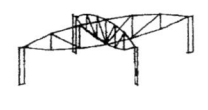

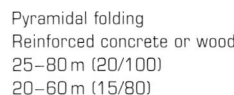

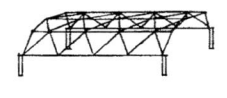

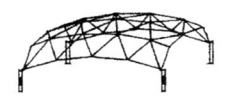

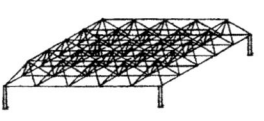

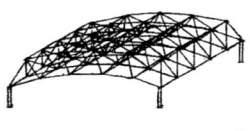

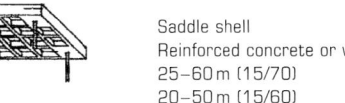

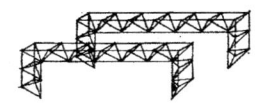

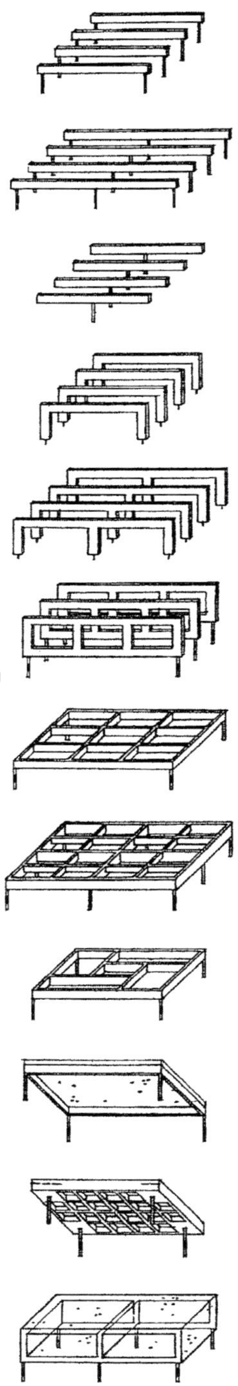

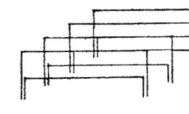

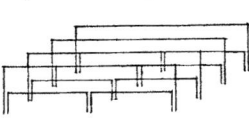

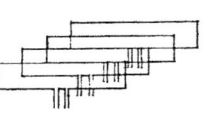

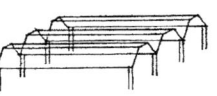

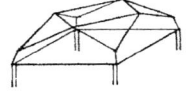

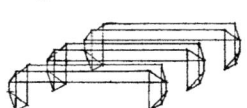

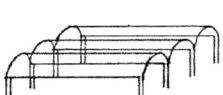

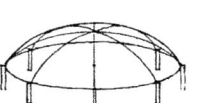

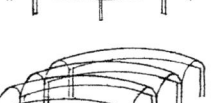

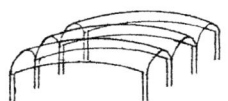

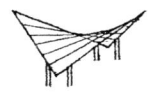

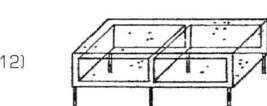

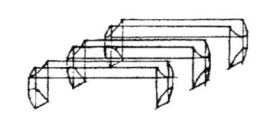

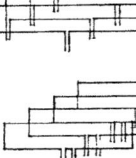

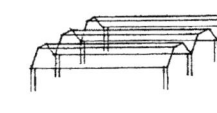

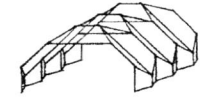

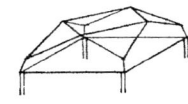

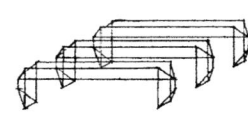

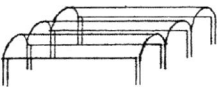

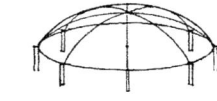

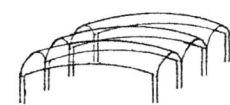

Linear systems

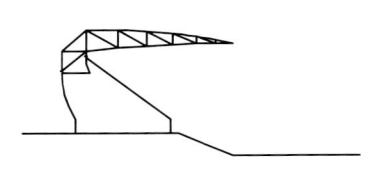

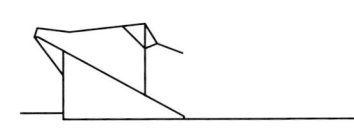

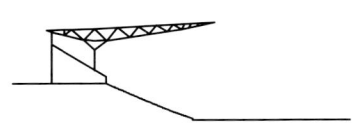

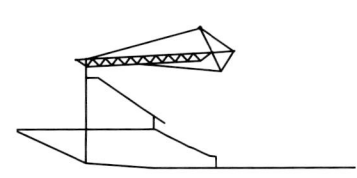

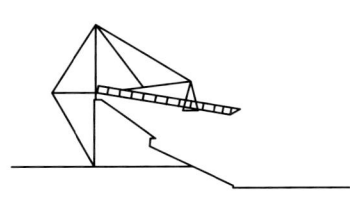

In order to carry out a classification of stadia or arena roofings, the question which type of bearing system to use as foundation should be clarified first.

As a rule, primary structures are distinguished from ancillary or roof structures; the first are primarily responsible for the load transfer of dead weight, snow and wind force.

Thus, all roof coverings could be defined as 'second-level' structures (secondary structural systems) serving either as simple cover for the stands or shielding the entire stadium from adverse weather by completely closing the roof. Frequently superimposed as hybrid structural systems, they sometimes cannot be clearly classified. Project definitions are mostly based on a design-specific innovation of the roof area (cf. 'Façade roof', Munich).

Linear systems are divided into two groups: *radial* and *axial systems*.

Mainly as cantilever-beam structures, the single trusses of radial systems are arranged circulating and at the rear of the terrace body, projecting toward the pitch. Three types of radial cantilever-beam systems may be distinguished.

Type A   simple cantilever systems
Type B   with support/holding
Type C   with tie-back or restraining cables

The cantilever moment of large spans must be dispersed via an equivalent construction height at the corner areas. If the building provides this option, a cantilever is a very simple static solution.

Because a single beam is not reliant on ring forces, beams may be installed successively; an advantage that comes to bear fruit for conversion measures during regular football operation.

If the truss (bar) cannot sustain the corner moment (Type A), the statically required height can be reduced by an additional support/holding (Type B) or by elongating the roof truss toward the rear (Type C) to form a force couple out of a tensile and compression bar.

If the horizontal member is also incapable of absorbing the corner moment arising from the cantilever span, it can be strengthened by elongating the vertical redirectional pole of the cable support (pylon + cantilever beam).

Type A: simple cantilever systems

Allianz Arena, Munich, has a complete roof cover as steel-cantilever structure:
a) cantilever with 96 main double beams (48 x 2), cantilever length ca. 60 m, cantilever-height 8–12 m, weight up to 106 t, rig depth $a = 10$ m as tension/compression support, equal distance at interior roof edge.
b) ring framework from 7 steel-frame ring trusses / purlin rings
c) supported pneumatic structure
d) retractable / fixed subroof

A lozenge-shape for the secondary structural system of the roof is difficult to realize from a static point of view. As an independent system, the soft structure of supported pneus behaves differently from the primary bearing structure of the main trusses. Therefore, 3 x 96 radial spring elements are fixed under the roof, serving to compensate the deformation of the lozenge-shaped grid (+/– 13 mm in every field). For better lighting of the turf, a subroof as retractable ceiling at S/W and a fixed subceiling at N/E made of a 60-m steel-membrane structure are retracted wave-like toward the roof edge via a cable polygon.

Type B: with support/holding

The roof above the Olympic Stadium Berlin is a cantilever structure supported around the quarter point. Components of the primary structure:
a) steel-pipe tree-like supports (20 x) at 32–40 cm intervals with circular solid bar and four supporting members on cast steel nodes.
b) radial double-boom truss (76 x) from circular hollow sections with circulating three-boom truss plus secondary beams.
c) roof edge from reinforced-concrete compression ring on 135 hinged pillars for tieback of cantilever structure to the roof parapet, 'narrow' sheet-lined roof edge.

The secondary roof structure is made of doubly curved membrane saddle-shape surfaces which are placed on seven tangential steel-pipe arches. Every arch is cable-supported. The roof material consists of PTFE coated fibre-glass fabric. For better lighting of the natural turf (15° roof cantilever, first row), a point-fixed glazed inner ring with laminated safety glass from 2 x 10 mm partially tensioned safety glass (5 x 2.40 m) is fixed to the interior roof edge.

In the bottom truss layer of the two-boom truss, an approx. 28,000 m large PTFE-membrane structure is accommodated as fixed, open grid fabric. The reinforced concrete ring on the outside is minimized in height and sheet-lined, with a 5-m joint toward the roof parapet.

Depending on the radius of curvature, oval athletics geometries allow wheel or rotation frameworks which exploit the positive adhesive properties of these tensile/compression-ring structures.

Out of respect for the historic ensemble of the former Reichssportfeld, the preservation of the famous opening of the marathon gate with the central viewing axis toward the Langemarck

496 Structural system sectional drawings
of all twelve World Cup 2006 stadia
a. Munich
b. Kaiserslautern
c. Berlin
d. Leipzig
e. Dortmund
f. Nuremberg

clock tower behind the Maifeld became paramount. This decision was a matter of principle and would have excluded a ring closure. With the help of tree-like supports, cantilever length was decreased by 17 m to statically more preferable 48 m, which also enabled the gradual assembly of the roof during the regular football season.

Supports tend to cause viewing restrictions and should hence be generally avoided or permitted only in exceptional cases.

Type C: with tie-back or restraining cables

Frankenstadion in Nuremberg was completely covered during conversion works in 1991. Its geometry resembles an elongated octagon and the roof is constructed from crane-like structures of tied-back steel trusses. The two-truss structure bearing a glazed inner roof, contra-rotational to the outer roof with a strong ridge offset (weather-exposed), is inclined inwardly and encloses the bottom of the stand area (no horizontal view towards the reverse stand). Tensile steel rods keep the front roof element in position. Forces are transferred vertically into the tensile base 32 x via a redirectional spreader bar at the pylon head ($l = 7$ m, $h = 22.0$ m, pylon height).

Axial systems

Main bearing systems with a flexural truss or arch in longitudinal or axial direction parallel to the pitch make up the second group of linear systems, to which also belong cable systems supporting the main truss as cable-stayed bridges or hanger bridges.

New Wembley Stadium in London sports a main truss as an elevated and sign-like single arch, a significant landmark in the surrounding area.

In Zentralstadion in Leipzig, two spatially supported arched cable-trusses inclined by 26° span the entire complex 220 m lengthwise. Cable-stay bracings improve the rigidity of the roof plate. The compression arch is around 17/22 m high (top boom 1,016 m, middle boom 711 mm, lower boom cables 2 x 90 mm). The curved roof (short sides) rests on eight V-supports projecting from the edge truss.

A spatially curved three-boom truss as circulating exterior compression ring is propped via 64 articulated columns onto the reinforced terrace structure. The roof area itself (trapezoid metal sheet roofing) is constructed out of radial trusses with a purlin-ring structure above. After analyzing its shadow effects, a transparent plate was installed at the inner roof edge (ca. 1,000 m² polycarbonate plates). Out of cost considerations, plans for a movable interior roof were not realized.

The World Cup stadia in Kaiserslautern and Dortmund belong to the group of 'evolved' stadia which were modernized and complemented several times during the past few decades.

Before conversion measures were undertaken, Westfalen-Stadion had two main supports with three booms positioned close to the pitch in each corner area.

A 12-m high bending truss connected these supports lengthwise to the terrace so that a secondary purlin structure as one-bay truss was capable of spanning from the main truss to the elevated roof edge.

After conversion in 2005 and the completion of the corners, the static function of the four corner pylons was taken over by a guyed crane structure consisting of 4 x two cable-restrained pylons/booms as propped cantilever system. These were structurally integrated without dismantling the existing roof. The new primary structure is braced by the reinforced concrete frames of the four new corner stands as well as one staircase core per pylon.

Fritz-Walter-Stadion in Kaiserslautern has a main truss running lengthwise to the terrace. It rest on a 'mega' support ($d = 330$ m) in the corner area. This static solution arises from the grown structure of the stadium.

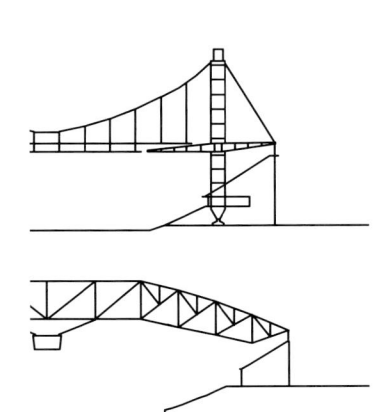

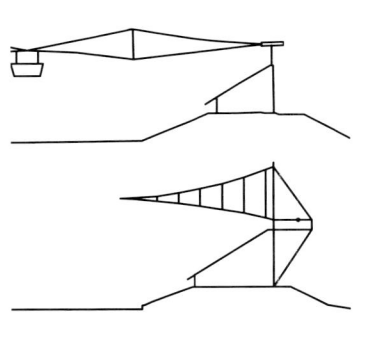

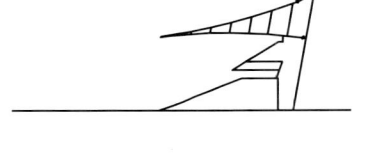

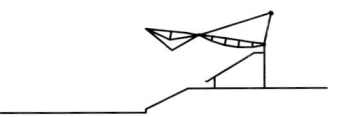

497 **Structural system sectional drawing**
of all twelve World Cup 2006 stadia
a. Cologne
b. Gelsenkirchen
c. Frankfurt
d. Hamburg
e. Stuttgart
f. Hanover

23 The planning of significant roof structures

'Simulated' spatial systems

A linear main and secondary framework spans as one-bay beam from outside with perforated double T-beams (h = 80 cm) to the top boom of the main truss. At the bottom boom, a bent single beam with 15 m cantilever is fitted which permits light into the rear part of the terrace through a glass roof at an 8-m offset. After completion, the horseshoe-shaped main roof was reinforced spatially. This basic shape arises from the historic role of the north stand as main stand.

Crosswise vertical hanger bridge

RheinEnergieStadion in Cologne is a self-anchoring hanger bridge structure on four pylons. Redirection of the main suspension cables is done by a trussed compression member (2.5 x 3 m, l = 170 / 215 m) spanning the whole stadium length, penetrating the pylon and continuing for about 20 m as cantilever. Per pylon there are two main directions of effective span and two tie-backs.

The secondary structure consists of linear trusses cantilevering from the compression bar to the stadium interior (secondary layer of double T-profiles). The exterior layer of trusses rests on 64 hinged pillars at an axial grid of 10 m, braced once at each side. The 'freely floating' roof plate is suspended from the four pylons to absorb tension from length changes in the cable structure. Therefore, the base point of the pylon is hinged (compare to a scout tent). This type of frame marks the transition between a simulated/true spatial system and a structurally independent mode of operation of all four sides and pylon stays.

If linear systems are laid out in two directions, these frameworks consisting of curved-beam, arch or cable intersections produce a spatial effect.

'Both appear to be largely decoupled; only the corner areas offer a possibility for load transfer in two different directions. From a structural point of view, it is not a true spatial system which requires spatial expansion for bearing purposes; it is a simulated spatial system' [Prof. U. Peil, 2005].[2]

A structure of intersecting trusses is clearly discernable at Guiseppe Meazza Stadium in Milan and crossed arches can be found at Estadio da Luz in Lissabon, erected for the occasion of EURO 2004 in Portugal.

True cable-suspended roofs, as employed at Estadio Municipal de Braga by architect Eduardo Souto de Moura, are rarely to be found in stadium construction. To the knowledge of the author, cable intersections as two-directional constructions remain as yet unbuilt. The example opposite merely shows a draft created at RWTH Aachen in 2005.

Reebok Stadium in Bolton, UK, employs four doubled corner pylons. The four linear main trusses are Jawerth trusses, according to the principle of stabilization by counter cable stay. A second rope underneath the main cable, reversely curved, ensures the tensioning of the suspended structure (planar or spatial). Main cable and counter cable stay are connected by ropes; to keep the system from rotating, it needs to be additionally fastened. (Reebok-Stadium employs pipes instead of cables).

498 **Intersecting arches**
Estadio da Luz, Lissabon Portugal
499 **Parallel longitudinal arches**
Estadio do Dragao, Portugal
500 **Cable-guyed pylon structure**
Manchester City Stadium, England
501 **Crosswise Jawerth trusses**
Reebok Stadium, Bolton, England

'True' spatial systems

We speak of true spatial systems only if the spatial effect also serves functional, i.e. load-bearing purposes. 'Closed' systems such as spoked-wheel or rotation frameworks do not require tie-back to the foundations since they balance out the acting forces with the ring connection and closed 'force bends', e.g. as tensile/compression-ring structures.

This applies also to spatial frameworks which, apart from horizontal wind forces, exclusively transfer vertical normal forces to the building. The second group of spatial systems are cable nets or true membrane structures in which the material (PTFE/PVC/PES) functions as statically effective primary structure and is not merely employed for weather-protection purposes. Lightweight stadium roofs are susceptible to vibration and can be agitated through wind (or earthquakes, rock concerts, etc.). The prestressed roof elements must therefore always be reversely curved (negative 'Gauß' curvature, U. Peil) and designed sufficiently strong as horizontal tensile forces approach the infinite and are, quite likely, not resisted by the material (cable net or membrane).

High (pylons) and low points should alternate; guy cables should continue the tensile direction of the form.

Saddle shapes are also hypar surfaces, which can be formed by two reversely spanned parallel cables in the direction of curvature. At evenly distributed loads these are parabolic or hyperbolic. The obligatory pretensioning of cable nets has to be secured by tensile foundations. We may assume that lightweight membranes or cable net structures bury a similar mass into the ground as needed for compression-stressed frameworks.

A hyperbolic paraboloid is created when a cylinder formed by parallel lines is twisted around its central axis. Such shapes can be formed by almost any type of material capable as straight-line generatrix (e.g. timber shells, etc.). The cables in a cable net structure run parallel to the edges.

Spoked-wheel structure

The large spans typical for stadium construction can be realized most economically as ring-cable roofs. The covering structure of Gottlieb Daimler Stadium in Stuttgart (1993), designed by engineering practice Schlaich Bergermann and Partner, has become a prototype worldwide, which has since been taken further and, in some instances, significantly altered. The principle of a horizontal spoked wheel was employed in the new-build and conversion of four of the twelve World Cup stadia in Germany. Normally, there is a pressure ring on the outside, which functions like a bicycle rim with a hub at the centre taking up all bearing and stabilizing cables.

Two types A/B are distinguished: Stabilizing cable *above/below* the suspension cable.

Since precise horizontal forces may become infinitely strong, the suspension cables have to run at an optimized pitch. If the main cable is above the stabilizing rope, it runs outward at the top and inward at the bottom. The geometry is thus elevated on the outside and shallow on the inside (cf. Stuttgart, Hanover, Hamburg). Frequently, weather protection is required only for the spectator areas; the tensile ring is designed as a hollow hub, leaving the stadium interior uncovered.

With the suspension rope in a low position, the structure is flat on the outside and high on the inside. The stabilizing cable for reversed loads (suction) runs above and is fixed to the hub (cf. Frankfurt). A spoked-wheel structure is capable of covering large areas without columns placed

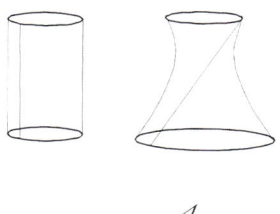

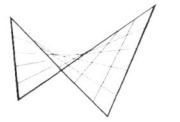

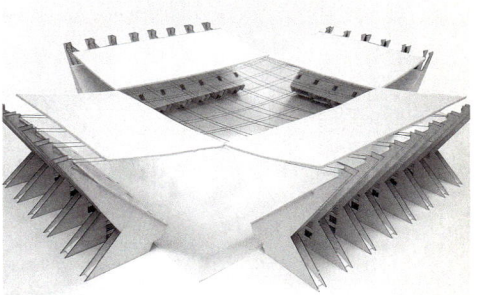

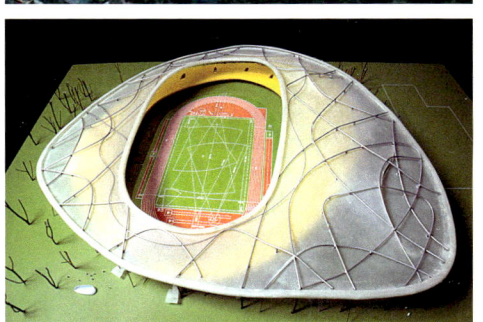

502 **Schematics**
Hyperbolic paraboloid/cable net (U. Peil)

503 **Intersection of trusses**
Guiseppe-Meazza-Stadium Milan, Italy

504 **Crossed suspension cables**
RWTH-Aachen, Taha Anwar, summer term 2005

505 **Cable-stayed pneumatic roof structure**
Tokyo Dome, 'Big Egg'

506 **Spatial framework on diagonal supports**
Draft: athletics stadium Chemnitz, 1995

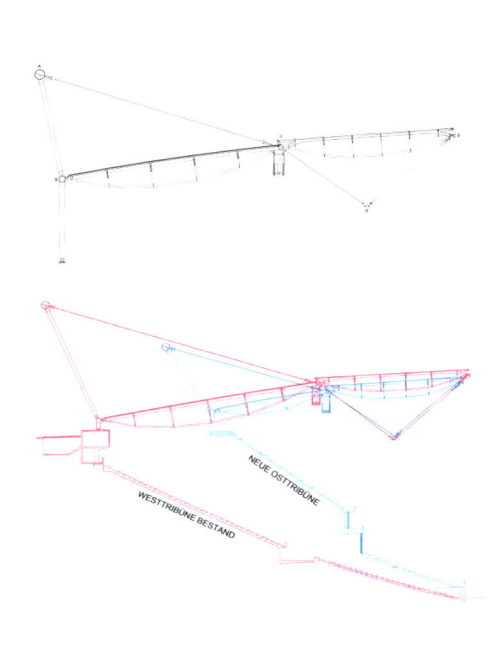

within the spectator zones. The tensile/compression ring structure in World Cup Stadium Frankfurt rests on the upper edge of the stand. There is a hinged column in each of the 44 main axes. An open hollow hub provides the option for installation of a retractable interior roof. The geometric height of a two-part tensile ring, which is separated at the main axes by cable spreader bars (air supports), enables another interior radial-cable structure bringing together all cables at point, plus a retractable PVC membrane at the lower layer of the stabilizing cable.

Equivalent to the rim of a spoked wheel, the optimum form of a pressure ring is a circle. The stadium structure tends to adapt to the rectangular shape of the playing field or the approximate oval basic shape of the athletics track. Frequently, the architectural design attempts an affine superposition of roof and stadium shape; on the other hand, an almost rectangular stadium form is unsuited for a pressure ring. In Frankfurt, for this reason, a pressure-ring was chosen which resembles the stadium shape but manages to round it off so as to conceal the asynchronous form with the 9 m wide roof-edge membrane and keep the necessary dimensions within economical bounds.

The long sides of World Cup stadium Stuttgart are a good example for the strengthening of an – in geometrical terms – insufficient pressure ring: the boxed steel ring circulating at the top around the centre-line area is expanded into a horizontal framework.

Because of spatial limitations, a further expansion of the overall shape and an optimization of the pressure ring are unfeasible. As stated on the stadium in a German magazine on steel construction, 'the membrane is supported in every section by seven tangential arches with tie rods. Load transfer to the lower cable takes places at coupling point to the upper cable. As they not only bear the membrane but are also stabilized by it against bending, a diameter of 200 mm is sufficient for these arches, in spite of a 20-m span. All membrane fields are entirely prefabricated.'[3]

An optimization of the compression ring has also been performed at World Cup Stadium Hamburg with a tensile ring made out of eight cables and exterior round pipe pylons (40 x) with suspension cable guyed at the head, a horizontal compressive spreader bar at the level of the terrace upper edge (bearing the circulating compression ring) and two horizontal diagonals, which form the tie-back plus cable stay to the base point.

The special roof structure at World Cup Stadium Hanover is a tensile ring with air supports which are bearing the inner part of the roof covered by an ETFE membrane. Normally, the cables are first laid out with the help of an auxiliary gantry above the seating rows, raised with the help of a strand jack and, via restraining cables, bolted in at the node. In Hanover, the cast elements of the air-support bearings (32 x) with the preassembled radial trusses are fixed to ring cable *D*.

This bearing ring has to be secured by cable stays and six bracings, due to its kinematic behaviour. The top booms of the two exterior and interior ancillary trusses are combined by compression bars and diagonals into trusses which are running horizontally at roof level, in order to receive the tensile forces of the cables guyed to the ETFE membrane. A special form of the spoked-wheel system is represented by the 'Geiger System', which as a tensegre structure is pulled up from the floor to its final height by successive pre-stressing. Frequently, several layers are superimposed and completed into a rotationally symmetric dome.

Curved spatial framework

Veltins Arena, Gelsenkirchen, consists of a curved spatial framework with circulating lattice truss structure (40 x = 7 x diagonal and 5 x lengthwise, intense play of shadows on the green). A diagonal-pipe structure is spatially interconnected through tangential pipe purlins and cast nodes and cross-braced by cables. F 30 fire-coating (forming insulating layers) was applied to the steel members with a distance to the terrace of less than 6 m. The roof covering is a double-layer membrane span and is guyed down at the pitch centre into the roof structure. Above the bearing layer, an open-air lattice is spanned for noise protection. The stadium interior roof is rectangular, as six single trusses are united in one nodal point at the corner area. The retractable roof structure with two movable parts runs on a rail system above the circulating trusses (running time 30 min, all-year, total weight 560 t).

FIFA requirements for the roof

Which type of stand covering is selected as weather protection (rain, snow, hail, sunshine) depends on the prevalent climate and spectators' demands for comfort.

In certain regions, protection against the sun may be more relevant than against rain or snow, as in more temperate climate zones. There are multiple options for coverings available, tried and tested, like most structural systems. According to FIFA World Cup 2006™ requirements, all VIP and media seats must be covered; roofing is generally expected for all seats.

Note: This request reflects the rising demands on comfort, compared to 1995 when FIFA Technical Recommendations deemed covered seats for all spectators merely 'desirable' for weather protection and shadowing purposes. As stated in the UEFA Technical Recommendations, all seats have to be covered for EURO 2008.

Definition of a roof overhang (new)

Based on empirical analysis of built examples from World Cup 2006, an average angle of 8° to the vertical above the first seating row may be assumed as minimum standard for planning of a roof overhang toward the stadium interior.
One recommendation, however, states a value of 12°.

Note: Experiences from matches with regard to weather conditions and severity of rain and wind have come to show that the first rows of the east stand (lower tier) will unavoidably be affected by weather despite overhanging cover.

Table M Overhang	[°]
01 Berlin	16
02 Dortmund	5
03 Frankfurt	7
04 Gelsenkirchen	11
05 Hamburg	9
06 Hanover	0
07 Kaiserslautern	8
08 Cologne	7
09 Leipzig	12
10 Munich	0
11 Nuremberg	12
12 Stuttgart	4
Average	8
Minimum	0
Maximum	16

page 252:
507 **System sectional view of roof framework**
AWD Arena, Hanover
508 **Adaptation of the asymmetric roof**
AWD Arena, Hanover
509 **Guyed cable structure**
World Cup 2002, Stadium Jeonju, Korea
510 **Impression of the roof structure at night**
World Cup 2002, Stadium Jeonju, Korea
511 **Table M: Roof overhang toward stadium interior**
Overview of World Cup stadia
512 **Membrane-roof structure**
King Fahd National Stadium, Saudi Arabia
513 **Aerial view of the entire complex**
King Fahd National Stadium, Saudi Arabia

FIFA scoreboard

As stated in the IAKS Planning Principles from 1993, glare-free scoreboards are to be installed for the information of spectators (item 2.3.7, p. 76). The main purpose of the alphanumerical information systems is to serve the display of competition results; matrix information systems can display texts and images via computer.

FIFA Technical Recommendations request at least one computer-controlled scoreboard (video board). As a rule, two scoreboards are arranged diagonally opposite one another, optimum viewing provided and without seat loss. IAKS recommends that the horizontal viewing angle should not exceed 120°.

In the World Cup stadia in Frankfurt and Gelsenkirchen, scoreboards are arranged on the four sides (ca. 7–8 m) of a video cube suspended over the pitch centre. The two roof structures bear the 30-t broadcasting technology.

In Frankfurt the cube structure also serves as storage space for the interior roof. Its lower edge is suspended around 27 m above the field.

Since sports associations have so far not defined a minimum height for video cubes, we may assume a measure of $h = 25$ m to be sufficient. Hence, the present minimum height (point of focus) of 15.0 m should be raised.

Note: The issue of shadow thrown by a cube situated at the pitch centre should be clarified in the run-up of planning with the responsible parties.

Visual communication system

Visitors should be able to find their orientation easily within the stadium complex. A clever layout of entries, lateral gangways and vertical circulation stairs are the preconditions for the smooth running of an event.

When the structural options for guiding streams of spectators through the building have been exhausted, a reliable information system for locating tiers, blocks, rows and seat becomes obligatory.

Signage

Since visitors to large places of congregation cannot be expected to know their way around, they are reliant on signage provided by universally intelligible pictograms on signposts throughout the extended area of a stadium complex.

Colour codes on tickets serve to identify parking areas. At the external areas around the main entrances, large stadium plans and information booths facilitate the further orientation of visitors. Current location, access routes, sectors and blocks are to be signed adequately.

An information system with video monitors should be placed in the players' recreational areas and dressing rooms, in areas for VIPs and media, and in control rooms, foyers and milling areas for spectators (UEFA Technical Recommendations, EURO 2008).

514 **Video scoreboard**
Olympic Stadium Berlin
515 **Video Cube (pitch centre)**
Commerzbank Arena, Frankfurt
516 **Video scoreboard (diagonal layout)**
RheinEnergieStadion, Cologne

BOOK RECOMMENDATION

Terms for Civil Engineers

Over 37.000 technical terms!

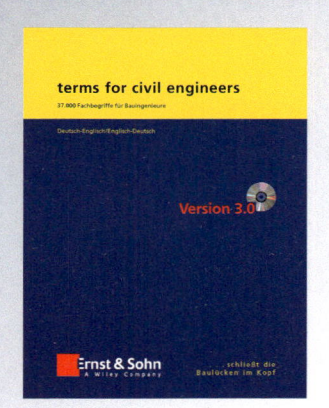

Ernst & Sohn

Terms for Civil Engineers

Deutsch-Englisch/
English-German CD-ROM Version 3.0, 2003.
€ 99,–/sFr 158,–
Update price
€ 60,–/sFr 96,–
ISBN 978-3-433-01714-2

Ernst & Sohn
Verlag für Architektur und
technische Wissenschaften
GmbH & Co. KG

www.ernst-und-sohn.de

For order and customer service:
Verlag Wiley-VCH
Boschstraße 12
69469 Weinheim
Germany
Tel.: +49(0) 6201 / 606-400
Fax: +49(0) 6201 / 606-184
E-Mail: service@wiley-vch.de

„Beton" is German for „concrete", but what is „Ortbeton"? A good knowledge of German is usually not enough when it comes to technical terms. This is where „Terms for civil engineers" comes in.

The program may be used to look up terms whose definitions are unclear, and users can extend and alter the definitions found in the dictionary for themselves. A simple mouse click on the self-explanatory symbol bar switches between German-English and English-German.

There are two ways to find the translation of a word:

1. Typing in the first few letters of the word brings the user to the relevant position in the dictionary, while the letters entered remain visible in a window.

2. A search for all entries containing part of a word returns a list of hits.

* In EU countries the local VAT is effective for books and journals. Postage will be charged. Whilst every effort is made to ensure that the contents of this leaflet are accurate, all information is subject to change without notice. Our standard terms and delivery conditions apply. Prices are subject to change without notice.

007587116_my

---

duraskin® – the durable skin for stadium roofs.

Foto: Manfred Storck

VERSEIDAG
COATING AND COMPOSITE

PVC coated PES fabrics and PTFE coated Glass fabrics supplied by Verseidag-Indutex GmbH have been covering stadiums around the world.

The high-level of our quality and our experience as weaver and coater for architectural textiles are convincing architects, engineers and investors since more than 35 years.

We would be glad to support you on your next project.

duraskin® Perfect solutions worldwide

VERSEIDAG-INDUTEX GmbH, Industriestr. 56, D-47803 Krefeld, www.vsindutex.de

Chapter 24

Grandstand profiles of built examples
A comparison of the stadia for
FIFA World Cup 2006™ in Germany

517–519 **Views of stadium interior areas**
from left to right:
World Cup Stadium **Berlin**
World Cup Stadium **Dortmund**
World Cup Stadium **Frankfurt**
World Cup Stadium **Gelsenkirchen**
World Cup Stadium **Hamburg**
World Cup Stadium **Hanover**
World Cup Stadium **Kaiserslautern**
World Cup Stadium **Cologne**
World Cup Stadium **Leipzig**
World Cup Stadium **Munich**
World Cup Stadium **Nuremberg**
World Cup Stadium **Stuttgart**

World Cup Stadium Berlin

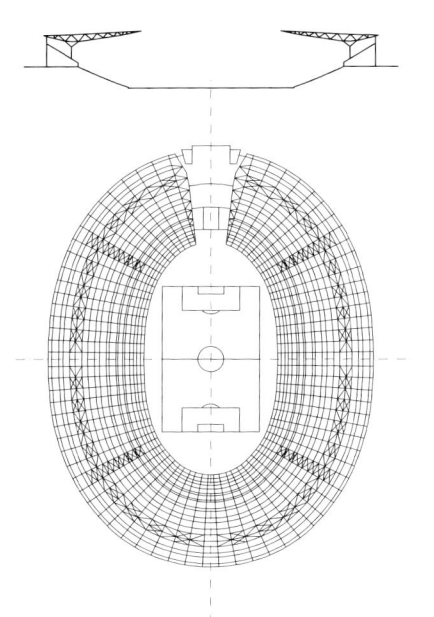

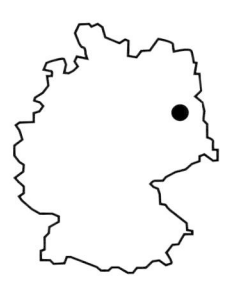

520 **Roof framework** (sectional view)
Olympic Stadium Berlin
521 **Roof framework** (top view)
Olympic Stadium Berlin
522 **Historic aerial view**
Olympic Stadium Berlin
523 **Geographic location** of the city of Berlin

page 261:
524 **Aerial view** (2006)

In 1913, Otto March started planning for the 'German Stadium' as multifunctional stadium accommodating almost all sports in one single edifice. Because of the First World War, the 1916 Olympics were cancelled.

For the XI Olympic Summer Games, the former 'German Stadium' at the Reichssportfeld was converted and in part newly erected by the architect's sons Walter and Werner March.

A second phase of conversion followed for the Football World Cup in 1974. A MERO-system roofing of the main and opposite stand plus floodlighting system were supplemented.

In 1998, gmp architects were awarded the 1st prize in an international competition with 10 participants. The Berlin senate allocated a construction and utilization concession for 13 to 21 years to Walter Bau AG. In the summer of 2000, the complete renovation and modernization of the Berlin Olympic Stadium began.

Based on the protection requirements for historical monuments, the construction was marked by cautious renovation with only gentle interventions into the general appearance. Roofing for spectator seats was provided and modern service facilities installed. Conversion took place with 48 months from 07/2000 to 12/2004 during the regular season (minimum capacity 55,000 seats, DFB Cup Final 70,000 seats).

The two-tier multipurpose stadium is used for athletics, football and open-air concerts and has a total capacity of around 74,200 seats, with around 5,000 executive seats. It is the largest of all twelve World Cup stadia. Clockwise, beginning at the marathon gate, in three construction phases (14 substages), lower and upper tier were dismantled and newly erected. Next, the integration of the load-bearing structure was carried out for the upper-tier roofing. No documents were available on the original structure from 1936. The framework, steppings and natural stone façades were restored. The lowering of the playing field by 2.65 m created two extra seating rows, improved the sightline and enabled the set-up of a new reporters' moat (1,600 additional places).

First, the outer columns and tree-like supports were erected to form the new roof structure, then binders and tangentials followed by the membrane structure (06/2002 – 05/2004). Next, the VIP area and the catering/merchandise points were inserted into the grandstand structure as well as the underground car parks south and north.

Quite exceptionally, the new tartan tracks are blue, the colour of home club Hertha BSC.

60% of spectators (around 44,500 places) are situated within an optimum viewing circle for athletics and only 22.5% of spectators (around 16,700 places) are seated within the best viewing radius of 90 m for football.

Due to the oval geometry, the circulating hospitality box tier is located near the halfway line at best viewing radius. Yet, around 3,000 spectators sit beyond the maximum admissible athletics distance of 230 m.

The large distance of the first row is typical for sports stadia with integrated athletics tracks. Football enthusiasts tend to assess more critically the rather 'expansive' atmosphere of an athletics stadium, a fact that certainly does not apply to the historic grandstand of the Berlin Olympic Stadium.

HIGHTEX

The World of Membranes

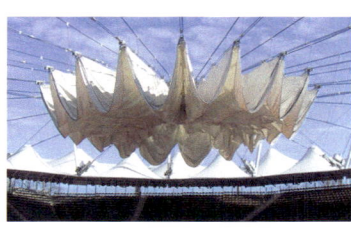

| Ascot Racecourse | Busan Main Stadium | Gottlieb-Daimler Stadium Stuttgart | Olympic Stadium Berlin | Tennis Stadium Rothenbaum |

Design | Engineering | Fabrication | Installation | Inspection | Maintenance

Hightex Group plc, London, UK
Hightex International AG, Switzerland
Hightex GmbH, Germany
Hightex Ltd., UK
Hightex Americas LLC, USA
Hightex Pty. Ltd., Australia
Hightex Structures Pty., South Africa

www.hightexworld.com

TECHNICAL LITERATUR FOR CIVIL ENGINEERS

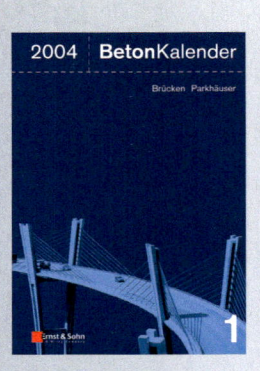

**Beton-Kalender 2004
(Concrete Structural Design)**
Bridges and multi-storey
car parks

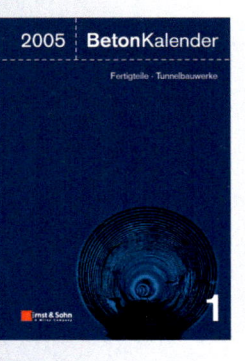

**Beton-Kalender 2005
(Concrete Structural Design)**
Precast elements and
tunnel structures

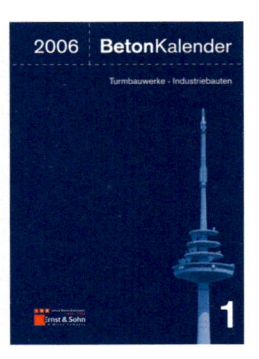

**Beton-Kalender 2006
(Concrete Structural Design)**
Tower structures,
industrial buildings

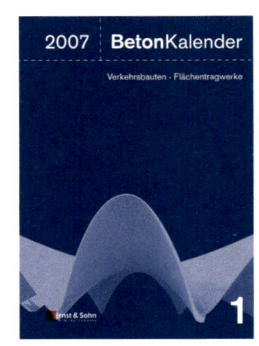

**Beton-Kalender 2007
(Concrete Structural Design)**
Traffic structures,
shell structures

**Beton-Kalender 2008
(Concrete Structural Design)**
Earthquake-proof construction,
hydraulic engineering

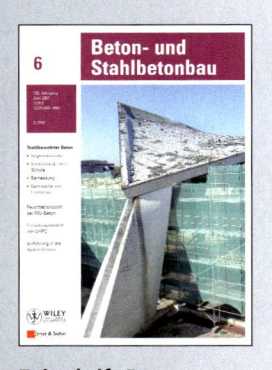

Selected sample chapters from our books under www.ernst-und-sohn.de!

**Zeitschrift Beton-
und Stahlbetonbau
(Concrete and
Reinforced Concrete
Strucutres)**

**FEM in Concrete
Construction**

**Structural Dynamics
in Practice**

**Stahlbau-Kalender 2007
(Steel Construction Design)**
Materials

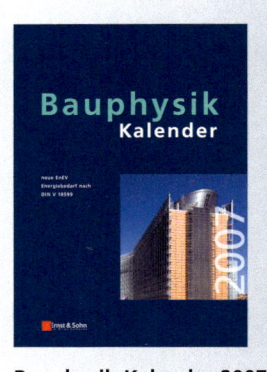

**Bauphysik-Kalender 2007
(Building Physics)**
Energy performance
of buildings

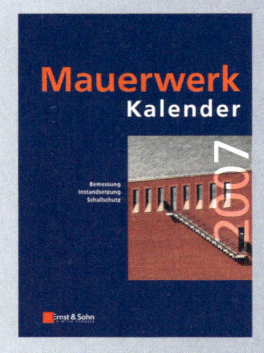

**Mauerwerk-Kalender 2007
(Masonry Year book)**

Ernst & Sohn
Verlag für Architektur
und technische Wissenschaften
GmbH & Co. KG

Ernst & Sohn
A Wiley Company
www.ernst-und-sohn.de

For order and customer service:

Verlag Wiley-VCH
Boschstraße 12
69469 Weinheim
Germany

Tel.: +49(0) 6201 / 606-400
Fax: +49(0) 6201 / 606-184
E-Mail: service@wiley-vch.de

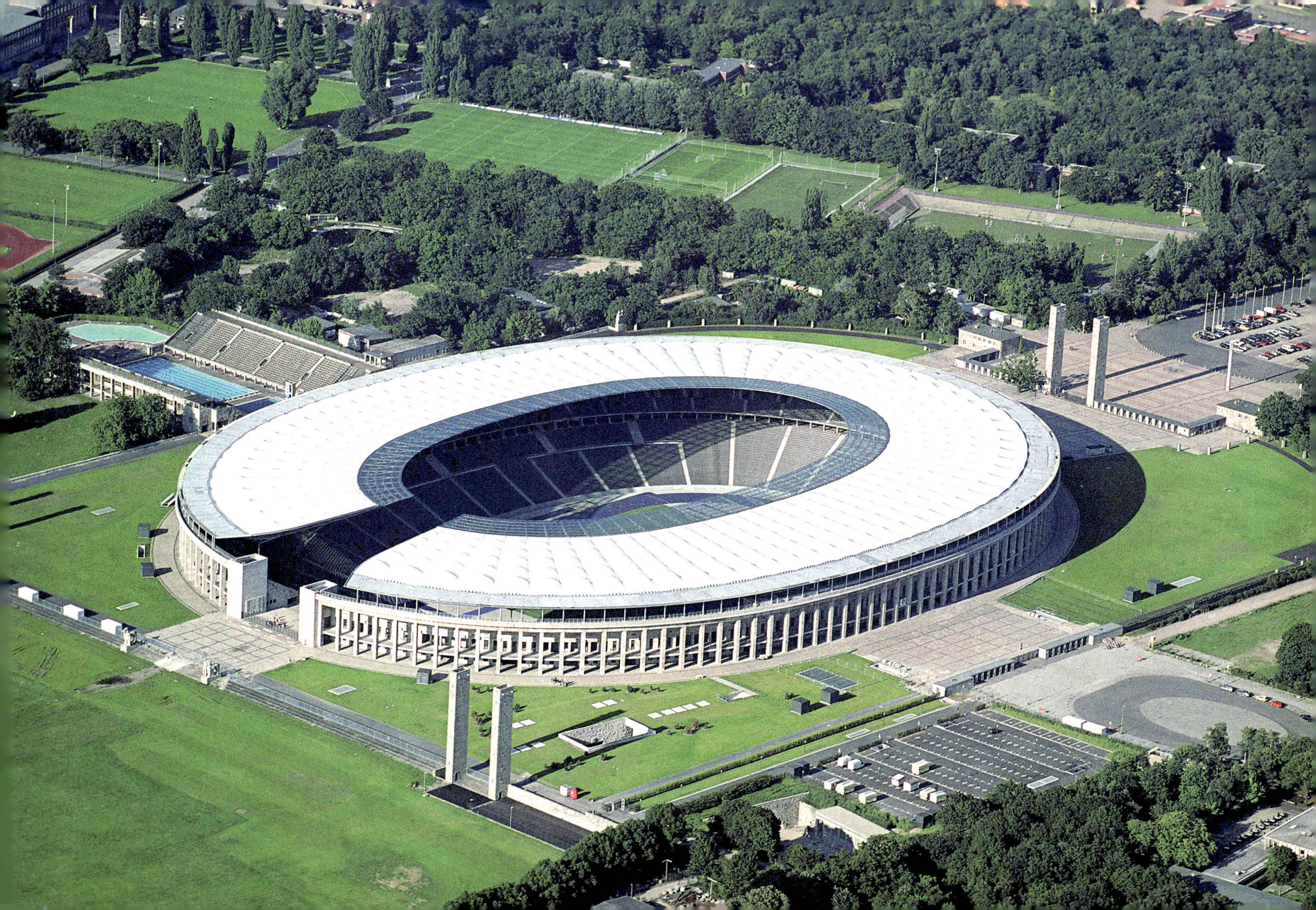

24 Grandstand profiles of built examples

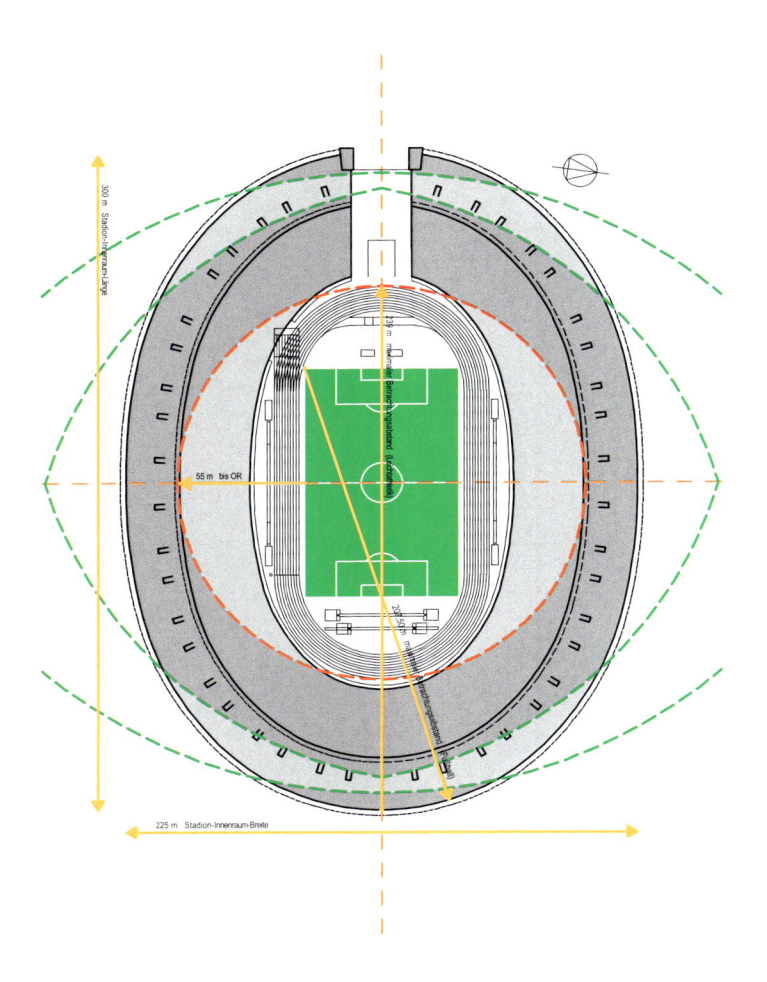

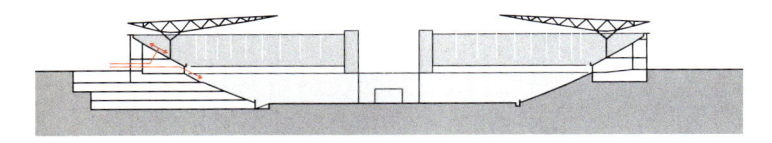

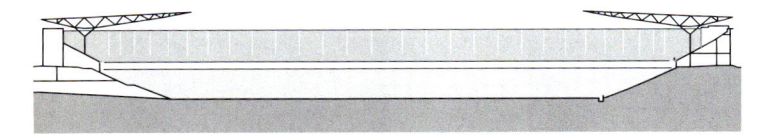

Olympic Stadium Berlin – Two-tier stadium (athletics)
Open oval geometry with circulating private box tier

Capacities

Gross capacity (international)	74,200 places [1]	100%
Gross capacity (national)	74,200 places [2]	
Standing places	–	
Seats available for World Cup 2006 (net)	63,400 places [1]	
Within opt. 90-m viewing circle (football)	16,700 places	22.5%
Within opt. 130-m viewing circle (athletics)	+44,500 places	60.0%
Up to the max. recomm. 230-m distance	+10,000 places	3.5%
Beyond 230-m distance	+ 3,000 places	4.0%
Lower tier	approx. 37,800 places	
Upper tier	approx. 36,400 places	
VIP guests (total)	5,653 places [1]	approx. 9,750 m²
Executive seats (World Cup 2006)	4,758 seats [1]	
Number of private boxes (total)	74 private boxes (+ 13 Sky-Boxes)	
Box seats, honorary tribune	895 box seats	
Media representatives (World Cup 2006)	200 commentators' positions (south)	
	400 television observers	
	1,000 press seats (with desk)	
	1,000 press seats (without desk)	
Wheelchair places	130 places [1]	

Dimensions (approx.)

	225 x 300 m (total)
Stadium interior	116 x 189 m = 22,000 m²
Width of roof	$b = 68$ m max.
Height above pitch	$h = 40$ m
Height above terrain (eaves)	$h = 21$ m
Roof area	$A = 42,000$ m²
Spans	77 spans (6 – 11 m)
Membrane roof	= 27,000 m²
Glass roof edge	= 6,000 m²
Concrete roof edge	= 9,000 m²
Drainage	Slope toward exterior = 3° or 5%

1) according to LOC World Cup 2006, Frankfurt (status: October 2005)
2) according to www.stadionwelt.de (status: October 2005)

525 **Viewing circle and dimensions** (plan)
   Olympic Stadium Berlin, Scale 1 : 3,000
526 **Structure and circulation**
   (transverse / longitudinal section)

   page 265:
527 **Picture of the stadium interior** (2006)

BOOK RECOMMENDATION

Kurrer, K.-E.

The History of the Theory of Structures
From Arch Analysis to Computational Mechanics

2008. Approx 800 pages with approx 650 figures. Hardcover.
Approx € 119.–*/sFr 188.–
ISBN: 978-3-433-01838-5

Ernst & Sohn
A Wiley Company
www.ernst-und-sohn.de

Ernst & Sohn
Verlag für Architektur und technische Wissenschaften GmbH & Co. KG

The History of the Theory of Structures

This major work is about much more than the origins of statics and its use in building and bridge engineering since the late sixteenth century. It is also about the very ideas of „statics" and „mechanics of materials" and how they came to be an integral part of the engineer's life; how they were developed into an academic discipline; how they became the subject of growing numbers of technical books and periodicals; and, ultimately, how the epistemology of the subject developed.

Drawing on a long series of specialized articles and more than two decades of study, the author begins each chapter with a personal reflection on his involvement with the subject under discussion. He demonstrates how engineers thought, far from being abstractly objective, is imbued with the character of its thinkers, their teachers, and their pupils.

For order and customer service:
Verlag Wiley-VCH
Boschstraße 12
69469 Weinheim
Germany
Tel.: +49(0) 6201 / 606-400
Fax: +49(0) 6201 / 606-184
E-Mail: service@wiley-vch.de

* In EU countries the local VAT is effective for books and journals. Postage will be charged. Whilst every effort is made to ensure that the contents of this leaflet are accurate, all information is subject to change without notice. Our standard terms and delivery conditions apply. Prices are subject to change without notice.

A Membrane Roof on Berlin's Olympic Stadium Ensures a Clear View of the Final Match

The newly refurbished Olympic Stadium in Berlin, Germany gives 74,000 spectators an unobstructed view of the soccer field. The new lightweight roof construction, made of woven fiberglass membranes coated with Dyneon™ PTFE and Dyneon™ Fluorothermoplastics, ensures that everyone is protected from varied weather conditions.

The conditions associated with the construction of the new roof of the Berlin Olympic Stadium were extremely challenging for the Berlin architects Gerkan, Marg and Partners: the entire structure of the roof was to stay within the boundaries of the ground plan of the existing stadium because as a historical monument its original appearance was to remain as unaffected as possible. At the same time, the spectators' view of the playing field was to be as unobstructed by supporting structures and columns as possible. To make things even more complicated, all of the reconstruction work was to be accomplished without interrupting regular-season matches.

The architects decided on a lightweight cantilevered steel construction for the U-shaped roofing over the stands. It does not form a closed ring but remains open in the middle and also open in front of the historical Marathon Gate Arch in order to preserve the view from the stadium seats to the Maifeld (May Field) and the Glockenturm (Bell Tower). In the upper regions of the stands, 132 outer steel posts and 20 slender steel posts positioned in the stands, having a pedestal diameter of only 250 mm, are essentially all that is supporting the steel construction of the roof.

The German company Hightex GmbH of Rimsting, specialized in textile architecture, covered the outside surface of the supporting structure with 27,000 m? of coated woven fiberglass membrane in 77 membrane sectors. These membranes have a tensile strength of up to several tons per square meter, and they themselves weigh only one to one-and-a-half kilograms per square meter. The coating with Dyneon™ PTFE and Dyneon™ Fluorothermoplastics is what gives this material these all-important properties for heavy-duty use in architectural applications: the surface of the coating is very smooth and has a long-lasting resistance to varied weather conditions. In addition, Dyneon PTFE possesses a nearly universal chemical resistance and exceptionally good mechanical properties.

A key advantage of PTFE coatings is that they require neither softeners nor stabilizers, which can evaporate over time and can cause the coatings to become brittle. The PTFE allows the membranes to remain elastic and smooth so that even after many years of service, dirt and contamination are unable to find cracks to settle into, and rain showers are all that is needed to clean the roof. The translucency of the woven material always guarantees that the lighting conditions inside the stadium are ideal for both spectators and players.

Hightex GmbH also lined the lower side of the roof construction with a Dyneon PTFE coated, open-grid woven fiberglass matting. This lower side serves as maintenance gangways for the glare-free floodlight system installed under the roof, as well as for the sound system. The open-structured fabric of the lower side allows sound and light to pass through, by ensuring ideal pressure compensation through to the upper membranes under windy conditions, thus also helping to reduce the static load on the steel construction of the roof.

Dyneon is a trademark of 3M Company

dyneon
a 3M company

3M

Contact:
Dyneon GmbH & Co. KG
www.dyneon.com
dyneon.europe@mmm.com

Geotechnical Engineering Handbook

Ed.: Ulrich Smoltczyk

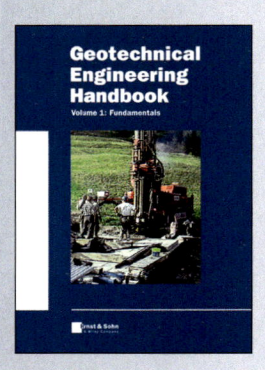

Volume 1: Fundamentals
2002.
829 pages, 616 fig.
Hardcover.
€ 179,–*/ sFr 283,–
ISBN 978-3-433-01449-3

This is the English version of the Grundbau-Taschenbuch - a reference book for geotechnical engineering. The first of three volumes contains all information about the basics on the field of geotechnical engineering. The book is written by authors from Germany, Belgium, Sweden, the Czech Republic, Australia, Italy, U.K., and Switzerland.

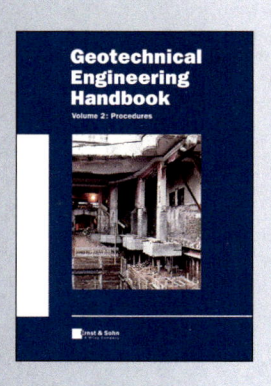

Volume 2: Procedures
2002.
679 pages, 558 fig.
Hardcover.
€ 179,–*/ sFr 283,–
ISBN 978-3-433-01450-9

Volume 2 of the Geotechnical Engineering Handbook covers the geotechnical procedures used in manufacturing anchors and piles as well as for improving or underpinning the foundations, securing existing structures, controlling ground water, excavating rocks and earthworks. It also treats such specialist areas as the use of geotextiles and seeding.

Volume 3: Elements and structures
2002.
646 pages, 500 fig.
Hardcover.
€ 179,–*/ sFr 283,–
ISBN 978-3-433-01451-6

Volume 3 of the Geotechnical Engineering Handbook deals with foundations. It presents shallow foundations starting with basic designs right up the necessary calculations. There is comprehensive coverage of the possibilities for stabilizing excavations, together with the relevant area of application, while another section is devoted to the useful application of trench walls. The entire book is an indispensable aid in the planning and execution of all types of foundations found in practice, whether for researches or practitioners.

Special Set Price (three volumes)
€ 477,–* / sFr 754,–
ISBN 978-3-433-01452-3

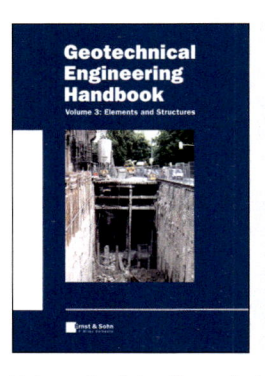

Fax-No. +49 (0)30 47031 240 – Ernst & Sohn Berlin, Germany

Number	Order-No.	Titel	Price
	978-3-433-01449-3	Volume 1: Fundamentals	179,– €*
	978-3-433-01450-9	Volume 2: Prodcedures	179,– €*
	978-3-433-01451-6	Volume 3: Elements and structures	179,– €*
	978-3-433-01452-3	Special Set (three volumes)	477,– €*

Company

Contact person | USt-ID Nr/VAT-ID No.

Street/No. | E-Mail

Country _ Zip code | Location

Date | Signature

Ernst & Sohn
A Wiley Company
www.ernst-und-sohn.de

Ernst & Sohn
Verlag für Architektur
und technische Wissenschaften
GmbH & Co. KG

For order and customer service:
Verlag Wiley-VCH
Boschstraße 12
69469 Weinheim
Germany

Tel.: +49(0) 6201 / 606-400
Fax: +49(0) 6201 / 606-184
E-Mail: service@wiley-vch.de

* In EU countries the local VAT is effective for books and journals. Postage will be charged. Whilst every effort is made to ensure that the contents of this leaflet are accurate, all information is subject to change without notice. Our standard terms and delivery conditions apply. Prices are subject to change without notice.

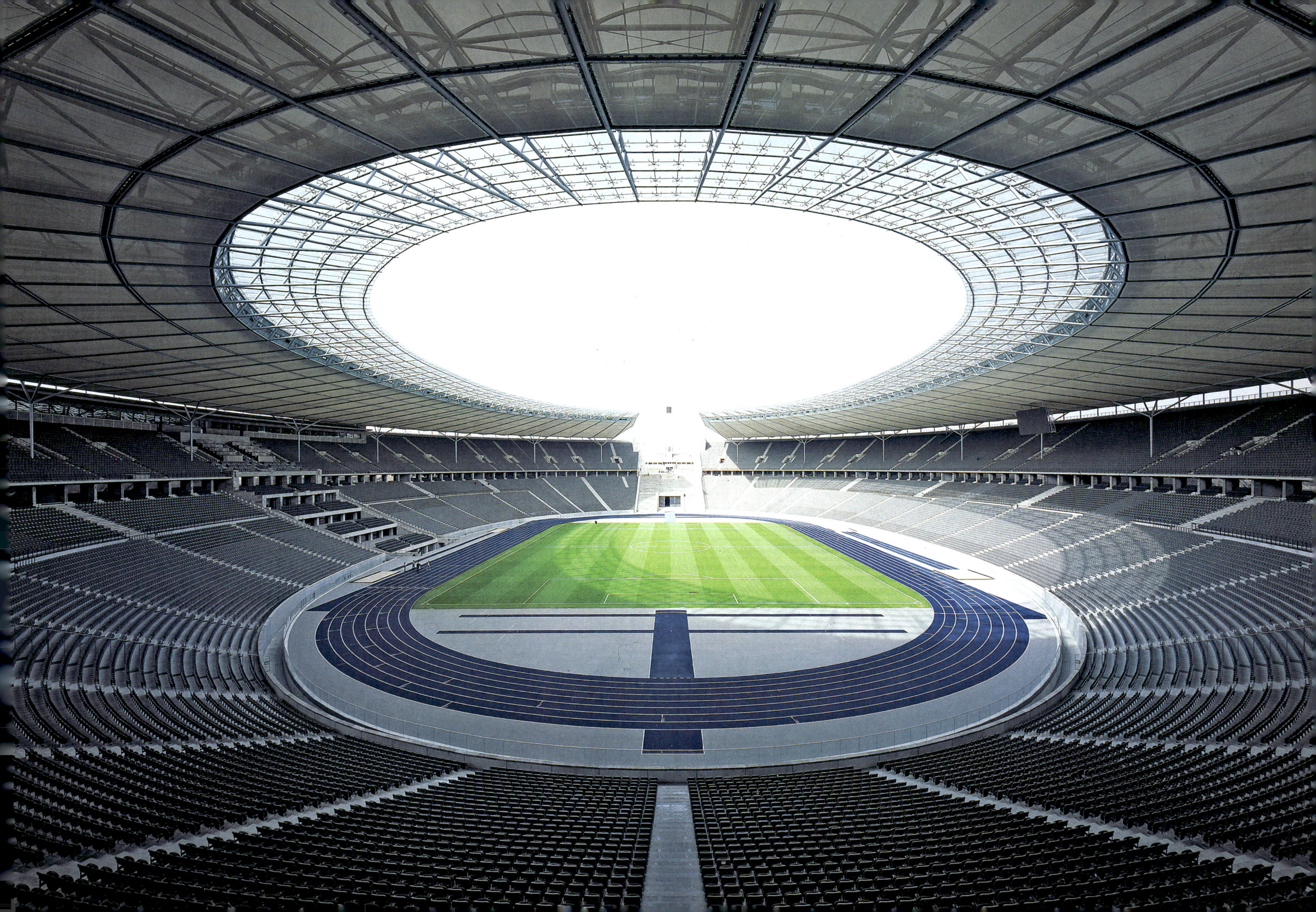

24 Grandstand profiles of built examples

01. PTFE membrane roof
02. Glass roof (interior ring)
03. Reinf.-conc. compression ring (exterior)
04. Tree-like support
05. Three-boom girder (radial)
06. Trussed girder (tangential)
07. Hinged pillar

Upper tier

Row number	31 rows
Tread depth	75 cm
Rise	39.6 – 51.0 cm
Stand inclination (average)	25°
Sightline elevation C	17.1 – 19.8 cm

Distance of stand (front edge)

Distance to field	24.20 m

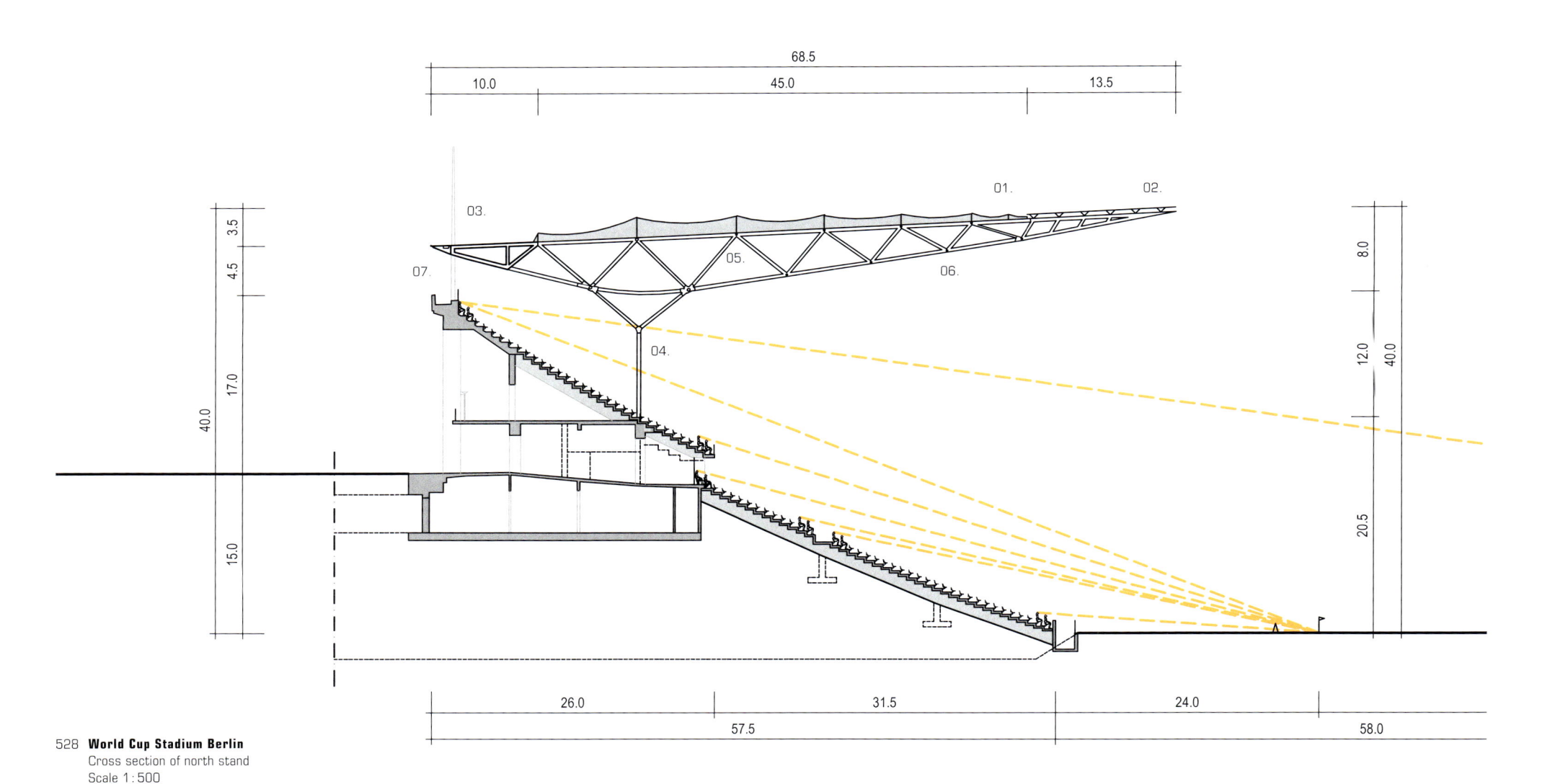

Lower tier

Row number	42 rows
Tread depth	78 cm
Rise	25.5 – 37.0 cm
Stand inclination (average)	23°
Sightline elevation C	19.7 – 15.0 cm

Viewing distance (to the point of focus)

Maximum horizontal distance	78.90 m
Maximum vertical distance	31.00 m

Distance of hosp. boxes (touch line) 55.00 m

Viewing angle (front touch line)

Upper tier	4.5° – 15.0°
Lower tier	18.0° – 21.5°

528 **World Cup Stadium Berlin**
Cross section of north stand
Scale 1:500

Lower tier

Row number	42 rows
Tread depth	78 cm
Rise	22.1–23.3 cm
Stand inclination (average)	16°
Sightline elevation C	22.1–23.3 cm

Upper tier

Row number	31 rows
Tread depth	75 cm
Rise	29.5–50.1 cm
Stand inclination (average)	25°
Sightline elevation C	11.7–24.8 cm

Distance of stand (front edge)

Distance to field	42.0 m (west)
	44.0 m (east)

Viewing distance (to the point of focus)

Maximum horizontal distance	96.75 m
Maximum vertical distance	31.00 m

Viewing distance (maximum)

Football	207.50 m
Athletics	239.00 m

Viewing angle (front goal line)

Lower tier	2.5°–11.5°
Upper tier	14.0°–17.0°

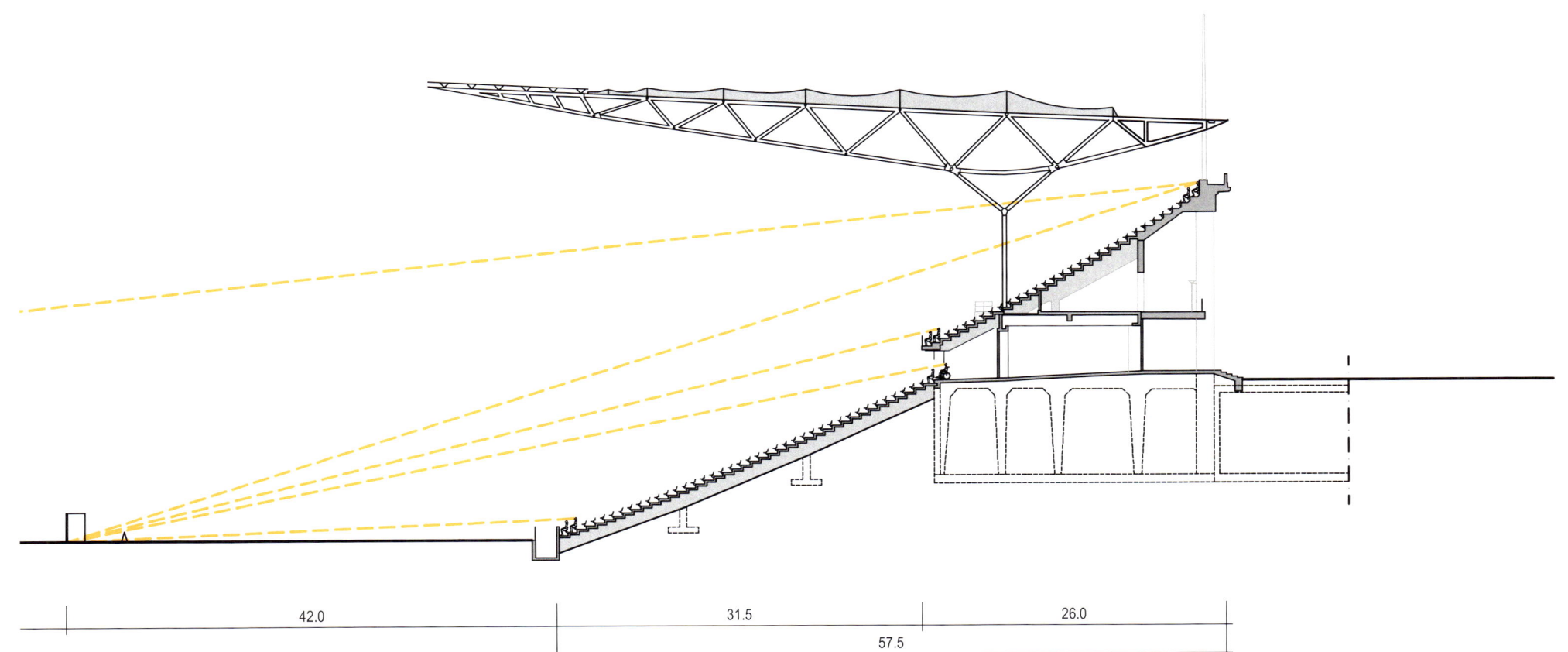

World Cup Stadium Berlin 267

529 **World Cup Stadium Berlin**
Longitudinal section of east stand
Scale 1:500

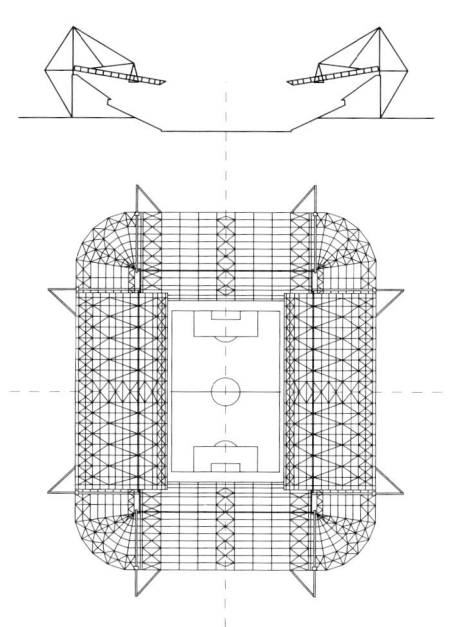

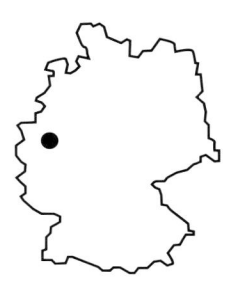

World Cup Stadium Dortmund

The Westfalenstadion was newly erected for the Football World Cup 1974, directly adjacent to the old 'Kampfbahn Rote Erde' from 1960, and had since then been the stage for many international matches. It is a pure football stadium (no concerts, athletic events or American Football).

From 1995, it was modernized in three phases. The stadium thus belongs to the group of 'evolved' stadia and therefore has no uniform tier geometry as it was developed in different stages during three decades.

1st construction phase 1995–1996: conversion of standing into seating places up to a capacity of 55,000 spectators; expansion of east and west stand (each plus 6,000 seats) and insertion of a 1,800-m² service/VIP level.

2nd construction phase 1998–1999: extension of the north stand plus 13,000 seats (during Bun-desliga: 68,600); conversion of the roof structure and expansion of the north stand: FAN 'Borussia Park' with 1,700 m² and new façade design; expansion of south stand to 25,000 spectators. It is said to be the largest standing-accommodation complex in Europe.

3rd construction phase 2002–2003: general contractor HochTief AG was commissioned with the modernization and extension of the complex to semi-final capacity for the World Cup.

In 2002, the corner areas were completed (plus 14,500 seats) as well as the expansion of the VIP area 'Stammtischebene' (700 m²) and the addition of 11 box-units (160 m²). An upscale in total capacity to 66,000 seats (81,250 incl. standing places) makes it the second largest of the twelve World Cup stadia. Conversion took place within 16 months during regular operation (05/2002 to 09/2003).

Shell construction began at all four corners simultaneously. After setting up the pylons, a load transfer of the existing roof structure was carried out.

Through tensioning, the roof was slightly lifted by 2 cm via a cable end-tensioning, to allow taking away the former corner columns. After this, works on the lounge corners began.

With 47,5 % of the total capacity, the new Signal Iduna Park (2006) as pure football stadium has 31,300 seats within the 90-m best viewing radius; the second largest capacity for optimum-viewing seats. Only about 4 % (2,600 seats) are outside the maximum admissible viewing distance for football matches of 190 m.

The stands of this rounded rectangular geometry connect without considerable height offset directly to the playing field (5.65/6.40 m). Securing measures during league operation are maintained through fencing at the short ends.

The stands have a linear rise of 40/44/54 cm and at the upper tier even a riser height of 60 cm and 80 cm tread width. 'C' value of sightline elevation decreases continuously in the lower tier of the east stand. The adjacent upper-tier values continue from 14.4 to 9.0 cm.

After 70 rows, the sightline profile of the south stand at the terrace front edge still has a 'C' value of 13.6 cm with a 54/80 cm rise.

The atmosphere in this very dense grandstand area has a good reputation in the Bundesliga. The south stand with its densely packed spectators and the vicinity to the goal represents the 'twelfth man' of the playing team.

530 **Roof framework** (sectional view)
Signal Iduna Park, Dortmund
531 **Roof framework** (top view)
Signal Iduna Park, Dortmund
532 **Historic aerial view**
Westfalenstadion and 'Kampfbahn Rote Erde'
533 **Geographic location** of the city of Dortmund

page 269:
534 **Aerial view** (2006)

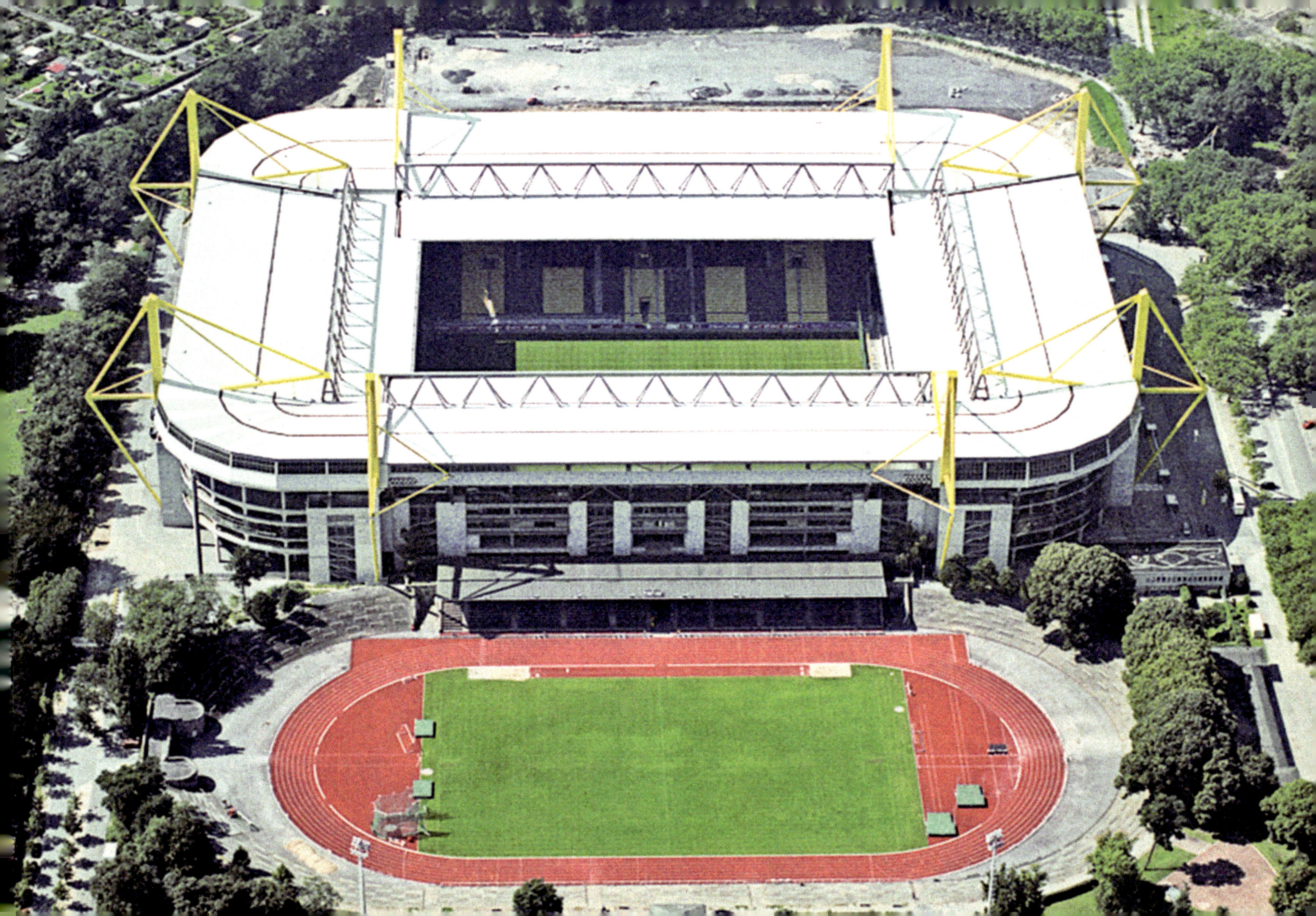

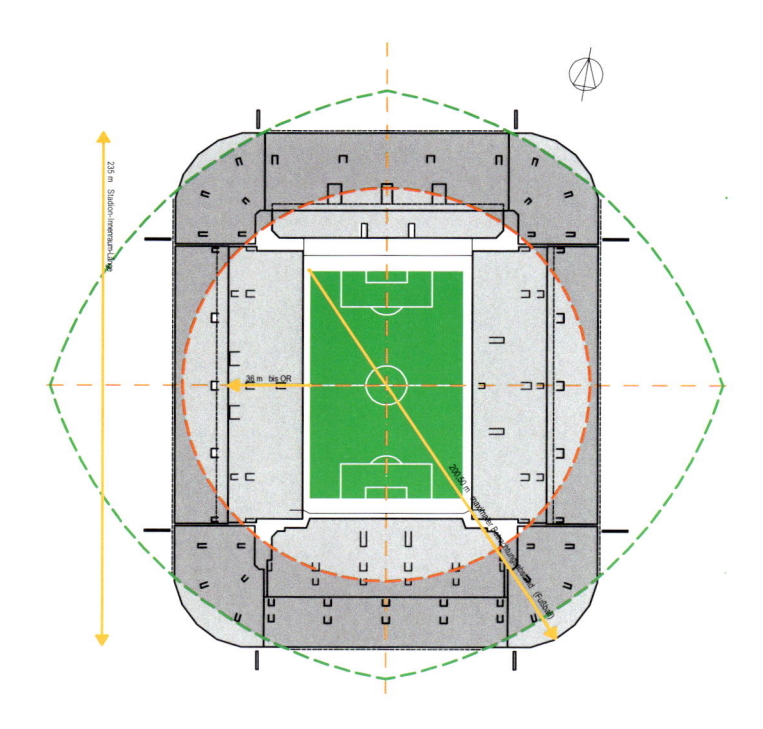

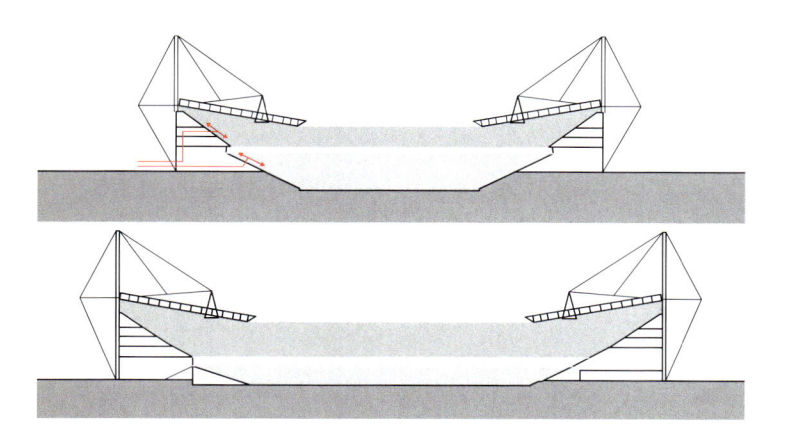

Signal Iduna Park, Dortmund – Two-tier stadium (football)
Rounded rectangular geometry with boxes on two sides

Capacities

Gross capacity (international)	65,900 places [1]	100%
Gross capacity (national)	81,250 places [2]	
Standing places	27,500 places [2]	
Seats available for World Cup 2006 (net)	56,400 places [1]	
Within opt. 90-m viewing circle (football)	31,300 places	47.5%
Up to the max. recomm. 190-m distance	+ 32,000 places	48.5%
Beyond 190-m distance	+ 2,600 places	4.0%
VIP guests (total)	1,946 places [1]	
Executive seats (World Cup 2006)	1,784 places [1]	
Number of private boxes (total)	15 private boxes	
Box seats, honorary tribune	162 box seats	
Media representatives (World Cup 2006)	200 commentators' positions (south)	
	400 television observers	
	1,000 press seats (with desk)	
	1,000 press seats (without desk)	
Wheelchair places	70 places [1]	

Dimensions (approx.) 190 x 230 m (total)
Stadium interior 79 x 118 m = 9,300 m²

Orientation of playing field Minus 9° in north-south direction

Width of roof $b = 62$ m max.
Height above pitch $h = 31$ m
Height above terrain (eaves) $h = 34$ m

Roof area $A = 34,000$ m²
Drainage Slope toward interior = 11°

1) according to LOC World Cup 2006, Frankfurt (status: October 2005)
2) according to www.stadionwelt.de (status: October 2005)

535 **Viewing circle and dimensions** (plan)
Sigal Iduna Park, Dortmund, Scale 1 : 3,000
536 **Structure and circulation**
(transverse / longitudinal section)

page 271:
537 **Picture of the stadium interior** (2006)

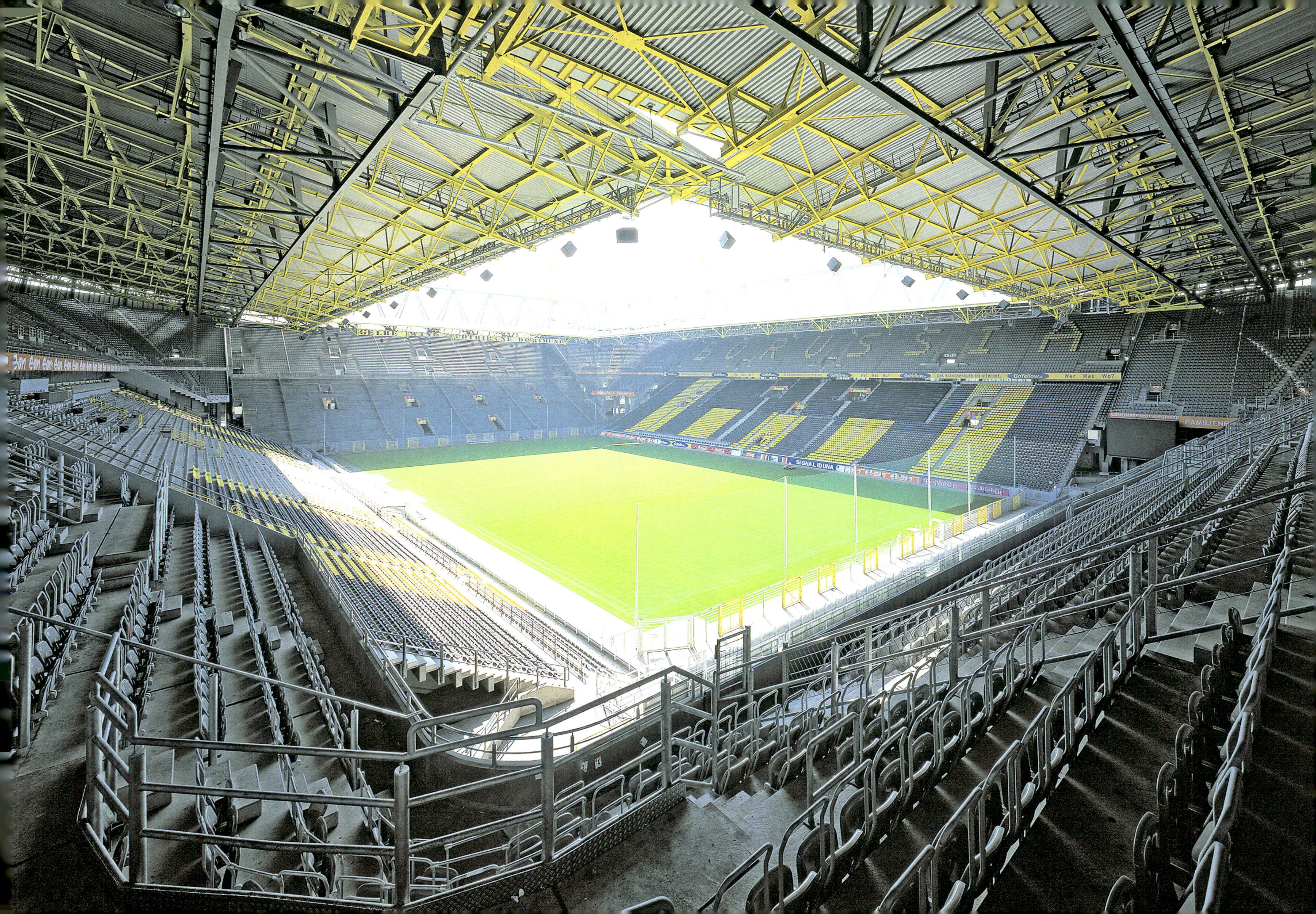
BORUSSIA
Wer Was Wo?
SIGNAL IDUNA
FAMILIEN

24 Grandstand profiles of built examples

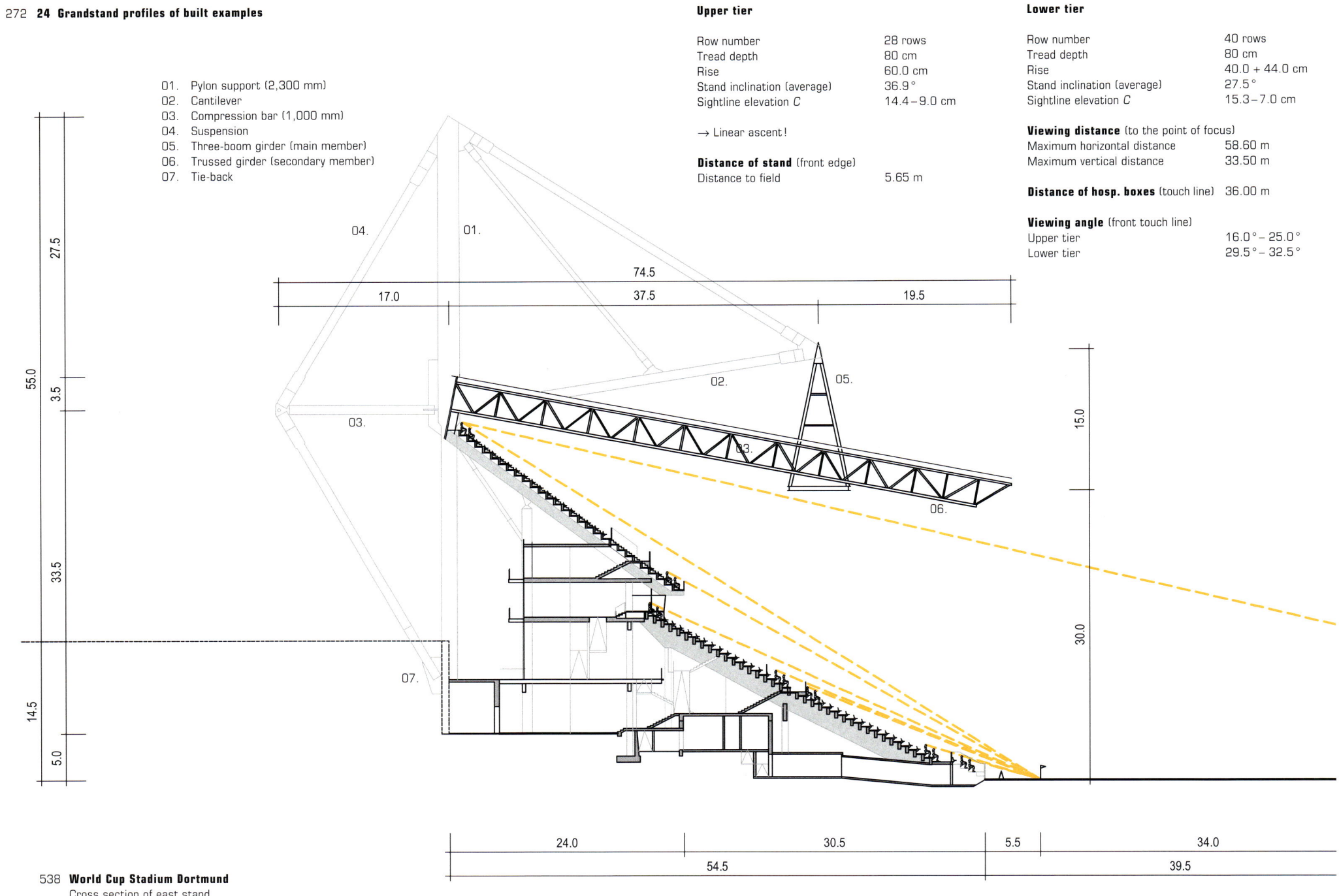

01. Pylon support (2,300 mm)
02. Cantilever
03. Compression bar (1,000 mm)
04. Suspension
05. Three-boom girder (main member)
06. Trussed girder (secondary member)
07. Tie-back

Upper tier

Row number	28 rows
Tread depth	80 cm
Rise	60.0 cm
Stand inclination (average)	36.9°
Sightline elevation C	14.4 – 9.0 cm

→ Linear ascent!

Distance of stand (front edge)
Distance to field   5.65 m

Lower tier

Row number	40 rows
Tread depth	80 cm
Rise	40.0 + 44.0 cm
Stand inclination (average)	27.5°
Sightline elevation C	15.3 – 7.0 cm

Viewing distance (to the point of focus)
Maximum horizontal distance   58.60 m
Maximum vertical distance   33.50 m

Distance of hosp. boxes (touch line)   36.00 m

Viewing angle (front touch line)
Upper tier   16.0° – 25.0°
Lower tier   29.5° – 32.5°

538 World Cup Stadium Dortmund
Cross section of east stand
Scale 1:500

Lower tier

Row number	43 rows
Tread depth	80 cm
Rise	40.0 + 44.0 cm
Stand inclination (average)	27.5 °
Sightline elevation C	20.4 – 18.0 cm

→ Two linear ascents !

Upper tier

Row number	27 rows
Tread dept	80 cm
Rise	54.0 cm
Stand inclination (average)	34 °
Sightline elevation C	15.5 – 13.6 cm

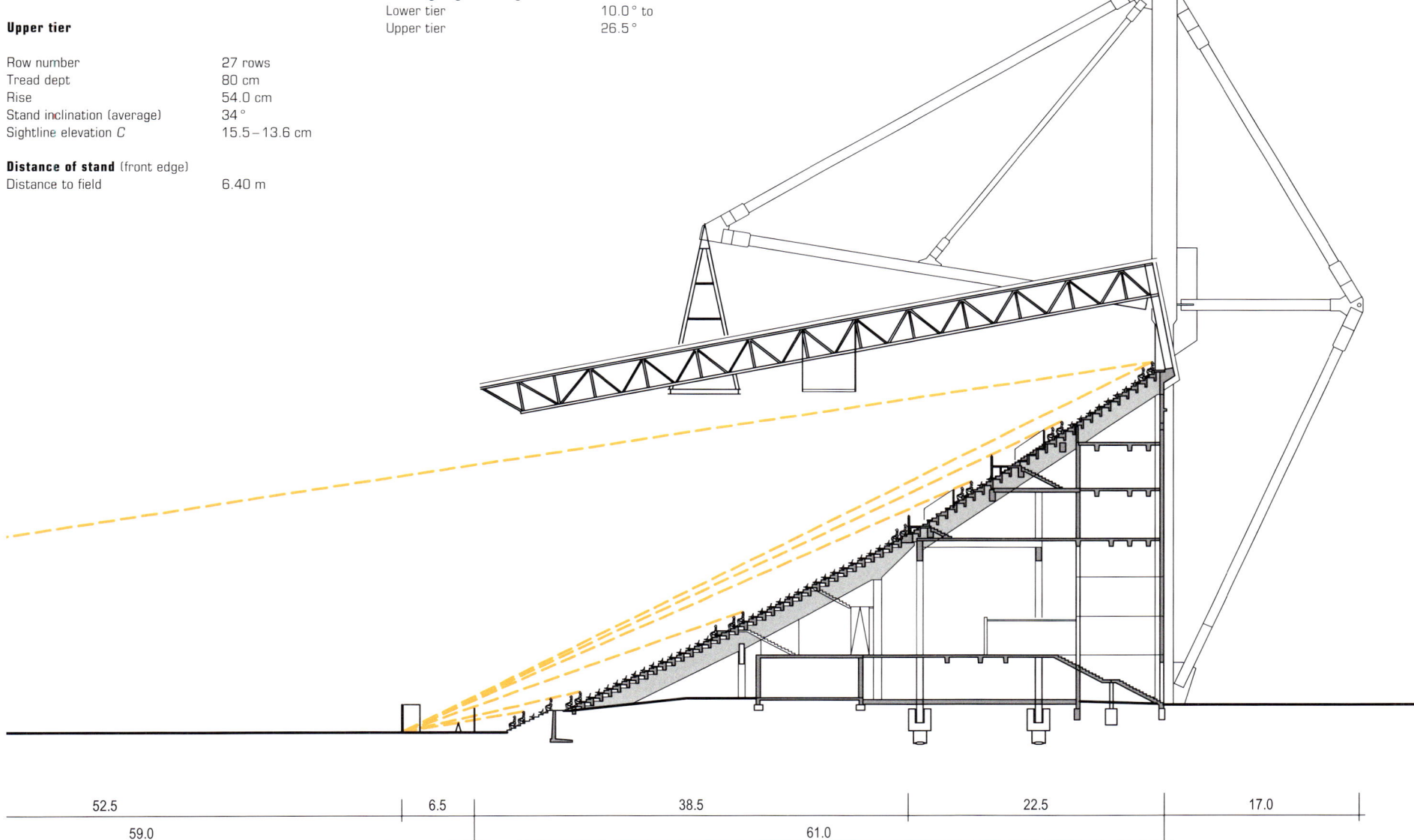

Distance of stand (front edge)

Distance to field	6.40 m

Viewing distance (to the point of focus)

Maximum horizontal distance	66.30 m
Maximum vertical distance	33.50 m

Viewing distance (maximum)

Football	200.50 m

Viewing angle (front goal line)

Lower tier	10.0 ° to
Upper tier	26.5 °

World Cup Stadium Dortmund

539 World Cup Stadium Dortmund
Longitudinal section of east stand
Scale 1 : 500

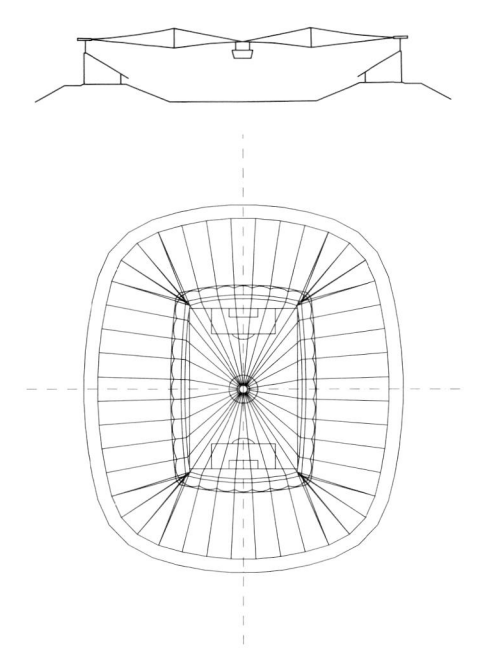

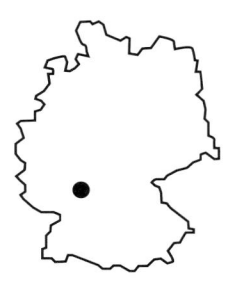

World Cup Stadium Frankfurt

The former athletics arena 'Waldstadion' was erected in the 1920s and completely renovated in 1954. Changed requirements by DFB instigated the conversion of standing into seating places (70,000 / 87,000) in 1960.

In 1974, a second alteration was executed for the Football World Cup. The main and reverse stand were newly covered (World Cup capacity: 61,000 seats). A third refurbishment took place for the European Championship in 1988.

The existing structure was deemed outmoded and not economically viable. A new build/complete conversion into a pure football stadium with multipurpose utilization was decided on. All construction works were to be executed during running operation. Max Bögl GmbH with gmp Architects, Berlin were commissioned with construction of a 'multifunctional arena' with a closing interior roof; an all-weather event venue mainly for football, open-air concerts and conferences.

Since 1998, American Football has been part of the sports programme.

The existing stadium was gradually dismantled and rebuilt completely anew. Commerzbank Arena was constructed in five phases in 36 months (06/2002 to 06/2005) with a minimum capacity of 30,000 seats.

The playing field remained untouched during construction. A major change in utilization came with the loss of the athletics track. Roof and stand geometry were co-ordinated perfectly: a convertible membrane interior roof with a membrane-spanned cable framework on a two-tier stand, with two circulating box tiers accommodating 74 boxes and special-use rooms.

The integration of the historic complex had the highest priority. From afar, the two media and service towers (65 m) accentuate the stadium build as urban landmark.

The location of the sports ground in the direct vicinity of Frankfurt city centre allows a view of the skyline from an elevated position on the upper tier.

Holding capacity is around 48,400 seats (52,000 incl. standing places).

The entire auditorium is within the maximum viewing distance of 190 m. 25,400 seats (52,5% of the capacity) are within the optimum viewing circle.

The risers are gradually elevated, which generates good 'C' values. The compact radial geometry out of three circular segments (radius of curvature: 285 / 165 / 49.50 m) guarantees a sufficient exposition of seats in the corner areas of the grandstand. The rounded profile of the steppings aims to align the individual seat as much as possible to the convergence or kick-off spot and, at the same time, keep sightline difference as low as possible.

120 wheelchair places with seats for helpers are located in an elevated position at the end of the lower tier, overlooking the touch line and goal line without restrictions.

Standing places on short ends are secured by inserted fence components. Behind the goals, retaining nets are suspended from the roof edge.

540 **Roof framework** (sectional view)
Commerzbank Arena, Frankfurt
541 **Roof framework** (top view)
Commerzbank Arena, Frankfurt
542 **Historic aerial view**
Sportpark Waldstadion, Frankfurt
543 **Geographical location** of the city of Frankfurt

page 277:
544 **Aerial view** (2006)

Givingscopeforenthusiasm*

*"Sports is ideal
 to abreact any emotions –
 be it happiness or angriness."
 Karl-Heinz "Kalle" Rummenigge

www.max-boegl.de

Building Construction

Civil Engineering

Steel Construction

Traffic Route Building

Turn-key Building

Tunnel Construction

Bridge Building

Environmental Technology

Slab Track Systems

Transrapid Guideway

MAX BÖGL

Progress is built on ideas.

P. O. Box 1120 · D-92301 Neumarkt, Germany
Phone +49 9181 909-0 · Fax +49 9181 905061
info@max-boegl.de

JOURNALS AND NEW PUBLICATIONS

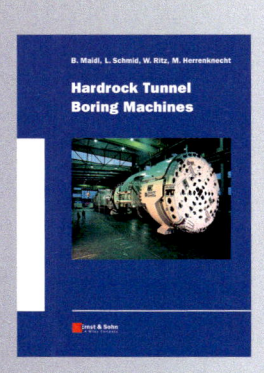

Hardrock Tunnel Boring Machines

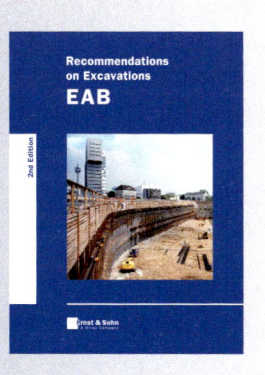

Recommendations on Excavations EAB

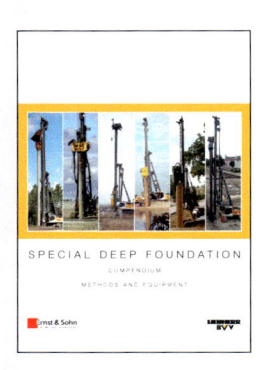

Special Deep Foundation Compendium Methods and Equipment

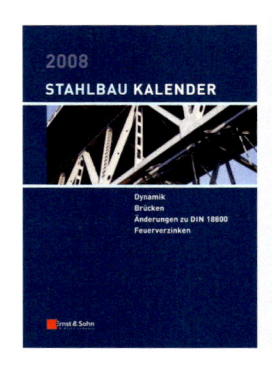

Stahlbau-Kalender 2008 (Steel construction design) Dynamics, Bridges

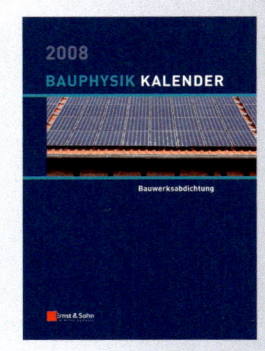

Bauphysik-Kalender 2008 (Building Physics) Sealings

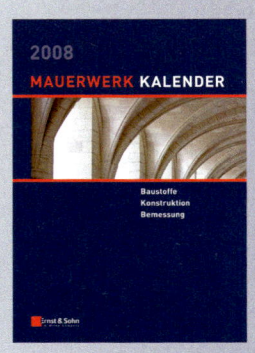

Mauerwerk-Kalender 2008 (Masonry Year book)

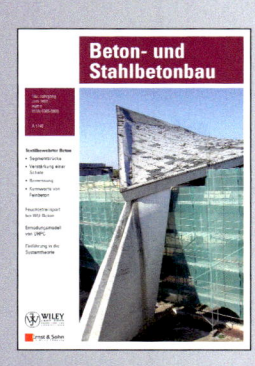

Zeitschrift Beton- und Stahlbetonbau (Concrete and Reinforced Concrete Strucutres)

Zeitschrift Stahlbau (Structural Steelwork)

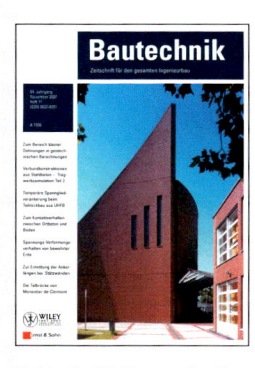

Zeitschrift Bautechnik (Structural Engineering)

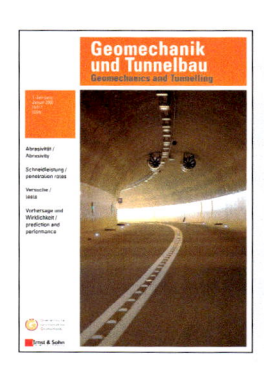

Zeitschrift Geomechanik und Tunnelbau (Geomechanics and Tunnelling)

Zeitschrift Bauphysik (Building Physics)

Zeitschrift Mauerwerk (Masonry)

Ernst & Sohn
Verlag für Architektur
und technische Wissenschaften
GmbH & Co. KG

Ernst & Sohn
A Wiley Company
www.ernst-und-sohn.de

For order and customer service:

Verlag Wiley-VCH
Boschstraße 12
69469 Weinheim
Germany

Tel.: +49(0) 6201 / 606-400
Fax: +49(0) 6201 / 606-184
E-Mail: service@wiley-vch.de

Free sample copy under www.ernst-und-sohn.de/zeitschriften

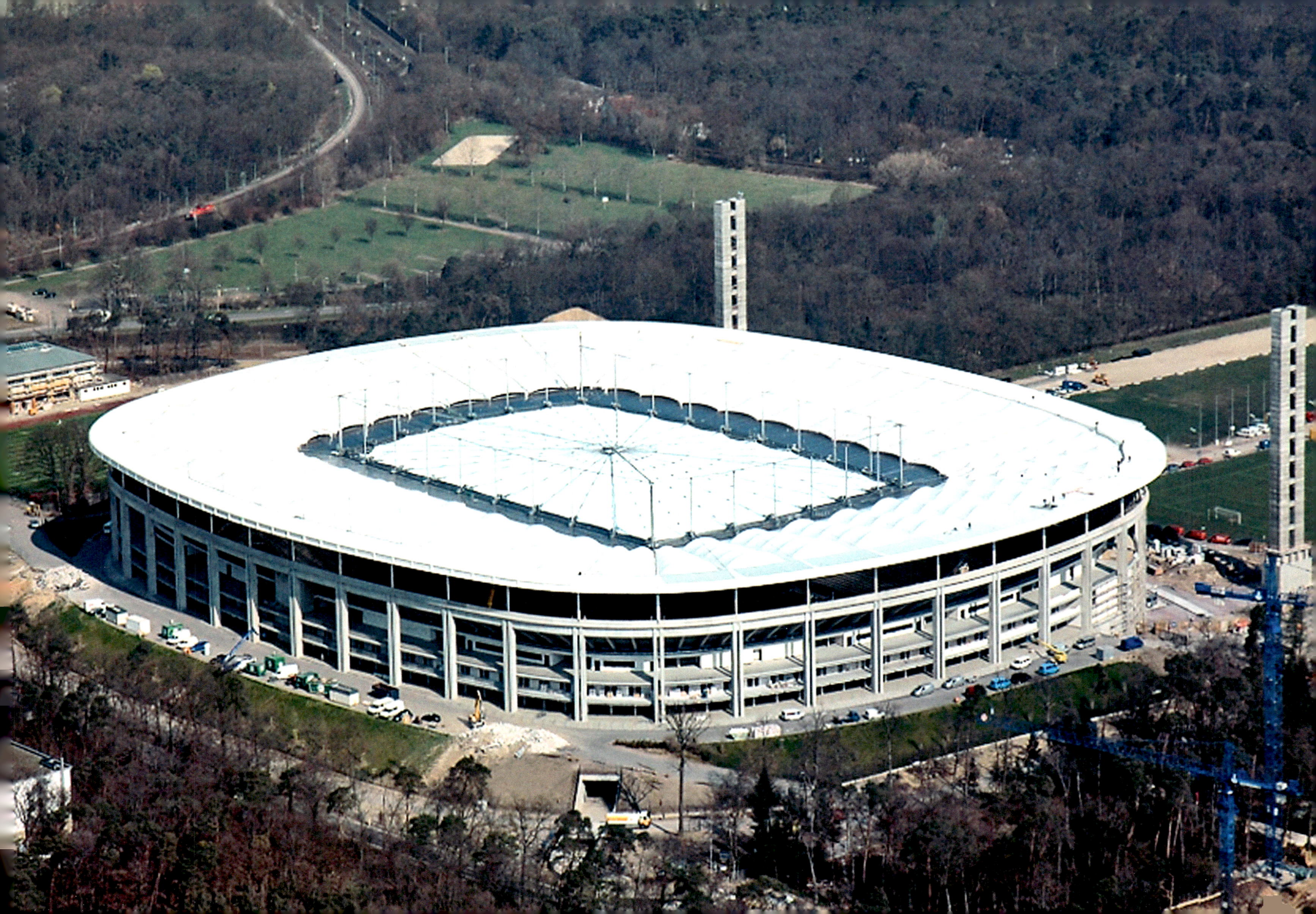

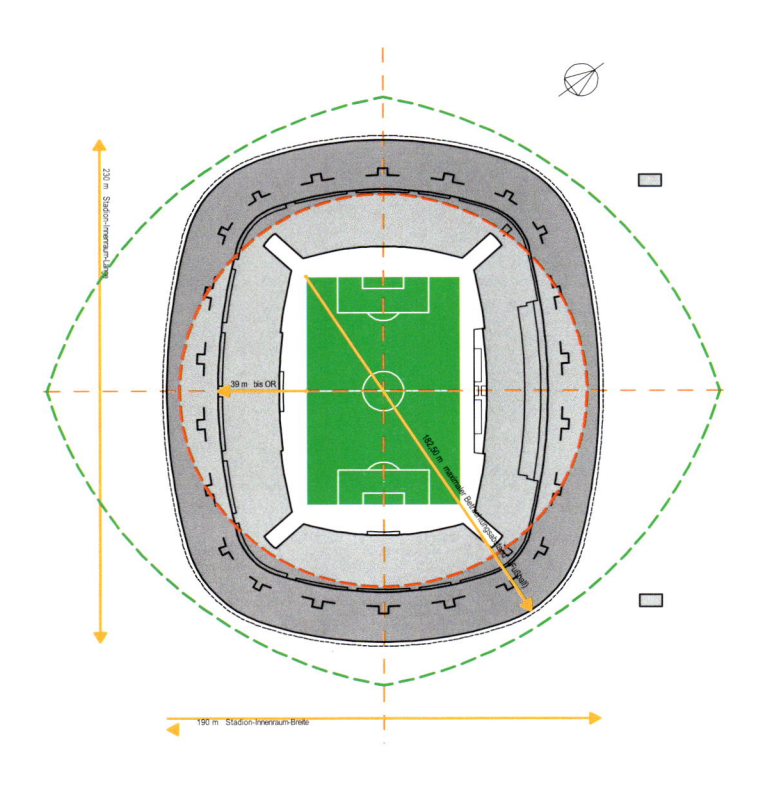

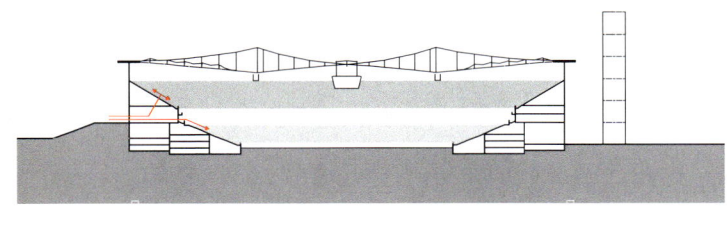

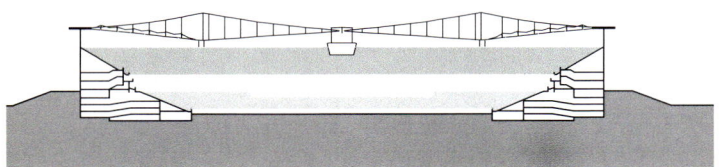

Commerzbank Arena, Frankfurt – Two-tier stadium (football)
Radial geometry of three circular segments and two circulating box tiers

Capacities

Gross capacity (international)	48,400 places [1]	100 %
Gross capacity (national)	52,100 places [2]	
Standing places	9,300 places [2]	
Seats available for World Cup 2006 (net)	41,100 places [1]	
Within opt. 90-m viewing circle (football)	25,400 places	52.5 %
Up to the max. recomm. 190-m distance	+23,000 places	47.5 %
VIP guests (total)	3,120 places [1]	
Executive seats (World Cup 2006)	2,218 places [1]	
Number of private boxes (total)	74 private boxes	
Box seats, honorary tribune	902 box seats	
Media representatives (World Cup 2006)	150 commentators' positions (south)	
	200 television observers	
	500 press seats (with desk)	
	100 press seats (without desk)	
Wheelchair places	120 places [1]	

Dimensions (approx.) 190 x 230 m (total)
Stadium interior 91 x 132 m = ca. 12,100 m²

Orientation of playing field Minus 53° in north-south direction

Air support $h = 11 – 13$ m
Width of roof $b = 59$ m max.
Height above pitch $h = 34$ m
Height above terrain (eaves) $h = 29$ m

Roof area $A = 27,000$ m²
Weight per unit area $G = 8$ kg/m²
Adaptable interior roof 2 x 39.5 m / 57.5 m
Drainage High performance roof drainage system / 'full-flow' vacuum drain
  Slope toward interior = 6°

1) according to LOC World Cup 2006, Frankfurt (status: October 2005)
2) according to www.stadionwelt.de (status: October 2005)

545 **Viewing circle and dimensions** (plan)
  Commerzbank Arena, Frankfurt, Scale 1 : 3,000
546 **Structure and circulation**
  (transverse / longitudinal section)

  page 279:
547 **Picture of the stadium interior** (2006)

adidas
Coca-Cola
i'm lovin' it
HYUNDAI
Fly Emirates
Postbank
PHILIPS
TOSHIBA
YAHOO!

24 Grandstand profiles of built examples

01. Compression ring (reinf. concrete)
02. Hinged pillars
03. Air support with upper and lower tensile ring
04. Central node
05. Stabilizing cable (top)
06. Structural cable (bottom)
07. Suspension cable
08. PTFE membrane roof
09. Video cube

Upper tier

Row number	25 rows
Tread depth	80 cm
Rise	48.2–49.5 cm
Stand inclination (average)	31.5°
Sightline elevation C	12.7–8.7 cm

Distance of stand (front edge)

Distance to field	11.80 m

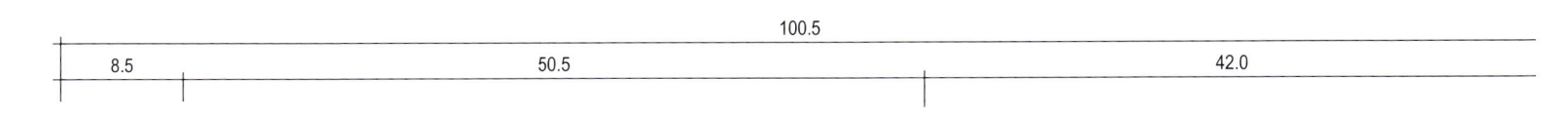

Lower tier

Row number	28 rows
Tread depth	80 cm
Rise	31.6–38.2 cm
Stand inclination (average)	23.5°
Sightline elevation C	18.6–11.8 cm
(upper private box level 3 rows)	7.9 cm

Viewing distance (to the point of focus)

Maximum horizontal distance	62.70 m
Maximum vertical distance	31.00 m

Distance of hosp. boxes (touch line) 39.00 m

Viewing angle (front touch line)

Lower tier	7.0°–15.5°
Middle tier (upper priv. box level)	19.5°
Upper tier	24.5°–27.0°

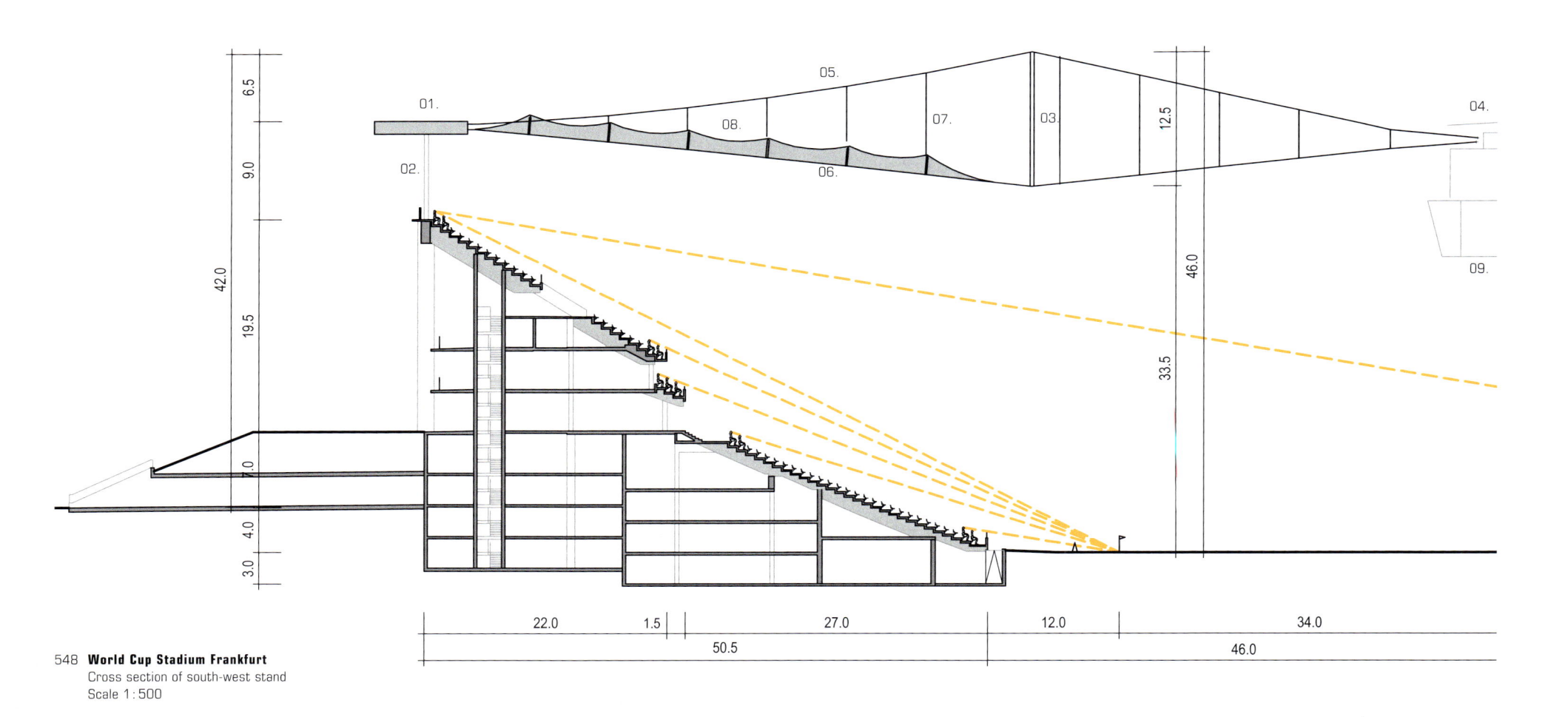

548 World Cup Stadium Frankfurt
Cross section of south-west stand
Scale 1:500

Lower tier

Row number	25 rows
Tread depth	85 cm
Rise	33.5–38.2 cm
Stand inclination (average)	23°
Sightline elevation C	18.6–10.4 cm

Upper tier

Row number	25 rows
Tread depth	80 cm
Rise	48.5–49.5 cm
Stand inclination (average)	31.5°
Sightline elevation C	13.3–9.3 cm

Distance of stand (front edge)

Distance to field	13.70 m

Viewing distance (to the point of focus)

Maximum horizontal distance	62.20 m
Maximum vertical distance	31.50 m

Viewing distance (maximum)

Football	182.50 m

Viewing angle (front goal line)

Lower tier	10.0°–17.0°
Upper tier	24.0°–26.5°

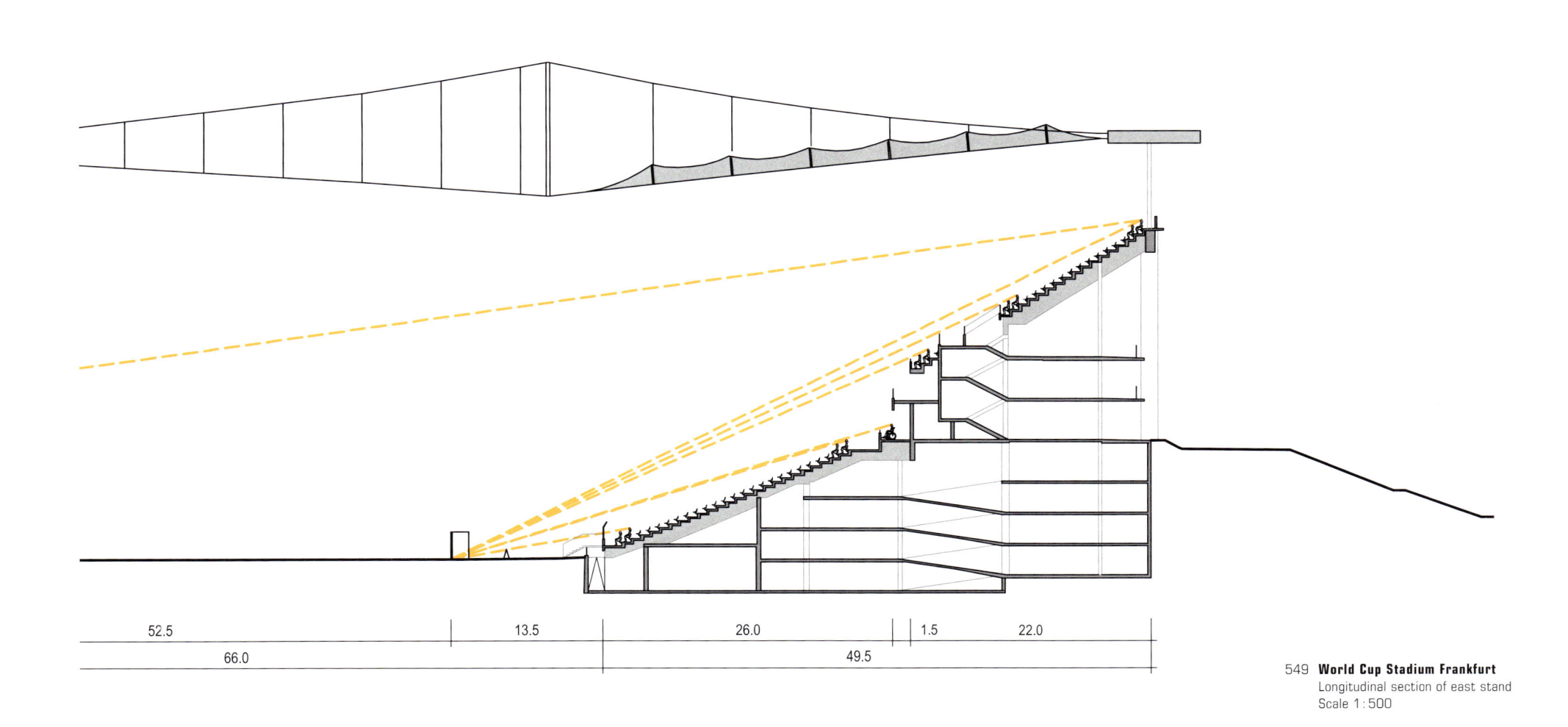

549 **World Cup Stadium Frankfurt**
Longitudinal section of east stand
Scale 1:500

World Cup Stadium Gelsenkirchen

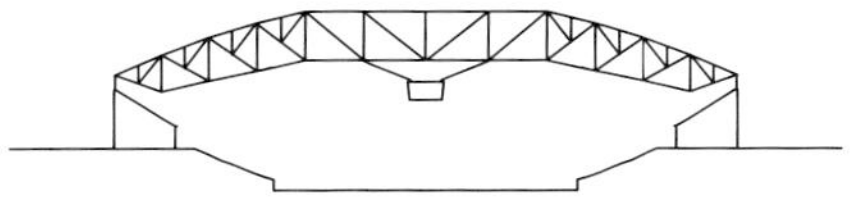

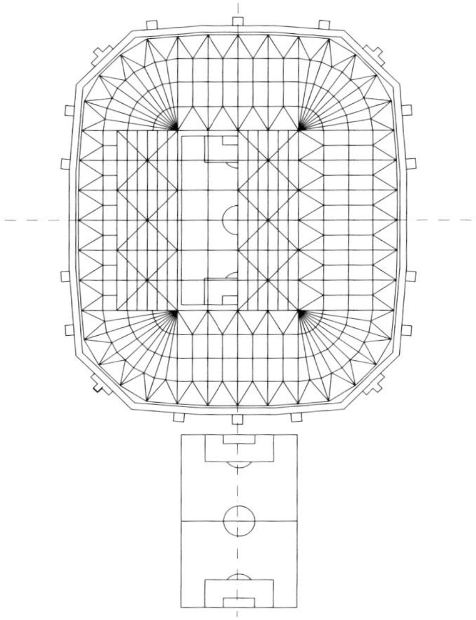

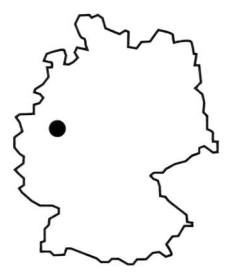

The new build of a multi-purpose arena in the direct vicinity of 'Parkstadion' (erected 1973) was commissioned in an 'open-book procedure' with maximum price guarantee in autumn 1998 to HBM Stadien- und Sportstättenbau GmbH. The new arena is rotated by 42° east due to the specific terrain in this mining area above two former coal seams.

Not a classic football stadium but an all-weather, consistently usable sports and event venue was supposed to become the new home of the 'Schalker Knappen'. Construction was financed completely through private funds, altogether 192 mio. €: 33,9 mio. € private equity capital, 12.8 mio. € from proprietary companies, 9.0 mio. € from Ruhrkohle / HBM, 8.5 mio. € outside capital, 115 mio. € from a bank syndicate and a 12.8 mio. € loan from the general contractor.

From 2005–2010, the name rights have changed from AOL Arena AufSchalke for 4–6 mio. € per year to Veltins Arena.

Due to mining foundation works, the conversion took place within 33 months (1,000 days) from 11/1998 to 08/2001 in 15 phases.
Roof erection (9 months for steel works): First, placement of quarter roof over eight tree-like supports and every third framework, then lowering of the 5000-t roof after assembly via hydraulic jacks on the mounting columns; just-in-time pre-fab components and installation.

A particular feature of the arena is the secondary floor as mobile grass pitch with parking position south of the complex. The lower tier south can be retracted as stand bridge to move the pitch tray in order to create additional space for the erection of a stage. The membrane-spanned steel structure is fitted with a sliding roof.

The entrance building west and a circulating internal route are dedicated for VIP guests. The 'Berger Feld' with its attractive leisure facilities and various new builds will be complemented as an ensemble in the coming years.

With an international capacity of around 53,600 seats (UEFA Five-Star stadia), and a total capacity of 61,500 including 16,300 standing places, it is the fourth largest of the World Cup stadia. 40 % of spectators are seated within the optimum viewing circle. No seats lie beyond the maximum recommended distance of 190 m to the corner flag. The 72 hospitality boxes and 36 special suites are located in a circulating two-level box tier in the main stand.

The majority of boxes is within the best viewing radius. Securing of the playing field is ensured by a combination of moat and elevation of the auditorium by 2.45 + 0.45 m = 2.90 m (first seating row). Sightline determination in this multipurpose arena has followed the height level of the grass pitch, lowered by 1.50 m for events with the grass surface parked outside. In this case, the interior area is 'activated' for the audience (maximum capacity of Veltins Arena ca. 78,400 seats). With the point of focus lowered by 1.50 m, sightline quality is diminished at the same ground-plan spot.

The first eye-point height follows the position of the advertising hoarding; the sightline follows the determined sightline quality 'C'. Therefore, when the grass pitch is parked outside, the point of focus moves by around 8.0 m to the centre.

550 **Roof framework** (sectional view)
Veltnis Arena, Gelsenkirchen
551 **Roof framework** (top view)
Veltnis Arena, Gelsenkirchen
552 **Historic aerial view**
Parkstadion Gelsenkirchen
553 **Geographic location** of the city of Gelsenkirchen

page 285:
554 **Aerial view** (2006)

Can a Stadium Roof Sway in the Wind?

Clearly, the answer is "yes". The increasingly complex roof designs of multi-functional stadia must withstand the most extreme environmental conditions. In some cases, they are even required to be earthquake-proof. The bearings are subject to weights of many tons. Our ELGES spherical plain bearings maintain their mobility throughout, even in the severe heat and cold.

Irrespective of whether they are used in the State Hockey Centre in Sydney, the Stade de France in Paris, the AufSchalke Arena or the Gerry Weber Stadium, our products offer a high degree of reliability and safety, allowing you to relax and concentrate on what is important, whatever the wind and weather conditions.

www.ina.com · www.fag.com

SCHAEFFLER GROUP
INDUSTRIAL

Beton-Kalender –
Basics, Examples, Engineer Standards

Since 2003, each issue of the compendium for concrete structures has been focussing on different subjects.
Editors: Konrad Bergmeister, Johan-Dietrich Wörner

BETON-KALENDER (CONCRETE STRUCTURAL DESIGN)

Hydraulic engineering, seismic design

Beton-Kalender 2008
2007. 1160 pages with
745 figures. Hardcover.
€ 165.–*/sFr 261.–
Series price:
€ 145.–*/sFr 229.–
ISBN: 978-3-433-01839-2

Under the main topic of hydraulic engineering the Yearbook „Concrete Structural Design" deals with design and construction of underwater foundation structures and protection structures for coastal areas and inland waterways, with notes relating to concrete under specific exposure and concrete repair.
Other sections cover seismic design, specifically dimensioning of reinforced and prestressed concrete structures according to DIN 4149 and Eurocode 8 and under dynamic loads.
Following the introduction of DIN 1055, highly topical notes provide insights into relevant aspects. The book contains reprints of DIN 1055 Parts 1, 3, 4, 5, 9, and 10.

Unter dem Schwerpunktthema Konstruktiver Wasserbau behandelt der Beton-Kalender Entwurf und Konstruktion von Gründungsbauwerken im Wasser sowie Schutzbauwerken an Küsten und Binnenwasserstraßen. Damit verbunden sind betone bei spezifischen Einwirkungen und die Betoninstandsetzung.
Das Erdbebensichere Bauen wird als weiterer Schwerpunkt in Kapiteln über die Bemessung der Stahlbeton- und Spannbetontragwerke nach DIN 4149 und Eurocode 8 bzw. unter dynamischen Beanspruchung erläutert.
Von hohen Aktualitätsgrad nach der bauaufsichtlichen Einführung snd die Hinweise zu Einwirkungen nach DIN 1055 mit Abdruck der Originalnormen DIN 1055 Teile 1, 23, 4, 5, 9, 100.

Traffic structures, shell structures

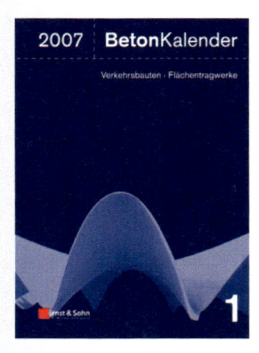

Beton-Kalender 2007
2006. 1100 pages with
1050 figures. Hardcover.
€ 165,–*/sFr 261,–
Series price:
€ 145,–*/sFr 229,–
ISBN: 978-3-433-01833-0

Traffic structures: The current state of the art in the construction of concrete roads, airports and rail tracks is described, together with road and rail supporting structures. Design aspects of highway traffic engineering are described taking into account the new generation of relevant standards.
Shell structures: all aspects of modelling, design and construction.
Standards: Up-to-date notes regarding design and construction according to DIN 1045-1, plus an outlook to relevant future European standards based on Eurocode 2.

Verkehrsbauten: Es wird der aktuelle Stand der Technik beim Bau von Betonstraßen, Flughäfen und Fester Fahrbahnen sowie von Stützbauwerken für Straßen- und Schienenwege vermittelt. Ebenso wird der Entwurf von Straßenverkehrsanlagen unter Berücksichtigung der neuen Normengeneration dargestellt.
Flächentragwerke: Alles zur Modellierung, Bemessung und Konstruktion der Tragwerke.
Normen: Aktuelle Hinweise zur Bemessung und Konstruktion nach DIN 1045-1. Mit DIN EN 1992-1-1 (Eurocode 2) und einer erläuternden Einführung wird ein Ausblick auf die zukünftig geltende europäische Normung vermittelt.

Tower structures, industrial buildings

Beton-Kalender 2006
2005. 1360 pages
with 1069 figures and
260 tables. Hardcover.
€ 165,–*/sFr 261,–
Series price:
€ 145,–* / sFr 229,–
ISBN: 978-3-433-01672-5

The 2006 anniversary edition covers tower structures and commercial and industrial buildings, with focus on design and execution, calculation, construction methods, special influences, and safety concepts. It covers new buildings as well as strengthening or re-utilization of existing structures.

In der Jubiläumsausgabe 2006 werden turmartige Bauwerke sowie Gewerbe- und Industriebauten umfassend behandelt. Dabei wird auf die Aspekte der Planung und Ausführung, die Berechnung, die Bauverfahren und die besonderen Einwirkungen und Sicherheitskonzepte eingegangen. Dies gilt sowohl bei Neubau als auch bei der Ertüchtigung oder Umnutzung der Bauwerke.

More informations:

www.ernst-und-sohn.de

All titles are published in German

Ernst & Sohn
A Wiley Company
www.ernst-und-sohn.de

Ernst & Sohn
Verlag für Architektur
und technische Wissenschaften
GmbH & Co. KG

For order and customer service:

Verlag Wiley-VCH
Boschstraße 12
69469 Weinheim
Germany

Tel.: +49(0) 6201 / 606-400
Fax: +49(0) 6201 / 606-184
E-Mail: service@wiley-vch.de

* In EU countries the local VAT is effective for books and journals. Postage will be charged. Whilst every effort is made to ensure that the contents of this leaflet are accurate, all information is subject to change without notice. Our standard terms and delivery conditions apply. Prices are subject to change without notice.

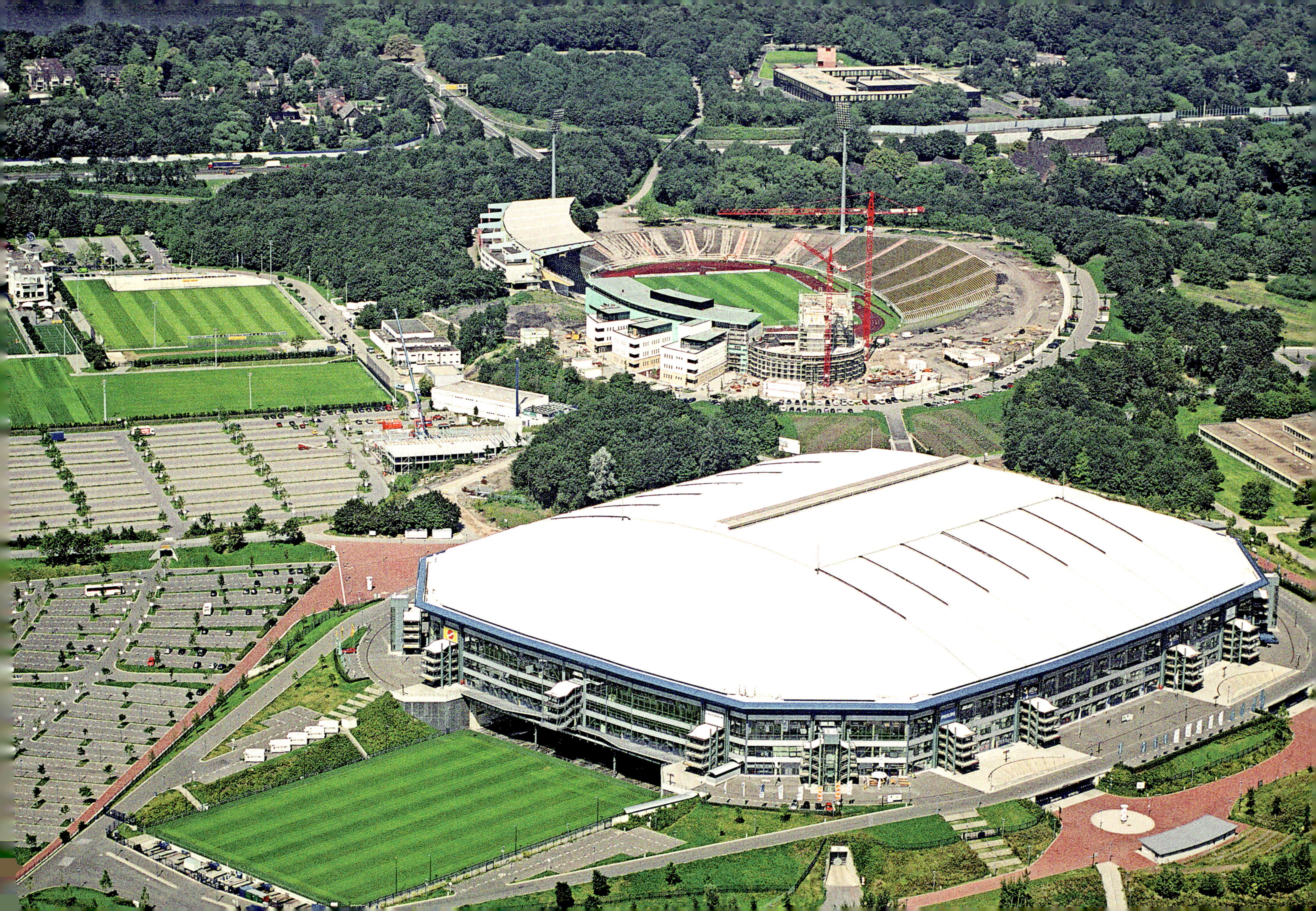

24 Grandstand profiles of built examples

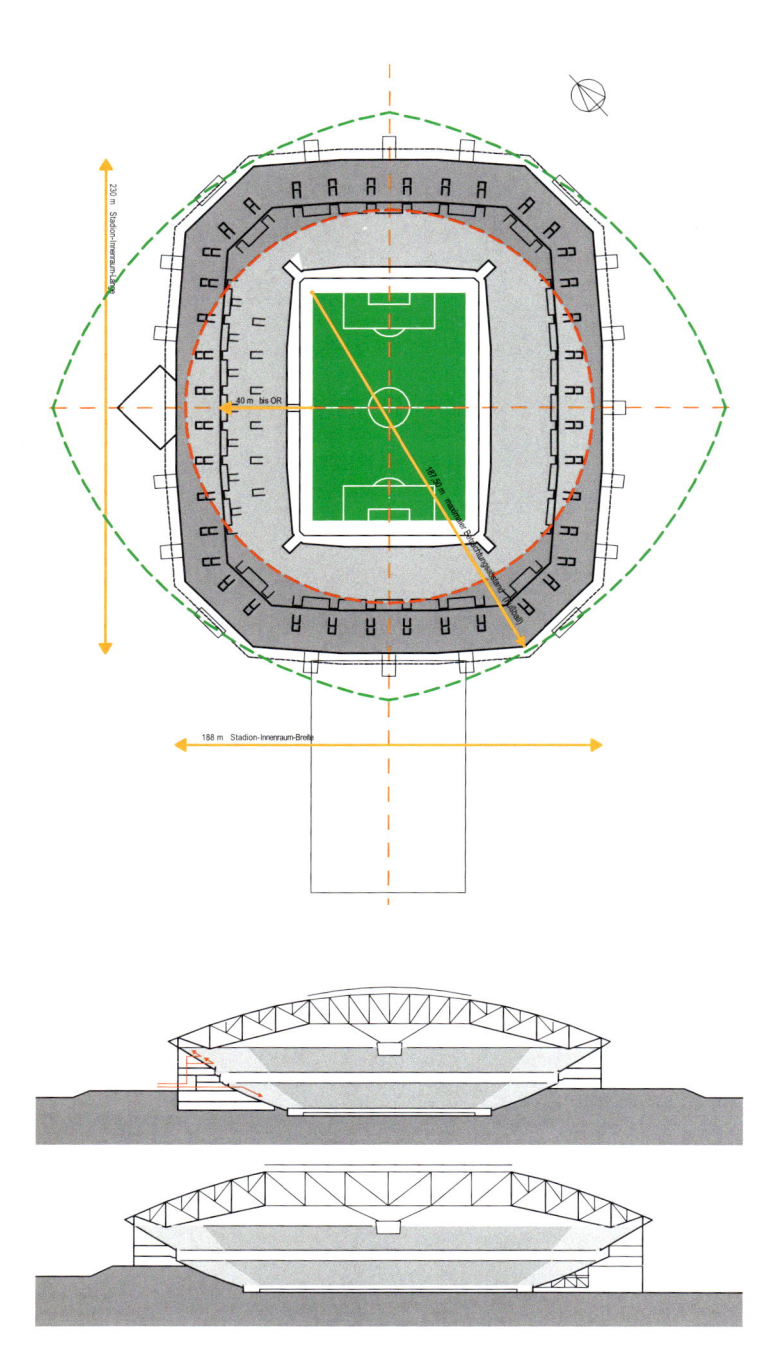

Veltins Arena, Gelsenkirchen – Two-tier arena (multipurpose hall)
Doubled octagonal geometry with a circulating box tier
(two-level box tier in the main stand)

Capacities

Gross capacity (international)	53,600 places [1]	100%
Gross capacity (national)	61,500 places [2]	
Standing places	16,300 places [2]	
Seats available for World Cup 2006 (net)	46,700 places [1]	
Within opt. 90-m viewing circle (football)	21,400 places	40.0%
Up to the max. recomm. 190-m distance	+ 32,200 places	60.0%
VIP guests (total)	3,255 places [1]	
Executive seats (World Cup 2006)	2,438 places [1]	
Number of private boxes (total)	72 box seats	
	36 private boxes	
Box seats, honorary tribune	817 box seats	
Media representatives (World Cup 2006)	150 commentators' positions (south)	
	200 television observers	
	500 press seats (with desk)	
	100 press seats (without desk)	
Wheelchair places	98 places [1]	

Dimensions (approx.) 188 x 230 m (total)
Stadium interior 90 x 128 m = 11,500 m²

Orientation of playing field plus 42° in north-south direction

Width of roof $b = 57$ m max.
Height above pitch $h = 43$ m
Height above terrain (eaves) $h = 21$ m

Roof area $A = 40,000$ m²
Convertible interior roof 70 x 108 m (two-part sliding roof)
Drainage Slope toward exterior = 15–22°

[1] according to LOC World Cup 2006, Frankfurt (status: October 2005)
[2] according to www.stadionwelt.de (status: October 2005)

555 **Viewing circle and dimensions** (plan)
Veltins Arena, Gelsenkirchen, Scale 1:3,000
556 **Structure and circulation**
(transverse/longitudinal section)

page 287:
557 **Picture of the stadium interior** (2006)

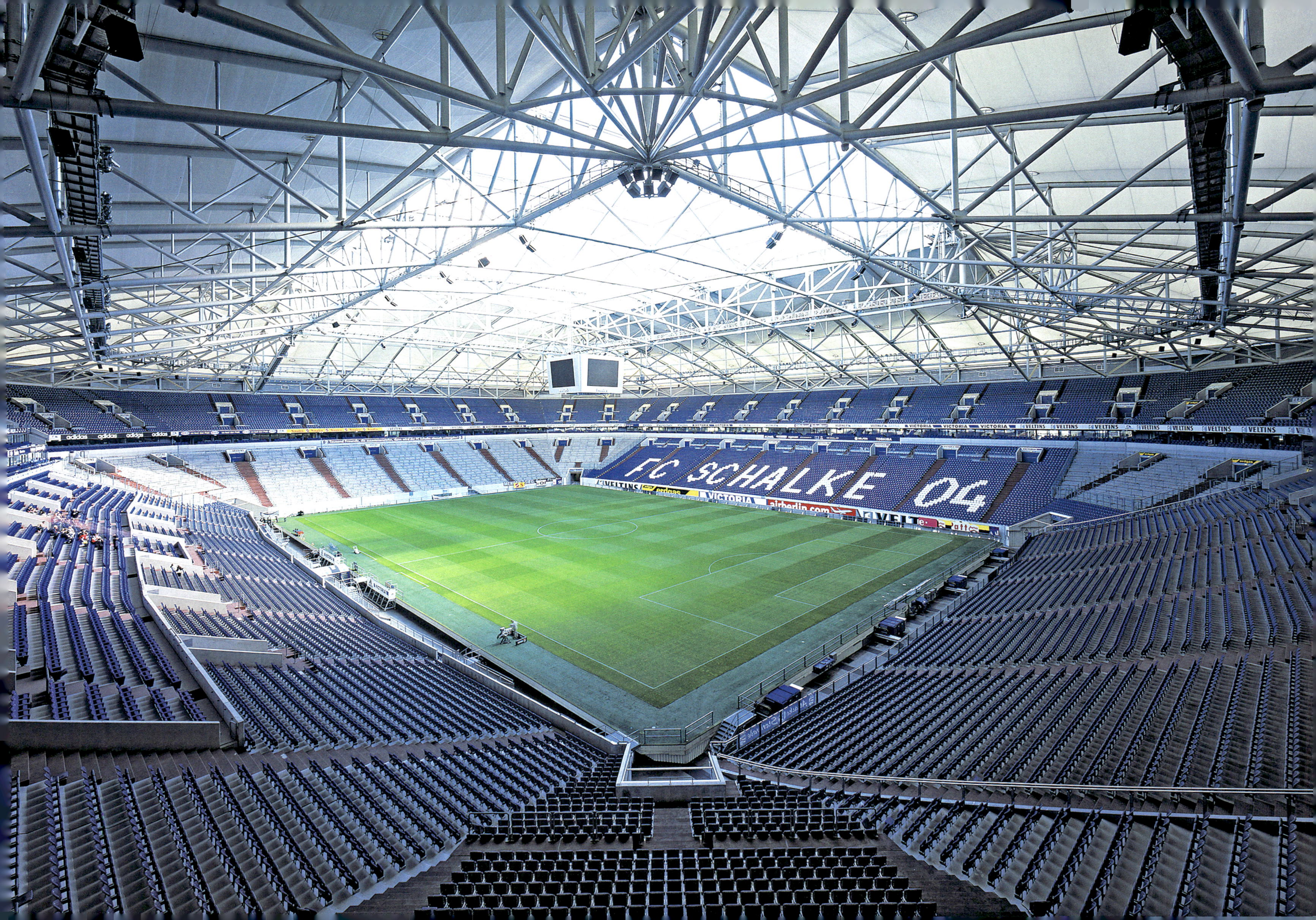
FC SCHALKE 04
VELTINS
VICTORIA

24 Grandstand profiles of built examples

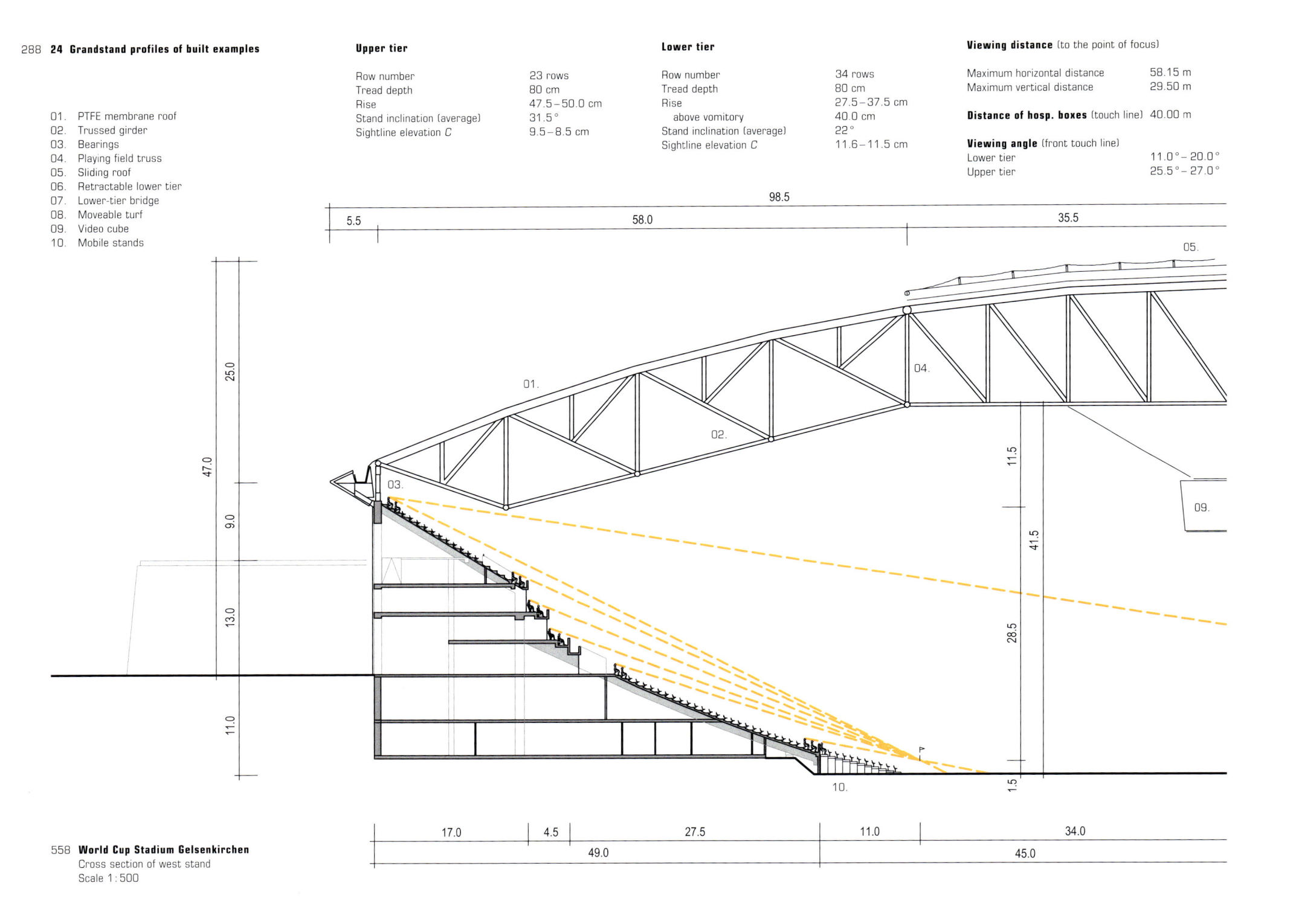

01. PTFE membrane roof
02. Trussed girder
03. Bearings
04. Playing field truss
05. Sliding roof
06. Retractable lower tier
07. Lower-tier bridge
08. Moveable turf
09. Video cube
10. Mobile stands

Upper tier

Row number	23 rows
Tread depth	80 cm
Rise	47.5–50.0 cm
Stand inclination (average)	31.5°
Sightline elevation C	9.5–8.5 cm

Lower tier

Row number	34 rows
Tread depth	80 cm
Rise	27.5–37.5 cm
above vomitory	40.0 cm
Stand inclination (average)	22°
Sightline elevation C	11.6–11.5 cm

Viewing distance (to the point of focus)

Maximum horizontal distance	58.15 m
Maximum vertical distance	29.50 m

Distance of hosp. boxes (touch line) 40.00 m

Viewing angle (front touch line)

Lower tier	11.0°–20.0°
Upper tier	25.5°–27.0°

World Cup Stadium Gelsenkirchen
Cross section of west stand
Scale 1:500

Lower tier

Row number	37 rows
Tread depth	80 cm
Rise	27.5–37.5 cm
above vomitory	40.0 cm
Stand inclination (average)	22°
Sightline elevation C	12.3–14.3 cm

Upper tier

Row number	22 rows
Tread depth	80 cm
Rise	47.5–50.0 cm
Stand inclination (average)	31.5°
Sightline elevation C	10.6–9.6 cm

Distance of stand (front edge)

Distance to field	11.50 m

Viewing distance (to the point of focus)

Maximum horizontal distance	58.65 m
Maximum vertical distance	29.50 m

Viewing distance (maximum)

Football	187.50 m

Viewing angle (front goal line)

Lower tier	11.0°–18.5°
Upper tier	25.0°–26.5°

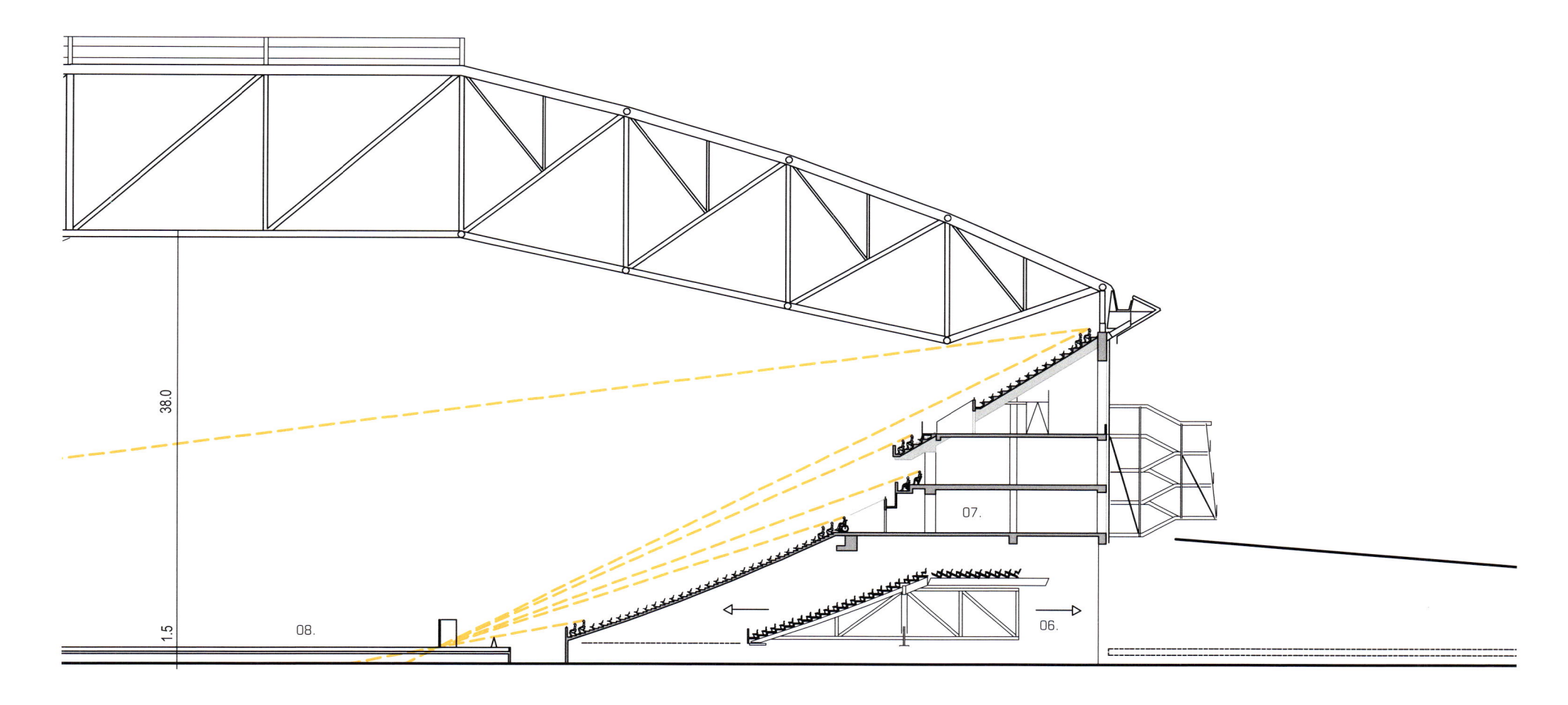

World Cup Stadium Gelsenkirchen

559 **World Cup Stadium Gelsenkirchen**
Longitudinal section of south stand
Scale 1:500

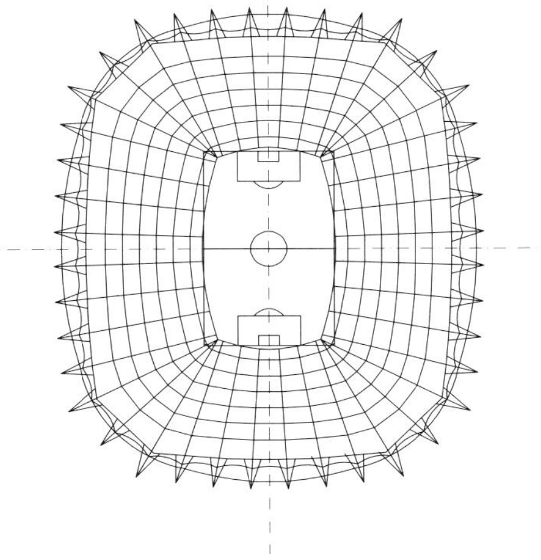

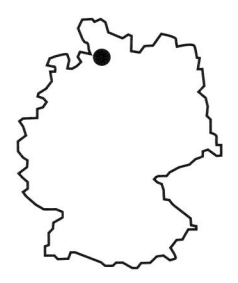

World Cup Stadium Hamburg

At the same location of the former Altonaer Stadion from 1925, the first Volksparkstadion was erected in 1937. The second conversion, including a running track, was carried out in 1953 on a two-storey earth tribune formed by war debris (75,000 spectators, of those 20,000 standing places).

For World Cup 1974, Volksparkstadion was refurbished and provided with a covered south stand, a floodlight system and scoreboard. After hosting several international matches and after another modernization it became one of the venues for the UEFA European Championship.

Based on the fact of its typically large pitch distance and the resulting bad viewing conditions, in 1997 the decision was taken to build a modern football stadium, a conversion during regular operations. Rotation of the playing field by 90° to the correct north/south direction was a vital measure.

Complete conversion into a multipurpose stadium mainly for football (venue for NFL Europe), with open-air concerts and conference use, was executed in eight stages in around 27 months (06/1998 to 09/2000). The new venue was officially opened with a concert by Tina Turner in July of the same year.

Redesign efforts and dismissal of the construction management/planning team increased the initial costs from 30.0 mio. € to 97.0 mio. €. An adequate roof structure had to be newly planned for the projected grandstand structure.

A rounded compression ring is supported by circulating pylons. Before construction of the roof and the set-up of the cables, reinforcement measures were carried out for the concrete structure; via strand grips the cable construction is spanned and raised in one assembly step. The open spoked-wheel structure as membrane roof rests on a free-standing stadium bowl made of reinforced concrete parts on a steel substructure. This pure football stadium (UEFA Five-Star stadia) bears the name AOL Arena. It offers accommodation for around 51,300 visitors, none of which will be placed outside the maximum viewing distance. Half of the audience, or 25,650 spectators, sit within the best viewing radius. The circulating grandstand follows an expanded octagon geometry.

The circulation system is marked by a two-part upper tier which is accessed by a circulating lateral gangway. This 2.0 m wide circulation ring divides the stand into two tiers (upper and middle tier).

AOL Arena is thus technically a three-tier stadium, although the term 'tier' generally implies a dedicated access system and not a concurrent distribution of the upper and lower part from the same internal gangway system.

Divided into three rise segments, the stand geometry does not follow the optimal parabolic rise but is divided into three linear rises with 34/90 LT, 51/90 MT and 70/100 UT. The rise/run ratio is fully exploited with four single steps per stepping row at the upper tier.[1]

The lateral gangway is cut into the stand geometry and thus necessitates short stairs within the grid of the ascending gangways to compensate the height difference.

560 **Roof framework** (sectional view)
    AOL Arena, Hamburg
561 **Roof framework** (top view)
    AOL Arena, Hamburg
562 **Historic aerial view**
    Volksparkstadion Hamburg
563 **Geographical location** of the city of Hamburg

page 291:
564 **Aerial view** (2006)

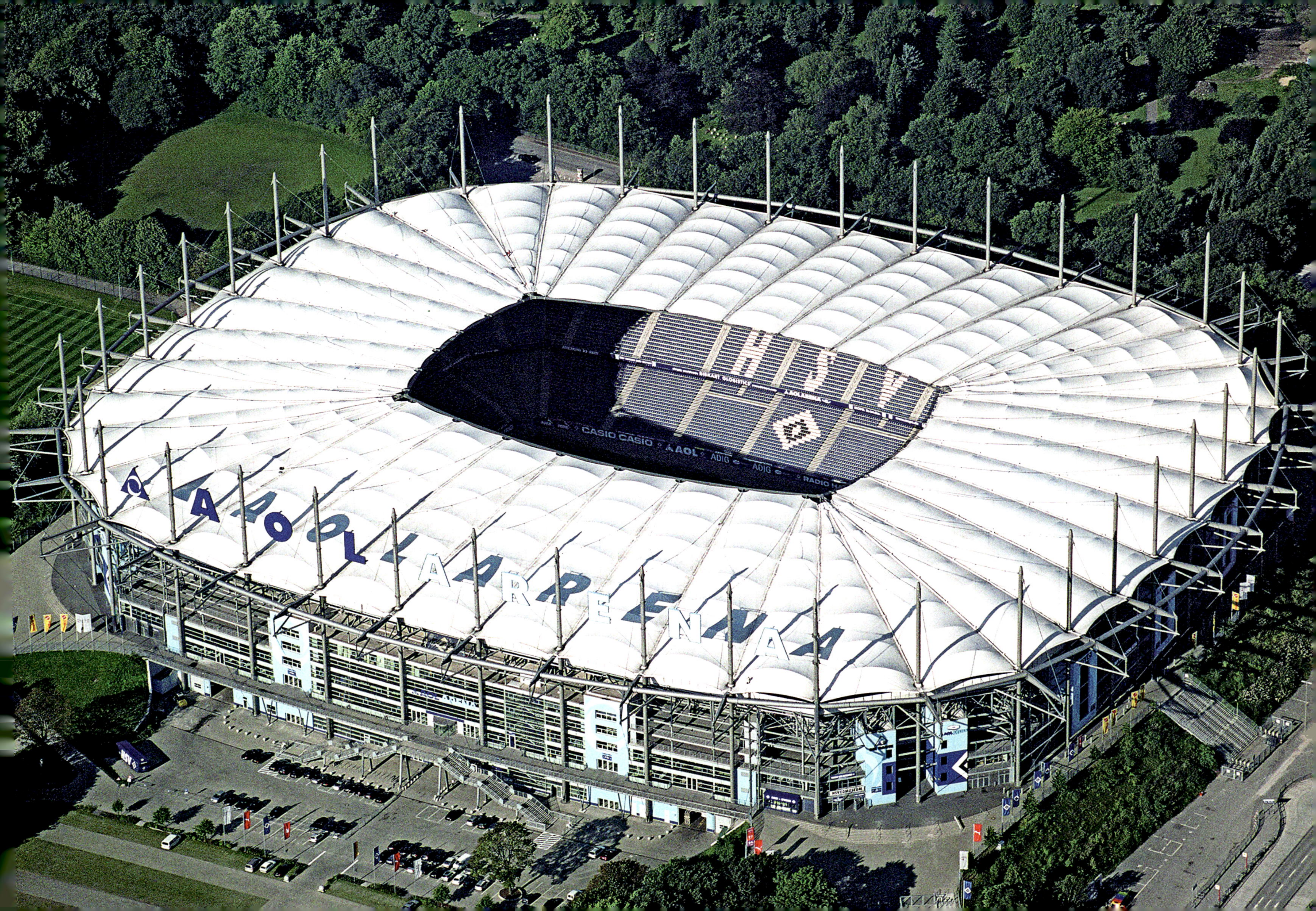
AOL ARENA
HSV
CASIO CASIO
AOL
ADIG

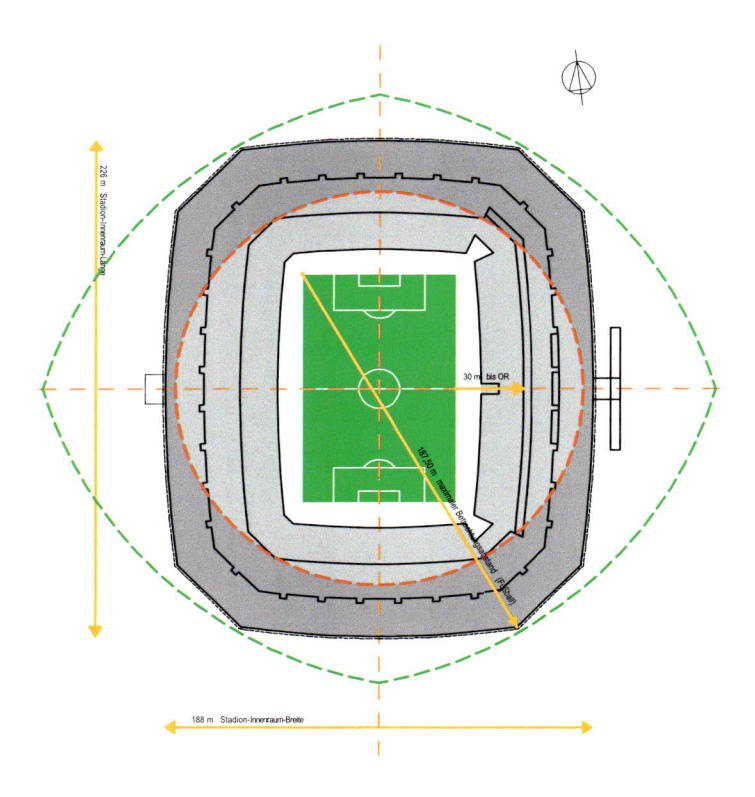

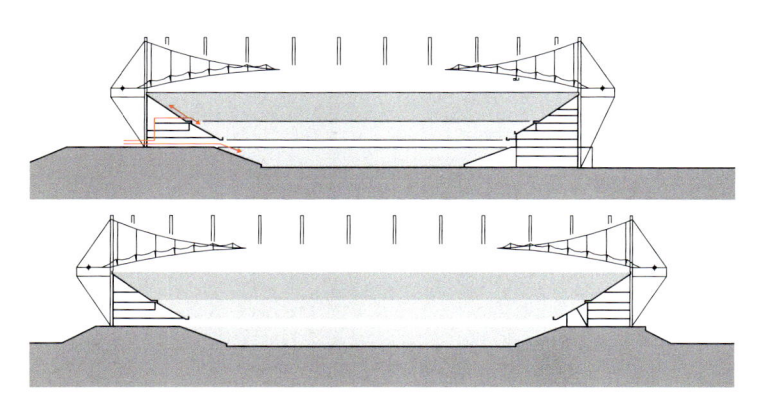

AOL Arena, Hamburg – Three-tier stadium (football)

Expanded octagonal geometry with circulating horizontal band (open) and with two-part upper tier

Capacities

Gross capacity (international)	51,300 places [1]	100%
Gross capacity (national)	56,800 places [2]	
Standing places	8,900 places [2]	
Seats available for World Cup 2006 (net)	43,300 places [1]	
Within opt. 90-m viewing circle (football)	25,650 places	50.0%
Up to the max. recomm. 190-m distance	+25,650 places	50.0%
Lower tier	17,500 places	
Middle tier	13,500 places	
Upper tier	20,300 places	
VIP guests (total)	2,759 places [1]	
Executive seats (World Cup 2006)	2,179 places [1]	
Number of private boxes (total)	50 private boxes	
Box seats, honorary tribune	580 box seats	
Media representatives (World Cup 2006)	150 commentators' positions (south)	
	200 television observers	
	500 press seats (with desk)	
	100 press seats (without desk)	
Wheelchair places	90 places [1]	

Dimensions (approx.)

	188 x 226 m (total)
Stadium interior	90 x 127 m = ca. 11,500 m²
Orientation of playing field	plus 7° in north-south direction
Pylon height	h = 58 m
Cantilever	h = 15 m
Width of roof	b = 59 m max.
Height above pitch	h = 44 m
Height above terrain (eaves)	h = 27 m
Roof area	A = 35,400 m²
Drainage	Slope toward exterior = 8°

1) according to LOC World Cup 2006, Frankfurt (status: October 2005)
2) according to www.stadionwelt.de (status: October 2005)

565 **Viewing circle and dimensions** (plan)
AOL Arena, Hamburg, Scale 1 : 3,000
566 **Structure and circulation**
(transverse/longitudinal section)

page 293:
567 **Picture of the stadium interior** (2006)

HSV
HSV
WERDE HSV-MITGLIED
OLYMPUS
RADIO HAMBURG
CASIO CASIO CASIO
ADIG
RADIO HAMBURG
SIGNAL IDUNA
OLYMPUS

294 24 Grandstand profiles of built examples

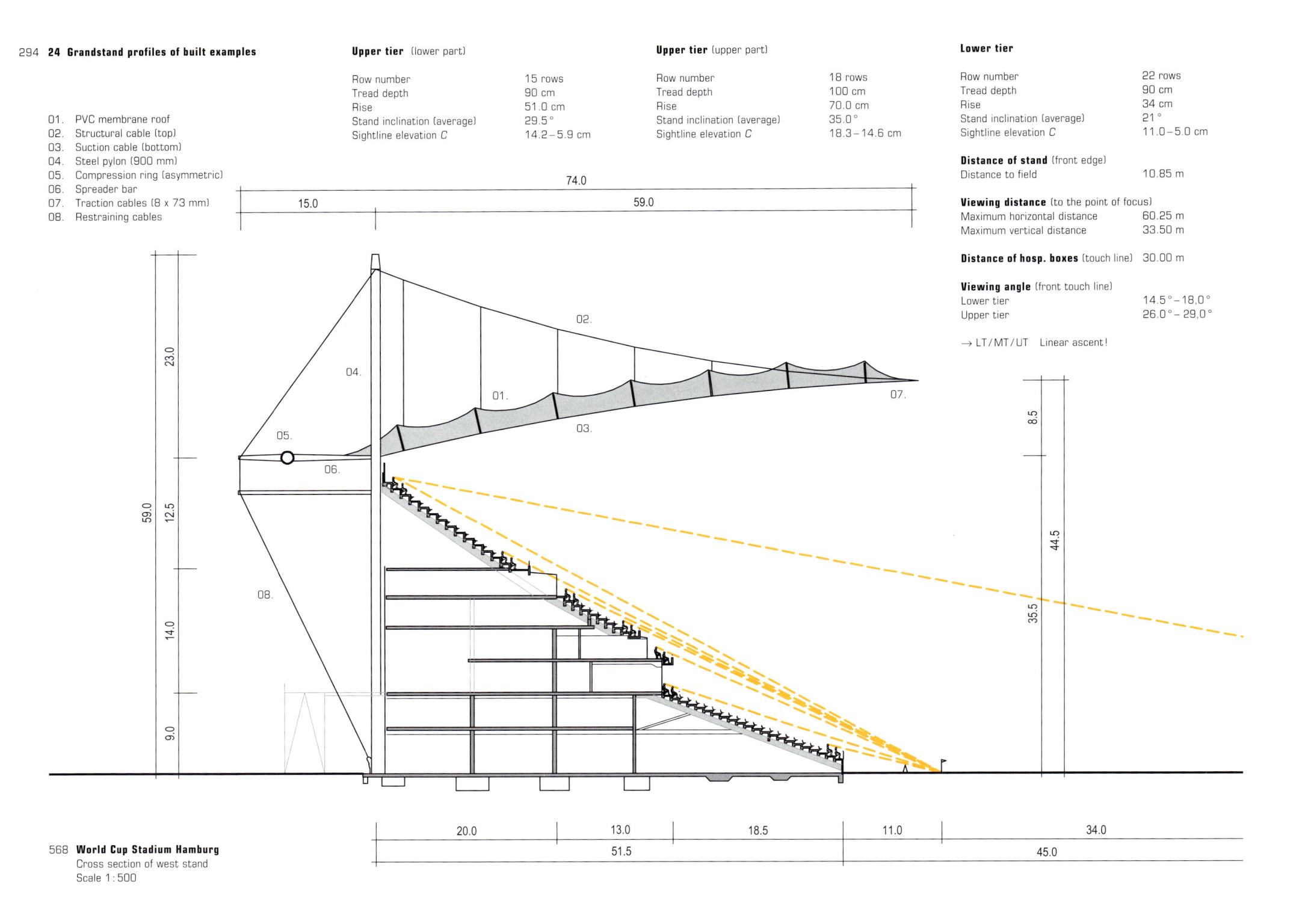

01. PVC membrane roof
02. Structural cable (top)
03. Suction cable (bottom)
04. Steel pylon (900 mm)
05. Compression ring (asymmetric)
06. Spreader bar
07. Traction cables (8 x 73 mm)
08. Restraining cables

Upper tier (lower part)

Row number	15 rows
Tread depth	90 cm
Rise	51.0 cm
Stand inclination (average)	29.5°
Sightline elevation C	14.2–5.9 cm

Upper tier (upper part)

Row number	18 rows
Tread depth	100 cm
Rise	70.0 cm
Stand inclination (average)	35.0°
Sightline elevation C	18.3–14.6 cm

Lower tier

Row number	22 rows
Tread depth	90 cm
Rise	34 cm
Stand inclination (average)	21°
Sightline elevation C	11.0–5.0 cm

Distance of stand (front edge)
Distance to field   10.85 m

Viewing distance (to the point of focus)
Maximum horizontal distance   60.25 m
Maximum vertical distance   33.50 m

Distance of hosp. boxes (touch line)   30.00 m

Viewing angle (front touch line)
Lower tier   14.5°–18.0°
Upper tier   26.0°–29.0°

→ LT/MT/UT   Linear ascent!

568 **World Cup Stadium Hamburg**
Cross section of west stand
Scale 1:500

Lower tier

Row number	23 rows
Tread depth	90 cm
Rise	34 cm
Stand inclination (average)	21°
Sightline elevation C	12.8 – 5.4 cm

Upper tier (lower part)

Row number	15 rows
Tread depth	90 cm
Rise	51.0 cm
Stand inclination (average)	29.5°
Sightline elevation C	9.6 – 6.7 cm

Upper tier (upper part)

Row number	18 rows
Tread depth	100 cm
Rise	70.0 cm
Stand inclination (average)	35.0°
Sightline elevation C	20.3 – 15.8 cm

Distance of stand (front edge)

Distance to field	11.30 m

Viewing distance (to the point of focus)

Maximum horizontal distance	60.70 m
Maximum vertical distance	33.50 m

Viewing distance (maximum)

Football	187.50 m

Viewing angle (front goal line)

Lower tier	13,5° – 17,5°
Upper tier	25,0° – 28,5°

→ LT / MT / UT   Linear ascent!

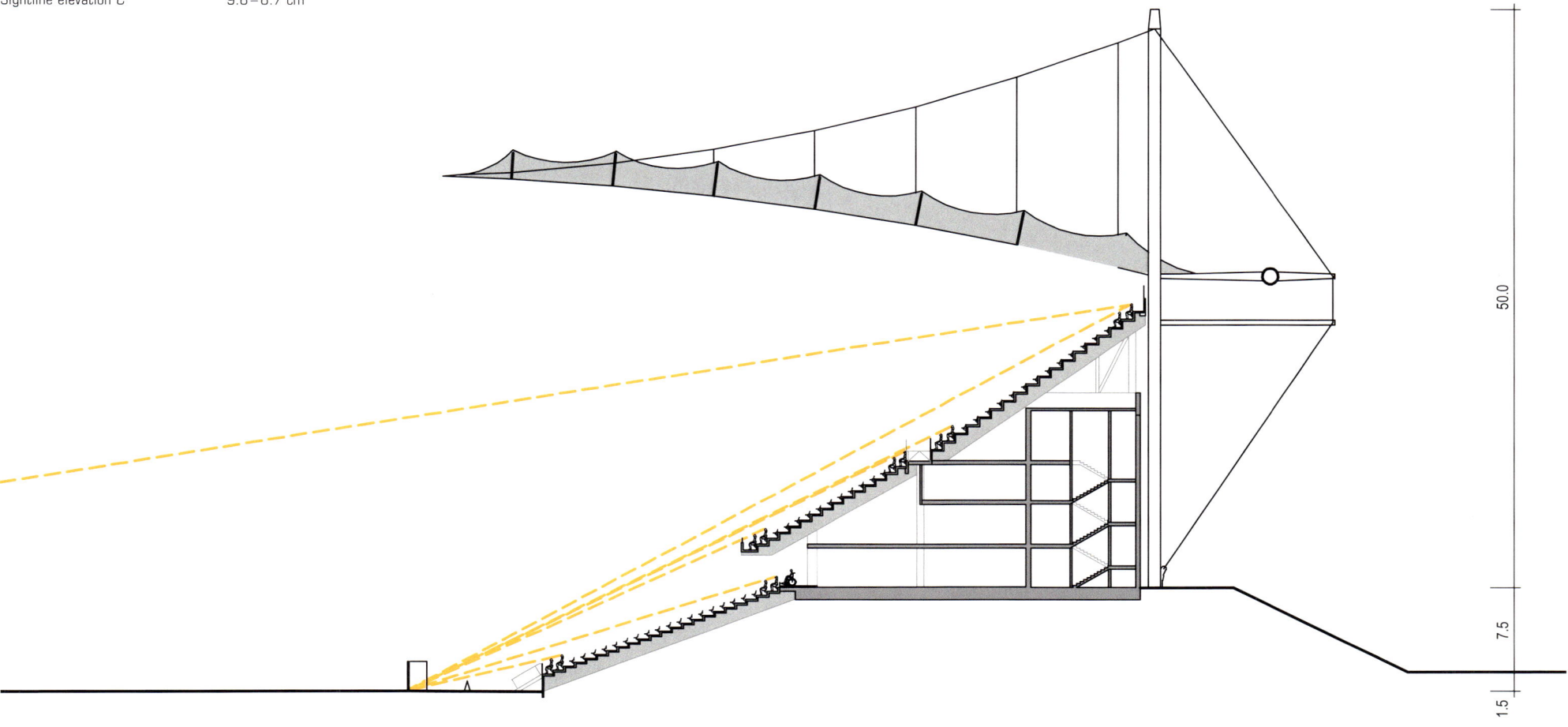

569 **World Cup Stadium Hamburg**
Longitudinal section of south stand
Scale 1 : 500

24 Grandstand profiles of built examples

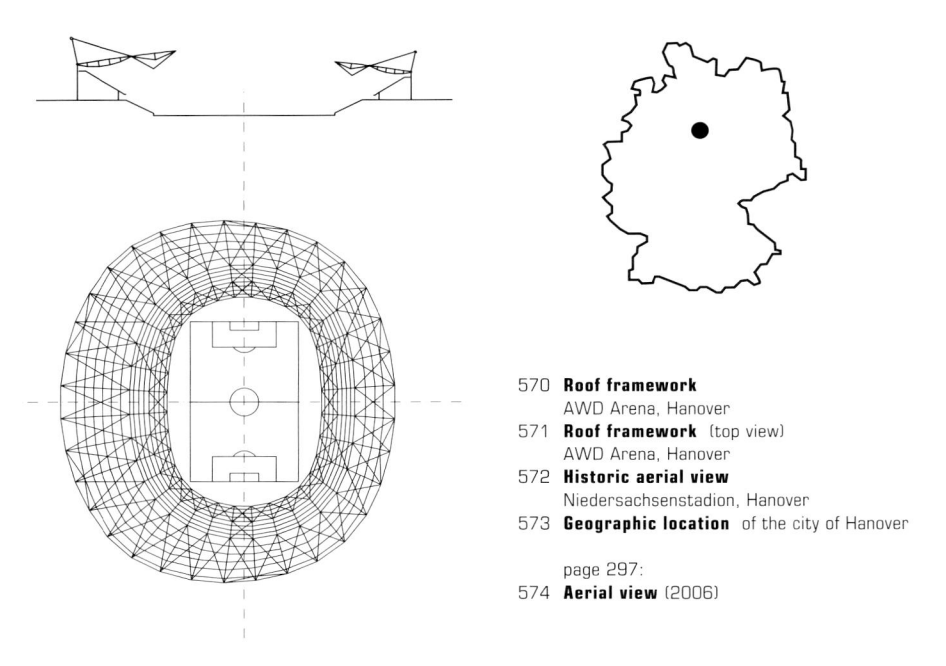

World Cup Stadium Hanover

Niedersachsenstadion Hanover was erected in 1954 as an earth-wall stadium on top of war debris and roofed at the west side for the World Cup 1974. The strong asymmetry was characteristic for this athletics stadium from the beginning.

Through a new construction of the east stand, elongated to the north and south, conversion into the new AWD Arena, now a pure football stadium, took place from 2003 to 2005.

Remaining as historical citation is the shallow west stand incline, typical for athletics stadia.

The design is marked by the difference of its two main stands, which today are incorporated by the circulating roof into the stadium body.

In a Europe-wide open architectural competition in the year 2000, a suitable design was chosen for the roof covering of the existing Niedersachsenstadion (first prize: Schulitz + Partner, Hanover).

After that, the city council of Hanover decided to modernize the entire stadium body, with full roof covering and addition of a new main stand east.

In a public-private-partnership investor competition (concession, operating concept), the Niedersachsenstadion Projekt-Betriebsgesellschaft-Hannover 96 was awarded the contract (during construction: Hannover 96 with 49% and Wayss & Freytag with 51%).

New build and conversion in three stages during running operation (minimum capacity 25,000 seats) in 21 months from 02/2003 to 12/2004. Completion of the new AWD Arena three months before planned finish date.

44,900 seats for international matches may be upscaled by 7,200 standing places to a total capacity of 49,800 places. 52,5% of spectators sit within the best viewing radius of 90 m.

The asymmetrical oval geometry creates different but good viewing conditions for all sides of the stadium.

The historic 'AVUS' as horizontal circulation band behind lower tier west is around 6.50 m wide. The height offset of the ascending upper-tier stand is sufficiently large to avoid viewing obstructions caused by people standing on the AVUS.

Stairs are redundant due to upper-tier access from above/behind via rear circulation areas across the debris wall.

The main design challenge lay in transforming the former geometry of an athletics grandstand into that of a pure football stadium.

New construction of the roofing in two steps:
1) The exterior ring is assembled in segments from the outside during running operation and provided with temporary columns.
2) The interior ring is mounted from inside during match-free time. Air supports were strategically mounted on a pre-stressed interior tension ring D.

Summer break 2004: roof construction, installation of new floodlighting and turf heating system.

Summer break 2003: rotation of the playing field in the direction of the west stand by 7 m and exchange of turf.

570 **Roof framework**
AWD Arena, Hanover
571 **Roof framework** (top view)
AWD Arena, Hanover
572 **Historic aerial view**
Niedersachsenstadion, Hanover
573 **Geographic location** of the city of Hanover

page 297:
574 **Aerial view** (2006)

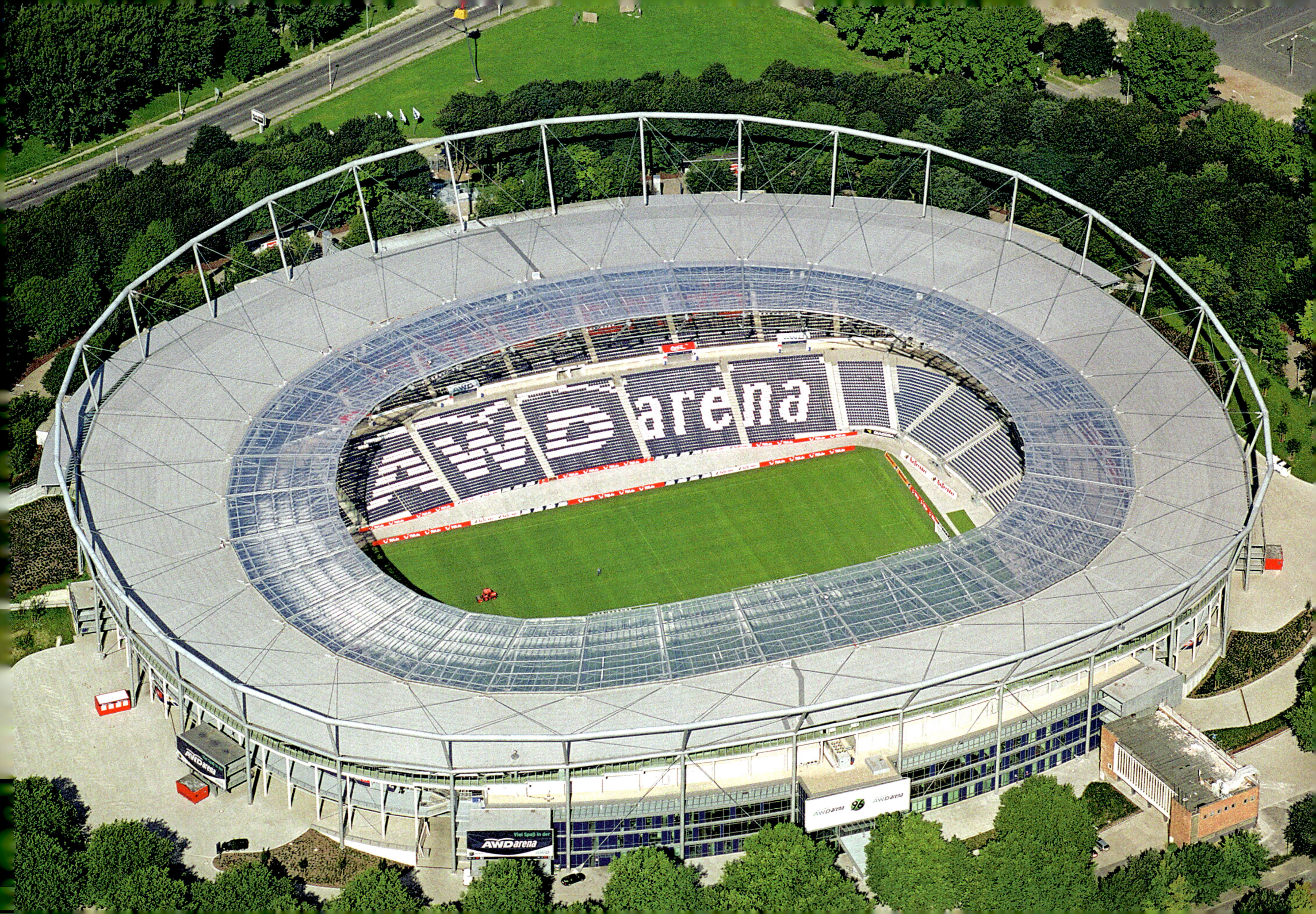
AWDarena

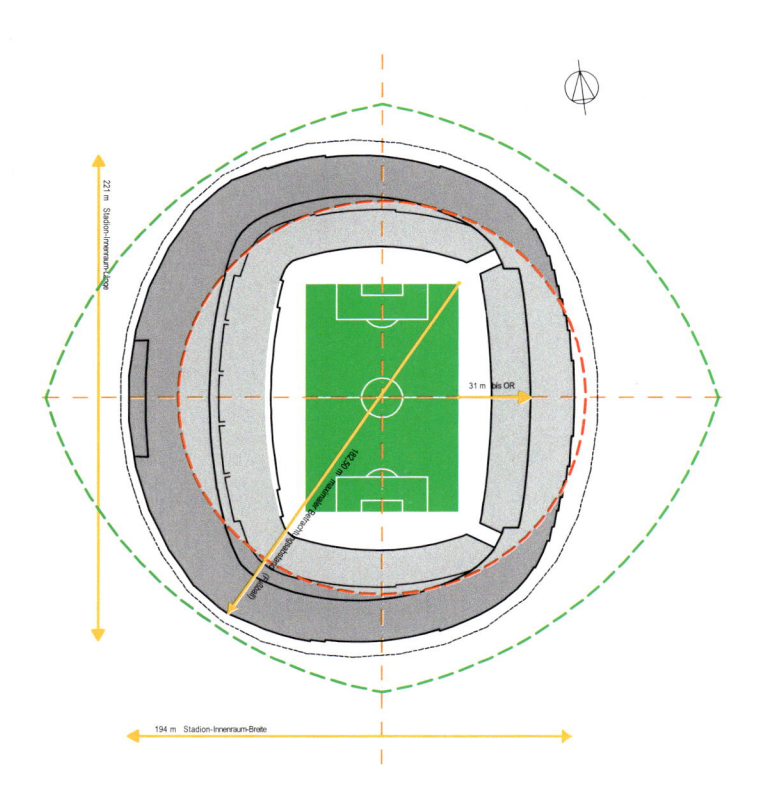

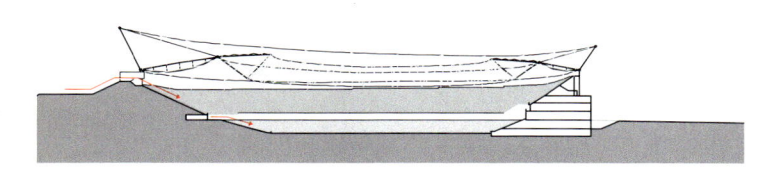

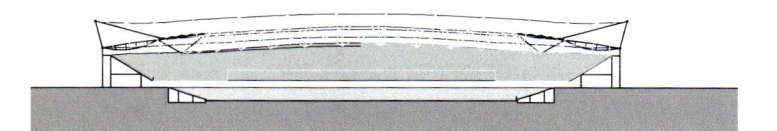

AWD Arena, Hanover – Two-tier stadium (football)
Asymmetric oval geometry with horizontal circulation band ('Avus')

Capacities

Gross capacity (international)	44,900 places [1]	100 %
Gross capacity (national)	49,800 places [2]	
Standing places	7,200 places [2]	
Seats available for World Cup 2006 (net)	38,800 places [1]	
Within opt. 90-m viewing circle (football)	23,600 places	52.5 %
Up to the max. recomm. 190-m distance	+ 21,300 places	47.5 %
Lower tier	17,500 places	
Middle tier	13,500 places	
Upper tier	20,300 places	
VIP guests (total)	1,551 places [1]	
Executive seats (World Cup 2006)	1,241 places [1]	
Number of private boxes (total)	39 private boxes	
Box seats, honorary tribune	310 box seats	
Media representatives (World Cup 2006)	150 commentators' positions (south)	
	200 television observers	
	500 press seats (with desk)	
	100 press seats (without desk)	
Wheelchair places	60 places [1]	

Dimensions (approx.)

	194 x 221 m (total)
Stadium interior	96 x 139 m = 13,400 m²
Orientation of playing field	plus 7° in north-south direction
Width of roof	$b = 39 – 57$ m max.
Height above pitch	$h = 30 – 34$ m
Height above terrain (eaves)	$h = 17.5$ m
Ring structure	5 x tensile/compressive ring A – E
Air support (34 x)	$h = 13.5$ m
Roof area	$A = 26,600$ m²
Trapezoid metal plate	$= 16,400$ m²
ETFE	$= 10,200$ m²
Drainage	Slope toward exterior = 8.5°

1) according to LOC World Cup 2006, Frankfurt (status: October 2005)
2) according to *www.stadionwelt.de* (status: October 2005)

575 Viewing circle and dimensions (plan)
AWD Arena, Hanover, Scale 1 : 3,000

576 Structure and circulation
(transverse/longitudinal section)

page 299:
577 Picture of the stadium interior (2006)

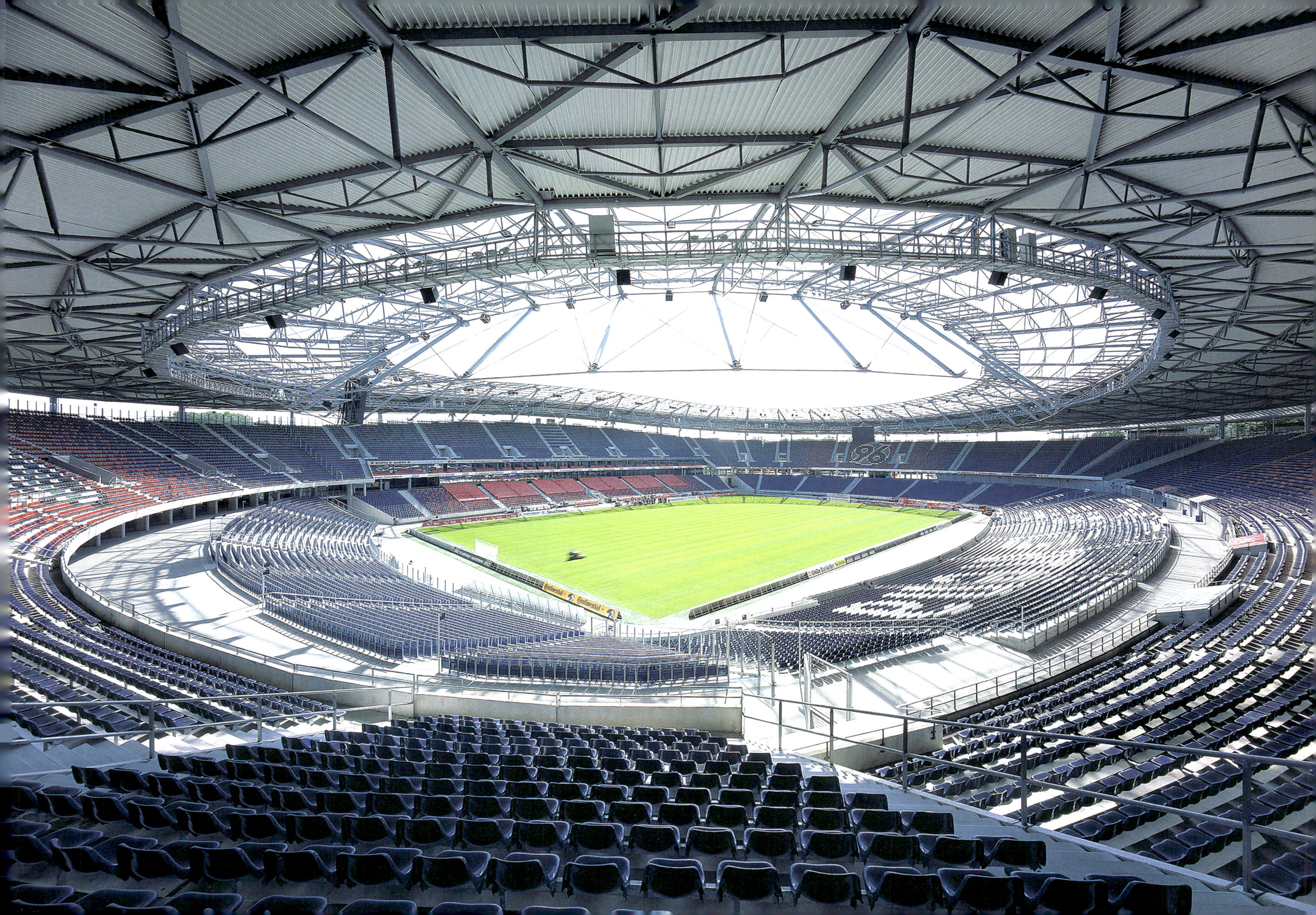

24 Grandstand profiles of built examples

01. ETFE membrane roof
02. Trapezoid metal plate roofing
03. Main compression ring A
04. Bearing ring B
05. Intermediate ring C
06. Structural ring D
07. Span ring E
08. Air support

Upper tier

Row number	23–33 rows
Tread depth	80/90 cm
Rise	56.0–57.0 cm
Stand inclination (average)	35.5°
Sightline elevation C	16.1–13.4 cm

→ Undulation!

Distance of stand (front edge)

Distance to field	14.05 m

Lower tier

Row number	15 rows
Tread depth	100 cm
Rise	41.3–44.4 cm
Stand inclination (average)	23°
Sightline elevation C	21.2–21.0 cm

Viewing distance (to the point of focus)

Maximum horizontal distance	50.00 m
Maximum vertical distance	26.50 m

Distance of hosp. boxes (touch line) 31.00 m

Viewing angle (front touch line)

Lower tier	11.5°–14.5°
Private box level	20.0°
Upper tier	24.5°–28.0°

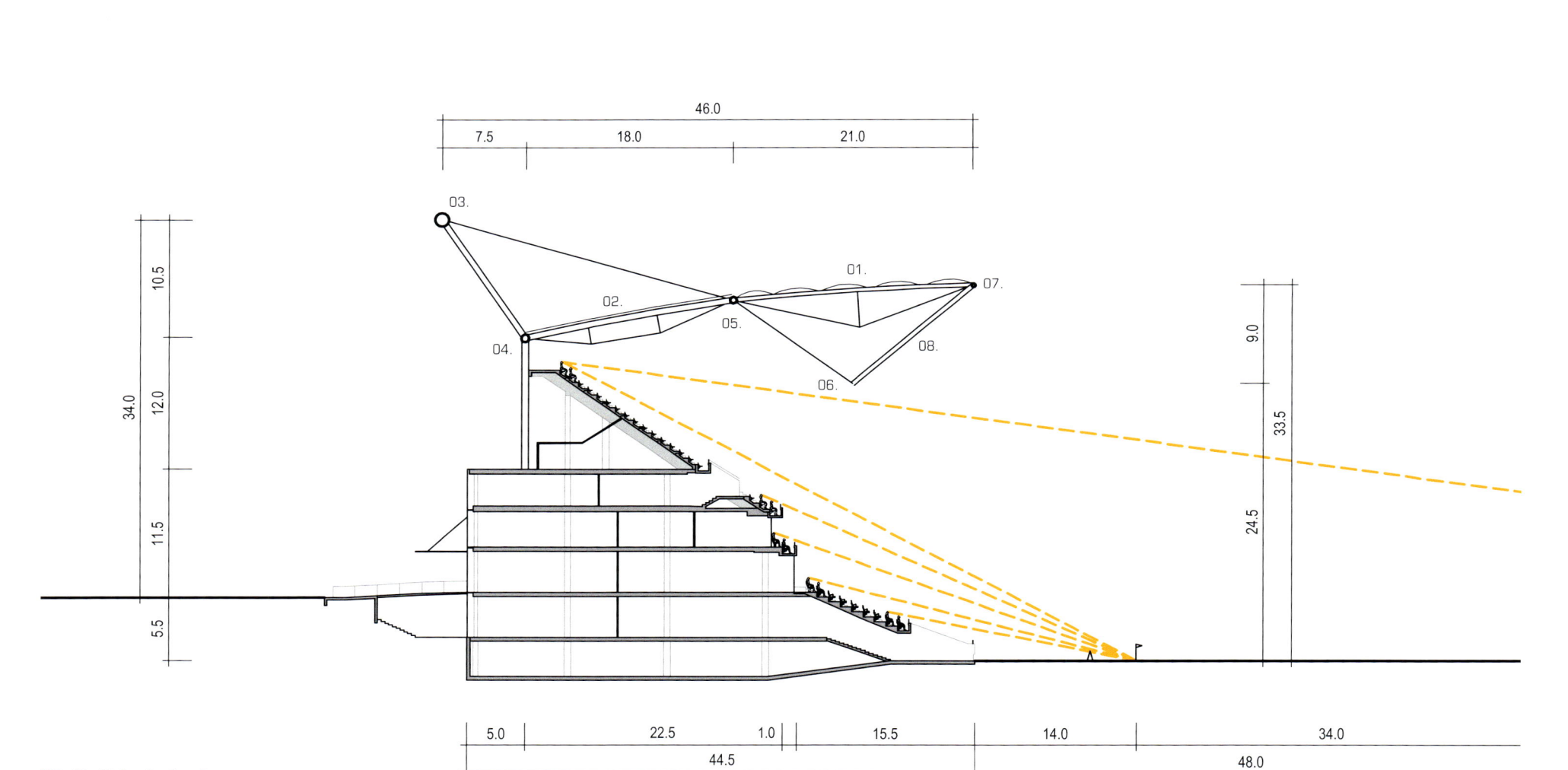

578 World Cup Stadium Hanover
Cross section of east stand
Scale 1:500

Lower tier

Row number	21 rows
Tread depth	40/80 cm
Rise	21.5–29.9 cm
Stand inclination (average)	21°
Sightline elevation C	11.7–12.4 cm

Upper tier

Row number	23–28 rows
Tread depth	80 cm
Rise	47.5–50.0 cm
Stand inclination (average)	31.5°
Sightline elevation C	26.5–21.4 cm

→ Undulation!

Distance of stand (front edge)

Distance to field	16.90 m

Viewing distance (to the point of focus)

Maximum horizontal distance	58.65 m
Maximum vertical distance	21.25 m

Viewing distance (maximum)

Football	187.50 m

Viewing angle (front goal line)

Lower tier	7.0°–12.5°
Upper tier	14.5°–20.0°

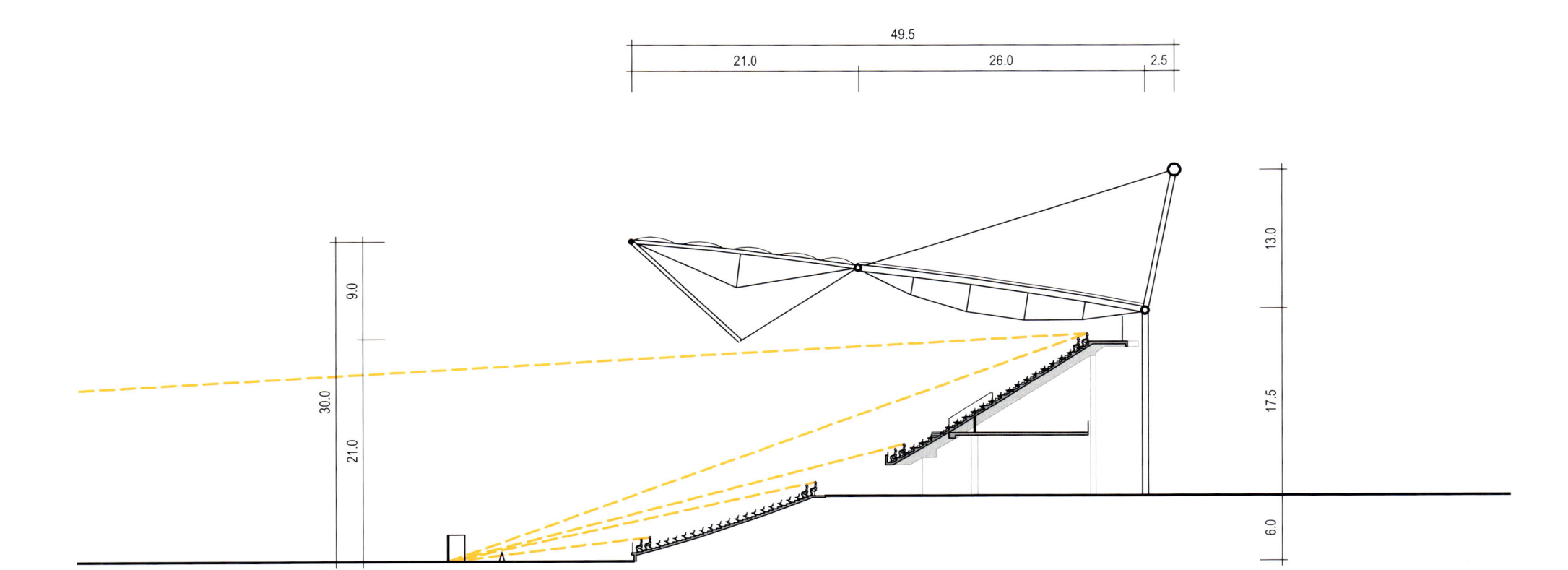

World Cup Stadium Hanover 301

579 **World Cup Stadium Hanover**
Longitudinal section of south stand
Scale 1:500

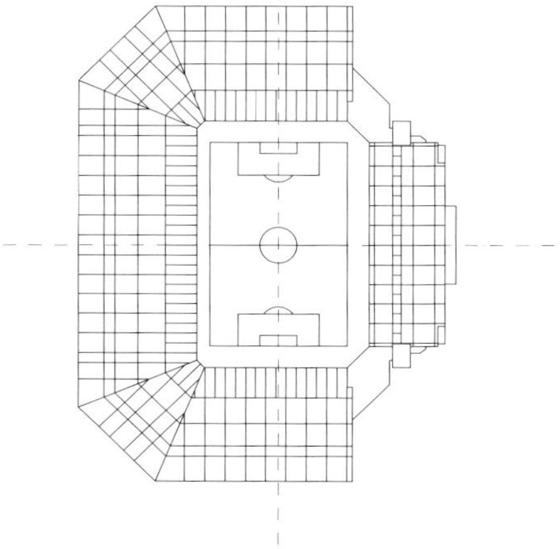

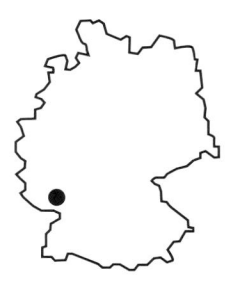

World Cup Stadium Kaiserslautern

The first stadium at Betzenberg was erected in 1920 and, after the first modernization in 1978, was time and again adapted in a modular fashion to the demands of a modern football stadium. Fritz-Walter-Stadion therefore belongs to the group of evolved stadia.

The main stand north has always been part of the city skyline, as the terrain on a stone quarry hillside slopes down rapidly ($h + 25$ m); due to the topography, the pitch is rotated by 90° to the east/west. The three-sided stadium is located only 3.5 km from the city, a 7-minute drive or 15 minutes walking.

In 1993, a two-storey catering plus VIP area was installed with conference and press facilities in the north stand. In 1998, an elongation of the south stand to the rear upscaled capacity.

In January 2000, the city council of Kaiserslautern decided to carry out all suitable conversion measures for the World Cup. A municipal stadium corporation was founded, the 'Objektgesellschaft Fritz-Walter-Stadion Kaiserslautern GmbH'.

In 2002, the east stand was expanded by 9,000 seats, and in 2003 the west stand by another 9,000 seats. In the same year, initial shell works at the hospitality and media towers next to the main stand north began. The final raise in capacity to 48,500 seats and the conversion of the north stand were also carried out during running operation in five stages from 11/2004 to 11/2005.

The classic horse-shoe form of the stadium was completed by a media tower at the north/west corner next to the main stand; supplementary media facilities were integrated into the north stand.

The erection of a hospitality tower at the north/eastern corner and the following modernization of the entire north stand completed the ring closure.

Lifting the roof structure by around 4.80 m enabled the expansion of the south stand by 13 rows and adaptation of the roof to the east and west stand. A new floodlight system at the roof of the south and north stand ascertains meeting new television and World Cup standards.
An international capacity of 47,700 can be raised to 48,500 by providing around 12,000 standing places.

The stadium is a one-tier grandstand, with exception of the main stand north and the two flanking media and hospitality towers, although from a circulation viewpoint the stands themselves are divided into two tiers.

Consequently, very large and visually dominant spectator stands emerge, meeting the demands on density and atmosphere of pure football stadia. The ambience 'auf dem Betze' is famous for being extremely challenging for the visiting teams and fans.

The new stands east/south employ again two linear rises (40/80 cm or 52/80 cm). Sightline elevation is decreased significantly at the lower but also upper tier due to the linear rise. Values range from 21.2 to 6.6 cm (lower part) and are continued at the upper part with 18.0 to 5.6 cm Distance to the playing field is around 10 m at the east/west and 13.60 cm at the main stand north, which is set up as a classic two-tier stand.

580 **Roof framework** (sectional view)
Fritz Walter Stadium, Kaiserslautern
581 **Roof framework** (top view)
Fritz Walter Stadium, Kaiserslautern
582 **Historic aerial view**
Betzenberg Stadium, Kaiserslautern
583 **Geographic location** of the city of Kaiserslautern

page 303:
584 **Aerial view** (2005)

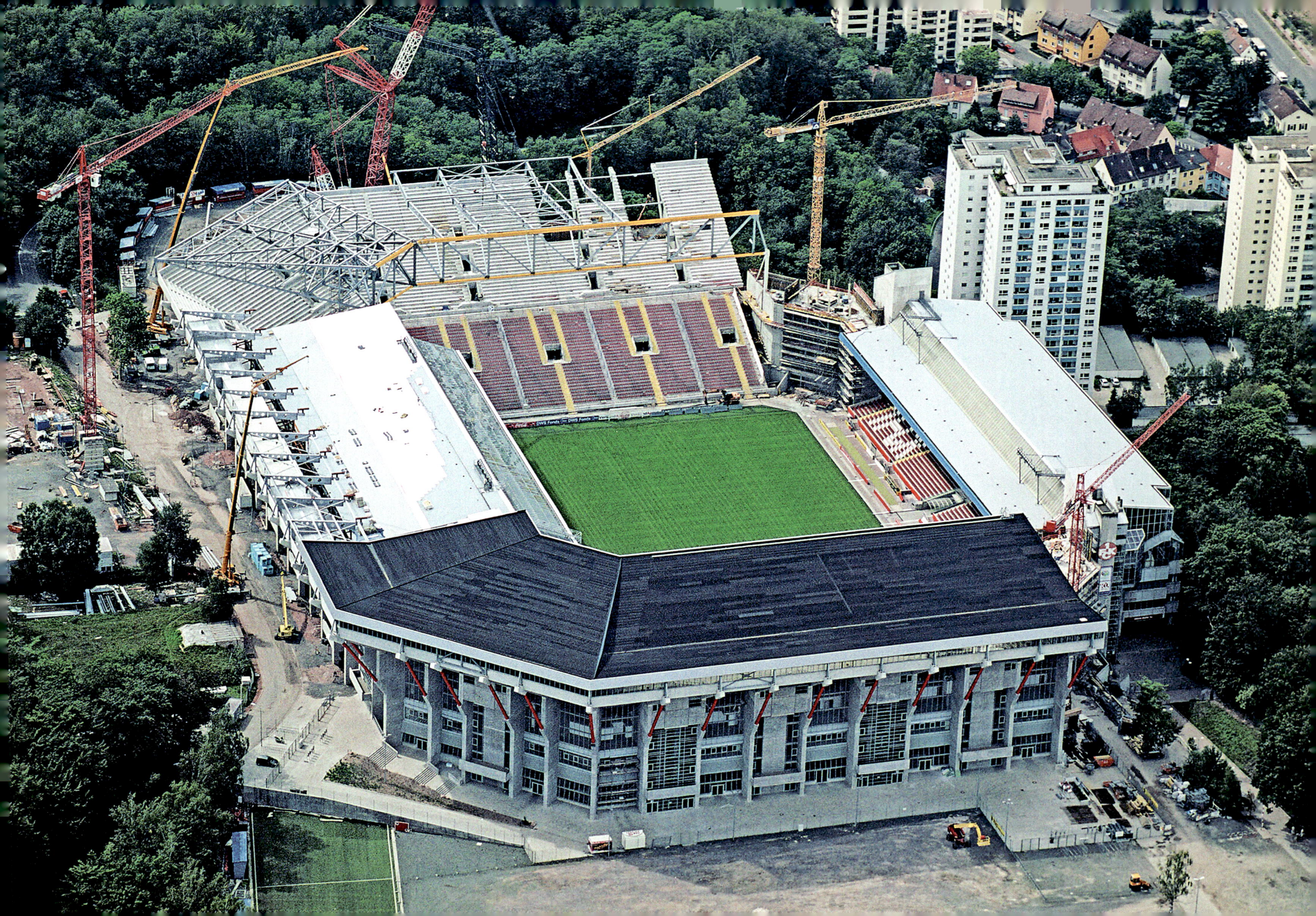

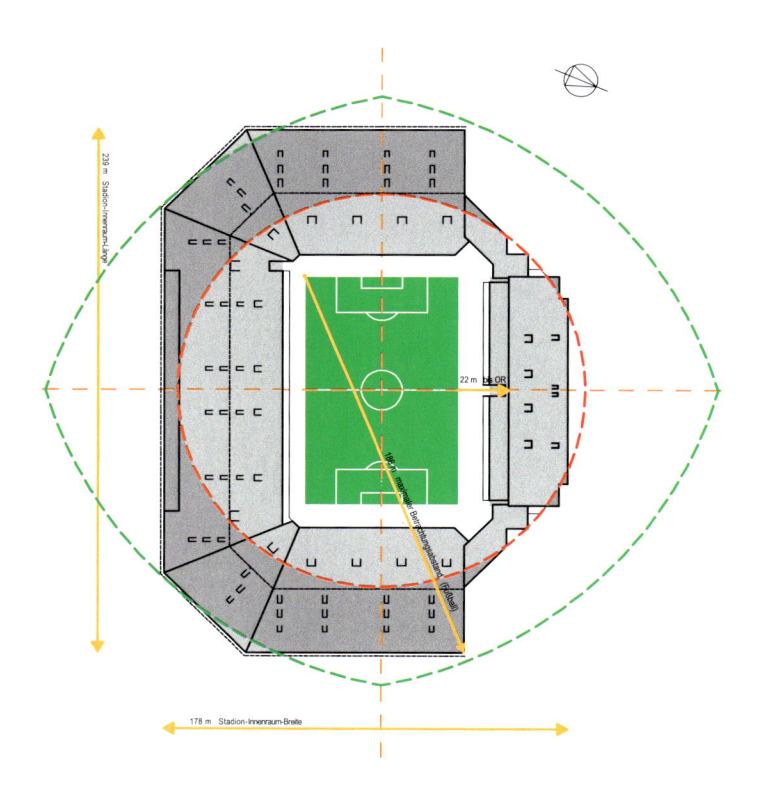

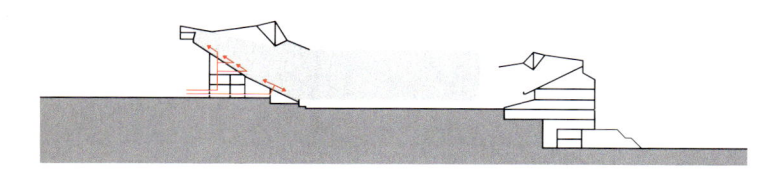

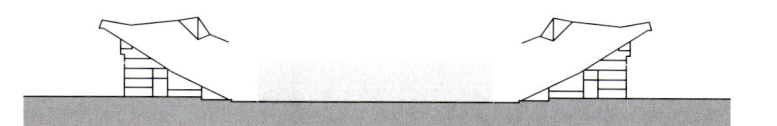

Fritz Walter Stadium, Kaiserslautern – One-tier stadium (football)
Asymmetric octagonal geometry with box tier on one side

Capacities

Gross capacity (international)	47,700 places [1]	100%
Gross capacity (national)	48,500 places [2]	
Standing places	12,000 places [2]	
Seats available for World Cup (net)	40,000 places [1]	
Within opt. 90-m viewing circle (football)	26,700 places	56.0%
Up to the max. recomm. 190-m distance	+ 20,000 places	42.0%
Beyond 190-m distance	+1,000 places	2.0%
Lower tier/upper tier (North)	1,500/2,600 places	
One tier (E/S/W)	43,600 places	
VIP guests (total)	1,327 places [1]	
Executive seats (World Cup 2006)	993 places [1]	
Number of private boxes (total)	28 private boxes	
Box seats, honorary tribune	334 box seats	
Media representatives (World Cup 2006)	150 commentators' positions (south)	
	200 television observers	
	400 press seats (with desk)	
	200 press seats (without desk)	
Wheelchair places	100 places [1]	

Dimensions (approx.)

	178 x 239 m (total)
Stadium interior	95 x 125 m = 11,900 m²
Width of roof	b = 57 m max.
Height above pitch	h = 18 – 25 m
Height above terrain (eaves)	h = 35 m
Roof area	A = 26,000 m²
Drainage	Slope toward interior = 22.5°

1) according to LOC World Cup 2006, Frankfurt (status: October 2005)
2) according to www.stadionwelt.de (status: October 2005)

585 **Viewing circle and dimensions** (plan)
Fritz Walter Stadium, Kaiserslautern, Scale 1 : 3,000
586 **Structure and circulation**
(transverse/longitudinal section)

page 305:
587 **Picture of the stadium interior** (2006)

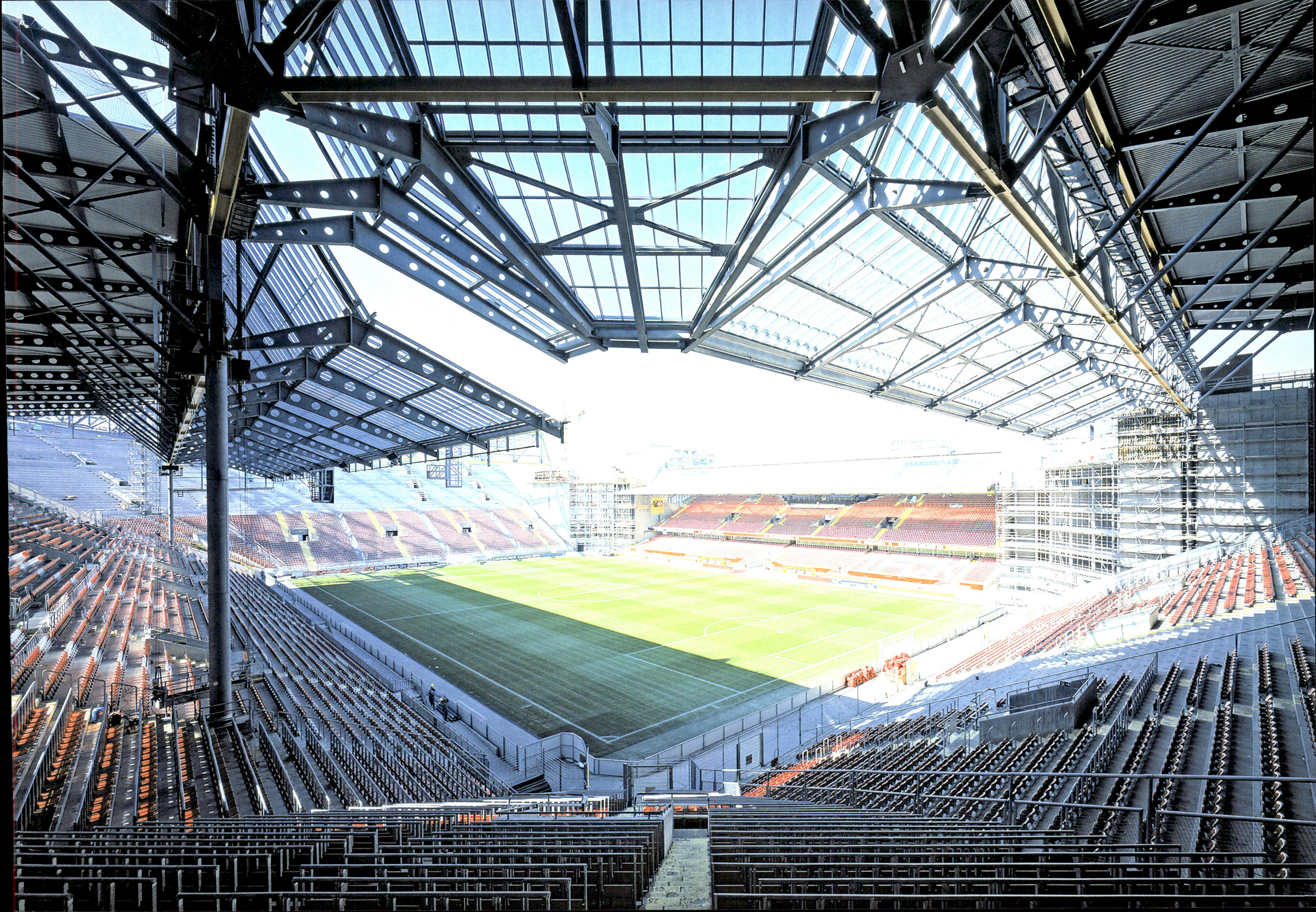

24 Grandstand profiles of built examples

Lower tier

Row number	15 rows
Tread depth	90 cm
Rise	24 cm
Stand inclination (average)	15°
Sightline elevation C	8.7 – 4.9 cm

→ Linear ascent!

Upper tier

Row number	31 rows
Tread depth	80 cm
Rise	40 cm
Stand inclination (average)	26.5°
Sightline elevation C	10.2 – 5.8 cm

→ Linear ascent!

Distance of stand (front edge)

Distance to field	13.60 m

Viewing distance (to the point of focus)

Maximum horizontal distance	47.10 m
Maximum vertical distance	20.10 m

Distance of hosp. boxes (touch line) 22.00 m

Viewing angle (front touch line)

Lower tier	9.5° – 12.0°
Upper tier	19.5° – 23.0°

588 World Cup Stadium Kaiserslautern
Cross section of north stand
Scale 1:500

Lower tier

Row number	33 rows
Tread depth	80 cm
Rise	40 cm
Stand inclination (average)	26.5°
Sightline elevation C	21.2–6.6 cm

→ Linear ascent!

Upper tier

Row number	35 rows
Tread depth	80 cm
Rise	52.0 cm
Stand inclination (average)	31.5°
Sightline elevation C	18.0–5.6 cm

→ Linear ascent!

Distance of stand (front edge)
Distance to field 10.10 m

Viewing distance (to the point of focus)

Maximum horizontal distance	58.65 m
Maximum vertical distance	34.50 m

Viewing distance (maximum)
Football 186.00 m

Viewing angle (front goal line)

Lower tier	13.5°
Upper tier	27.5°

World Cup Stadium Kaiserslautern

01. Main truss
02. Mega support (2x)
03. Secondary truss
04. Beam extension (new)
05. Compression column (outside)
06. Trapezoid metal plate roofing
07. Glass roof

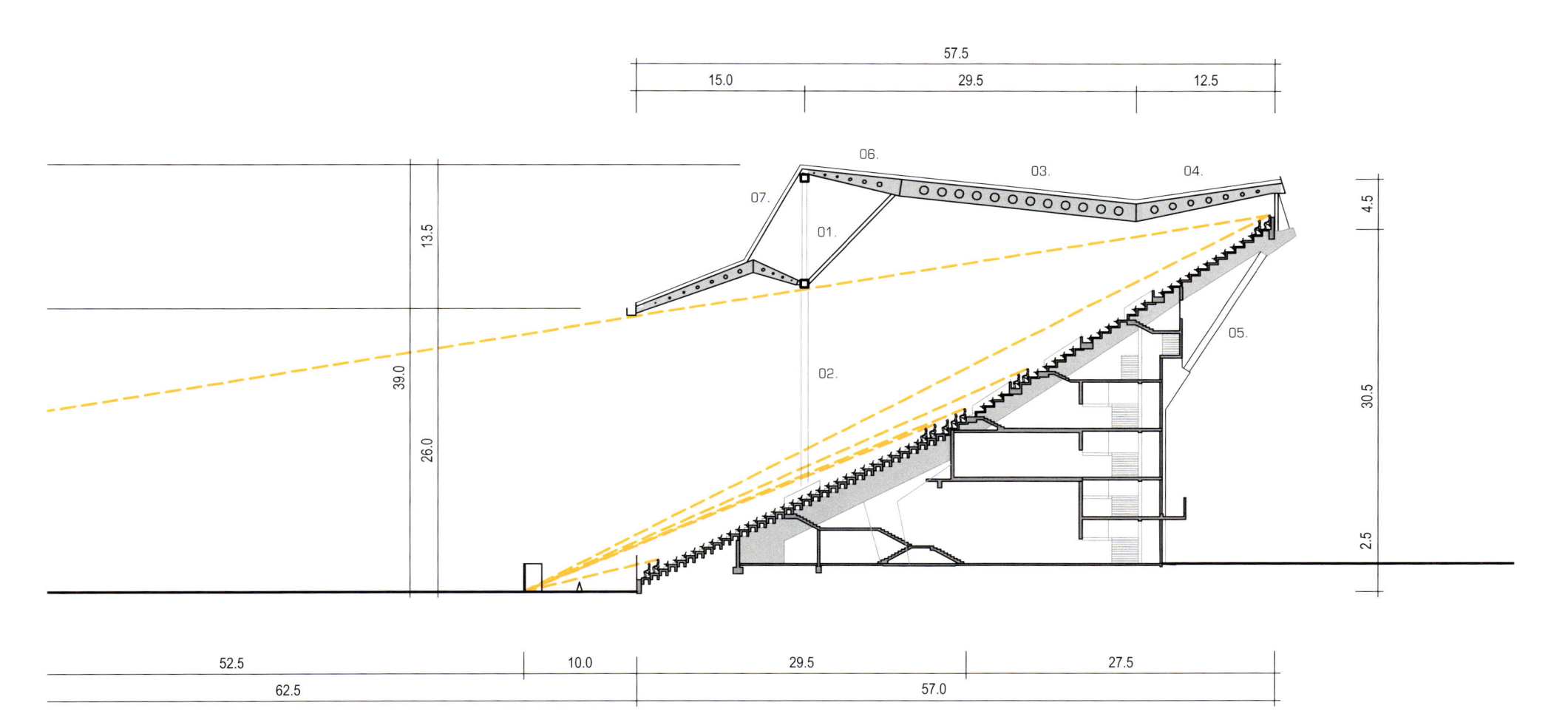

589 World Cup Stadium Kaiserslautern
Longitudinal section of east stand
Scale 1:500

World Cup Stadium Cologne

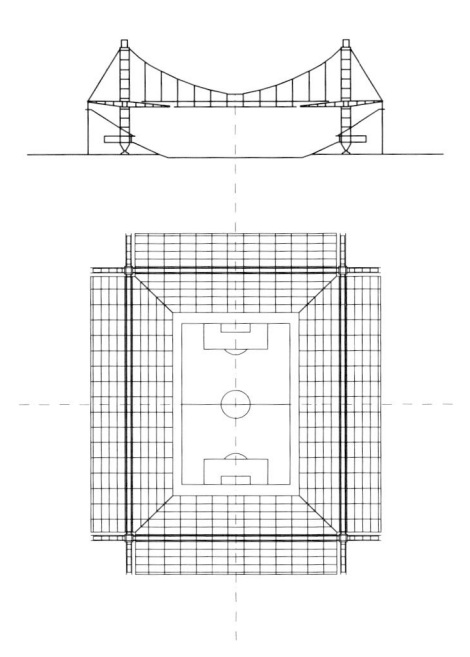

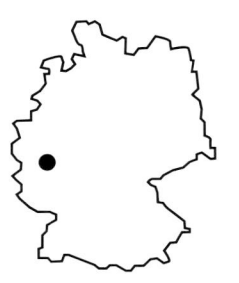

Müngersdorfer Stadion was finally completed in 1975, one year after the World Cup 1974. It was deemed one of the most modern multipurpose athletics stadia of its time. In time for World Cup 2006, conversion into a pure football stadium, now called RheinEnergieStadion, was completed in 2004.

In an open competition as combined procedure (GRW 1995) gmp Architects was awarded first prize in June 2001.

The anonymous process required a partnership of the design architect with a building contractor and the submission of a binding offer. In a parallel procedure (VOB part A) in September 2001, Max Bögl GmbH, Neumarkt, was commissioned as general contractor.

One week after the contract letting, the general contractor had to submit a binding cost offer via a notary for the conversion into a modern football stadium during running operation.

The complete conversion (new build) was inaugurated after 27 months (12/2001 to 03/2004) with a friendly match between Germany and Belgium on 31 March 2004. During construction, a minimum capacity of 30,000 had to be guaranteed. For logistical reasons, no fifth construction phase for installation of a more economical spoked-wheel structure was feasible. Thus, the clear rectangular geometry of the stadium responds to the playing field and the orthogonal garden architecture of the sports park. The suspended cable bridge enabled the dismantling, structural works and finishing of a quarter of the stadium, in half-year steps.

Of the 46,00 seats, around 60% (27,800) are within the optimum viewing circle, and the entire auditorium lies within the maximum viewing distance of 190 m. With a size of 215 x 170 m, the Cologne stadium was one of the most compact sports stadia of World Cup 2006.

All stands are defined by a parabolic sightline elevation; the proximity to the pitch mirrors British football stadia by exploiting FIFA minimum distances. Inclination follows the parameters of a best-possible sightline; hence, the first row is just high enough for spectators to view over the advertising hoarding, in a position recommended by FIFA.

The low base point of the stand of 75 cm is compensated with regards to security by vertically adjustable fencing, which in turn would affect viewing toward the touch and goal line for the entire lower tier.

Even though a circulating box tier with a clear height of 2.70 cm was integrated into the parabolic rise, 'C' values remain within the admissible range of 12–9 cm.

The upper tier west is distinguished from the other grandstand parts as the desired number of 2 x 25 private boxes has to be accommodated on two levels at the main stand. Two listed tree-lined avenues to the left and right of the stadium prevent an expansion of the stadium complex. The historic axis along the green surfaces at Aachener Straße is continued into the stadium window (former marathon gate), allowing a view into the stadium interior, even on non-match days.

590 **Roof framework** (sectional view)
RheinEnergieStadion, Cologne
591 **Roof framework** (top view)
RheinEnergieStadion, Cologne
592 **Historic aerial view**
Sportpark Müngersdorf, Cologne
593 **Geographic location** of the city of Cologne

page 309:
594 **Aerial view** (2006)

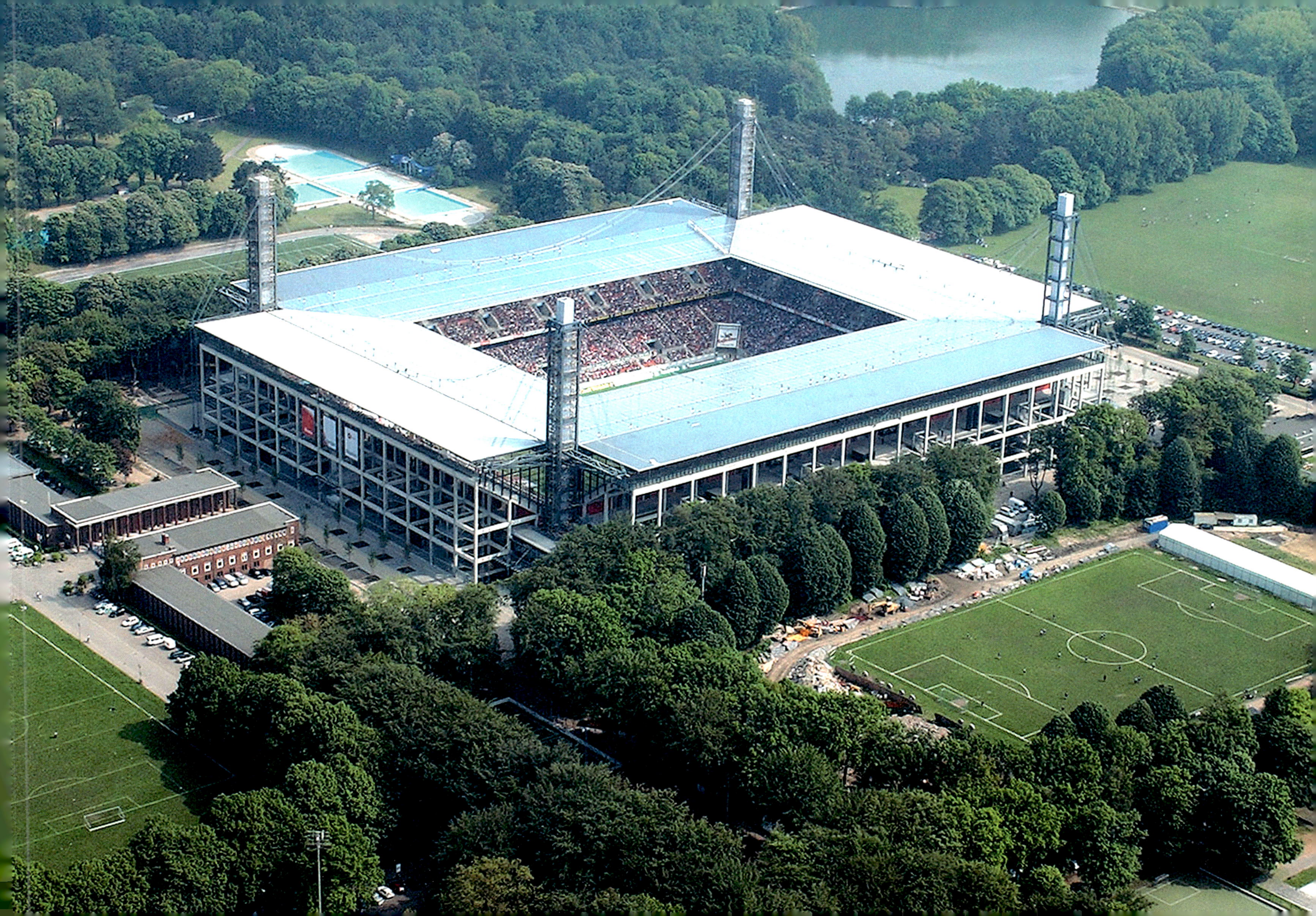

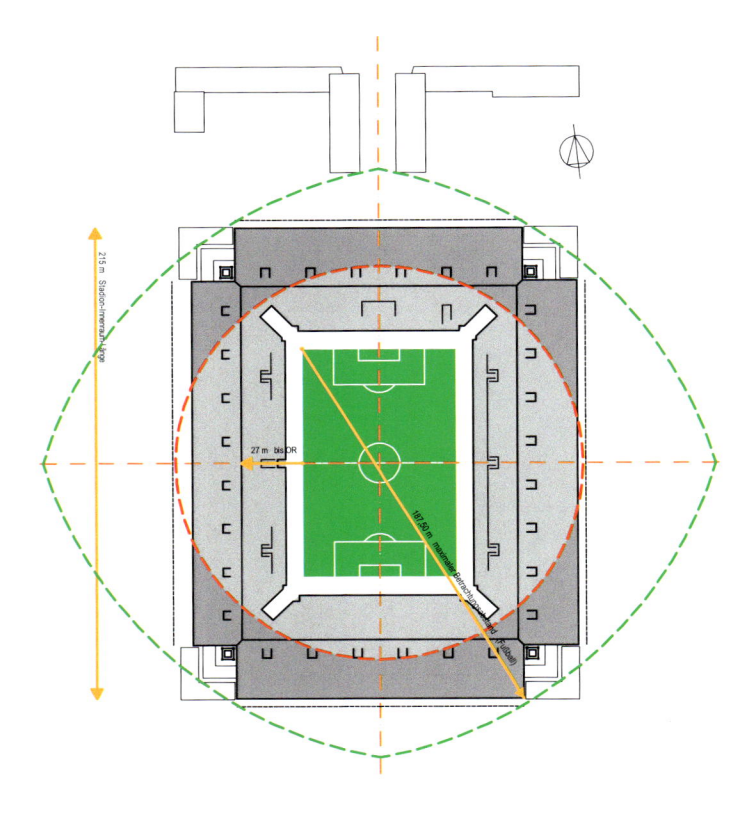

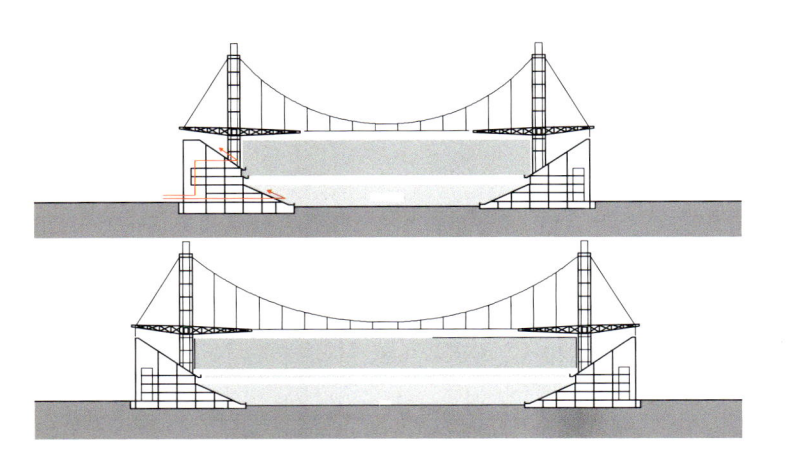

RheinEnergieStadion, Cologne – Two-tier stadium (football)
Open rectangular geometry with a circulating box tier
(two-level at the main stand)

Capacities

Gross capacity (international)	46,300 places [1]	100%
Gross capacity (national)	50,700 places [2]	
Standing places	9,200 places [2]	
Seats available for World Cup 2006 (net)	38,200 places [1]	
Within opt. 90-m viewing circle (football)	27,800 places	60.0%
Up to the max. recomm. 190-m distance	+18,500 places	40.0%
Lower tier	17,600 places	
Upper tier/balcony	28,700 (360) places	
VIP guests (total)	2,685 places [1]	
Executive seats (World Cup 2006)	2,089 places [1]	
Number of private boxes (total)	48 private boxes/2 special boxes	
Box seats, honorary tribune	596 box seats	
Media representatives (World Cup 2006)	150 commentators' positions (south)	
	200 television observers	
	400 press seats (with desk)	
	200 press seats (without desk)	
Wheelchair places	100 places [1]	

Dimensions (approx.)

	170 x 215 m (total)
Stadium interior	83 x 121 m = 10,000 m²
Orientation of playing field	plus 7° in north-south direction
Pylon height	$h = 66$ m + 6 m (light stele)
Width of roof	$b = 49$ m max.
Height above pitch	$h = 33$ m
Height above terrain (eaves)	$h = 33$ m
Roof area	$A = 28,400$ m²
Trapezoid metal plate	$= 13,000$ m²
Makrolon	$= 15,400$ m²
Drainage	Slope toward interior = 3° or 5%

1) according to LOC World Cup 2006, Frankfurt (status: October 2005)
2) according to www.stadionwelt.de (status: October 2005)

595 **Viewing circle and dimensions** (plan)
RheinEnergieStadion, Cologne, Scale 1:2,500
596 **Structure and circulation**
(transverse/longitudinal section)

page 311:
597 **Picture of the stadium interior** (2006)

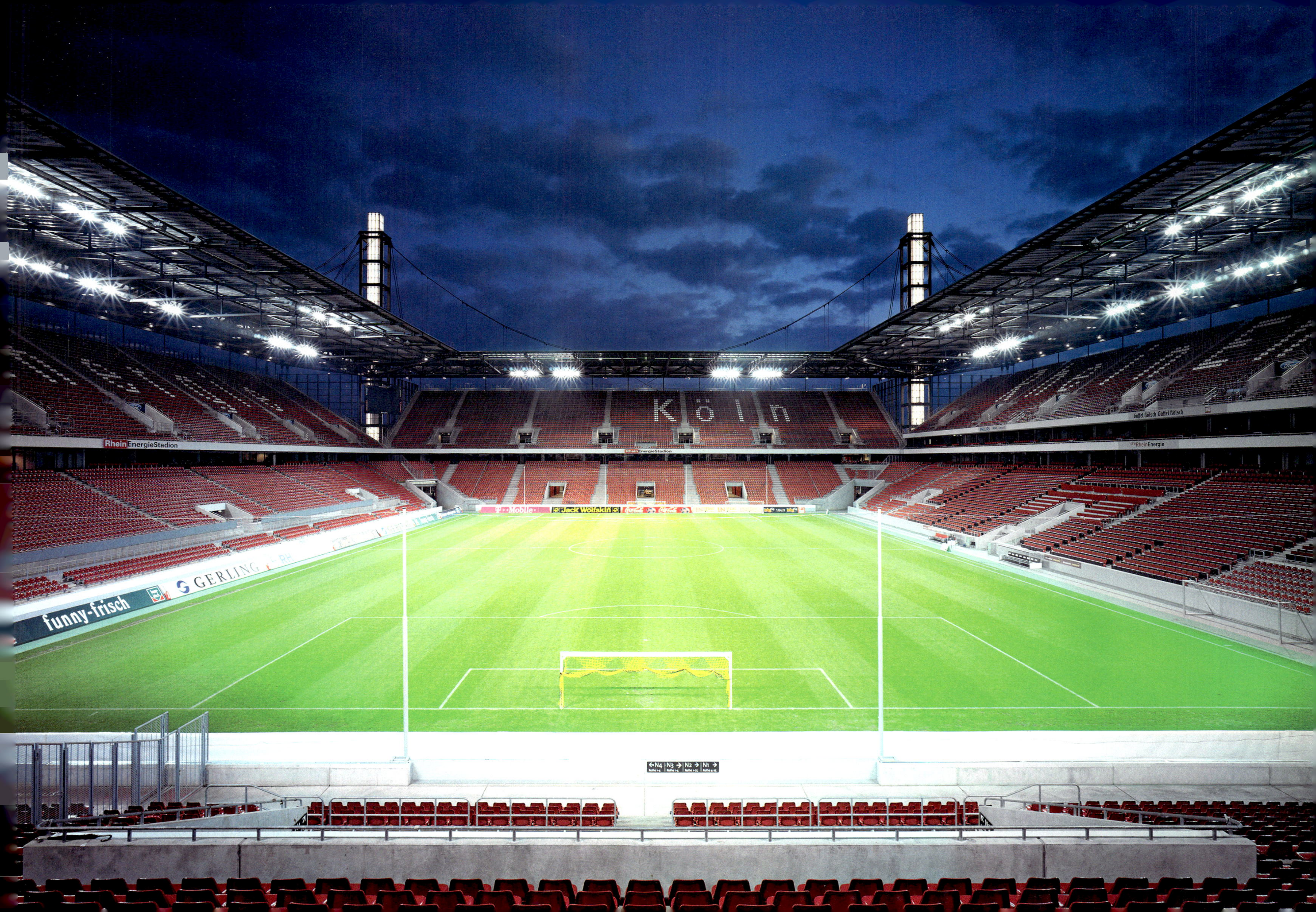
Köln
RheinEnergieStadion
GERLING
funny-frisch

24 Grandstand profiles of built examples

01. Pylon with light stele (4 x)
02. Structural cable (2 x 2)
03. Cantilever (force redirection)
04. Virendeel truss (compression bar)
05. Secondary truss
06. Trapezoid metal plate roofing
07. Glass roof

Lower tier

Row number	20 rows + 2 box rows
Tread depth	90 cm
Rise	34.9 – 46.6 cm
Stand inclination (average)	24.5°
Sightline elevation C	14.5 – 11.9 cm

Upper tier

Row number	24 rows
Tread depth	80 cm
Rise	55.0 – 57.0 cm
Stand inclination (average)	35.0°
Sightline elevation C	5.8 – 4.9 cm

Distance of stand (front edge)
Distance to field 7.20 m

Viewing distance (to the point of focus)
Maximum horizontal distance 56.10 m
Maximum vertical distance 31.00 m

Distance of hosp. boxes (touch line) 27.00 m

Viewing angle (front touch line)
Lower tier 13.5° – 21.0°
Private boxes level 27°
Upper tier 31.5° – 33.0°

598 World Cup Stadium Cologne
Cross section of west stand
Scale 1 : 500

Lower tier

Row number	24 rows
Tread depth	80 cm
Rise	30.3 – 42.1 cm
Stand inclination (average)	24.5°
Sightline elevation C	13.6 – 12.7 cm

Upper tier

Row number	31 rows
Tread depth	85 cm
Rise	55.0 – 56.8 cm
Stand inclination (average)	33°
Sightline elevation C	12.2 – 8.8 cm

Distance of stand (front edge)

Distance to field	8.15 m

Viewing distance (to the point of focus)

Maximum horizontal distance	54.80 m
Maximum vertical distance	31.00 m

Viewing distance (maximum)

Football	187.50 m

Viewing angle (front goal line)

Lower tier	11.5° – 20.0°
Upper tier	26.5° – 29.5°

World Cup Stadium Cologne 313

599 **World Cup Stadium Cologne**
Longitudinal section of north stand
Scale 1:500

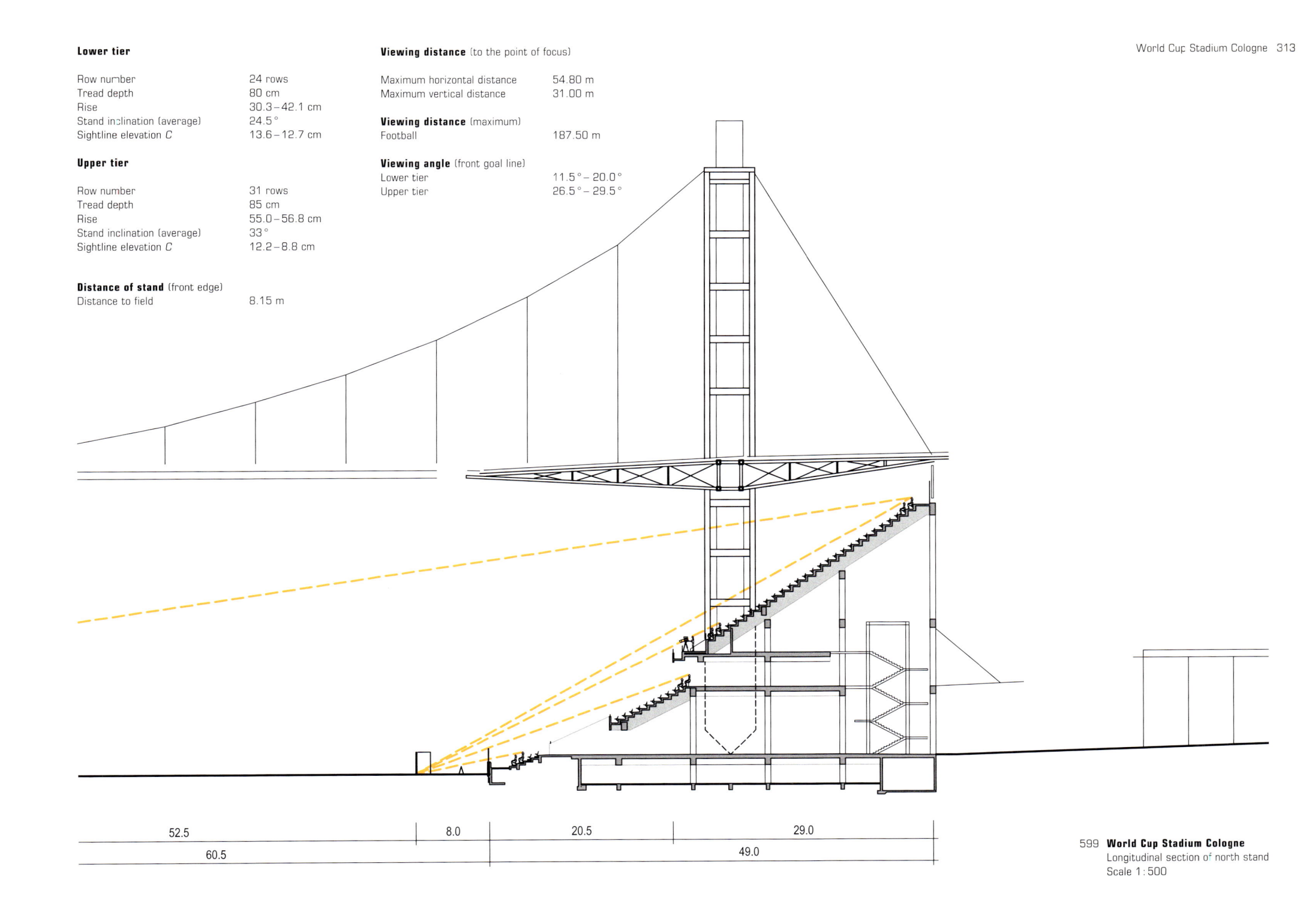

World Cup Stadium Leipzig

The old Zentralstadion at Sportplatz Leipzig (1892) was called 'Stadion der Hunderttausend' [stadium of the one hundred thousand] and was the largest stadium at that time in Germany. In 1920, the neighbouring Elster flood basin was completed, and in 1939 planning of the sports field by Werner March (cf. Berlin stadium) began.

Architectural practice Glöckner, Nürnberg, won the international competition in 1996. A multistage selection process for project development with investor bidding followed, which the city council of Leipzig decided on in October 1997.

The new Sportforum Leipzig covers around 55 ha and is only 2.5 km from the city centre; a 'sports and leisure park' realized on the meadows of the Elsterbecken.

The new build of a compact football complex with a capacity of 45,000 was to be integrated into the existing earth-wall system ($h = +22.0$ m) made of debris from World War II and located within the drainage area of the 'White Elster', the local river. Planning envisaged a 'stadium within a stadium', i.e. the new construction of the grandstand circle within the existing stadium bowl. For reasons of historical preservation it was attempted to incorporate the existing main building at the Sportforum, which is partly integrated into the wall.

In around 40 months (12/2000 to 03/2004) the following construction works were carried out: erection of the roof structure (building parts) with two arches; expansion and reinstatement of the main building; connecting the utilization areas of the stadium on levels 1/3/5/7; alteration of the entrance situation at Sportforum east.

There was no parallel match activity. The Zentralstadion is the only World Cup hosting venue without a home team. With 44,300 seats it has the lowest capacity of all World Cup venues. Therefore, most spectators (70%) sit within the optimum viewing area (31,000 seats). A typical feature of the Zentralstadion Leipzig is its special topographical situation in the centre of the old stadium circle, and the open short sides, as commonly found in stadia in Portugal and Korea/Japan. The grandstand has therefore only two upper tiers at its long sides.

In the curved segments and on the straights the upper edge of the lower tier circulates with an oval geometry at a curvature radius of 365/370 m, or 26 m in the corner areas, and at equal height around the playing field.

Between the rounded profile and the rectangular geometry of the field, different numbers of rows (29/34) and distances emerge: distance of the first row 14.7/13.9 m in the pitch axes and a minimum distance of approx. 7.50 m at the corner flag area.

The approximate circular form of the stand's rear edge causes an 'undulating' ascent so that the number of rows increases from 28 to 33. With a maximum vertical distance of 38.50 m, Leipzig, the smallest of the World Cup stadia, has the second largest stand height at maximum stand inclination.

The event profile of Sportforum Leipzig is complemented by the Arena Leipzig (a multifunctional hall). Utilization: 12,000 places for concerts, 4,000 places for indoor athletics with a 200-m indoor track, or as general or judo sports hall (300 places). The Festwiese Leipzig (a fairground with 200 x 200 m) for 80,000 spectators lies directly adjacent to the Zentralstadion.

600 **Roof framework** (sectional view)
Zentralstadion, Leipzig
601 **Roof framework** (top view)
Zentralstadion, Leipzig
602 **Historic aerial view**
„Stadion der Hundertausend", Leipzig
603 **Geographic location** of the city of Leipzig

page 315:
604 **Aerial view** (2006)

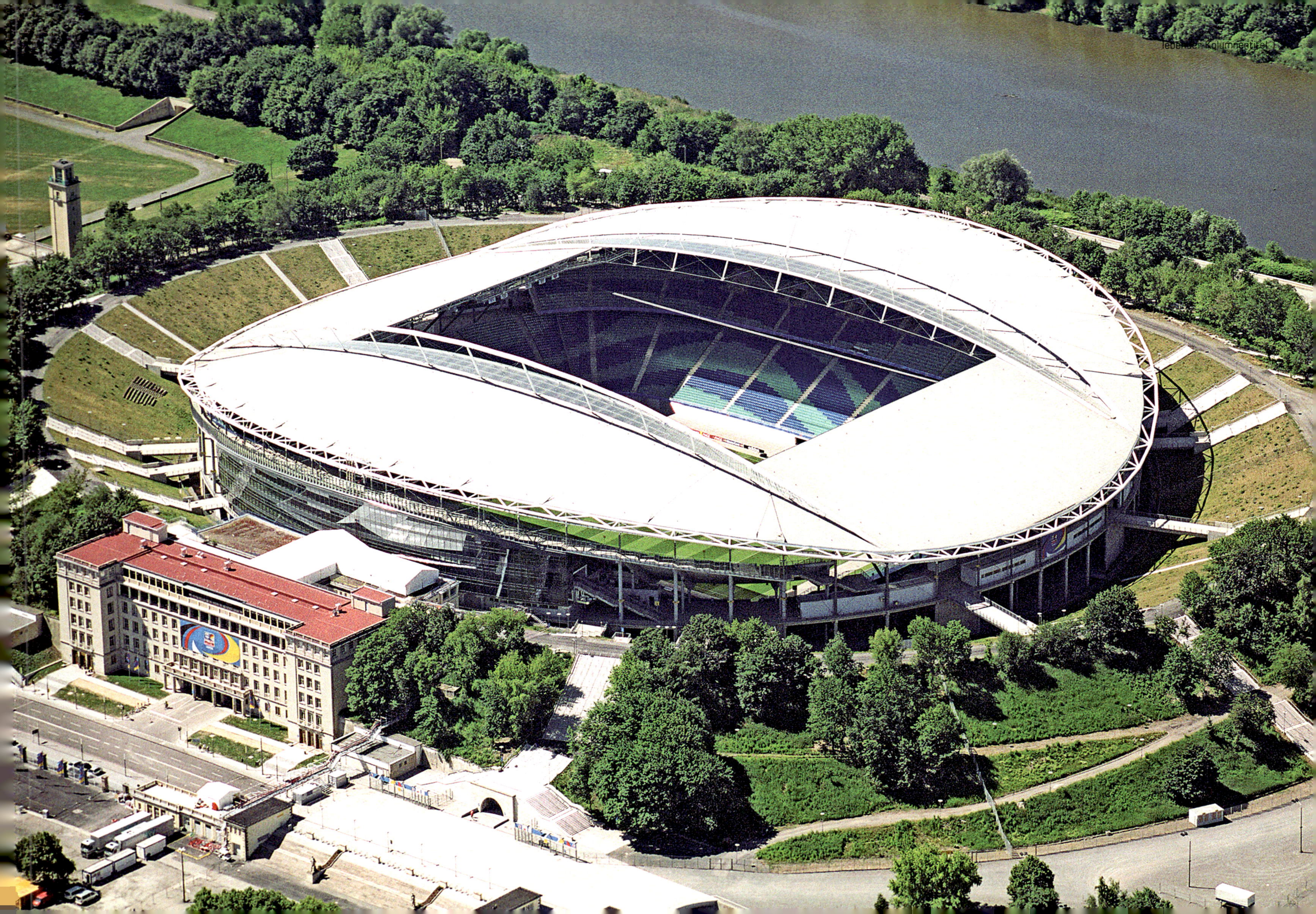

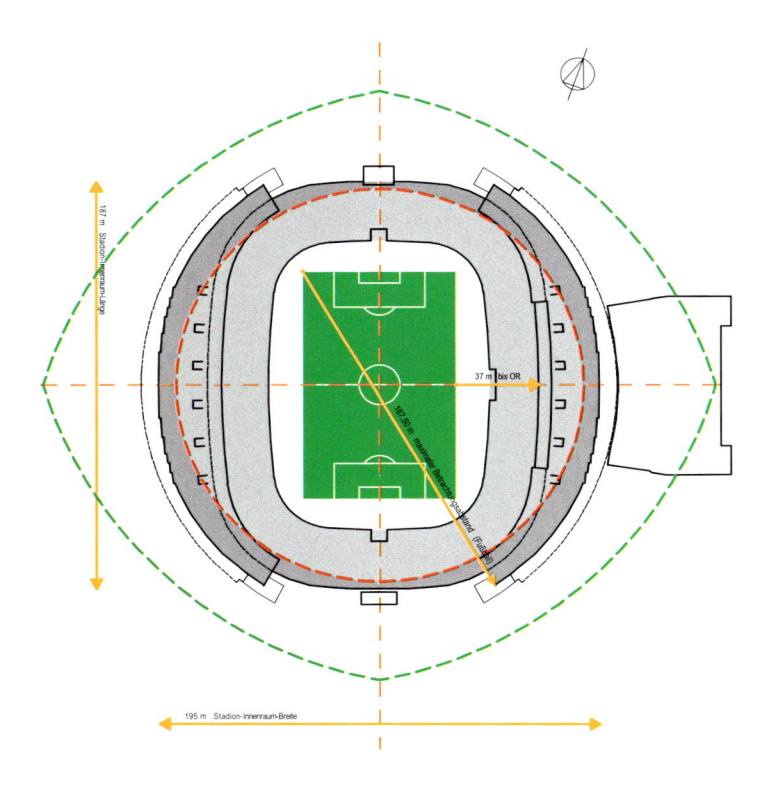

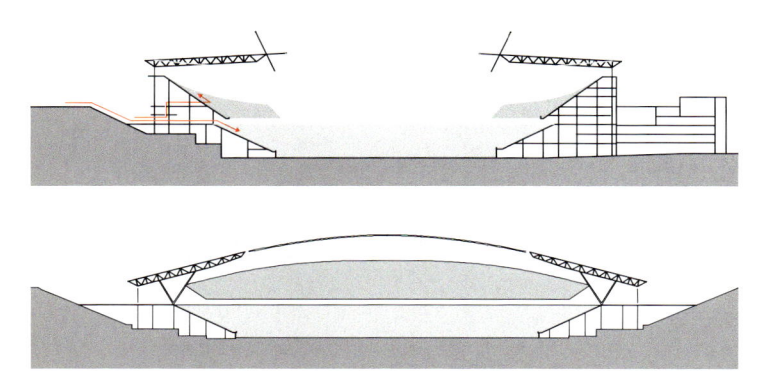

Zentralstadion, Leipzig – Two-tier stadium (football)
Asymmetric radial geometry with upper tier stand on both sides

Capacities

Gross capacity (international)	44,300 places [1]	100%
Gross capacity (national)	44,300 places [2]	
Standing places	. /.	
Seats available for World Cup 2006 (net)	37,100 places [1]	
Within opt. 90-m viewing circle (football)	31,000 places	70.0%
Up to the max. recomm. 190-m distance	+13,300 places	30.0%
Lower tier	27,250 places	
Upper tier	17,250 places	
VIP guests (total)	2,969 places [1]	
Executive seats (World Cup 2006)	2,660 places [1]	
Number of private boxes (total)	16 private boxes	
Box seats, honorary tribune	309 box seats	
Media representatives (World Cup 2006)	150 commentators' positions (south)	
	200 television observers	
	400 press seats (with desk)	
	200 press seats (without desk)	
Wheelchair places	139 places [1]	

Dimensions (approx.) 187 x 195 m (total)
Stadium interior 97 x 133 m = 12,900 m²

Orientation of playing field minus 20° in north/south direction

Width of roof b = 62 m max.
Height above pitch h = 40–47 m
Height above terrain (eaves) h = 12–28 m (Level 5)

Roof area A = 28,000 m²
Arched roof = 19,000 m²
Curved roof = 9,000 m²
Drainage Slope toward exterior = 3° or 5%

1) according to LOC World Cup 2006, Frankfurt (status: October 2005)
2) according to www.stadionwelt.de (status: October 2005)

605 **Viewing circle and dimensions** (plan)
Zentralstadion, Leipzig, Scale 1:3,000
606 **Structure and circulation**
(transverse/longitudinal section)

page 317:
607 **Picture of the stadium interior** (2006)

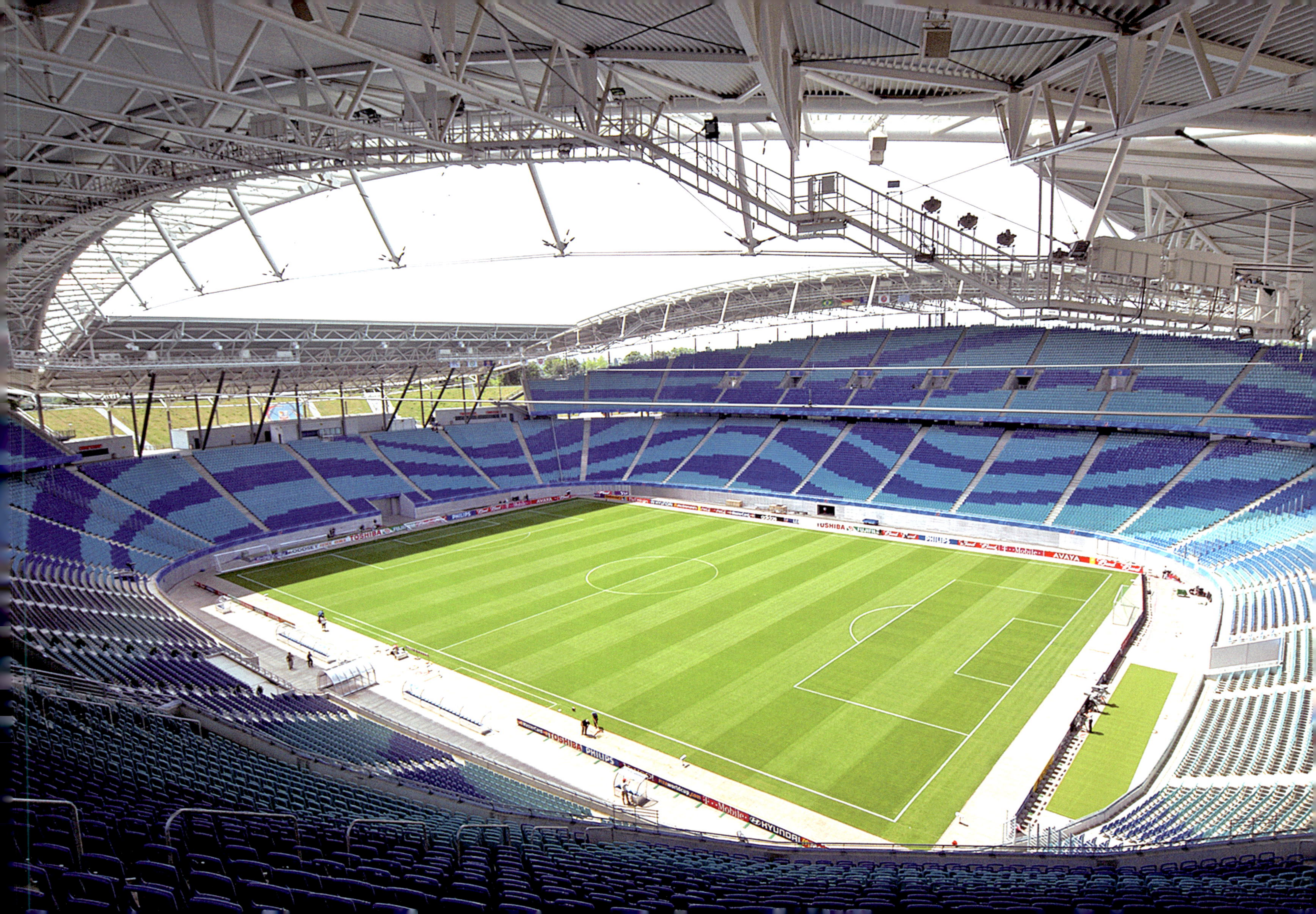
TOSHIBA
PHILIPS
AVAYA
TOSHIBA PHILIPS
HYUNDAI

24 Grandstand profiles of built examples

01. Arched girder (2 x)
02. Three-boom ring truss
03. Lattice radial truss
04. Restraining cables
05. Glass roof, shield

Lower tier

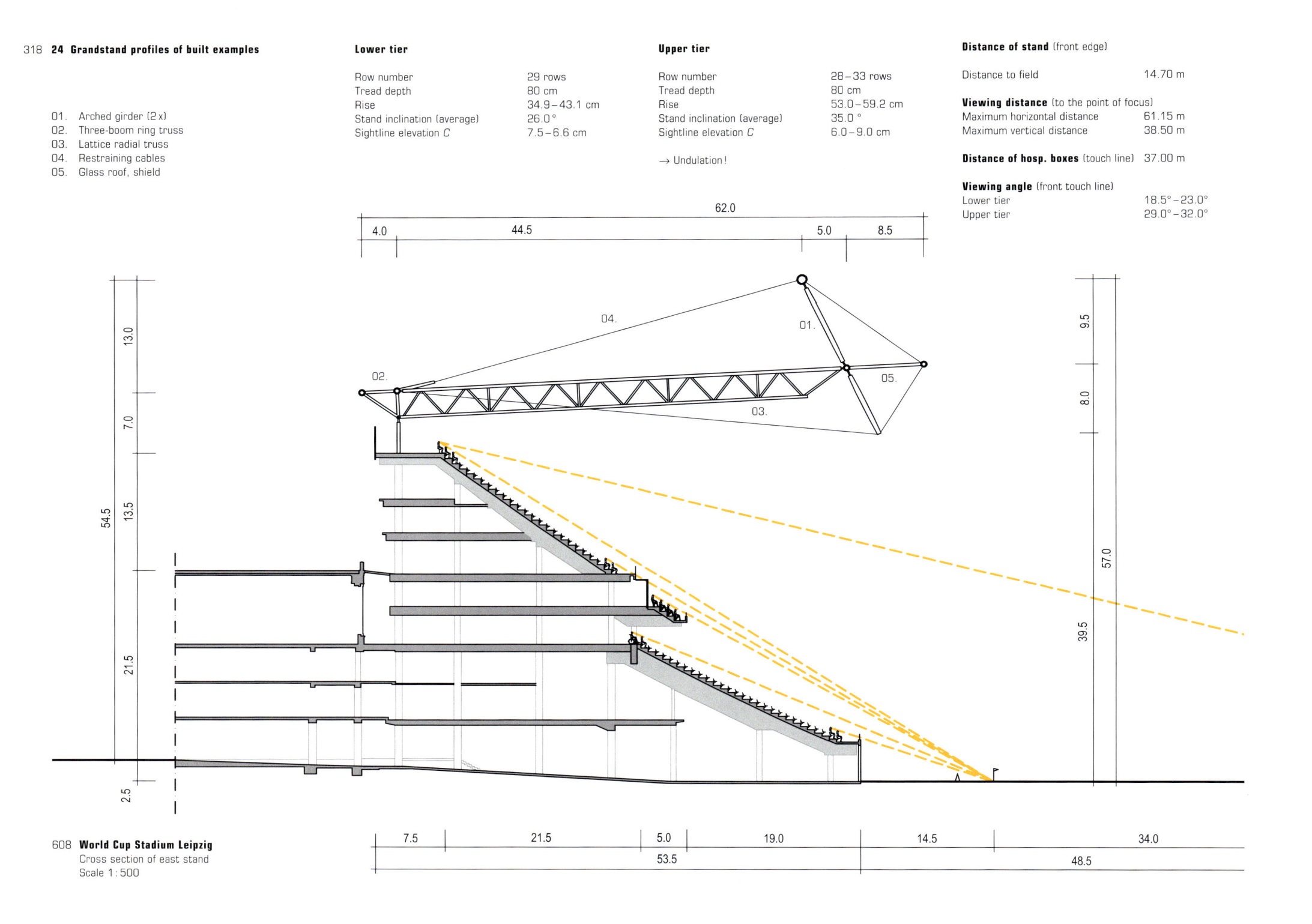

Row number	29 rows
Tread depth	80 cm
Rise	34.9 – 43.1 cm
Stand inclination (average)	26.0°
Sightline elevation C	7.5 – 6.6 cm

Upper tier

Row number	28 – 33 rows
Tread depth	80 cm
Rise	53.0 – 59.2 cm
Stand inclination (average)	35.0°
Sightline elevation C	6.0 – 9.0 cm

→ Undulation!

Distance of stand (front edge)

Distance to field	14.70 m

Viewing distance (to the point of focus)

Maximum horizontal distance	61.15 m
Maximum vertical distance	38.50 m

Distance of hosp. boxes (touch line) 37.00 m

Viewing angle (front touch line)

Lower tier	18.5° – 23.0°
Upper tier	29.0° – 32.0°

World Cup Stadium Leipzig
Cross section of east stand
Scale 1 : 500

Lower tier

Row number	34 rows
Tread depth	80 cm
Rise	32.4 – 43.1 cm
Stand inclination (average)	25.0°
Sightline elevation C	10.7 – 11.3 cm

→ Short side without upper tier

Distance of stand (front edge)

Distance to field	13.90 m

Viewing distance (to the point of focus)

Maximum horizontal distance	57.60 m
Maximum vertical distance	16.60 m

Viewing distance (maximum)

Football	167.50 m

Viewing angle (front goal line)

Lower tier	14.5° – 21.5°

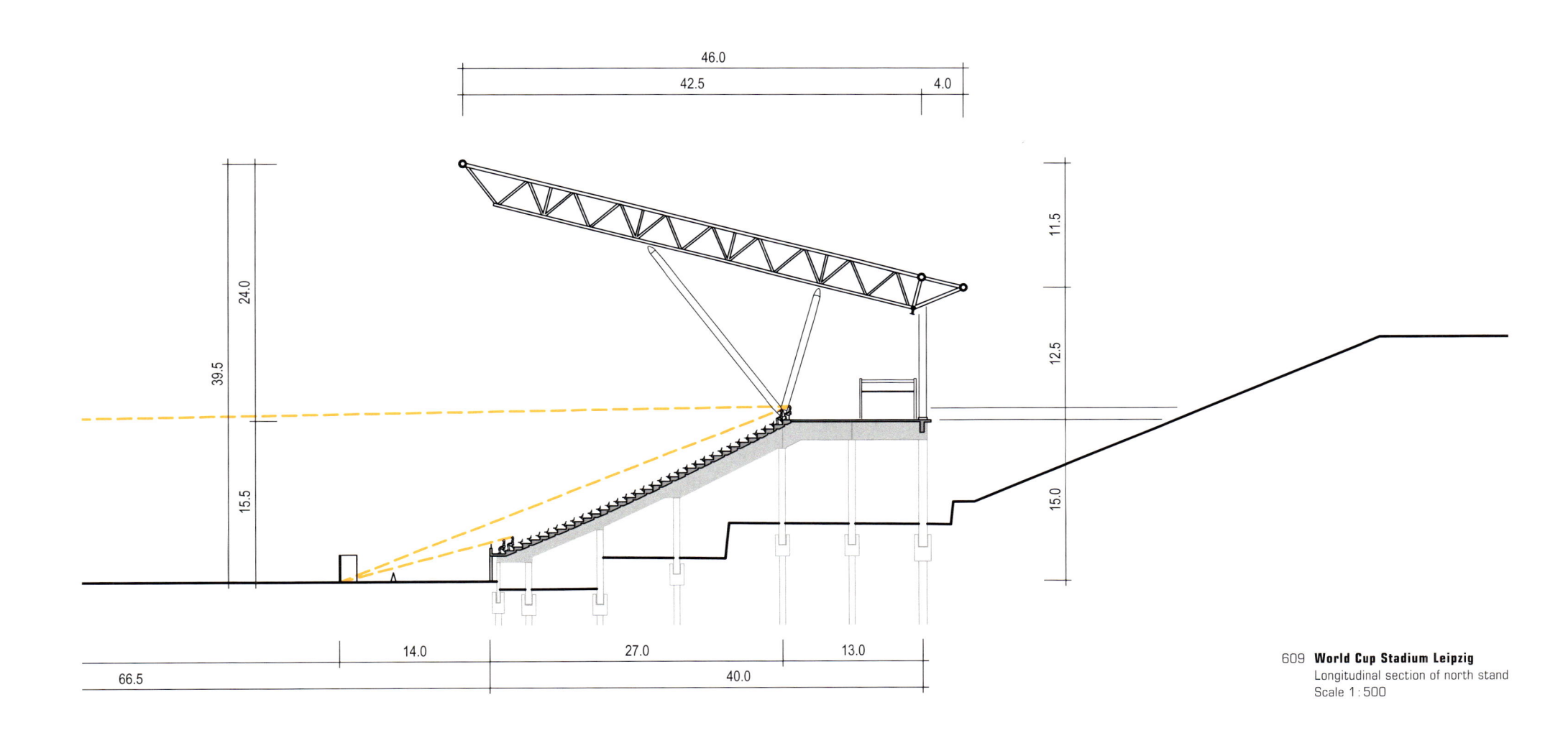

World Cup Stadium Leipzig 319

609 **World Cup Stadium Leipzig**
Longitudinal section of north stand
Scale 1 : 500

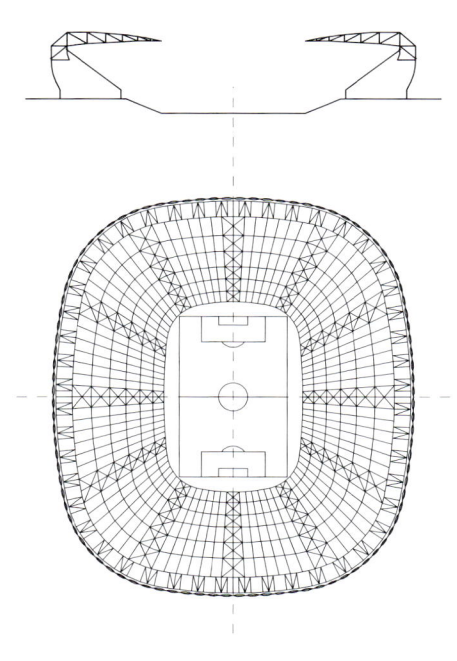

World Cup Stadium Munich

In March 1997, the city of Munich examined the conversion of the Munich Olympic Stadium without reaching an unanimous vote. In 2001, FC Bayern Munich and TSV 1860 decided to build their own stadium and began the location search for a new-build pure football stadium after the British model.

In October 2001, a local referendum confirmed the decision of the city council for the new stadium site at Fröttmaning. Parallel, the practices of Herzog de Meuron, Basle, and gmp Architects, Aachen, were selected for the second round of the architectural competition, with participation of a general contractor. The design by HdM was selected for the new Allianz Arena, with Alpine Co. (Austria) as building contractor.

After 32 months (10/2002 to 5/2005), without simultaneous match operation, the opening ceremony took place on 30/31 May 2005.

In eight stages with 12 building parts, first the concrete structure (components 01–08, with four straights and four corners each), then the roof framework (first the 'shoulders' or tension/compression supports, then two main trusses each) and finally the internal finish works were executed.

A spoked-wheel solution was rejected quite early for deadline reasons. Instead, a simple cantilever frame was planned as primary structure and a lozenge-pneu roof façade as secondary structure.

Besides the actual stadium build, extensive infrastructure works were carried out: a large 'esplanade' with several parking levels, 'FAN-Canyons' with shops and cash desks on an inclined circulation level as well as construction of an urban train station Fröttmaning and motorway access.

The compact stadium offers space for 66,000 spectators, the third-largest capacity of the Bundesliga. The grandstand itself is divided into three tiers; the stand elevation is marked by two hospitality box tiers.

At the main circulation level, the 'grand promenade', lies the first circulating structural joint, through which visitors enter the stadium and from which they may overlook the stadium interior.

The entire auditorium lies within the maximum viewing distance of 190 m, and 48.8 % of seats (ca. 32,000) are within the optimum viewing circle, the largest figure of all World Cup stadia.

The distance of 8.0 m (stands to pitch) on the long and short sides and the projecting terraces allow spectators to sit close to the action on the pitch even in the upper tier (maximum horizontal distance to the point of focus: circa 64.2 m).

The elevation of the first seating row to 0.80 m seat height does not correspond to the 2.0 m required by DFB in 2004 (securing of the pitch), but slip-on fence elements in the affected spectator areas provide a solution. Originally, a lowering of the stand base point was planned to achieve a height of 2.0 m flush with the upper edge of the barrier. The lower tier ensures a sufficient sightline elevation. Due to the different elevational profile of the terrace rear edge from the stepping geometry, rows 22–28 undulate.

A remarkably dense atmosphere evolves from the uniformity of the stadium interior and the calmness of the three tiers. Aided by the complete roof covering, a compact sense of space is created.

610 **Roof framework** (sectional view)
Allianz Arena, Munich
611 **Roof framework** (top view)
Allianz Arena, Munich
612 **Historic aerial view**
Olympiastadion Munich (1972)
613 **Geographic location** of the city of Munich

page 321:
614 **Aerial view** (2006)

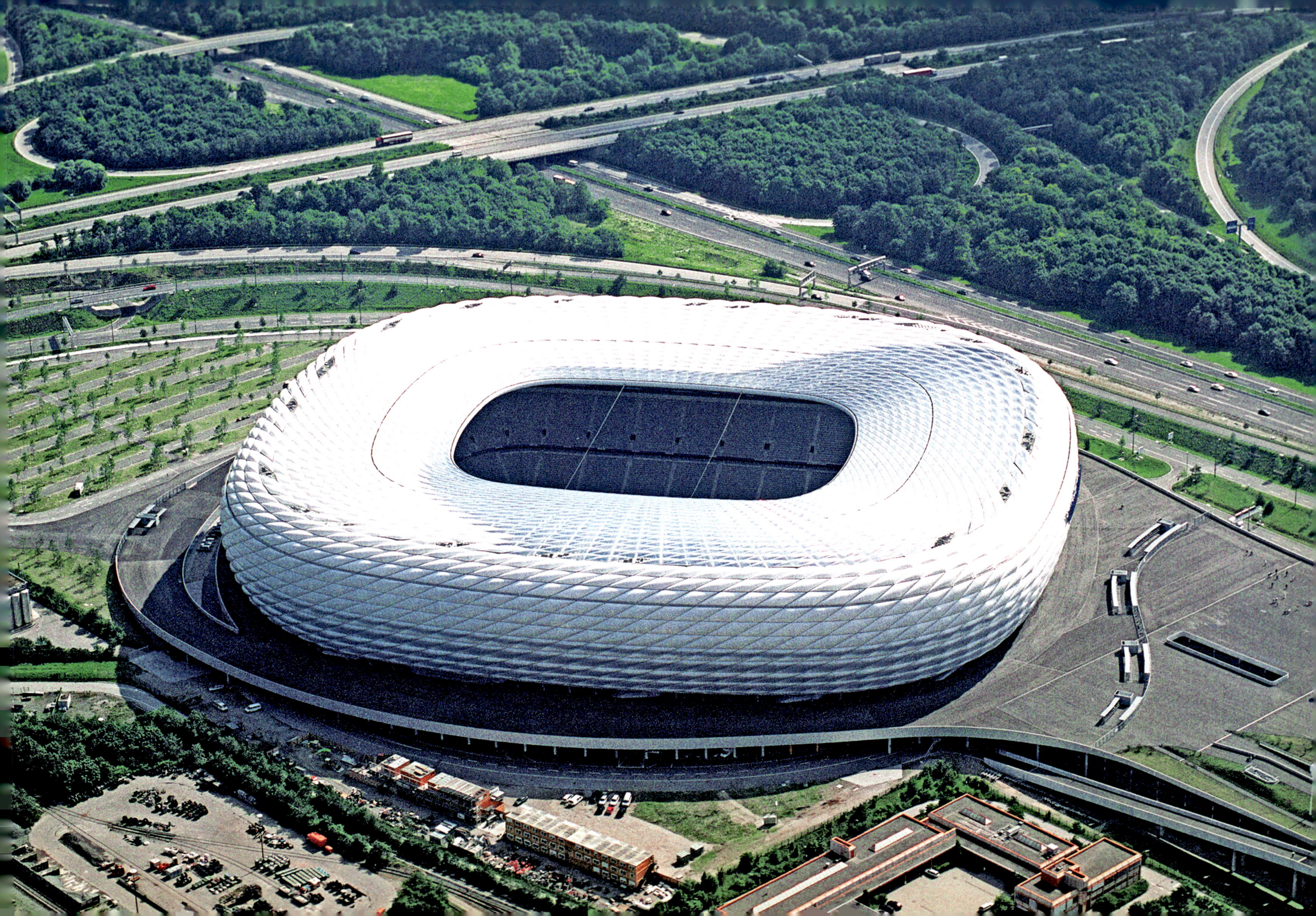

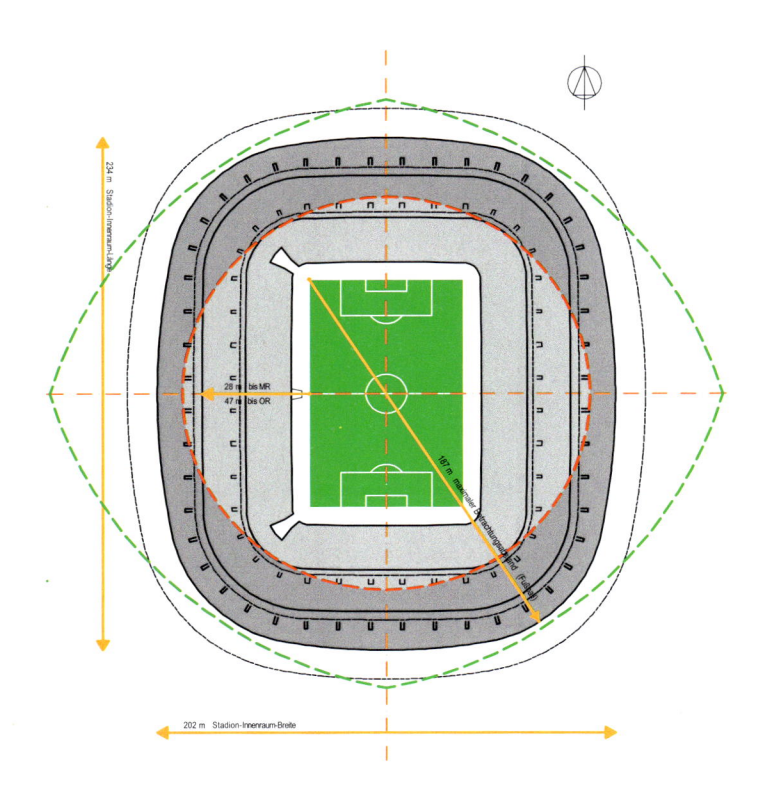

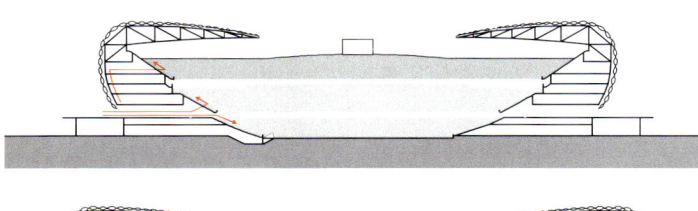

Allianz Arena, Munich – Three-tier stadium (football)
Rounded / expanded rectangular geometry
with two circulating box / horizontal bands

Capacities

Gross capacity (international)	65,600 places [1]	100 %
Gross capacity (national)	66,000 places [2]	
Standing places	20,000 places [2]	
Seats available for World Cup 2006 (net)	55,800 places [1]	
Within opt. 90-m viewing circle (football)	31,800 places	48.5 %
Up to the max. recomm. 190-m distance	+33,800 places	52.5 %
Lower tier	20,000 places	
Middle tier	24,000 places	
Upper tier	22,000 places	
VIP guests (total)	3,449 places [1]	
Executive seats (World Cup 2006)	2,152 places [1]	
Number of private boxes (total)	100 private boxes / 8 special boxes	
Box seats honorary tribune	1,297 box seats	
Media representatives (World Cup 2006)	200 commentators' positions (south)	
	400 television observers	
	1,000 press seats (with desk)	
	1,000 press seats (without desk)	
Wheelchair places	200 places [1]	

Dimensions (approx.)
Stadium interior

202 x 234 m (total)
84 x 121 m = 10,100 m²

Orientation of playing field minus 14° in north/south direction

Width of roof $b = 71$ m max.
Height above pitch $h = 45$ m
Height above terrain (eaves) $h = 35$ m

Roof area $A = 64,000$ m²
Roof $= 38,000$ m²
Façade $= 26,000$ m²
Drainage Slope toward interior $= 7.5°$

1) according to LOC World Cup 2006, Frankfurt (status: October 2005)
2) according to www.stadionwelt.de (status: October 2005)

615 **Viewing circle and dimensions** (plan)
Allianz Arena, Munich, Scale 1:3,000
616 **Structure and circulation**
(transverse / longitudinal section)

page 323:
617 **Picture of the stadium interior** (2006)

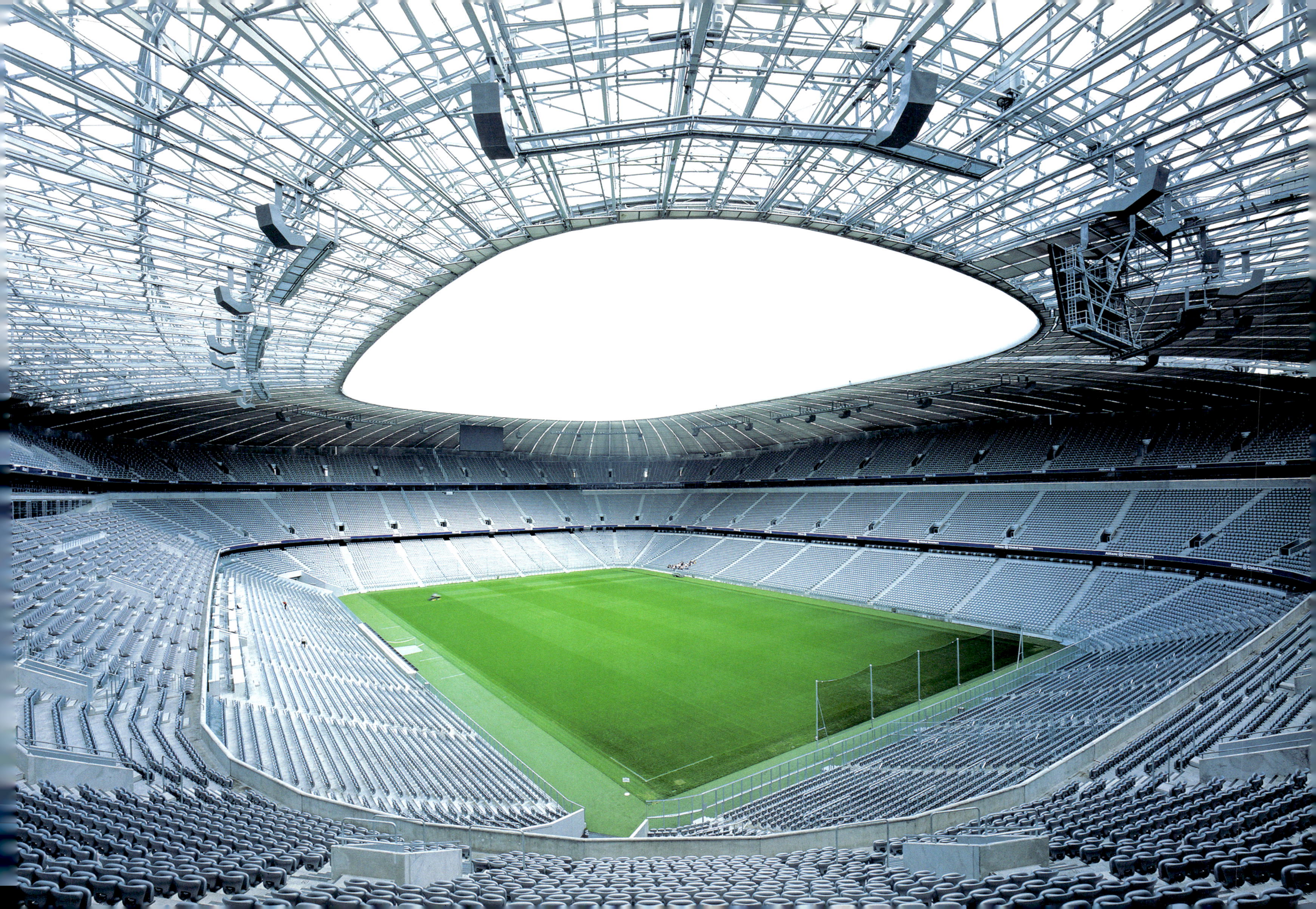

24 Grandstand profiles of built examples

01. Pneumatic PTFE membrane cushions
02. Hinged pillar (+ spring member)
03. Primary cantilever truss
04. Lattice ring truss
05. Compressive/tensile bearings
06. Cantilever-beam façade
07. Subroof (partly retractable)

Lower tier

Row number	23 rows + disabled
Tread depth	80 cm
Rise	26.6 – 39.3 cm
Stand inclination (average)	22.5°
Sightline elevation C	10.3 – 10.1 cm

Middle tier

Row number	25 rows + 2 extra rows
Tread depth	80 cm (105 cm)
Rise	43.6 cm (67 cm)
Stand inclination (average)	28.5° (32.5°)
Sightline elevation C	8.4 – 7.0 cm (23.2 cm)

Upper tier

Row number	22 – 28 rows
Tread depth	80 cm
Rise	54.6 – 58.0 cm
Stand inclination (average)	35°
Sightline elevation C	8.3 – 8.2 cm

→ Undulation!

Distance of stand (front edge)

Distance to field	8.00 m

Viewing distance (to the point of focus)

Maximum horizontal distance	66.65 m
Maximum vertical distance	40.65 m

Distance of boxes (touch line) 47.00 m

Viewing angle (front touch line)

Lower tier	11.0° – 19.0°
Middle tier	23.5° – 27.0°
Upper tier	30.0° – 31.5°

618 World Cup Stadium Munich
Cross section of west stand
Scale 1:500

Lower tier

Row number	28 rows
Tread depth	80 cm
Rise	25.8 – 39.3 cm
Stand inclination (average)	22°
Sightline elevation C	10.3 – 10.6 cm

Middle tier

Row number	29 rows
Tread depth	80 cm
Rise	43.8 – 51.6 cm
Stand inclination (average)	31°
Sightline elevation C	8.4 – 7.5 cm

Upper tier

Row number	22 – 28 rows
Tread depth	80 cm
Rise	54.6 – 57.0 cm
Stand inclination (average)	35°
Sightline elevation C	8.4 – 7.5 cm

→ Undulation!

Distance of stand (front edge)

Distance to field	8.00 m

Viewing distance (to the point of focus)

Maximum horizontal distance	64.20 m
Maximum vertical distance	39.00 m

Viewing distance (maximum)

Football	187.50 m

Viewing angle (front goal line)

Lower tier	11.0° – 19.0°
Middle tier	23.5° – 27.0°
Upper tier	30.0° – 31.5°

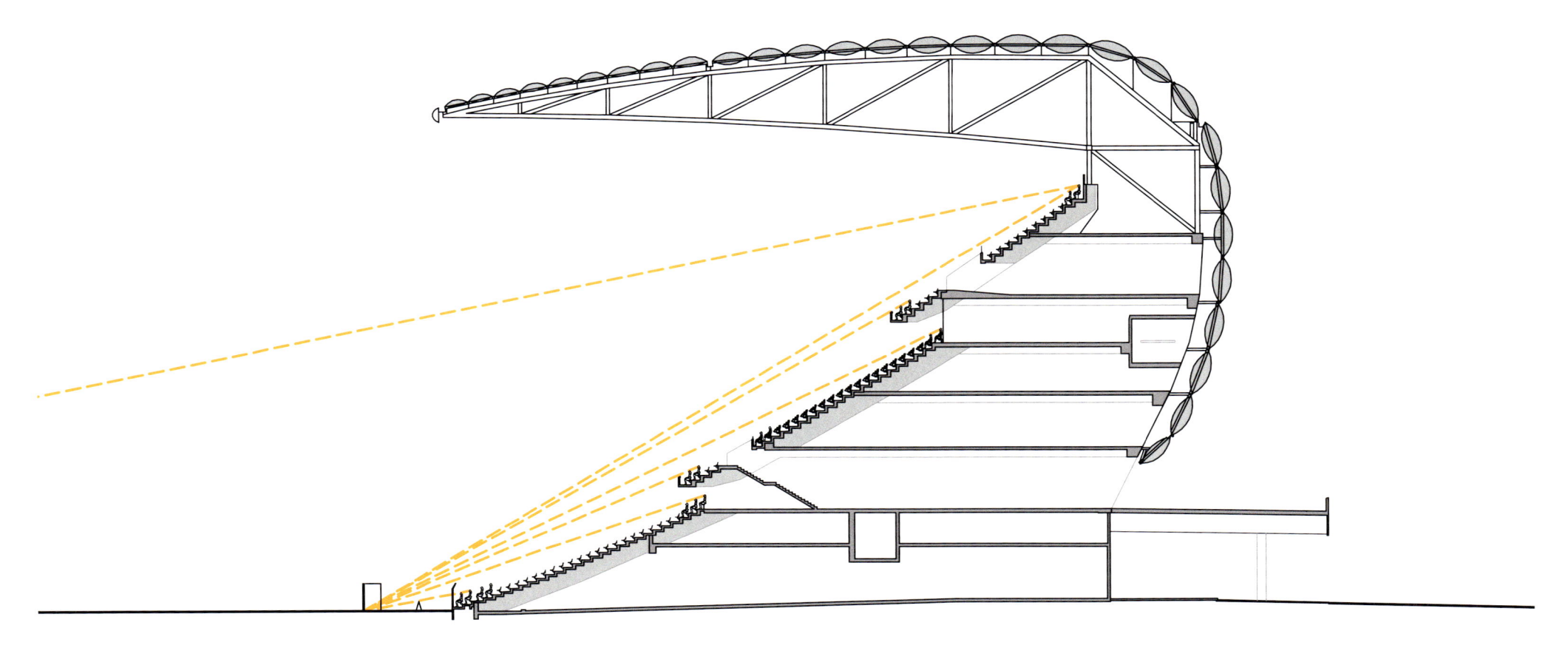

619 World Cup Stadium Munich
Longitudinal section of north stand
Scale 1:500

World Cup Stadium Nuremberg

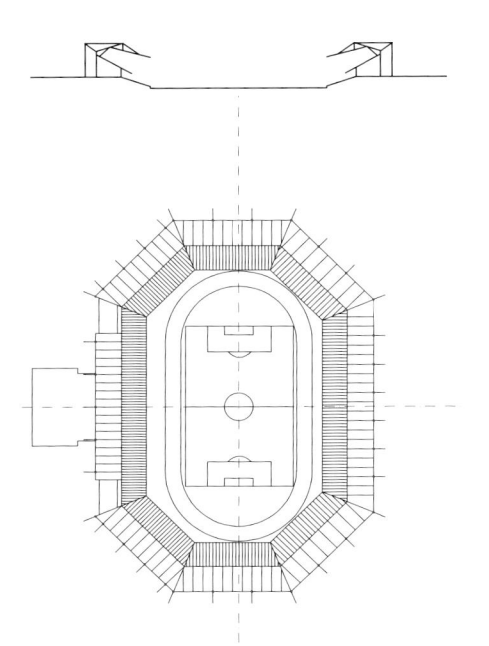

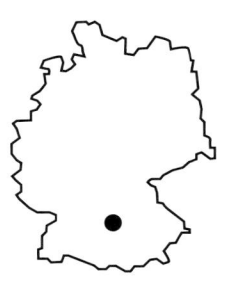

620 **Roof framework** (sectional view)
World Cup Stadium Nuremberg
621 **Roof framework** (top view)
World Cup Stadium Nuremberg
622 **Historic aerial view**
World Cup Stadium Nuremberg
623 **Geographic location** of the city of Nuremberg

page 327:
624 **Aerial view** (2006)

The 'Städtisches Stadion Nürnberg' [a municipal stadium] was inaugurated in 1928 as an exemplary model of Bauhaus architecture by Otto Ernst Schweizer. In 1963, additional reinforced concrete stands were erected with a total capacity of 56,500.

In 1986, an idea competition was initiated for the conversion into a modern multipurpose arena for 60,000 spectators. After the stadium disasters of Brussels and Bradford, the conversion and installation of temporary seats was to reduce the 7,910 standing places by half.

In November 2000, the city council confirmed the modernization and new construction of a supplementary functional building during the regular football league. Additionally, the construction works comprised an expansion of the SW and NW stands to a capacity of 44,300, including the improvement of viewing conditions for 4,000 spectators in the lower seating rows through lowering the pitch by 1.30 m. Fan facilities like vending kiosks, WC utilities and a fan hall were also newly installed. Other measures included: a new functional building with VIP area, catering facilities, commentators' headquarters and offices for the LOC 2006; the conversion of the main stand for media facilities, mixed zone, press stand, VIP area and security headquarters; technical services such as power supply, floodlights, video walls, sound system; exterior areas in the stadium surroundings and outer security ring with parking spaces for media representatives and VIPs. In around 17 months (11/2003 to 04/2005) the construction work was completed up to Confederations Cup 2005.

The new draft took up the historical form of an elongated rectangle, supplementing a second framed-on tier on the circulating wall including cover and service ring. The existing running track was preserved and the capacity raised to 43,800 spectators, 39% of which (17,000) placed within the optimum viewing circle. There are no seats beyond the maximum 190-m area. The largest viewing distance at 187 m is quite comparable to pure football geometries. The stadium interior dimensions, however, are 108 x 178 m, which makes it the smallest of the three athletics stadia (Berlin/Stuttgart/Nuremberg). The total number of visitors remains within the 130-m viewing circle which is considered optimal for athletics (cf. EN/DIN 13200-1).

Viewing restrictions in the stands arise only when a standard advertising hoarding is erected at the minimum distance (FIFA). In this case, the view toward the touch line is restricted in the entire lower tier. In spite of a linear rise of 34,7/85 cm, a very comfortable 'C' value of 21.5–18.2 cm on 11 + 2 rows of the upper tier remains. The large athletics distance of 19.85/36.55 m causes the 'C' value in the lower tier to fall only to 17.5–11.0 cm (linear rise 44.4/80.0 cm).

In the private box area at main stand west, the upper tier projects by 5.0 m, which limits the clear height of the box tier to circa 1.55 m. Due to the inclined stand there are no problems with headroom, but standing spectators have, immediately after stepping out of the box, no unrestricted view to the pitch centre, at a height of 15.0 m. All boxes lie within the 90-m best viewing radius. Viewing conditions of the long sides differ slightly as the playing field is not located centrally within the stadium bowl.

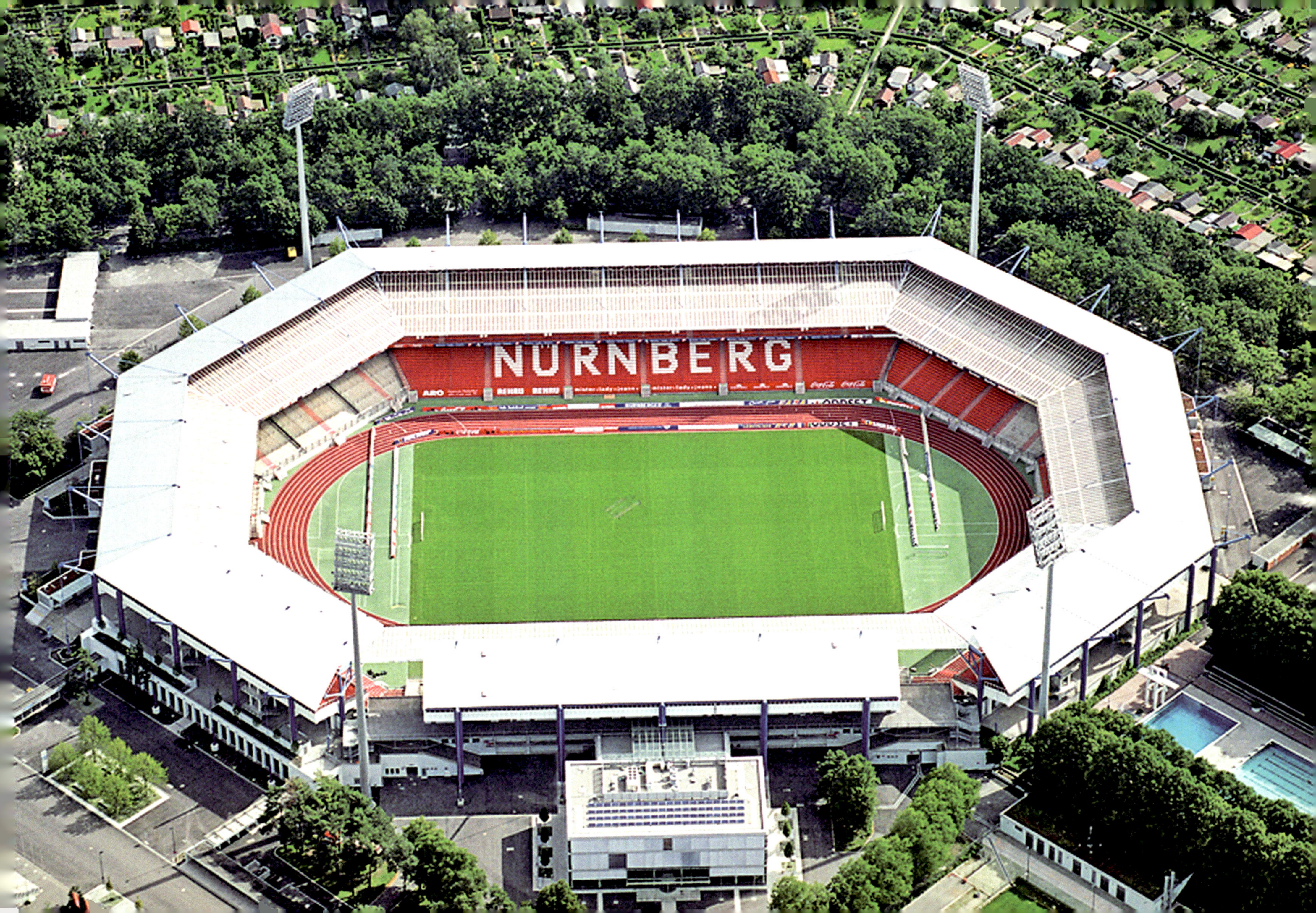
NÜRNBERG

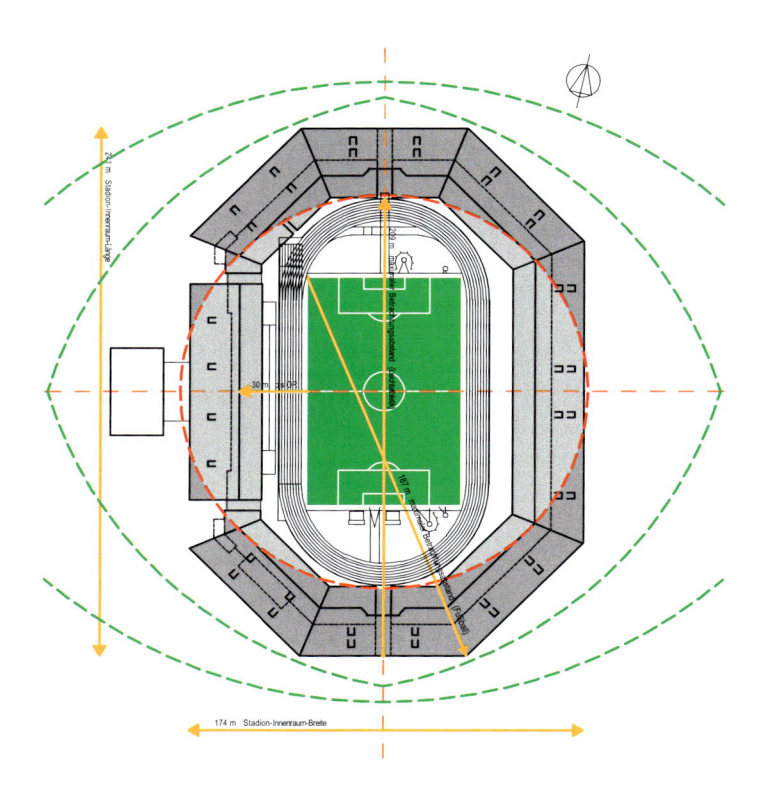

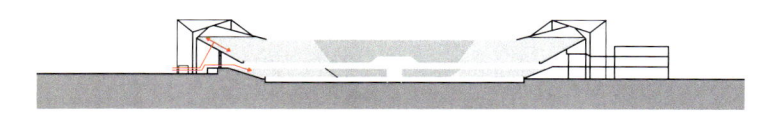

World Cup Stadium Nuremberg – Two-tier stadium (athletics)
Oblong octagonal geometry with a circulating horizontal band

Capacities

Gross capacity (international)	43,800 places [1]	100%
Gross capacity (national)	46,800 places [2]	
Standing places	7,800 places [2]	
Seats available for World Cup 2006 (net)	39,300 places [1]	
Within opt. 90-m viewing circle (football)	17,000 places	39.0%
Up to the max. recomm. 190-m distance	+ 26,800 places	61.0%
VIP guests (total)	621 places [1]	
Executive seats (World Cup 2006)	441 places [1]	
Number of private boxes (total)	14 private boxes	
Box seats, honorary tribune	180 box seats	
Media representatives (World Cup 2006)	150 commentators' positions (south)	
	200 television observers	
	400 press seats (with desk)	
	200 press seats (without desk)	
Wheelchair places	83 places [1]	

Dimensions (approx.)
Stadium interior — 272 x 207 m (total)
108 x 178 m = 19,100 m²

Orientation of playing field — minus 13° in north/south direction

Width of roof — $b = 32$ m max.
Height above pitch — $h = 19.5$ m
Height above terrain (eaves) — $h = 16.5$ m

Roof area — $A = 42,400$ m²
Drainage — Slope toward interior/exterior
Interior roof = 14°
Exterior roof = 16°

1) according to LOC World Cup 2006, Frankfurt (status: October 2005)
2) according to www.stadionwelt.de (status: October 2005)

625 **Viewing circle and dimensions** (plan)
World Cup Stadium Nuremberg, Scale 1:3,000
626 **Structure and circulation**
(transverse/longitudinal section)

page 329:
627 **Picture of the stadium interior** (2006)

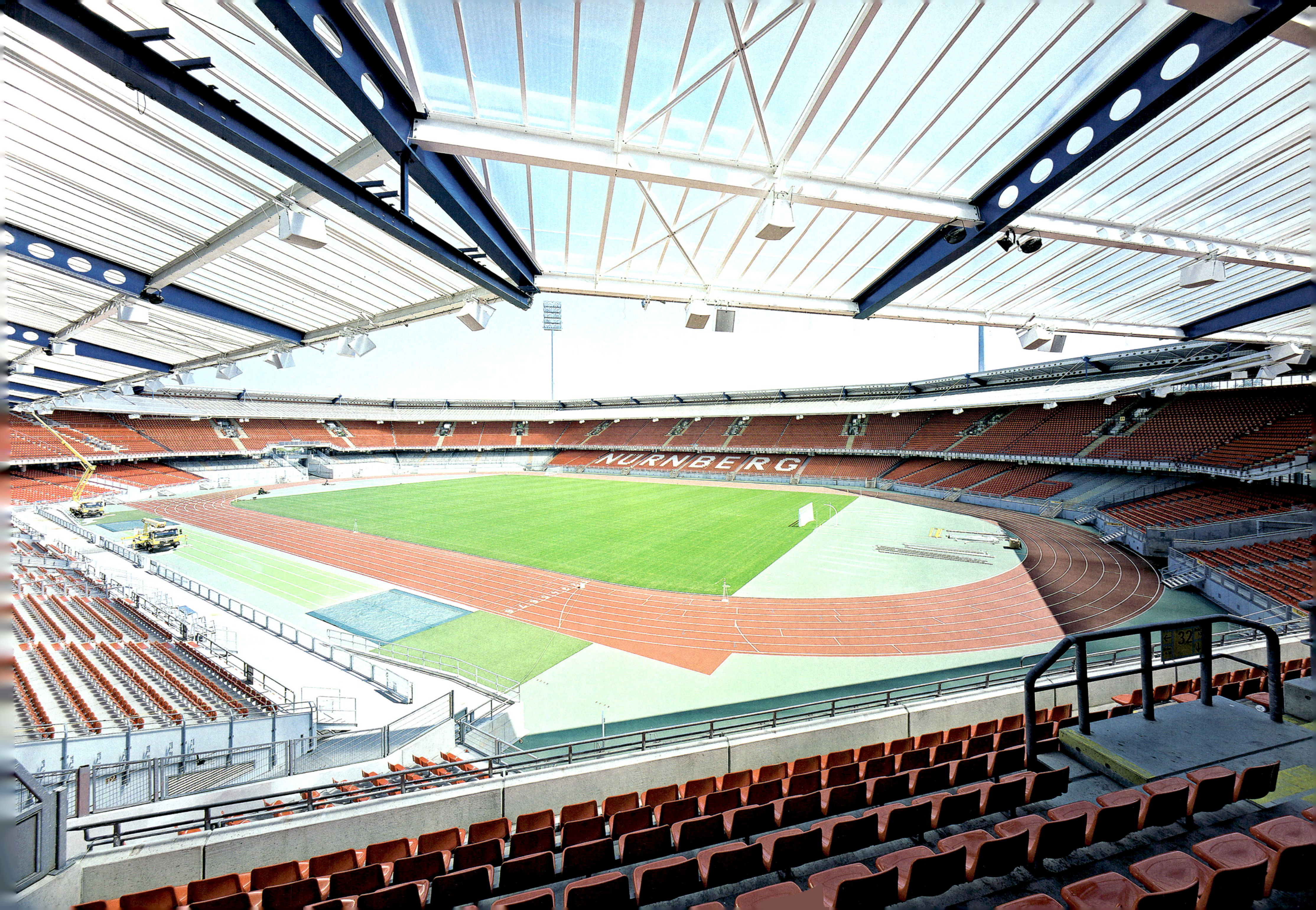
NURNBERG

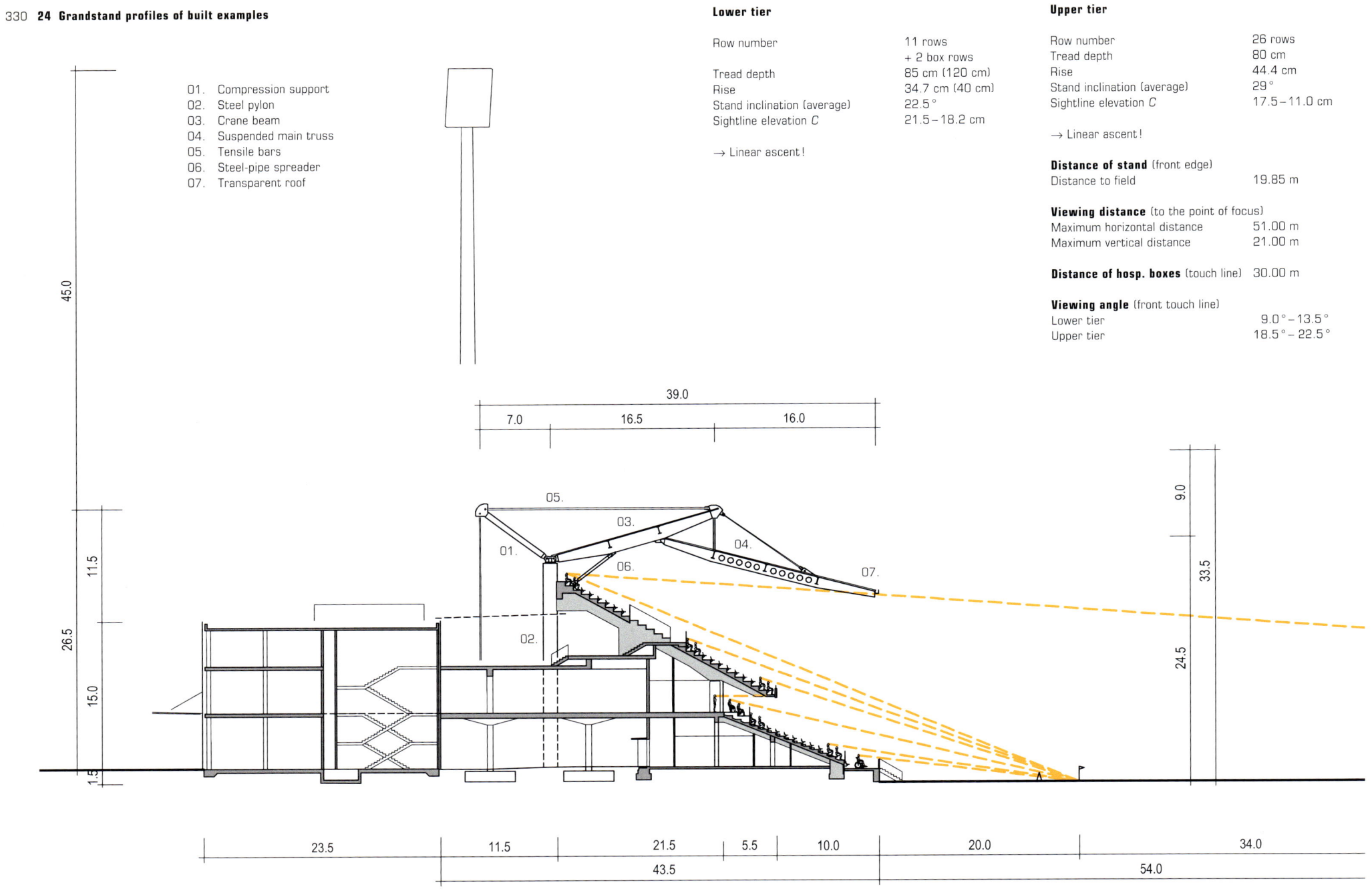

330 **24 Grandstand profiles of built examples**

01. Compression support
02. Steel pylon
03. Crane beam
04. Suspended main truss
05. Tensile bars
06. Steel-pipe spreader
07. Transparent roof

Lower tier

Row number	11 rows + 2 box rows
Tread depth	85 cm (120 cm)
Rise	34.7 cm (40 cm)
Stand inclination (average)	22.5°
Sightline elevation C	21.5 – 18.2 cm

→ Linear ascent!

Upper tier

Row number	26 rows
Tread depth	80 cm
Rise	44.4 cm
Stand inclination (average)	29°
Sightline elevation C	17.5 – 11.0 cm

→ Linear ascent!

Distance of stand (front edge)
Distance to field — 19.85 m

Viewing distance (to the point of focus)
Maximum horizontal distance — 51.00 m
Maximum vertical distance — 21.00 m

Distance of hosp. boxes (touch line) — 30.00 m

Viewing angle (front touch line)
Lower tier — 9.0° – 13.5°
Upper tier — 18.5° – 22.5°

628 **World Cup Stadium Nuremberg**
Cross section of west stand
Scale 1 : 500

Lower tier

Row number	19 rows
Tread depth	78 cm
Rise	26 cm
Stand inclination (average)	18°
Sightline elevation C	19.7–15.0 cm

→ Linear ascent!

Upper tier

Row number	26 rows
Tread depth	85 cm
Rise	44.4 cm
Stand inclination (average)	29°
Sightline elevation C	26.9–19.2 cm

→ Linear ascent!

Distance of stand (front edge)

Distance to field	36.55 m

Viewing distance (to the point of focus)

Maximum horizontal distance	66.75 m
Maximum vertical distance	21.00 m

Viewing distance (maximum)

Football	187.00 m
Athletics	209.00 m

Viewing angle (front goal line)

Lower tier	4.5°–8.0°
Upper tier	12.5°–17.5°

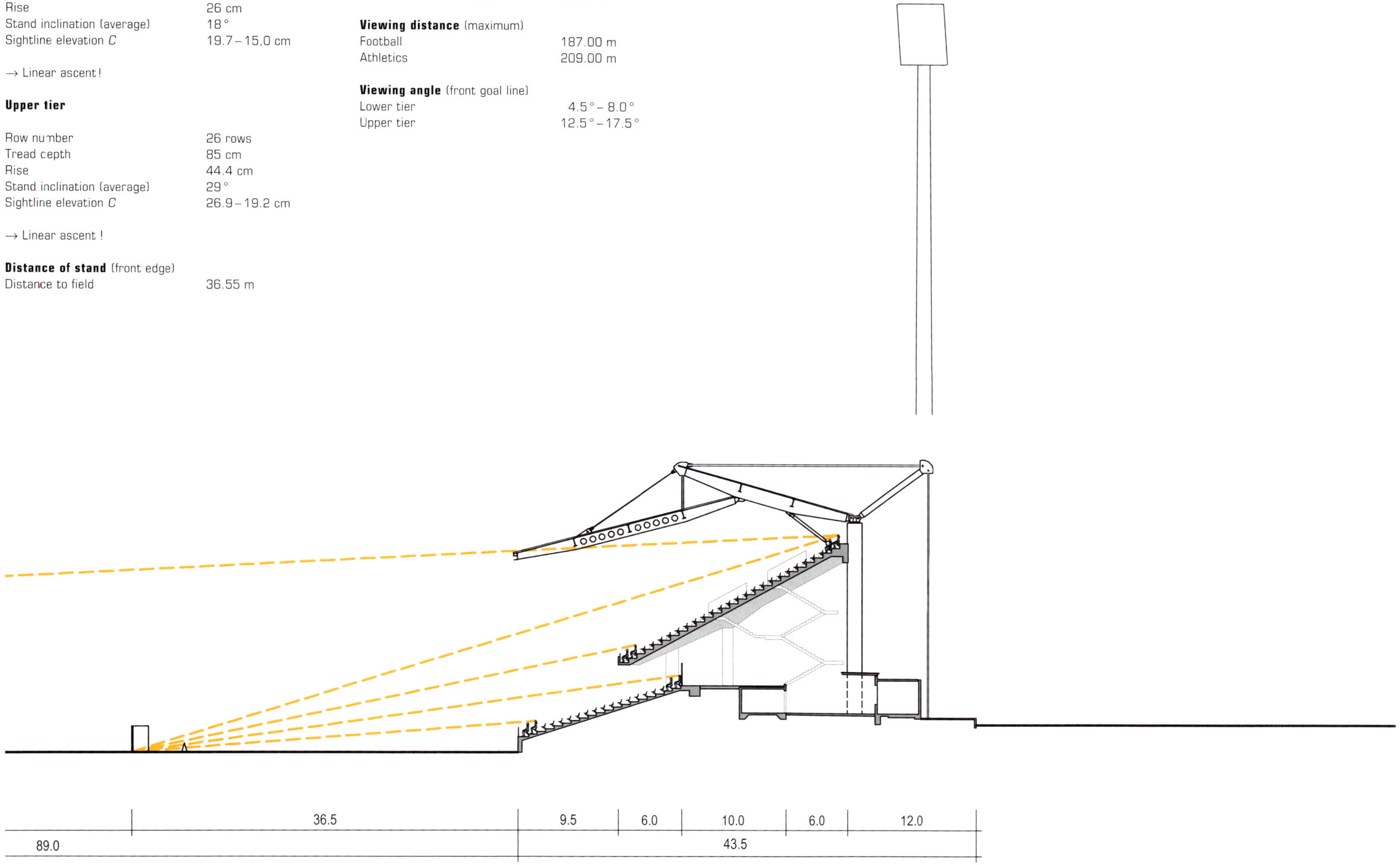

629 World Cup Stadium Nuremberg
Longitudinal section of south stand
Scale 1:500

24 Grandstand profiles of built examples

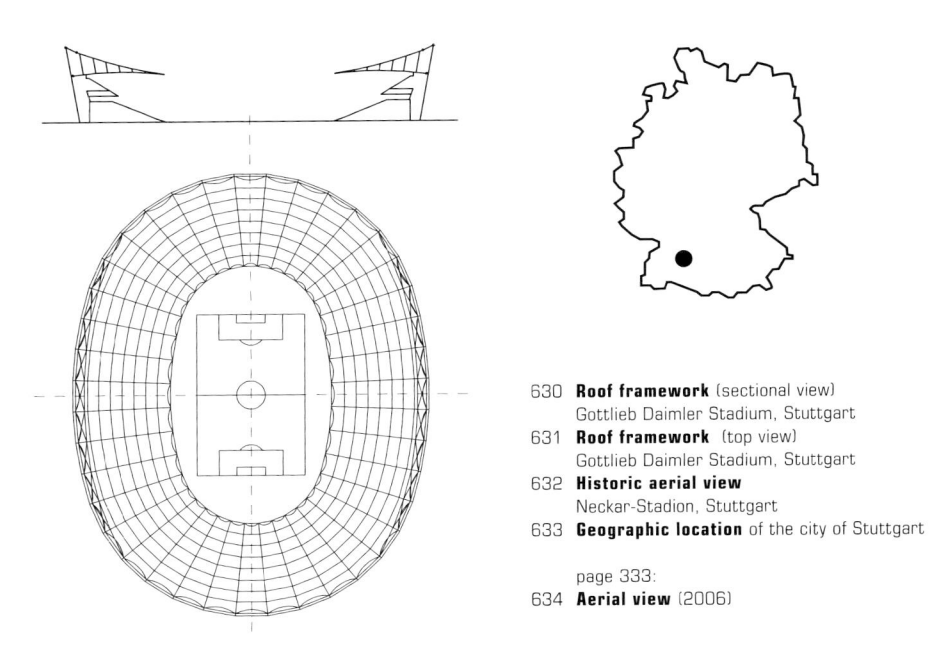

World Cup Stadium Stuttgart

630 **Roof framework** (sectional view)
Gottlieb Daimler Stadium, Stuttgart
631 **Roof framework** (top view)
Gottlieb Daimler Stadium, Stuttgart
632 **Historic aerial view**
Neckar-Stadion, Stuttgart
633 **Geographic location** of the city of Stuttgart

page 333:
634 **Aerial view** (2006)

The former 'Neckar-Stadion' was built in 1933 according to the plans of architect Paul Bonatz. Various conversions since 1949 expanded the stadium several times and adapted it to the changed demands of a multipurpose sports complex (first floodlight system in 1963, Video full-matrix scoreboard.)

Since 1993, the stadium carries the name Gottlieb-Daimler Stadium. It was renovated and made World-Cup-compatible in three stages.

Conversion stage 1 1992–1993: complete membrane covering, planning and execution in 18 months, construction time for cable net circa 3 weeks.

Conversion stage 2 1999–2003: completion of a second tier (5,600 persons) at the main stand and new construction of a business centre; conversion of bench seats into single tip-up seats; 44 private boxes; 1,500 executive seats and a multi-storey car park. With the installation of two upper-tier stands on the long sides the original one-tier stadium is raised in gross capacity to 53,100. The short sides remain open.

With the first stadium membrane covering in 1993, the cable net structure set standards. Enough space was left under the roof for a later supplementation of the upper tier stands, which follows the height profile of the cable net membrane of the roof, adapting its elevation to it. Thus, the number of rows in the upper tier undulates from 16 to 29. Due to the large terrace distances of 17.75/38.70 m, only one third of the spectators sit in the 90-m optimum viewing circle. For athletics geometries larger viewing distances are admissible. Only 5% of the stadium occupants are outside this 130-m viewing circle, comparable in number to the Berlin Olympic Stadium. The stands have linear rises (33/80 cm or 52/82 cm), which allow an economical factory-built construction. The higher the starting value for sightline elevation, the later the limit for lowering of the 'C' value is reached.

For the upper tier of the main stand, in spite of a maximum seating height of 28.40 m, 'C' value is lowered from 23.8 cm to only 16 cm as the horizontal distance with 65.70 m is quite large. This is a clear advantage of an athletics stadium compared to pure football geometries, although at a certain expense in atmosphere, which can only be compensated by raising capacity (as in Berlin with 74,000 spectators).

The shallow base visual angle of around 2–5° (2–9° Berlin/Nuremberg) requires an adjustment of the position of the advertising hoarding as a 13° visual angle is necessary at the distance of 4.0 m (long side) recommended by FIFA to view over it freely toward the point/line of focus.

Extensive modernization and conversion measures during running operations were carried out within 24 months (01/2004 to 12/2005).

At a desired World Cup capacity of 48,000 seats (national), the third phase was divided into two stages: dismantling and new construction of the reverse stand (2,150 additional seats in the upper tier with improved circulation) as well as installation of catering and functional rooms, façade cleaning and crowd-control system, new payment and access areas, extensive renovation including sanitary facilities, new catering facilities and attractive milling areas.

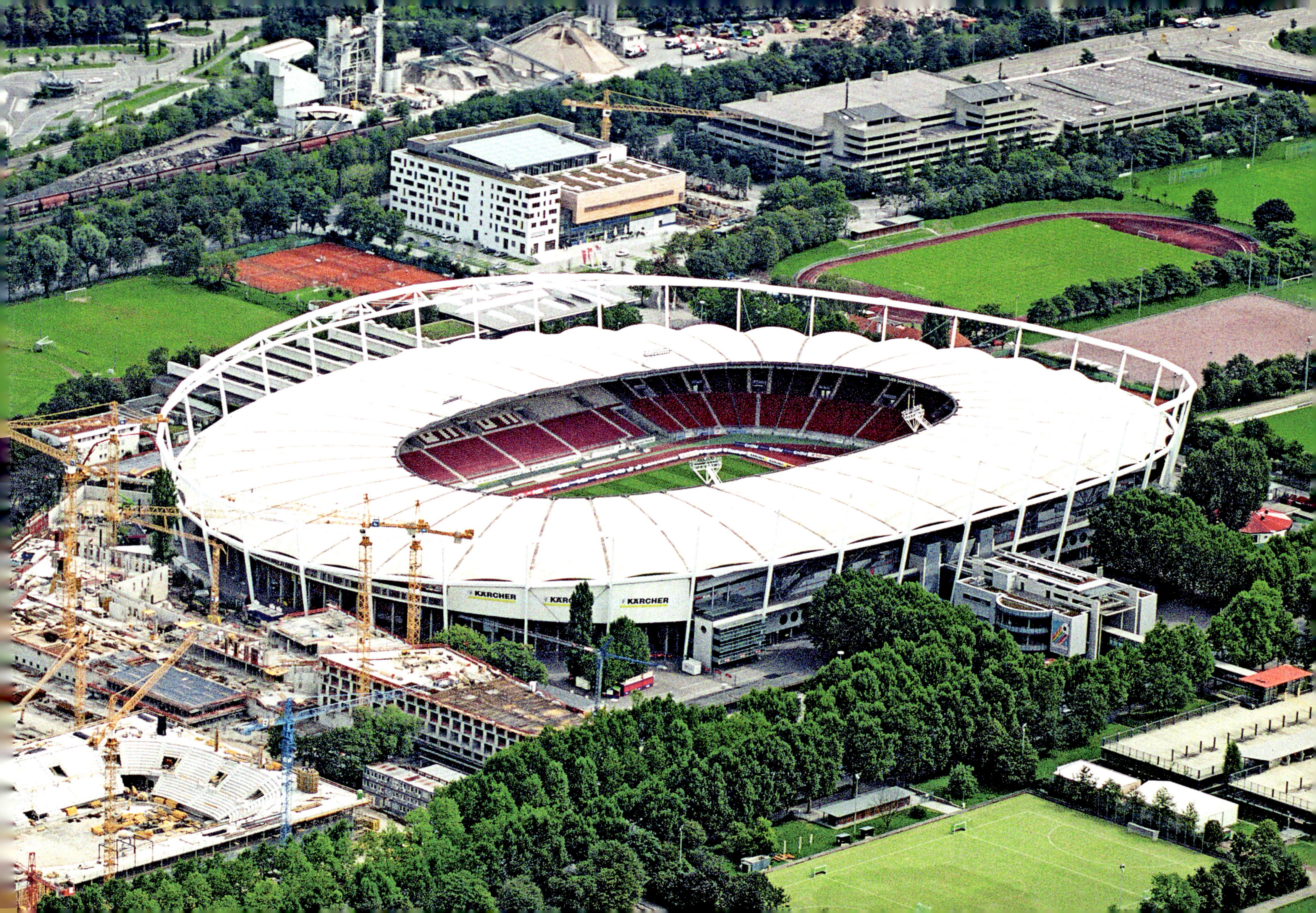
KÄRCHER
KÄRCHER

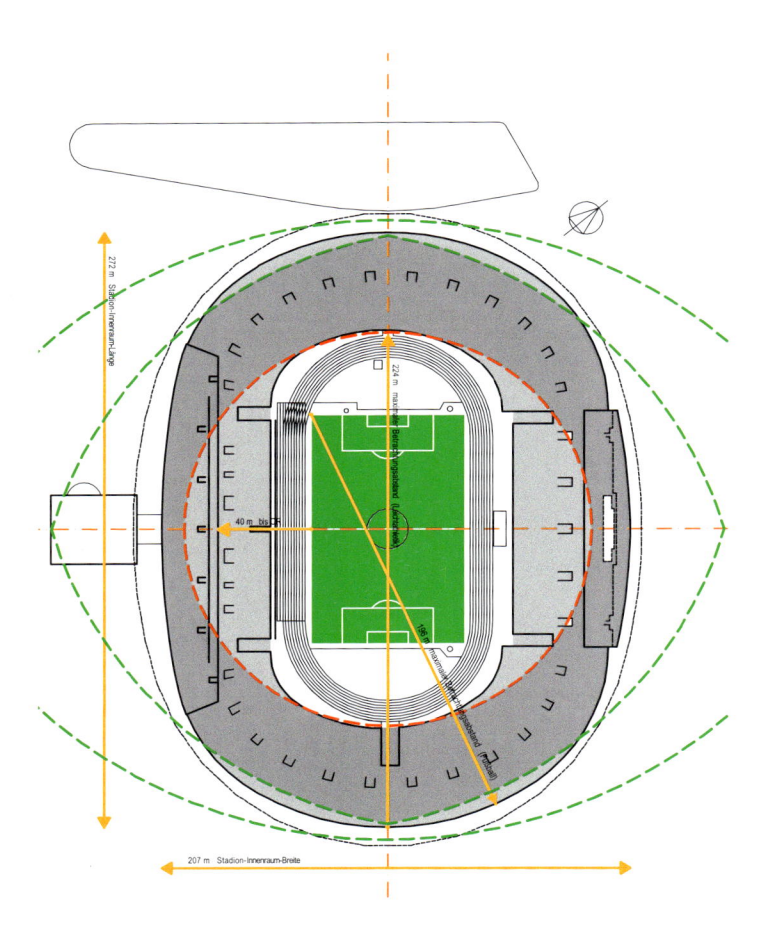

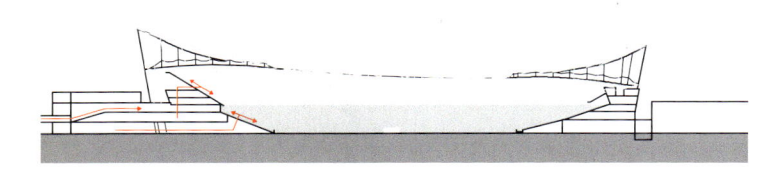

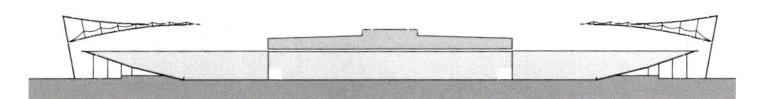

Gottlieb Daimler Stadium, Stuttgart – Two-tier stadium (athletics)
Asymmetric semicircular geometry with two-sided upper tier stand

Capacities

Gross capacity (international)	53,100 places [1]	100%
Gross capacity (national)	57,000 places [2]	
Standing places	4,300 places [2]	
Seats available for World Cup 2006 (net)	37,100 places [1]	
Within opt. 90-m viewing circle (football)	14,900 places	28.0%
Within opt. 130-m viewing circle (athletics)	+50,500 places	95.0%
Up to the max. recomm. 230-m distance	+2,600 places	5.0%
VIP guests (total)	2,182 places [1]	ca. 4,400 m²
Executive seats (World Cup 2006)	1,504 places [1]	
Number of private boxes (total)	44 private boxes / 4 special boxes	
Box seats, honorary tribune	678 box seats	
Media representatives (World Cup 2006)	150 commentators' positions (south)	
	400 television observers	
	700 press seats (with desk)	
	700 press seats (without desk)	
Wheelchair places	198 places [1]	

Dimensions (approx.) 207 x 272 m (total)
Stadium interior 103 x 182 m = 18,900 m²

Orientation of playing field minus 51° in north/south direction

Width of roof $b = 57$ m max.
Height above pitch $h = 22.5$ m
Height above terrain (eaves) $h = 20 – 29$ m

Roof area $A = 34,000$ m²
Weight per unit area $G = 12$ kg/m²
Drainage Slope toward exterior = $3 – 15°$

[1] according to LOC World Cup 2006, Frankfurt (status: October 2005)
[2] according to www.stadionwelt.de (status: October 2005)

635 **Viewing circle and dimensions** (plan)
Gottlieb Daimler Stadium, Stuttgart, Scale 1:3,000
636 **Structure and circulation**
(transverse/longitudinal section)

page 335:
637 **Picture of the stadium interior** (2006)

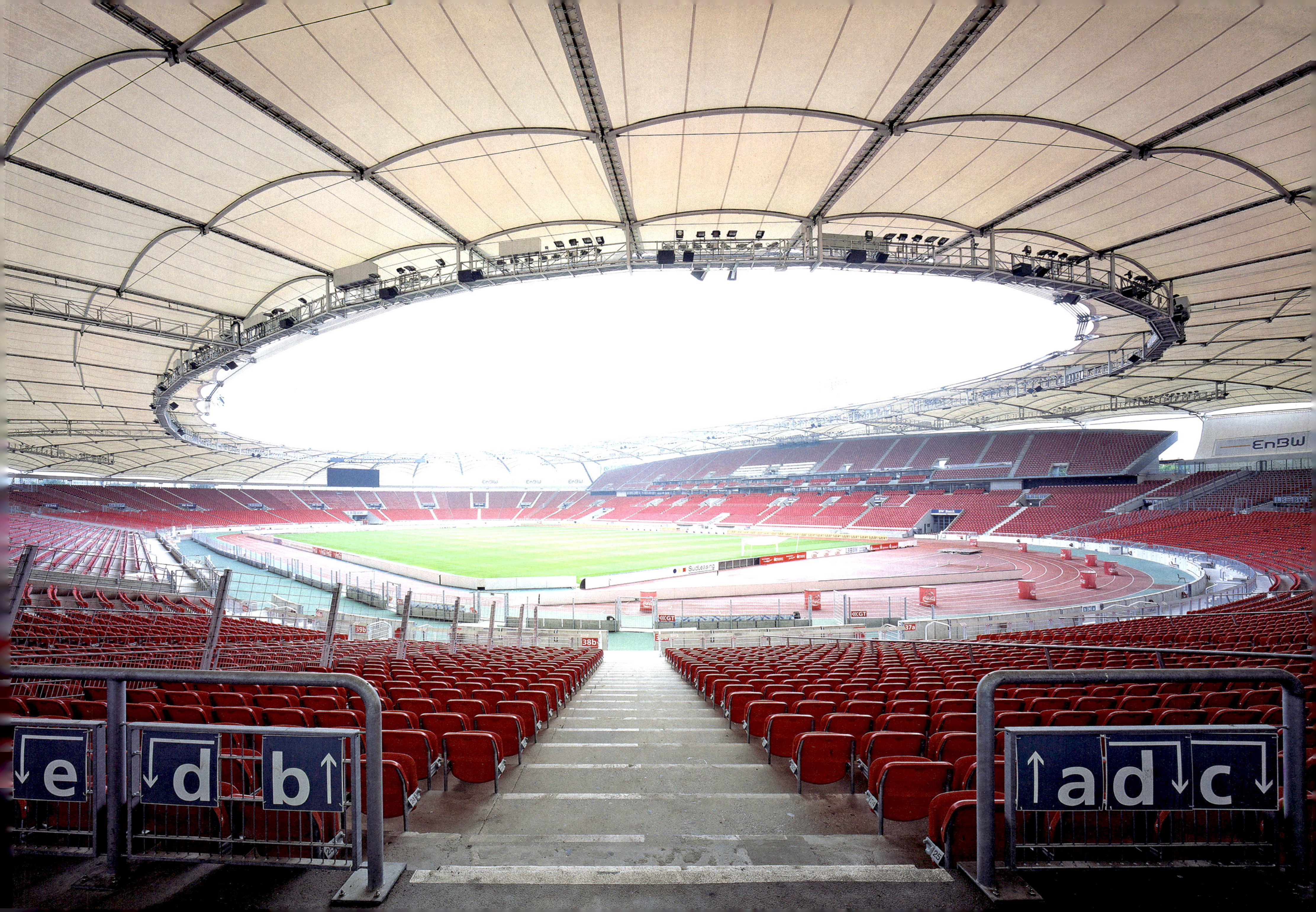
EnBW
Sudleasing
e
d
b
a
d
c

24 Grandstand profiles of built examples

01. Compression ring (top)
02. Bracing (long side)
03. Steel pylon
04. Compression ring (bottom)
05. Cable truss
06. PVC membrane roof
07. Steel-pipe arched girder
08. Ring traction cable (inside)
09. Tree-like support (tier extension)

Lower tier

Row number	25 rows + 2 box rows
Tread depth	85 cm below vomitory 80 cm above vomitory
Rise	20.0 cm 1–3 rows 35 cm below vomitory 33 cm above vomitory
Stand inclination (average)	14°/22.5°
Sightline elevation C	12.3/16.0/9.7 cm

→ 3 Linear ascents!

Upper tier

Row number	16–29 rows
Tread depth	82.0 cm
Rise	52.0 cm
Stand inclination (average)	33°
Sightline elevation C	23.8–16.0 cm

→ Linear ascent!
→ Undulation!

Distance of stand (front edge)

Distance to field 17.75 m

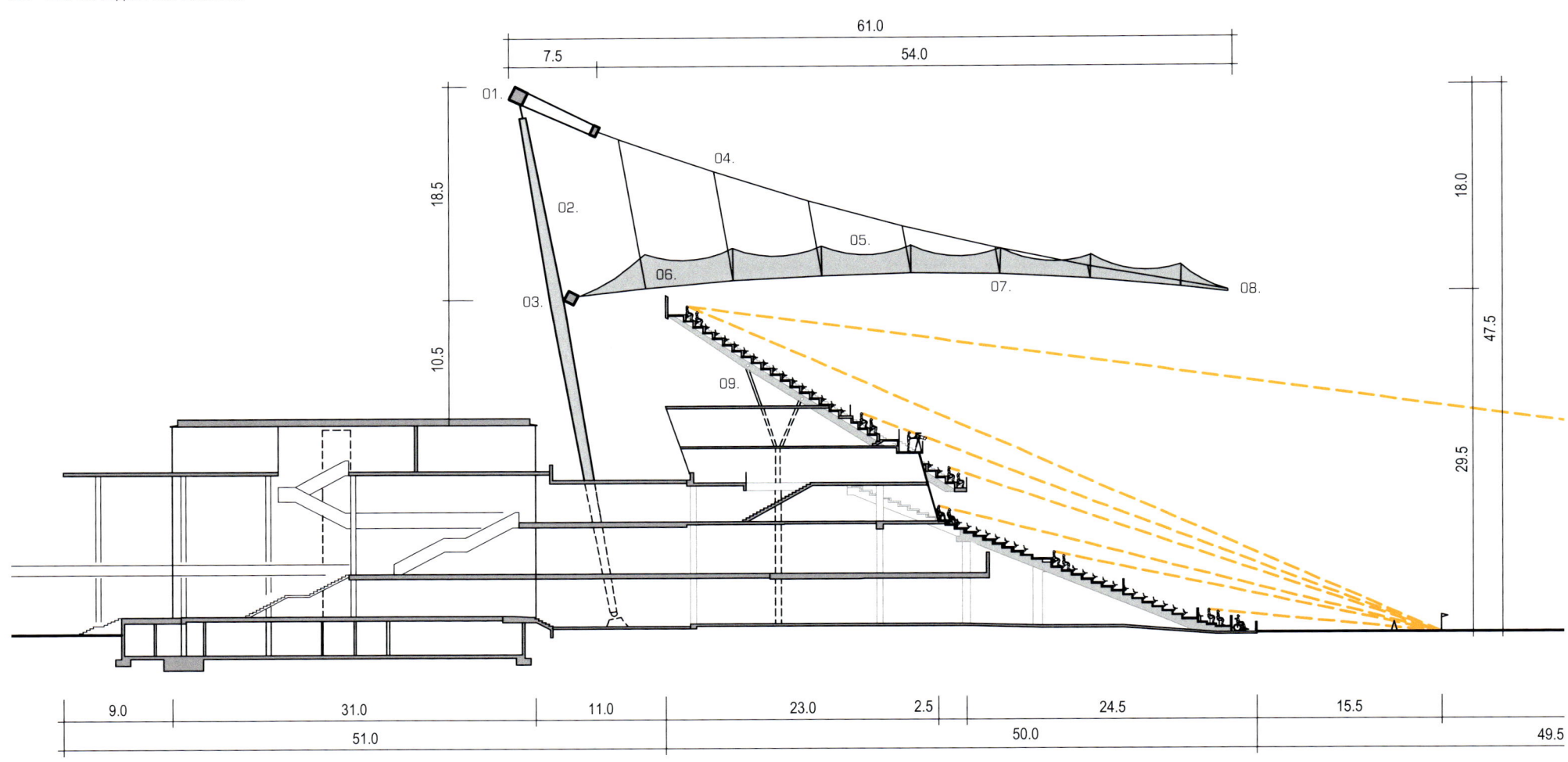

Viewing distance (to the point of focus)

Maximum horizontal distance 65.70 m
Maximum vertical distance 28.40 m

Distance of hosp. boxes (touch line) 40.00 m

Viewing angle (front touch line)
Lower tier 5.5°–14.5°
Upper tier 19.0°–24.0°

World Cup Stadium Stuttgart
Cross section of west stand
Scale 1:500

Lower tier

Row number	25–27 rows
Tread depth	85/95/78 cm
Rise	21.1–28.8 cm
Stand inclination (average)	17.5°
Sightline elevation C	10.4–14.9 cm

→ Undulation!

Distance of stand (front edge)

Distance to field	38.70 m

Viewing distance (to the point of focus)

Maximum horizontal distance	82.40 m
Maximum vertical distance	14.10 m

Viewing distance (maximum)

Football	207.50 m
Athletics	239.00 m

Viewing angle (front goal line)

Lower tier	2.0°
Upper tier	9.5°

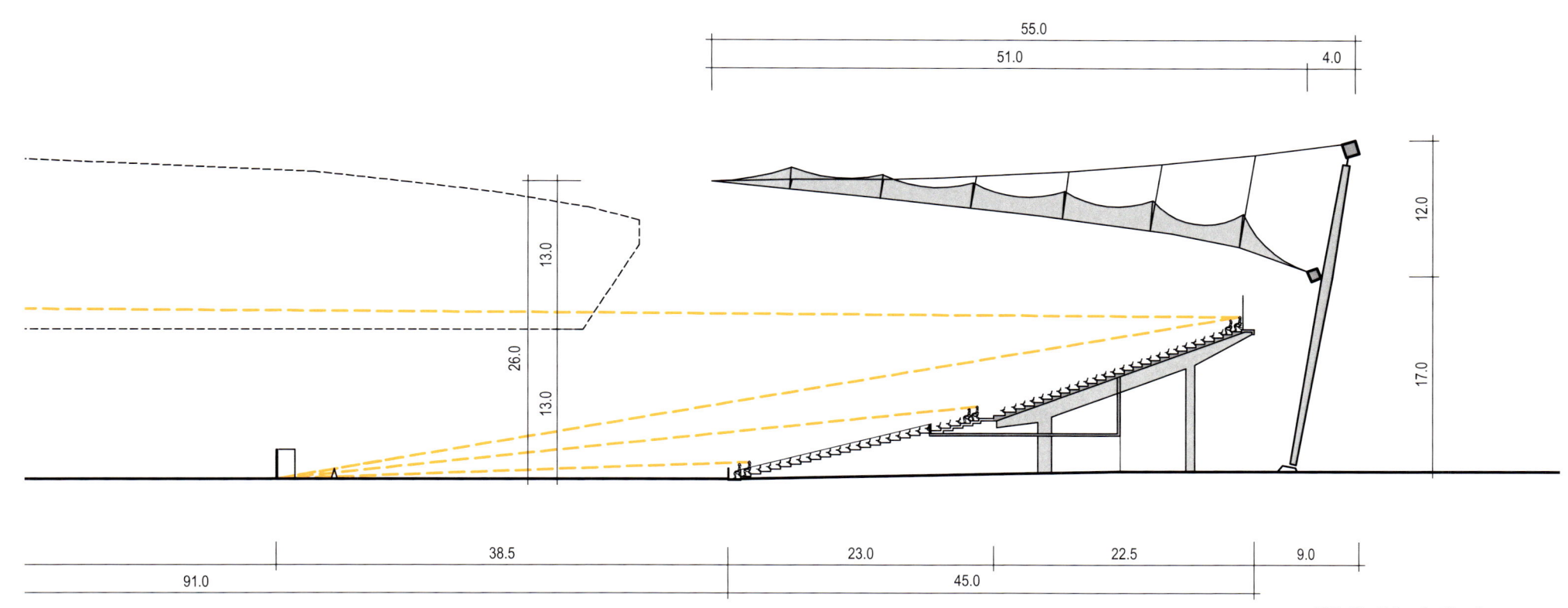

World Cup Stadium Stuttgart

639 **World Cup Stadium Stuttgart**
Longitudinal section of north stand
Scale 1:500

Chapter 25

Table N
World Cup capacities

	International		National (total)	Seats		VIP seats			Percentage of seats	
	Seats only	Standing places	With standing places	Usable (net)	Usable (%)	Total VIP seats	Executive seats	Priv. box tribune	Honorary tribune	Priv. box tribune
01 Berlin	74,200	./.	74,200	63,400	85.4%	5,653	4,758	895	6.4%	1.2%
02 Dortmund	65,900	27,500	81,250	56,400	85.6%	1,946	1,784	162	2.7%	0.2%
03 Frankfurt	48,400	9,300	52,300	41,100	84.9%	3,120	2,218	902	4.6%	1.9%
04 Gelsenkirchen	53,600	16,300	61,500	46,700	87.1%	3,255	2,438	817	4.5%	1.5%
05 Hamburg	51,300	8,900	55,800	43,300	84.4%	2,759	2,179	580	4.2%	1.1%
06 Hanover	44,900	7,200	49,800	38,800	86.4%	1,551	1,241	310	2.8%	0.7%
07 Kaiserslautern	47,700	12,000	48,500	40,000	83.9%	1,327	993	334	2.1%	0.7%
08 Cologne	46,300	9,200	50,300	38,200	82.5%	2,685	2,089	596	4.5%	1.3%
09 Leipzig	44,300	./.	44,300	37,100	83.7%	2,969	2,660	309	6.0%	0.7%
10 Munich	65,600	20,000	66,000	55,800	85.1%	3,449	2,152	1,297	3.3%	2.0%
11 Nuremberg	43,800	7,800	46,800	39,300	89.7%	621	441	180	1.0%	0.4%
12 Stuttgart	53,100	4,300	57,000	45,400	85.5%	2,182	1,504	678	2.8%	1.3%
Total	639,100	122,500	687,750	545,500	./.	31,517	24,457	7,060	./.	./.
Average	53,258	12,250	57,313	45,458	85.4%	2,626	2,038	588	3.7%	1.1%
Minimum	43,800	4,300	44,300	37,100	./.	621	441	162	./.	./.
Maximum	74,200	27,500	81,250	63,400	./.	5,653	4,758	1,297	./.	./.

640 **Table N: Overview of stadium capacities national / international**

Final evaluation and compendium
An assessment summary of the twelve German World Cup stadia and selected stadia in Korea/Japan, Portugal, Austria/Switzerland, South Africa

Stadium evaluation

Best viewing radius

Every stadium draft has to provide for an efficient management of seating capacities, independent of form and geometry.

Hence, the following percentage values represent gross reference figures for 'international' seating capacities in relation to the optimum 90-m viewing circle.

A comparison of the figures shows that multi-purpose stadia with running tracks have a significantly lower share of their capacities (22.5 to 39%) placed within the best viewing radius, as compared to football stadia with 40 to 70%.

Athletics complexes require on average around 6,200 m² more central surface area compared to football stadia with an average value of 11,200 m².

Note: These extra space requirements were often the main reason for football-oriented stadium operators to decide against athletics tracks, thus relinquishing the option of hosting athletics competitions.

EN/DIN 13200-1 (B.1) determines the viewing distance in relation to the type of sports. For athletics events (group A): 190 m (recommended) and 230 m (maximum admissible value).

Note on athletics: Complementing the optimum viewing circle of 90 m for football geometries amounts to a 130-m best viewing radius for athletics stadia.

From an athletics perspective, World Cup Stadium Berlin has 60% instead of 22.5% of seats placed within the optimum viewing circle (Stuttgart: 95% instead of 28%). Only 4–5% of seats are outside of the maximum viewing distance of 230 m.

This clearly shows that stadia can only be adequately compared in the context of the relevant disciplines or stadium types. A primary assessment of viewing qualities with reference to football is recommended, based on the majority of such utilizations of the stadium interior.

Distance to the playing field

The distances of the stands to the pitch in football stadia range from 6 m to 15 m; only half that of athletics stadia with 17.5 m to 24 m.
The distance to the goal line is between 6.5 m to 17 m, whereas they are 2.5 times higher for athletics stadia with 36.5 m to 42 m.

The omission of the running track at World Cup stadium Cologne allowed bringing the 'Südtribüne' football enthusiasts on the short side around 34 m closer to the action on the pitch.

Note: Sightline factor D is further quantified in the framework of the parameter studies.

page 339:
641 **Comparison of all form types** (World Cup 2006)
(red) 90-m viewing radius (marking)
(green) 190-m circle for maximum football viewing distance
(green) 230-m circle for maximum athletics viewing distance

Stadium evaluation 339

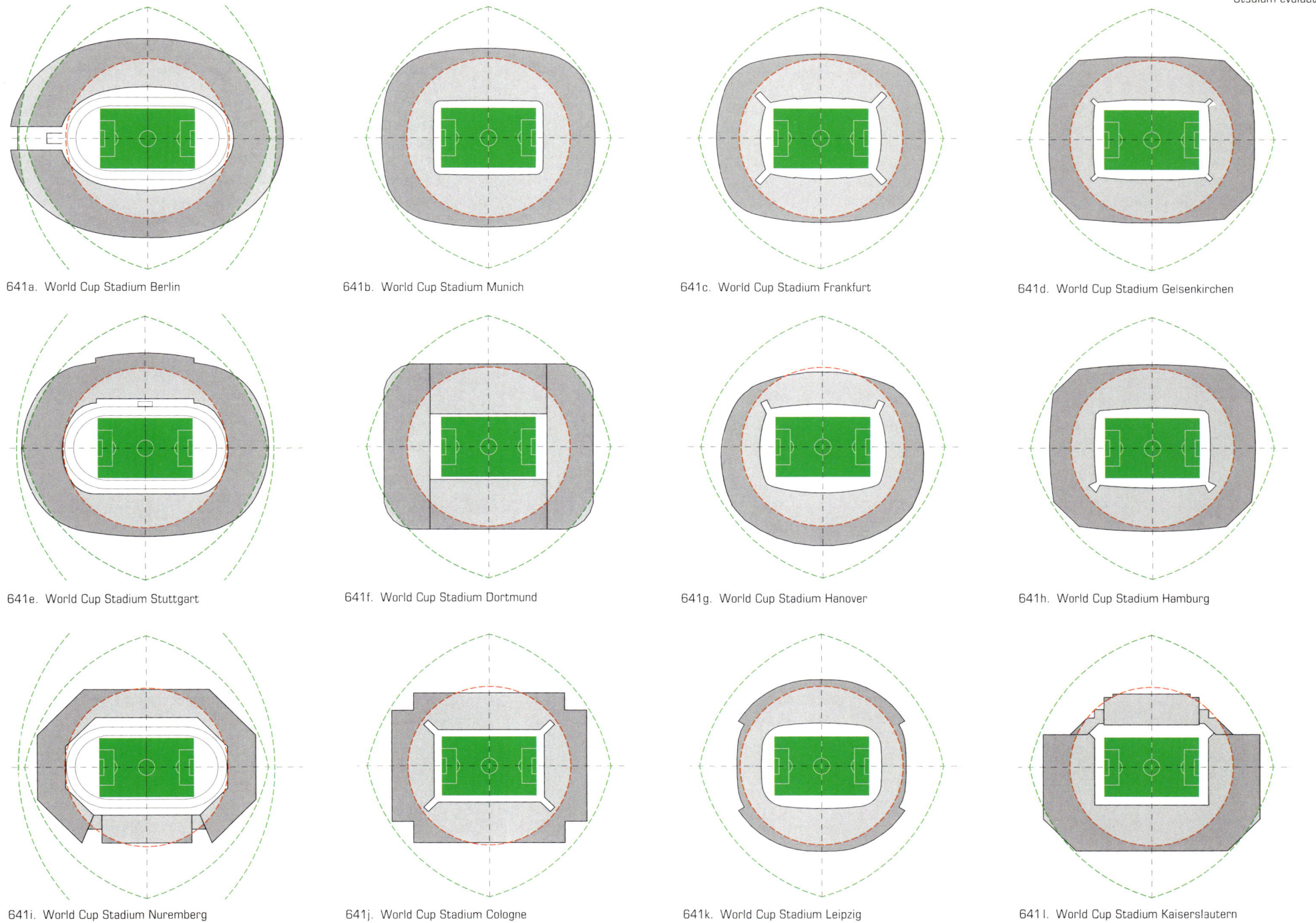

641a. World Cup Stadium Berlin

641b. World Cup Stadium Munich

641c. World Cup Stadium Frankfurt

641d. World Cup Stadium Gelsenkirchen

641e. World Cup Stadium Stuttgart

641f. World Cup Stadium Dortmund

641g. World Cup Stadium Hanover

641h. World Cup Stadium Hamburg

641i. World Cup Stadium Nuremberg

641j. World Cup Stadium Cologne

641k. World Cup Stadium Leipzig

641l. World Cup Stadium Kaiserslautern

Number of rows (block definition)

An analysis of the twelve stadia of World Cup 2006 shows that lower tiers generally consist of 13 to 43 rows, and upper tiers accommodate between 18 and 35 rows (44 rows in Dortmund, corner areas).
Lower-tier average: 28 rows
Upper-tier average: 26 rows

The number of rows might vary for the upper tier of stadia with an undulating terrace edge, i. e. when the profile of steppings is non-affine to the overall shape, depending on the difference of their non-affine image.

The more circular the boundary of the rear terrace edge in comparison to the rectangular pitch or the straight/curved profile of the step front edges, the more rows can be accommodated.
Stuttgart: long side 14 rows
Hanover: long side 10 rows
Leipzig: long side 6 rows
Munich: long side 6 rows

Block definition: MVStättV 2005 (§ 10 cl. 4) defines the size of a seating block with a maximum of 30 rows. Between and behind the blocks, gangways with a minimum width of 1.20 m are to be provided. In 20 % of all World Cup grandstands up to 42/44 rows can be found, which is made possible by adequate fire protection (Berlin), reasonable deviations from specifications or by a structural separation into two blocks (Dortmund) with separate circulation systems. Multiplying a value of 30 rows with a maximum of 40 seats between two gangways (uncovered) amounts to a maximum of 1,200 persons within one block. Setting this reference figure as maximum value, then 40 rows are feasible, inasmuch as the number of people seated in one row is limited to 30 persons. (Interpolation of intermediate values is possible.)

Note on block formation: The built examples illustrate that the block size should not be limited by the number of seating rows but by a maximum number of 1,200 persons to be accommodated therein.

Tread depth

An analysis of the World Cup stadia has come up with two reference sizes for minimum tread depth:
– 80 cm for regular seating accommodation
– 90 cm for special seating accommodation such as media or executive areas
– 85 cm as expansion measure for fittings such as safety stirrups or crush barriers

At times, the measures for honorary guest areas can be higher (100/105 cm); a decision that largely depends on space available and the demands on comfort on the part of the stadium operator.

Rises

The steppings of a lower-tier stand at the examined World Cup stadia are between 22 and 44 cm high; the largest rises for athletics geometries are around 33 cm. Upper-tier risers in athletics stadia start from 29.5 cm up to a maximum of 52 cm.

The starting value for upper-tier football geometries is on average 53.5 cm, which is higher than the maximum value for athletics. Football grandstands commonly have a limit value for safety stirrups of 50 cm. Depending on their capacity, steep terraces tend to exploit the maximum rise.[1]

Note: The built examples are generally limited to a rise of 57 cm, with maximum values of 59 cm in exceptional cases. This indicates a general feasibility of gangway rises of up to 20 cm, per step x 3 = 60 cm maximum. Standing places are almost exclusively located in the more shallow-inclined lower tier.

The link between sightline and rises has been pointed out sufficiently; required capacities of large sports venues are limited significantly by a maximum rise of 19 cm per step. When proximity to the pitch is desired, sightline quality in the upper seating rows can hardly be maintained. Moreover, large distances produce larger grandstands. On the other hand, profitability aspects tend to prohibit excessive construction volumes; therefore, compact stadium structures are generally preferred.

Stand inclination

The vertical profile of a stand arises from the ratio of the selected tread depth and the riser height determined by the sightline geometry. Average lower-tier gradients of 18° for athletics grandstands differ significantly from those of football geometries, due to a considerably increased distance to the playing field (long sides: 20/10 m, short sides: 40/11 m). Lower tiers in football stadia have an average gradient of 23.5° (ex.: 40/80 cm = 0.5 = arc tan 26.5°). The median gradient of an upper tier varies from 27° (athletics) to 34° (football).

The following values can be used for predimensioning purposes:
Lower tier: approx. 20° to 24°, maximum 27°
Upper tier: approx. 30° to 34°, maximum 36°

The planning principles of IAKS (1993) for embankments state that inclined terrain which is

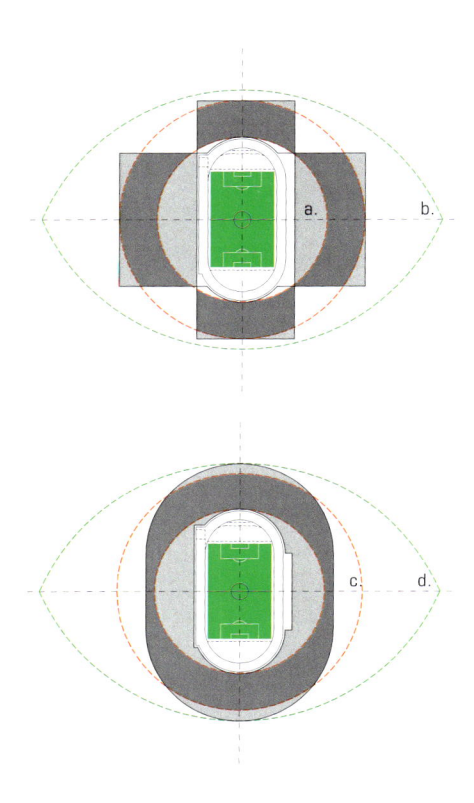

642 **Circles of viewing distances**
 a. Football geometry – recommended: 150-m circle (corner flag) = 90-m circle at centre spot
 b. Football geometry – maximum admissible: 190-m circle (corner flag)
 c. Athletics geometry – recommended: 190-m circle (corner flag) = 130-m circle at centre sport
 d. Athletics geometry – maximum admissible: 230-m circle (corner flag)

occasionally used as spectator facility should not have a slope in excess of 30°.

Sightline elevation

Values for lower-tier terraces average between 4.9 cm to 21 cm for the transverse section; improved by 1 to 2 cm at the curved areas of the longitudinal section. This illustrates the possible effects of an average increase of merely one metre in distance to the goal line compared to the halfway line. Athletics (football) stadia with a 16-cm (11.6 cm) transverse section, or a 17.5-cm (13.2 cm) longitudinal section are much better from a sightline perspective than denser football geometries; another indication for the interrelation of distance and sightline quality.

Note: Here, mean values were compared because individual stadia exhibited very extreme sightline values; for example the transition of tiers without sufficient elevation of the next row. If such values were to be compared, a distorted picture would emerge.

Only at the upper tier do geometric difficulties emerge with regard to maintaining a clear sightline profile. For the upper tier, which generally rises above the private boxes and at a structural height offset, the starting and end points of a sightline are relevant:
Transverse section:
First row, upper tier 5.8 cm – 23.8 cm
Last row, upper tier 4.9 cm – 19.8 cm
Longitudinal section:
First row, upper tier 8.4 cm – 26.9 cm
Last row, upper tier 5.6 cm – 24.8 cm

In all four cases, the highest value relates to an athletics geometry. For almost half of all football geometries, the last rows of an upper tier in the longitudinal section are below the admissible limit of 9 cm. This applies to almost two thirds of all upper-tier stands for the transverse section close to the pitch.

Note: The relevant undercuttings are comprehensible from a planning perspective and are within a tolerance of up to 5 mm. This fact clearly points out that a new evaluation of the maximum limits for rises seams reasonable, in order to maintain a consistent sightline quality, or 'C' value, also for the upper tier stands.

Viewing distances

Limit values to be observed for maximum viewing distances are stated in EN/DIN 13200-1 (B.1). Initially, this makes no reference to actual viewing distances.

Nowadays, proximity to the playing field is generally deemed a quality criterion for a stadium, providing the live experience and the possibility to directly follow the match activity. In the twelve World Cup stadia the horizontal maximum distance to the point of focus (diagonal corner flag) ranges between 167.5 m and 207 m (average value: 188.3 m). Analysis has confirmed the admissible DIN requirement of a maximum of 190 m, independent of the stadium form and capacity. The maximum values for athletics usage are accordingly higher (209 m to 239 m, measured in the longitudinal axis of the running track). The vertical height above the playing field in athletics stadia is significantly lower with 21 m to 31 m compared to pure football stadia with 26.5 m to 40.5 m.

Note: These figures might become relevant in the context of necessary vertical distribution and terrace circulation systems.

Distance to the kick-off spot

The activity surface on the playing field has a size of 7,140 m². The centre spot is chosen, vicariously, for the evaluation of spectator distance to the activity area. For the examined football stadia, the minimum distance ranges from 39 m to 50 m and the maximum distance from 105 m to 120 m. For the three athletics stadia, maximum distances are increased by around 15 % and minimum distances even by around 25 %.

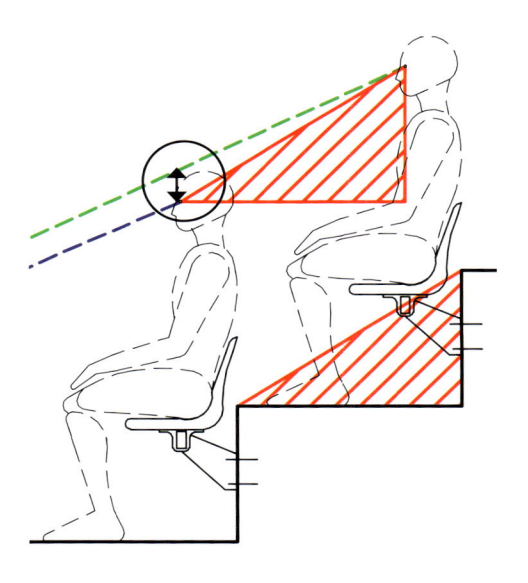

643 Schematic sketch of sightline elevation
The perpendicular distance between the sightlines of two successively seated spectators is measured above the eye of the person sitting in front. This value reflects the sightline quality of a spectator stand.

Visual angle

The effective binocular field of vision (driving regulations) is screened by the involuntary eye movements without head movements:

25° upwards
30° to the right/left
40° downwards

The last aspect on sightlines examines the gradient of the visual angle since it applies for the entire duration of the match, depending on the location (height) of the spectator.

Also in this context, transverse and longitudinal sections are distinguished. A spectator seated in the first row of a stand must tilt his view to the goal line by between 6.3° (athletics) to 12.5° (football).

In the last row, the view has to be slanted twice or three times as much by 14.3° to 18.7°, which means that spectators in the upper tier have to constantly incline their heads by around 10°. Values are around 2–3° lower in relation to the goal line.

As a matter of fact, tilting of the head has determined the viewing habits of stadium or amphitheatre visitors for the last 2,000 years.

644–645 **Comparison of sightline profiles**
2 x 6 World Cup stadia (distinguished by colour)

644. Gelsenkirchen
Leipzig
Berlin
Cologne
Hanover
Nuremberg

page 343:
645. Munich
Hamburg
Frankfurt
Dortmund
Stuttgart
Kaiserslautern

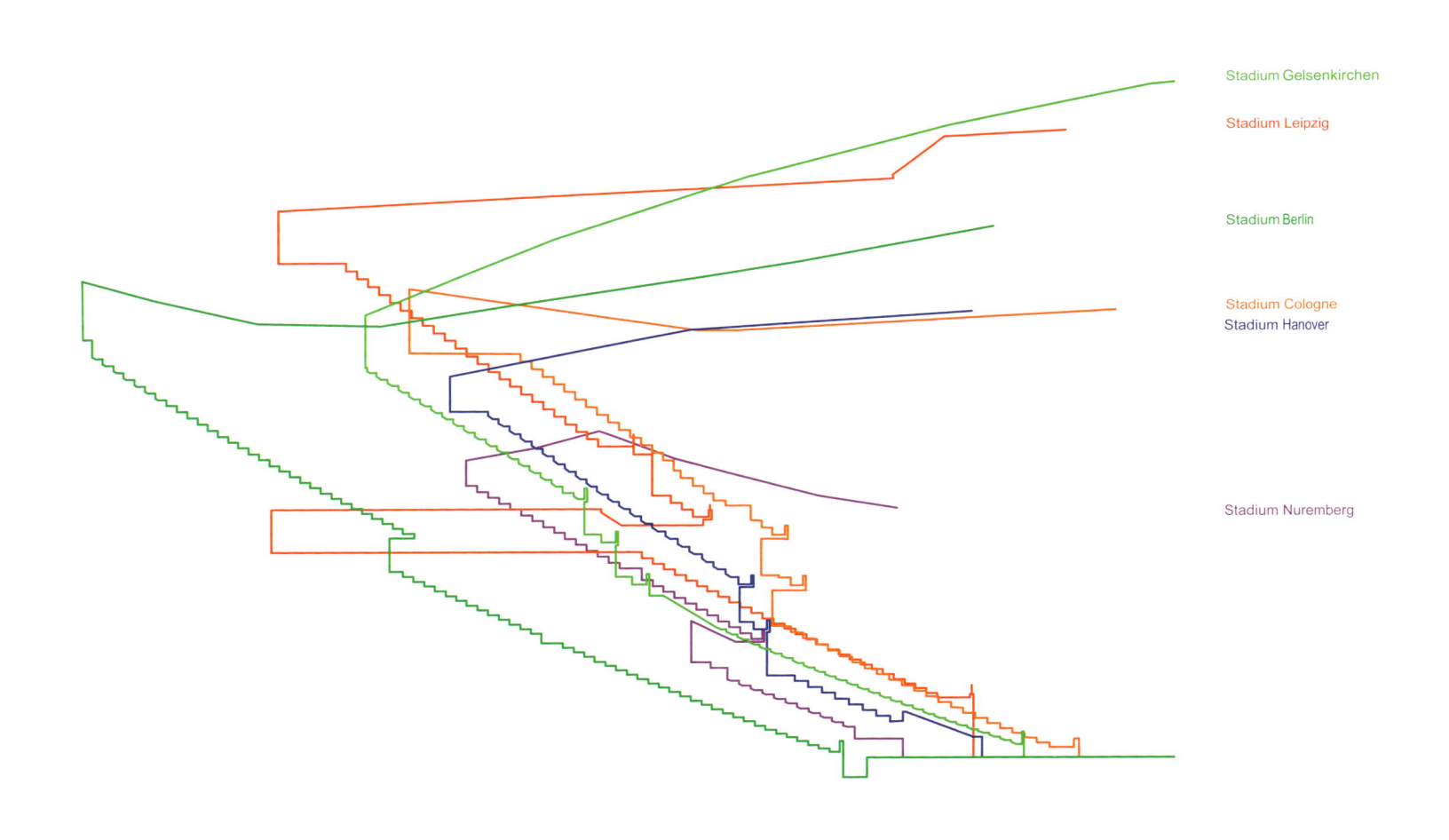

Stadium Gelsenkirchen
Stadium Leipzig
Stadium Berlin
Stadium Cologne
Stadium Hanover
Stadium Nuremberg

Stadium evaluation 343

The roof structure should never be so dominant as to inhibit a relaxed 'wandering' gaze. Spectators should be enabled to experience the entire stadium bowl and truly partake in its atmosphere and comforting effect.

The World Cup stadia in Nuremberg, Dortmund and Kaiserslautern have very low roof structures. Their obvious advantage of improved cover in bad weather is counteracted by almost forcing the view of spectators downward onto the pitch, affecting the overall sense of space. A final evaluation of this particular issue should naturally be left to each individual planner.

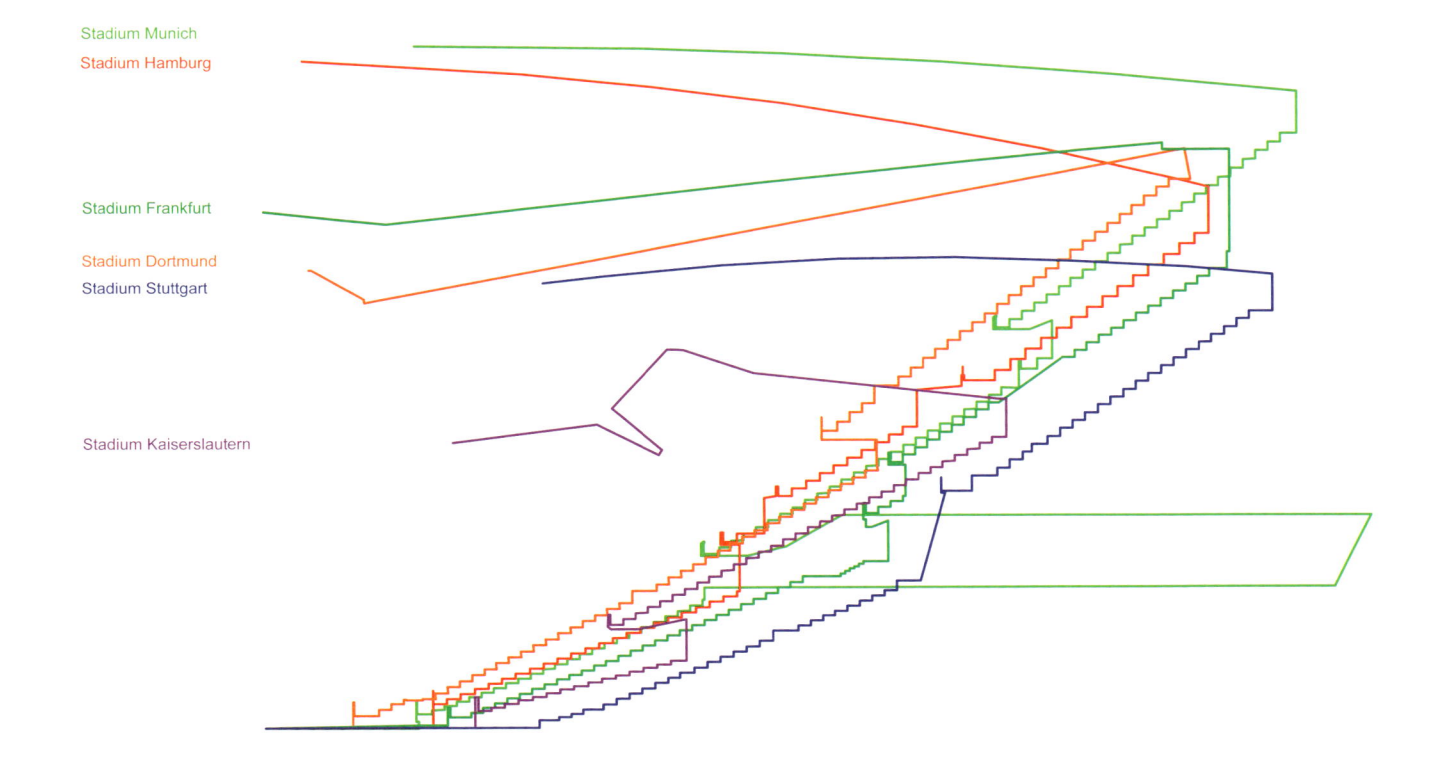

Stadium Munich
Stadium Hamburg
Stadium Frankfurt
Stadium Dortmund
Stadium Stuttgart
Stadium Kaiserslautern

page 344, 345:
646–647 **Table 0.1 / 0.2: Values derived from built World Cup sightline geometries**

01	BRL	World Cup Stadium Berlin
02	DOR	World Cup Stadium Dortmund
03	FRA	World Cup Stadium Frankfurt
04	GEL	World Cup Stadium Gelsenkirchen
05	HAG	World Cup Stadium Hamburg
06	HAN	World Cup Stadium Hanover
07	KAI	World Cup Stadium Kaiserslautern
08	CGN	World Cup Stadium Cologne
09	LEI	World Cup Stadium Leipzig
10	MUC	World Cup Stadium Munich
11	NUR	World Cup Stadium Nuremberg
12	STU	World Cup Stadium Stuttgart

Commentary on comparative table 0.1 / 0.2

The table is subdivided into ten categories, equivalent to the foregoing section on sightline geometries. All values are adjusted as mean values for better systematic comparison.
Inasmuch as maximum and minimum values are important for statements made in the text, they shall be considered. Due to the stark geometric deviations between athletics and football stadia, the according figures are summarized in separate tables.

The respective extreme values are marked in red (minimum value: green). They are not included in the overall evaluations, but are merely accepted as geometric sightline phenomena. The data record refers to an exemplary tribune on the longitudinal (short side) and transverse section (main stand).

In Frankfurt the circulating balcony in front of the upper box tier is not considered as an independent middle tier.

In Gelsenkirchen and Cologne the executive seats at the upper end of the lower tier are included into the total number of rows of the lower tier.

In Hanover, likewise, only the main and the south stand are analyzed, due to the asymmetric geometry.
In Kaiserslautern, vicariously, the east stand as new build is included in the comparison.

In Leipzig there is no upper tier on the short sides of the stadium.

The numbers 'from / to' refer to the number of rows of an upper tier and apply in this case to the increasing number of seating rows produced by the 'undulating' upper tier edge.

Table 0.1 Cross section Main Tribune		01 BRL	02 DOR	03 FRA	04 GEL	05 HAG	06 HAN	07 KAI	08 CGN	09 LEI	10 MUC	11 NUR	12 STU
Pitch distance													
Across	[m]	24.20	5.65	11.80	11.00	10.85	14.05	13.60	7.20	14.70	8.00	19.85	17.75
Lower tier													
Rows	[rows]	42	40	28	34	22	15	15	22	29	24	13	27
Tread depth	[cm]	78	80	80	80	90	100	90	90	80	80	85	85
Ascent from	[cm]	25.5	40 / 44	31.6	27.5	34.0	41.3	24.0	34.9	34.9	31.6	34.7	20 / 33
to	[cm]	37.0	linear	38.2	37.5	linear	44.4	linear	46.6	43.1	39.3	linear	linear
Stand inclination (average)	[°]	23.0	27.5	23.5	22.0	21.0	23.0	15.0	24.5	26.0	22.5	22.5	14.0
Sightline elevation C from	[cm]	19.7	15.3	18.6	11.6	11.0	21.2	8.7	14.5	7.5	10.3	21.5	12.3
to	[cm]	15.0	7.0	11.8	11.5	5.0	21.0	4.9	11.9	6.6	10.1	18.2	9.7
Middle tier													
Rows	[rows]	./.	./.	./.	./.	15	./.	./.	./.	./.	27	./.	./.
Tread depth	[cm]	./.	./.	./.	./.	90.0	./.	./.	./.	./.	80.0	./.	./.
Ascent from	[cm]	./.	./.	./.	./.	51.0	./.	./.	./.	./.	43.5	./.	./.
to	[cm]	./.	./.	./.	./.	linear	./.	./.	./.	./.	./.	./.	./.
Stand inclination (average)	[°]	./.	./.	./.	./.	29.5	./.	./.	./.	./.	28.5	./.	./.
Sightline elevation C from	[cm]	./.	./.	./.	./.	14.2	./.	./.	./.	./.	8.4	./.	./.
to	[cm]	./.	./.	./.	./.	5.9	./.	./.	./.	./.	7.0	./.	./.
Upper tier													
Rows	[rows]	31	28	25	23	18	23 – 33	31	24	28 – 33	22 – 28	26	16 – 29
Tread depth	[cm]	75	80	80	80	100	80	80	80	80	80	80	82
Ascent from	[cm]	39.6	60.0	48.2	47.5	70.0	56.0	40.0	55.0	53.0	54.6	44.4	52.0
to	[cm]	51.0	linear	49.5	50.0	linear	57.0	linear	57.0	59.2	58.0	linear	linear
Stand inclination (average)	[°]	25.0	36.9	31.5	31.5	35.0	35.5	26.5	35.0	35.0	35.0	29.0	33.0
Sightline elevation C from	[cm]	17.1	14.4	12.7	9.5	18.3	16.1	10.2	5.8	6.0	8.3	17.5	23.8
to	[cm]	19.8	9.0	8.7	8.5	14.3	13.4	5.8	4.9	9.0	8.2	11.0	16.0
Sight													
Max. distance – horizontal	[m]	78.9	58.6	62.7	58.1	60.3	50.0	47.1	56.1	61.1	66.7	51.0	65.7
Max. distance – vertical	[m]	31.0	33.5	31.0	29.5	33.5	26.5	20.1	31.0	38.5	40.6	21.0	28.4
Distance													
Centre spot – max. approx.	[m]	150.0	120.0	115.0	117.0	121.0	120.0	133.5	124.5	105.5	115.5	126.0	135.0
Centre spot – min. approx.	[m]	59.0	39.0	47.0	46.0	45.0	49.0	42.5	43.5	50.0	42.5	57.5	53.0
Viewing angle													
LT touch line / first row	[°]	4.5	16	7.0	11.0	14.5	11.5	9.5	13.5	18.5	11.0	9.0	5.5
LT touch line / last row	[°]	15.0	25	15.5	20.0	18.0	14.5	12.0	21.0	23.0	19.0	13.5	14.5
UT touch line / first row	[°]	18.0	29.5	24.5	25.5	26.0	24.5	19.5	31.5	29.0	30.0	18.5	19.0
UT touch line / last row	[°]	21.5	32.5	27.0	27.0	29.0	28.0	23.0	33.0	32.0	31.5	22.5	24.0
Linear ascent			LT/UT			LT/UT		LT/UT				LT/UT	LT/UT

Stadium evaluation 345

Table 0.2 Longotudinal section / Short side		01 BRL	02 DOR	03 FRA	04 GEL	05 HAG	06 HAN	07 KAI	08 CGN	09 LEI	10 MUC	11 NUR	12 STU
Pitch distance													
Across	[m]	42.00	6.40	13.70	11.50	11.30	16.90	10.10	8.15	13.90	8.00	36.55	38.70
Lower tier													
Rows	[rows]	42	43	25	37	23	21	33	24	34	28	19	25 – 27
Tread depth	[cm]	78	80	85	80	90	80	80	80	80	80	78	85
Ascent	from [cm]	22.1	40 / 44	33.5	27.5	34.0	21.5	40.0	30.3	32.4	25.8	26.0	21.1
	to [cm]	23.3	linear	38.2	37.5	linear	29.9	linear	42.1	43.1	39.3	linear	28.8
Stand inclination (average)	[°]	16.0	27.5	23.0	22.0	21.0	21.0	26.5	24.5	25.0	22.0	18.0	17.5
Sightline elevation C	from [cm]	22.1	20.4	18.6	12.3	12.8	11.7	21.2	13.6	10.7	10.3	19.7	10.4
	to [cm]	23.3	18.0	10.4	14.3	5.4	12.4	6.6	12.7	11.3	10.6	15.0	14.9
Middle tier													
Rows	[rows]	./.	./.	./.	./.	15	./.	./.	./.	./.	29	./.	./.
Tread depth	[cm]	./.	./.	./.	./.	90	./.	./.	./.	./.	80	./.	./.
Ascent	from [cm]	./.	./.	./.	./.	51.0	./.	./.	./.	./.	43.8	./.	./.
	to [cm]	./.	./.	./.	./.	linear	./.	./.	./.	./.	51.6	./.	./.
Stand inclination (average)	[°]	./.	./.	./.	./.	29.5	./.	./.	./.	./.	31.0	./.	./.
Sightline elevation C	from [cm]	./.	./.	./.	./.	9.6	./.	./.	./.	./.	8.4	./.	./.
	to [cm]	./.	./.	./.	./.	6.7	./.	./.	./.	./.	7.5	./.	./.
Upper tier													
Rows	[rows]	31	27	25	22	18	23 – 28	35	31	./.	22 – 28	26	./.
Tread depth	[cm]	75	80	80	80	100	80	80	85	./.	80	85	./.
Ascent	from [cm]	29.5	54.0	48.5	47.5	70.0	47.5	52.0	55.0	./.	54.6	44.4	./.
	to [cm]	50.1	linear	49.5	50.0	linear	50.0	linear	56.8	./.	57.0	linear	./.
Stand inclination (average)	[°]	25.0	34.0	31.5	31.5	35.0	31.5	31.5	33.0	./.	35.0	29.0	./.
Sightline elevation C	from [cm]	11.7	15.5	13.3	10.6	20.3	26.5	18.0	12.2	./.	8.4	26.9	./.
	to [cm]	24.8	13.6	9.3	9.6	15.8	21.4	5.6	8.8	./.	7.5	19.2	./.
Sight													
Max. distance – horizontal	[m]	96.8	66	62.2	58.7	60.7	58.6	58.6	54.8	57.6	64.2	66.8	82.4
Max. distance – vertical	[m]	31.0	33.5	31.5	30.5	33.5	21.3	34.5	30.9	16.6	39.0	21.0	14.1
Distance													
Centre spot – max. approx.	[m]	150.0	120.0	115.0	117.0	121.0	120.0	133.5	124.5	105.5	115.5	126.0	135.0
Centre spot – min. approx.	[m]	59.0	39.0	47.0	46.0	45.0	49.0	42.5	43.5	50.0	42.5	57.5	53.0
Viewing angle													
LT touch line / first row	[°]	2.5	10	10.0	11.0	13.5	7.0	13.5	11.5	14.5	11.0	4.5	2.0
LT touch line / last row	[°]	11.5	./.	17.0	18.5	17.5	12.5	./.	20.0	21.5	19.0	8.0	9.5
UT touch line / first row	[°]	14.0	./.	24.0	25.0	25.0	14.5	./.	26.5	./.	30.0	12.5	./.
UT touch line / last row	[°]	17.0	26.5	26.5	26.5	28.5	20.0	27.5	29.5	./.	31.5	17.5	./.
Linear ascent			LT/UT			LT/UT		LT/UT				LT/UT	

25 Final evaluation and compendium

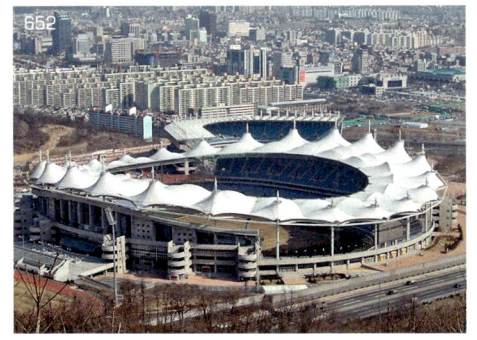

FIFA WorldCup 2002™

648 **Busan Asia Main Stadium, Korea**
Capacity: 55,982
Parking spaces: 3,098
Completion: July 2001

649 **Daegu World Cup Stadium, Korea**
Capacity: 68,014
Parking spaces: 3,550
Completion: May 2001

650 **Daejeon World Cup Stadium, Korea**
Capacity: 40,407
Parking spaces: 3,188
Completion: September 2001

651 **Gwangju World Cup Stadium, Korea**
Capacity: 42,880
Parking spaces: 4,248
Completion: September 2001

652 **Incheon Munhak Stadium, Korea**
Capacity: 52,179
Parking spaces: 4,559
Completion: December 2001

653 **Jeonju World Cup Stadium, Korea**
Capacity: 42,391
Parking spaces: 3,729
Completion: September 2001

654 **Jeju World Cup Stadium, Korea**
Capacity: 42,256
Parking spaces: 666
Completion: December 2001

655 **Seoul World Cup Stadium, Korea**
Capacity: 63,961
Parking spaces: 2,648
Completion: December 2001

656 **Suwon World Cup Stadium, Korea**
Capacity: 43,188
Parking spaces: 2,748
Completion: May 2001

657 **Munsu-Stadium (Ulsan), Korea**
Capacity: 43,550
Parking spaces: 3,963
Completion: May 2001

648–657 **Aerial views of World Cup Stadia 2002**
(from left to right)

Aerial views of World Cup Stadia 2002 347

FIFA WorldCup 2002™

658 **Ibaraki Kashima-Stadion, Japan**
Capacity: 42,000
Parking spaces: 12,000
Completion: May 2001

659 **Kobe Wing Stadium, Japan**
Capacity: 42,000
Completion: October 2001

660 **Miyagi Stadium, Japan**
Capacity: 49,000
Parking spaces: 7,000
Completion: March 2000

661 **Niigata Stadium Big Swan, Japan**
Capacity: 42,300
Parking spaces: 4,000
Completion: March 2001

662 **Oita Stadium Big Eye, Japan**
Capacity: 43,000
Completion: March 2001

663 **Osaka Nagai-Stadion, Japan**
Capacity: 50,000
Parking spaces: 2,500
Completion: May 1996

664 **Saitama Stadium, Japan**
Capacity: 63,000
Parking spaces: 2,500
Completion: July 2001

665 **Sapporo Dome, Japan**
Capacity: 42,000
Parking spaces: 1,700
Completion: May 2001

666 **Shizuoka-Stadion ECOPA, Japan**
Capacity: 50,600
Parking spaces: 3,695
Completion: March 2001

667 **International Stadium Yokohama, Japan**
Capacity: 70,000
Parking spaces: 3,500
Completion: October 1997

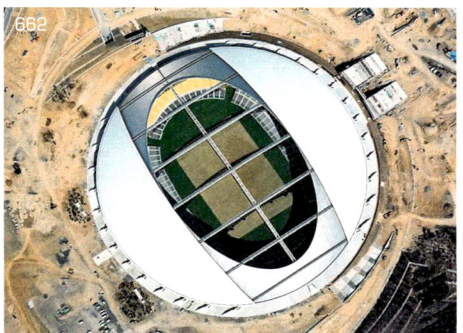

658–667 **Aerial views of World Cup Stadia 2002**
(from left to right)

25 Final evaluation and compendium

UEFA EURO 2004™

668 **Estadio Aveiro Municipal, Portugal**
Capacity: 31,498
Parking spaces: 3,798 (uncovered)
Completion: November 2003

669 **Estadio Municipal de Braga, Portugal**
Capacity: 30,360
Parking spaces: 5,224 (uncovered)
Completion: December 2003

670 **Estadio Cidade de Coimbra, Portugal**
Capacity: 30,216
Parking spaces: 542
Completion: September 2003

671 **Estadio Algarve, Faro Portugal**
Capacity: 30,305
Parking spaces: 5,000 (uncovered)
Completion: November 2003

672 **Estadio Afonso D. Henriques, Guimaraes Port.**
Capacity: 34,600
Parking spaces: 800 (underground)
Completion: July 2003

673 **Estadio Dr. Magalhaes Pessoa, Leiria Portugal**
Capacity: 30,000
Parking spaces: 542
Completion: November 2003

674 **Estadio Jose Alvalade XXI, Lisbon Portugal**
Capacity: 52,000
Parking spaces: 2,400 (underground)
Completion: August 2003

675 **Estadio da Luz, Lissabon Portugal**
Capacity: 65,272
Parking spaces: 1,410 (underground)
Completion: October 2003

676 **Estadio do Bessa Sculo XXI, Porto Portugal**
Capacity: 30,000
Completion: December 2001

677 **Estadio do Dragao, Porto Portugal**
Capacity: 50,106
Completion: November 2003

668–677 **Aerial views of EM Stadia 2004**
(from left to right)

page 351:
524–634 **Aerial views of World Cup Stadia 2006**
(from left to right)

general planners **agn**

www.agn.de

Stadium Alemannia Aachen | Stadium FC St. Pauli Hamburg | Sportsparc Cologne | Stadium VfL Osnabrück | Arena Artland Dragons | Stadium 1899 Hoffenheim

Lindapter® – THE Innovative Clamping System for Steelwork and Hollow Sections since 1934

The worldwide network of dedicated distributors coupled with free technical design service provided by a team of experienced Lindapter engineers guarantee quality solutions and hassle-free sourcing. Technical approvals offer additional peace of mind in applying the system's versatility.

Lindapter CF – High Friction Clamp

The main benefits of clamping steel together with the Lindapter system

- Reduces design time
  Quicker turnarounds
- No on-site drilling or welding necessary – Undamaged steelwork and most coatings
- 'Hot Working' not required
  Minimised fire risk
- Only hand tools required to install
  Lower equipment costs
- Site adjustable
  Easier & speedier re-alignment
- Power not required
  No inaccessible areas
- Speedier installation
  Quicker turnarounds – minimised labour costs

Aston Villa – High Friction Clamp

Engineered Solutions

Lindapter offers more than just a comprehensive clamping system, a team of experienced engineers supports specifiers and end-users alike in finding exactly the right solution. Within just a few hours customers are supplied with their solution drawing ready to be imported into the overall project plan as well as a parts list, no extra work or calculation required. Of course Lindapter supplies complete pre-assembled Girder Clamps, stools etc. upon request.

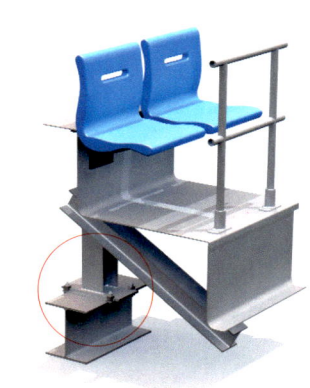

Seating secured with Lindapter Girder Clamp

Lindapter Hollo-Bolts were used to secure the rails for the 127 m long and 38 m wide sliding roof of **LTU Arena in Düsseldorf**. The protection against corrosion inside the hollow-sections was a particular challenge. The experienced Lindapter engineers co-operated closely with the specifier and contractor. Many other Lindapter products were used to secure the air-conditioning and further services.

Lindapter Hollo Bolt

Millenium Award granted to the Hollo-Bolt for outstanding design

- Blind fixing for hollow-sections or pipes
- Patented 5-part design in M16 and M20 guarantees high clamping forces
- Accepted by the UK Steel Construction Institute and included in the 'Green Book' for primary connections
- Approved by the DIBt since 2002
- Standard coating JS500 or hot dip galvanised
- A variation of bolt heads available
- Stainless steel available

Telstra Stadium Sydney – Seating fixed with Lindapter Hollo Bolt

References include

- Telstra Stadium Sydney (2000 Olympics)
- Velodrome Athens
- Allianz-Arena Munich
- Old Trafford Re-development (Manchester United)
- The Lawn Tennis Association Wimbledon
- New Wembley Stadium
- Emirates Stadium (Arsenal)
- Enron Field (Houston Astros) Texas
- Kentucky Speedway Circuit
- SAFECO Field (Seattle Mariners Baseball)
- Millennium Stadium Cardiff

Lindapter Girder Clamp

LINDAPTER GmbH
Tenderweg 11
45141 Essen
Germany
Tel.: 0201 / 21 47 78
Fax: 0201 / 29 06 14
E-Mail: info@lindapter.de
www.lindapter.de

Lindapter®
Verbindungs- und Klemmsysteme seit 1934

FIFA WorldCup 2006™

524 Berlin: 'Olympiastadion Berlin'
World Cup capacity: 74,200
Parking spaces: 815 VIP (underground)
Completion: December 2004

534 Dortmund: 'Signal-Iduna-Park'
World Cup capacity: 65,900
Parking spaces: 270 VIP (at the stand)
Completion: 2003/2005

544 Frankfurt: 'Commerzbank-Arena'
World Cup capacity: 48,400
Parking spaces: 1,800 VIP (underground)
Completion: June 2005

554 Gelsenkirchen: 'Veltins-Arena'
World Cup capacity: 53,600
Parking spaces: 2,500 VIP (uncovered)
Completion: August 2001

564 Hamburg: 'AOL-Arena'
World Cup capacity: 51,300
Parking spaces: 550 VIP (uncovered)
Completion: September 2000

574 Hanover: 'AWD-Arena'
World Cup capacity: 44,900
Parking spaces: 140 VIP (uncovered)
Completion: December 2004

584 Kaiserslautern: 'Fritz-Walter-Stadion'
World Cup capacity: 47,700
Completion: Nov. 2005 (planned)

594 Cologne: 'RheinEnergieStadion'
World Cup capacity: 46,300
Parking spaces: 650 VIP (underground)
Completion: March 2004

604 Leipzig: 'Zentralstadion'
World Cup capacity: 44,300
Parking spaces: 500 VIP (underground)
Completion: March 2004

614 Munich: 'Allianz-Arena'
World Cup capacity: 65,600
Parking spaces: 1,200 VIP (underground)
Completion: May 2005

624 Nuremberg: 'Frankenstadion'
World Cup capacity: 43,800
Parking spaces: 205 VIP (uncovered)
Completion: June 2005

634 Stuttgart: 'Gottlieb-Daimler Stadion'
World Cup capacity: 53,100
Parking spaces: 885 VIP (multi-storey)
Completion: December 2005

25 Final evaluation and compendium

UEFA EURO 2008™

678 **Basel: St. Jakobs-Park, Switzerland**
Costs (newbuild): 150 mio. Euro
Capacity: 42,500 (2008)
Completion: March 2001
Spring 2005

679 **Bern: Stade de Suisse, Wankdorf Switzerland**
Costs (newbuild): 200 mio. Euro
Capacity: 32,000
Completion: June 2005

680 **Genf: Stade de Genève, Switzerland**
Costs (newbuild): 63 mio. Euro
Capacity: 30,500
Completion: March 2003

681 **Zurich: Letzigrund Stadion, Switzerland**
Costs (newbuild): 237 mio. Euro
Capacity: 30,000
Completion: 2007 (planned)

682 **Innsbruck: Tivoli (new), Austria**
Costs for add. storeys: 37 mio. Euro
Capacity: 15,000
30,600 (2008)
Completion: 2006 (planned)

683 **Klagenfurt: Wörtherseestadion, Austria**
Costs (newbuild): 40 mio. Euro
Capacity: 12,000 + 18,000
Completion: 2007 (planned)

684 **Salzburg: Stadion Salzburg, Austria**
Costs for add. storeys: 15 mio. Euro
Capacity: 18,500
30,000 (2008)
Completion: March 2003

685 **Vienna: Ernst Happel-Stadion, Austria**
Modernization: 15 mio. Euro
Capacity: 53,000
Completion: 2005 (Five-Star Stadia)

Financing

In Austria mainly through public funds (one third of the costs respectively from national, federal state and municipal funds). Merely the adaptation of Happel Stadium is born in half by the Federal Republic and the city.

In Switzerland the construction of stadia is mainly backed by with private financing. Private syndicates, mostly lead by Credit Suisse, bear up to 96 percent of the costs. [2]

678–685 **Aerial views of stadia for EURO 2008**
(from left to right)

ARCHITECTURAL BOOKS FROM WILEY-BLACKWELL

Ernst & Peter Neufert
Architects' Data
2002. Third Edition.
648 pages 6200 ilustrations.
Softcover.
27,8 × 21,9 cm
€ 74,90* / sFr 120,–
ISBN: 978-0-632-05771-9

Architects' Data is an essential aid in the initial design and planning of a building project. It contains essentials of architectural planning.

Features:
- new in paperback
- classic reference for architects, known worldwide
- based on 35th edition of the original German „Neufert", first published in 1936 and translated into 16 languages

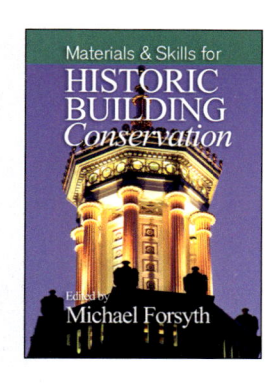

Edited by Michael Forsyth
Materials & Skills for Historic Building Conservation
2007. 256 pages. Hardcover.
€ 67,90* / sFr 109,–
ISBN: 978-1-405-11170-6

Here within a single volume is provided all the essential information on the properties of traditional building materials. Subjects covered include their availability and sourcing, the causes of erosion and decay, knowledge of the skills required for their application on conservation projects and the impact the materials have on the environment. Each material is considered in the constructional sequence: wall materials, roof coverings, wood and timber framing, metals and finishes.

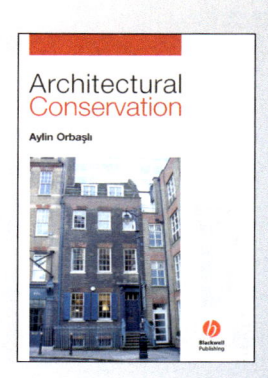

Aylin Orbasli
Architectural Conservation
2007. 280 pages. Paperback.
€ 47,90* / sFr 77,–
ISBN: 978-0-632-04025-4

This book provides an introductory text for students in built environment disciplines, as well as those who manage or own historic properties, and those embarking upon architectural conservation professionally. It is designed to give an understanding of the main principles, materials and problems in the field of conservation and it features a number of case studies.

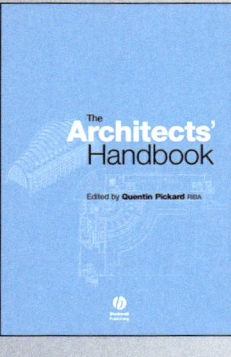

Edited by Quentin Pickard
The Architects' Handbook
2005. 484 pages.
Softcover.
29,7 × 21 cm
€ 49,90* / sFr 80,–
ISBN: 978-1-405-13505-4

This book examines each main building type, with a brief outline of the main design considerations, followed by a range of building examples, primarily in the form of plans.

Features:
- provides references to key technical standards and design guidance plus a comprehensive bibliography for most building types
- includes over 300 case studies
- complements Neufert's Architects' Data which concentrates on measurements and data

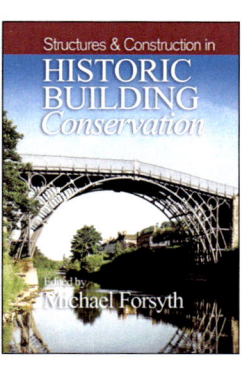

Edited by Michael Forsyth
Structures & Construction in Historic Building Conservation
2007. 248 pages. Hardcover.
24,6 × 18,9 cm
€ 67,90* / sFr 109,–
ISBN: 978-1-405-11171-3

Structures and Construction explains how structural principles have influenced the evolution of building forms and styles, and how structural failure can occur in historic buildings.

Features:
- gives key information in each area with where to go for more detailed guidance
- strong focus on engineering and craft solutions

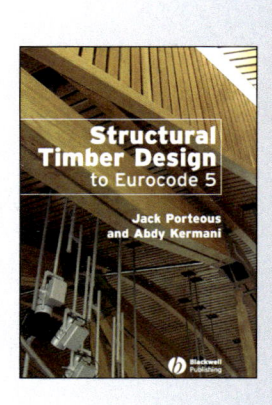

Jack Porteous, Abdy Kermani
Structural Timber Design to Eurocode 5
2007. 280 pages. Hardcover.
24,4 × 17,2 cm
€ 109,–* / sFr 172,–
ISBN: 978-1-405-14638-8

This book offers detailed guide to Eurocode 5 on structural timber design, providing a comprehensive source of information on how to design timber elements and structures to the new code. Numerous worked examples are provided, with the use of computers encouraged to carry out the mathematical calculations.

Ernst & Sohn
A Wiley Company
www.ernst-und-sohn.de

Ernst & Sohn
Verlag für Architektur
und technische Wissenschaften
GmbH & Co. KG

For order and customer service:
Verlag Wiley-VCH
Boschstraße 12
69469 Weinheim
Germany

Tel.: +49(0) 6201 / 606-400
Fax: +49(0) 6201 / 606-184
E-Mail: service@wiley-vch.de

* In EU countries the local VAT is effective for books and journals. Postage will be charged. Whilst every effort is made to ensure that the contents of this leaflet are accurate, all information is subject to change without notice. Our standard terms and delivery conditions apply. Prices are subject to change without notice.

seatbox

There are bleachers, mobile grandstands and luxury stadiums, but there is only one product that combines all three. Telescope-like extendable step elements, arranged in a supporting frame, can be erected and dismounted easily without requiring time-consuming moving of many individual parts. The Seatbox Container Grandstand system speaks for itself, through its easy transportability, minimal storage requirements, extremely short installation time and vast variability in individual constructions.

This concept offers unique, transportable spectator stands made of steel and aluminium. They have telescopic extensible ranks, which allow time-saving assembly and disassembly for all types of use, both indoors and outdoors. The system offers the possibility of a seating capacity of fewer than 100 seats up to several thousand seats. One of the greatest advantages of the concept compared with conventional grandstand construction is the minimal amount of time and personnel required for installation. For example, a spectator grandstand comprising six-row containers and approximately 700 seats can be assembled by two workers in only four to five hours and disassembled in another three to four hours.

Seatbox Aktiengesellschaft
Tel.: 0041 (0) 79 341 1068
Email: office@seatbox.com
www.seatbox.com

Aerial views of World Cup Stadia 2010 355

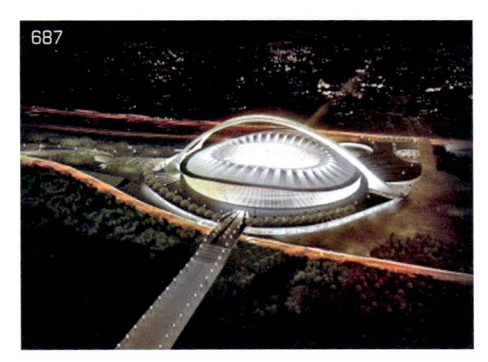

FIFA WorldCup 2010™

686 **Bloemfontein: Free State Stadium**
Capacity: 40,000
Completion (Extension): Initial inauguration 1952, June 2007 (planned)

687 **Durban: King Senzangakhona-Stadium**
Capacity: 60,000
Completion (new build): Initial Inauguration 1957, September 2007 (planned)

688 **Johannesburg: Ellis Park Stadium**
Capacity: 60,000
Completion (Extension): Initial inauguration 1927, September 2007 (planned)

689 **Johannesburg: 'Soccer City'**
Capacity: 94,700
Completion (extension): Initial inauguration 1987, April 2007 (planned)

690 **Kapstadt: Africain Rennaissance Stadium**
Capacity: 70,000
Completion (new build): Initial inauguration 1888, November 2007 (planned)

691 **Nelspruit: Mbombela-Stadium**
Capacity: 40,000
Completion (new build): June 2007 (planned)

692 **Polokwane: Peter Mokaba Stadium**
Capacity: 40,000
Completion (extension): Initial inauguration 1976, April 2007 (planned)

693 **Port Elizabeth: Port Elizabeth Stadium**
Capacity: 30,000 + 20,000
Completion (new build): June 2007 (planned)

694 **Pretoria: Loftus Versfeld Stadium**
Capacity: 50,000
Completion (extension): Initial inauguration 1903, April 2005

695 **Rustenburg: Royal Bafokeng Stadium**
Capacity: 40,000
Completion (extension): Initial inauguration 1999, November 2005

686–695 **Aerial views of World Cup Stadia 2010**
(from left to right)

Appendix

Epilogue

The present *StadiumATLAS* aims to clarify step-by-step the complex decision parameters for the construction of grandstand profiles of modern sports and event venues. First, the present guidelines for stadium construction are analyzed and geometric rules are derived by the examination of built examples of World Cup 2006™ as well as by gathering and sorting out available material. The brought out influence factors which are important for the planning of a spectator facility are not merely subject to the categories of beauty, truth, information or to the functional and technical.[1] Moreover, these aspects always involve a relevant group of decision makers and their motivations.

Many intentions have to be reconciled in the development of a sightline profile. Therefore, in the run-up of planning of a new build or the modernization of a sports and event venue, it is paramount to interrelate these strongly diverging aspects, to inform the participants about the effects of guidelines and to provide holistic answers to their questions.

The main basis of information for this book was the dissertation entitled *Sichtlinien und Sicherheit* [Sightlines and Safety] which was produced by this author during his PhD studies, supervised by Prof. Volkwin Marg, gmp architects, Hamburg and approved by the faculty for architecture 2006 in Aachen.

The strongest influence on the ascending profile of a stand, apart from the safety of circulation, is exerted by the safety measures of the first row, which are to be evaluated primarily in reference to their effects on sightline quality.

Stadia of the third generation, as constructed for World Cup 2006™ in Germany, possess accrue a certain orientation and recommendation character. They are stepping stones for further developments and the coming generations in stadium construction, representing a constantly evolving building type. The concept of multifunctionality will surely remain essential in the light of profitable utilization.

Beyond the basic demands on comfort with regards to weather protection, safety and service, the exploitation of technical and digital possibilities will determine the design of future stadium complexes. According to Prof Dr F.-J. Verspohl (Friedrich Schiller University, Jena) a strong trend toward 'adaptable and media-oriented stadia' may be anticipated, 'architecture as action object' in the 'framework of variability'.[2]

These venues will offer a broad spectrum of events and uses order to guarantee maximum utilization. Progress in telecommunication technology will entail the further fitting-out of viewing areas with media technology; additional measures of comfort that will further alter the rough but honest character of a traditional stand. Specialist stadia designed for a special type of event or sports represent a contrary development to the adaptable allrounder. Additional costs for adaptable roofing, terrace or playing field not incurred, are offset by a very restricted event spectrum. Apart from short-term use of temporary stands or mobile supplementary stands to maximize capacity (ex. 'Telstra Stadium' in Sydney, Olympic Games 2000), temporary stadia are also thinkable (ex. Bondi Beach Volleyball Stadium, Sydney Olympics 2000, Nüssli Co., CH). In such cases, investment costs for a singular event further entail permanent operation and maintenance costs.

On the occasion of an exhibition entitled 'The Stadium' that took place in Rotterdam in 2000, Verspohl cites curator Michelle Provoost: 'Stadia today have taken over the role of museums as walking ads for cities.' As contemporary witnesses and permanent point of identification, they are making an important cultural contribution to architectural history. Stadia appear to have become a particular means of self expression of a society.

Regardless of the development in typology or utilization, as long as people will gather to view an event live from inclined or stepped tribunes, the necessity to provide good viewing conditions will remain. The ultimate experience is the sense of community, as J.W. Goethe put it in his *Italian Journey*, 'making people feel at their best' and coming together in such a 'crater' created as plain as possible so that 'the public itself supplies its decoration'. It is the hope of this author to be able to make a contribution in facilitating planning efforts with this *StadiumATLAS*.

Notes

TC = Translated Citation (acronym for citations translated from a German publication into English).

Preface 5
1) Muthesius, Hermann. *Die Einheit der Architektur* (1908). TC.
2) Vitruvius. *De Architectura libri Decem* (1991). TC.
3) Alberti, Leon Battista. *De re aedificatoria* (1492). TC.
4) BauO NRW § 75 Baugenehmigung [German federal state building regulations, permit].

Chapter 01 10–23
1) VersammlungsstättenVO § 2.
2) *Brockhaus: die Enzyklopädie,* 20th edition (Leipzig/Mannheim, 1996). TC.
3) Marg, Volkwin. 'Regie und Selbsterfahrung der Massen – Beispiel Berlin', publ. in *Bauingenieur* (May, 2004), vol. 79, p. 201. TC.
4) Ibid.
5) Alberti, Leon Battista. *Zehn Bücher über die Baukunst.* German transl. by Max Theuer (Wiss. Buchgesellschaft, Darmstadt, 1975). TC.
6) Note: Here, 2.0 m refers to the height of the seats, which results in a barrier height of 2.5 m.
7) Vitruvius. *The Ten Books on Architecture.* Transl. by Morris Hicky Morgan (Dover Books, New York, 1960), p. 139.
8) Ibid.
9) Ibid., p. 97.
10) Cf. drawings in EN/DIN 13200–1, published by Deutsches Institut für Normung.
11) Alberti, L. B. *Zehn Bücher über die Baukunst* (1975), book 8, chapter 7, p. 447. TC.
12) Goethe, J. W. *The Italian Journey.* Transl. by W. H. Auden and Elizabeth Mayer (Penguin Books 1962), vol. 1, pp. 52–53.
13) Gussman, H. *Theatergebäude – Technik des Theaters* (Berlin, 1954), p. 23. TC.
14) Ibid. TC.
15) Gellinek, P. O. *Sichverhältnisse in 15 Zuschauerräumen von Theatern* (Hanover, 1956), pp. 13/21. TC.
16) Ibid.
17) Gussman, H. *Theatergebäude – Technik des Theaters* (Berlin, 1954), p. 23. TC.
18) Kallmorgen, W. *Was heißt und zu welchem Ende baut man Kommunaltheater?* (Darmstadt, 1955), p. 135. TC.
19) Burk, A. and R. *Checklisten der aktuellen Medizin - Augenheilkunde* (Stuttgart, 2005), p. 517. TC: all terms in quotation marks.
20) Verspohl, J. *Stadionbauten von der Antike bis zur Gegenwart* (Gießen, 1976), dust jacket text. TC.
21) Marg, Volkwin. *Olympia Stadion Berlin – Sanierung und Modernisierung 2002–2004* (Berlin, 2004), p. 12.
22) Cf. J. Verspohl (Gießen, 1976). TC.
23) Ibid., pp. 30–31.
24) Ibid., p. 32.
25) Marg, Volkwin. *Olympia Stadion Berlin – Sanierung und Modernisierung 2002–2004* (Berlin, 2004), pp. 9–13.
26) Ibid.
27) Lowe, B. 'Vom Spielfeld zur Sportstätte der Zukunft', *Zeitschrift für Baukultur* 3/2004, pp. 7–15. TC.
28) Drawn from Sheard, R. *The Stadium – Architecture for the New Global Culture* (Singapore, 2005), pp. 103–116.
29) Marg, Volkwin. *DETAIL-concept* 09/2005, p. 897. TC.
30) Augustin, E. *Fußball unser* (Munich, 2005), pp. 5–29. TC.
31) Merkt, A. *Fußballgott – Elf Entwürfe* (Cologne, 2006), p. 50. TC.

Chapter 02 24–31
1) A German business magazine, issue April 2005.
2) 'Faszination Stadion 2006', *Stadionwelt* (Cologne, 2005), p. 10. TC.
3) Request of UEFA for EURO 2008, completion December 2006.
4) According to IAKS (1993), ca. 75 % of the annual capital.
5) According to planning principles by IAKS (No. 33). TC.
6) German title: *Grundsätze und Richtlinien für Wettbewerbe auf den Gebieten der Raumplanung, des Städtebaus und des Bauwesens – GRW 1995*.
7) German regulations on allocation of construction works. [VOB/A – *Verdingungsordnung für Bauleistungen*].
8) German regulation on architects' and civil engineers' fees. [HOAI – *Honorarordnung für Architekten und Ingenieure*].
9) 18th BlmSchV – *Sportanlagenlärmschutzverordnung* § 1.
10) Drawn and translated from http://www.bmu.de/laermschutz/kurzinfo/doc/4022.php.
11) *Regeln für die Auslobung von Wettbewerben* (RAW 2004).
12) According to W. Steinhart. *Verkehrsberatung* [traffic consultants], Aachen.
13) Ibid. [definition according to BMVBW, the German Federal Ministry for Transport, Construction and Urban Planning].
14) GarVO NRW (1990) § 6. [Regulation on garages, North Rhine-Westphalia], *Einzelstellplätze und Verkehrsflächen*.
15) German title of recommendations: *Empfehlungen für Anlagen des ruhenden Verkehrs*.
16) MVStättV 2005 § 13 (1), § 10 (7).

Chapter 03 32–37
1) DFB safety guidelines § 5.
2) MVStättV 2005 § 30 para 1.
3) DFB safety guidelines § 32.
4) FIFA safety guidelines § 7 on guest block.
5) DFB safety guidelines § 5.
6) Versammlungsstättenverordnung § 30 (2).
7) MVStättV 2005 § 26 (3).

Chapter 04 38–51
1) EN/DIN 13200-1 section 5.1.3.
2) MVStättV 2005 § 10 (1) 'Bestuhlung'
3) MVStättV § 33
4) According to UEFA (2004).
5) As stipulated in MVStättV 2005 and DIN 13200-1 item 5.3.3, which in principle also still allows seats on benches.
6) Here, MVStättV 2005 § 1 section 3 takes over the regulations of VersammlungsstättenVO 1978.
7) DIN 13200-1 item 5.3.4 (minimum distance from backrest to backrest).
8) MVStättV 2005 § 10 section 3.
9) DIN 13200 item 5.3.2 + 5.5.3.
10) DIN 18065 'Gebäudetreppen' (table 2).
11) Not subject to ArbeitsstättenVO and according to LBO [federal state building regulations].
12) MVStättV 2005 § 11 clause 2.
13) DIN 18065 part 3 section 4.2.
14) MVStättV 2005 § 11 clause 3.
15) EN/DIN 13200-3/4.1.3.
16) During a match VStättVO applies; after a match ArbStättVO.
17) MVStättV 2005 § 11 clause 1.
18) EN/DIN 13200-3 section 4.1.1.
19) EN/DIN 13200-3 section 4.1.3, E.
20) MVStättV 2005, § 11, clause 2–3.
21) MVStättV 2005.
22) MVStättV 2005, § 11 section (1) clause 3, ARGEBAU, Fachkommission Bauaufsicht. TC.
23) MVStättV 2005 § 10 clause 3.
24) EN/DIN 13200-1 item 5.3.4.
25) MVStättV 2005 § 27 section 3.
26) IAKS planning guidelines 33.
27) DIN 13200-1 item 5.2.2.
28) MVStättV 2005 § 1.
29) In Germany this would be the department of *Wiederkehrende Prüfung, Brandschau*.
30) MVStättV 2005 § 10 section 8.
31) MVStättV 2005 § 28 section 1.
32) FIFA *Pflichtenheft Stadion 2006* item 1.3.
33) MVStättV 2005 § 11 clause 4.
34) Table A.1 in appendix A.
35) DIN 18030 'Barrierefreies Bauen – Planungsgrundlagen' (Berlin, draft Nov 2002). TC.
36) DIN 18030 (draft) item 4.4.1.
37) MVStättV 2005.
38) DIN 18030 'Barrierefreies Bauen'.
39) DIN 18030 item 5.3.1. TC.
40) MVStättV 2500 § 10 clause 7.
41) FIFA requirement as stated in item 1.6 of the Technical Recommendations.

Chapter 05 52–67
1) The term 'business lounge', as it is frequently used in German stadia.
2) http://fifaworldcup.yahoo.com/06/de/tickes/index.html. TC.
3) FIFA Technical Recommendations 2006, item 1.4.
4) Status as of 2005.
5) Descriptions are taken from the official Hospitality Programme of World Cup 2006™. TC.
6) ArbStättVO § 23, clause 2.
7) GarVO, Teil III für Mittel- und Großgaragen, NRW 1990 § 9 clause 4. TC.

8) EN/DIN 13200-1 'Spectator facilities', Normenausschuss Bauwesen (NABau) in DIN 2004, Appendix D, Note.
9) MVStättV 2005 § 1.

Chapter 06 68–75
1) DFB guidelines § 23.
2) FIFA safety guidelines § 19.
3) MVStättV 2005 § 9.
4) DIN 18030 (2002-11).
5) DIN 18035 'Sportplätze' item 4.5.3.4
6) 10 % of surface area, as required by ArbStättVO.
7) MVStättV § 12 'Toilettenräume'. The regulation follows the Muster-Gast-BauVO 1982 § 22.

Chapter 07 76–93
1) Marg, Volkwin. *Stadien und Arenen* (Hamburg, 2005), p. 17.
2) DIN 18035 'Sportplätze', Appendix A.
3) ArbeitsStättVO 2002 § 23 on room dimensions, section 2.
4) Ibid. TC.
5) ArbStättVO § 23, second section, first part.
6) All measures in approximation.
7) FIFA Technical Recommendations for FIFA World Cup 2006, 'Profile und Anforderungen'. TC.
8) Information received in a consultation with German film and television production company Michaelis GmbH, Film- und Fernsehproduktion Veranstaltungstechnik, Erkrath.
9) Interview published in *Stadionwelt* 2005. TC.

Chapter 08 94–97
1) FIFA Technical Recommendations of (1995), item 17.
2) UEFA specifications for EURO 2008 (5.13).

Chapter 09 98–99
1) DIN 18035 'Sportplätze'.
2) MVStättV 2005 § 21 'Werkstätten, Magazine und Lagerräume'.

Chapter 10 100–109
1) LBO NRW § 3.
2) BGB § 74.
3) *Nationales Konzept* 'Sport und Sicherheit', Ständige Konferenz der Länder-Innenminister.
4) *Richtlinien zur Verbesserung der Sicherheit bei Bundesligaspielen*.
5) BGB § 54 'Sonderbauten'. TC.
6) BGB § 68 'Vereinfachtes Genehmigungsverfahren'. TC.
7) Lukowski, S. *Bad Blankenburger Sportstättentagung* (press article BIS DSB 2004, Frankfurt), pp. 7/8. TC.
8) DIN 13200-1, item 5.1.4.
9) DFB safety guidelines § 25 (2004).
10) DIN 4844, part 1. Equivalent to § 7 FIFA safety guidelines.
11) MVStättV 2005 § 15, section 3.
12) Predtetschenski, M. M. and Milinski, A. I. *Personenströme in Gebäuden* (Staatsverlag der DDR, Berlin, 1971).
13) In comparison to no. 2.
14) Fachkommission Bauaufsicht. *Begründung und Erläuterung zur MVStättV 2005*, p. 14. TC.
15) Neufert, P. *Bauentwurfslehre* (33rd edition, 1992), p. 427. TC.
16) Cf. chapter 12: 'Entspannungstore'.

Chapter 11 110–121
1) MVStättV 2005 § 8.
2) DIN 18030 'Barrierefreies Bauen'. TC.
3) MVStättV § 7 clause 4 'Bemessung der RW'.
4) This solution goes against MVStättV § 8, clause 6.
5) This conforms to MVStättV 2005.

6) In the open, derived from § 7 MVStättV, clause 4.
7) HHP-West. *Brandschutzgutachten Stadion Köln* (Bielefeld 04/2004) item 4.1.7, Abweichungsantrag 31, p. 184. TC.
8) MVStättV 2005 § 10 (4).
9) For example VStättVO 1990, § 14 section 3.
10) Annex E (informative).
11) Compare to MVStättV 2005 § 10 section 4.
12) Values have been rounded for simplification.
13) MVStättV 2005 § 1 (see also section 'MF-Stehplätze').
14) IAKS planning guidelines no. 33.

Chapter 12 122–129

1) Panofsky, Erwin. *Meaning in the Visual Arts* (New York, 1955), p. 55.
2) Naredi-Rainer, Paul von. *Architektur und Harmonie* (1982), pp. 18 ff. TC.
3) Hausmann, A. *Der goldene Schnitt* (Aachen 2002), p. 11. TC.
4) Theuer, Max. *Leon Battista Alberti IX/5* (1912), p. 492. TC.
5) Bruchhaus, G. *Proportion und Harmonie – Maß und Zahl in Architektur* (Aachen, 1990), p. 80.
6) Kepler's principle work is *Harmonices mundi libri V* (Linz, 1619).
7) Le Corbusier. *The Modulor,* first edition (Faber & Faber, London, 1954).
8) Naredi-Rainer, Paul von. *Architektur und Harmonie* (1982), p. 101. TC.
9) Vitruvius. *De architectura libri Decem.* Transl. C. Fensterbusch (Darmstadt 1991). TC.
10) DIN 33402 'Körpermaße' (draft 01/2005), introduction, p. 5. TC.
11) DIN 33402 – 'Ergonomie – Körpermaße' (NormBau, Berlin, 2005) section 'Anwendungsbereich'. TC.

Chapter 13 130–141

1) Lachenmayr, B. *Auge, Brille, Refraktion* (Stuttgart, 1996), p. 12. TC.
2) http://www.medizinfo.de.
3) Gellinek, P. O. *Sichtverhältnisse in Zuschauerräumen von Theatern* (Hannover, 1956), p. 68. TC.
4) Wördenweber, Burkhard. *Physiologie des Sehens – Einführung in die Lichttechnik* (University of Paderborn) TC.
5) Reim, Martin (Director of eye clinic at RWTH Aachen). *Augenheilkunde,* 4th edition, (1993), p. 58. TC.
6) http://www.medizinfo.de/augenheilkunde. TC
7) Reim, Martin. *Augenheilkunde,* p. 31. TC
8) Grehn, Franz. *Augenheilkunde,* 29th edition (Heidelberg), pp. 13/14. TC.
9) DFB. *Fußballregeln des DFB 2005/2006* (Frankfurt, 2005), p. 10.
10) Cf. EN/DIN 33402-2 'Körpermaße'.
11) IAKS. *Planungsgrundlagen Sportplätze/ Sportstadien Nr. 33* (Köln, 1993), p. 71.
12) Ibid.
13) Gellinek, C. *Der Hörsaal im Hochschulbau* (Berlin, 1934), p. 32.
14) Neufert, P. *BEL – Bauentwurfslehre,* 33rd edition (Köln, 1992), p. 416. TC.

Chapter 14 142–149

1) *Planungsgrundlagen Sportplätze/Sportstadien Nr. 33,* IAKS (1993), p. 71. TC.
2) 3 x lower tier east and respectively 1 x south/ north; with the exception of World Cup Stadium Munich, where executive seats are located circular at the middle tier.
3) Brockhaus: *die Enzyklopädie,* 20th edition (Leipzig/Mannheim, 1996). TC.

4) Designed by this author for gmp Architekten, Aachen, Germany.
5) Cf. MVStättV 2005 § 10 (4).

Chapter 15 150–161

1) Equivalent to sightline elevation, according to G. Graubner, Hanover, 1968 (cf. Note 5).
2) The former regulation is the VStättVO 1990, compared to the new MVStättV 2005.
3) A reduction of up to 70 cm would be possible, according to EN/DIN 13200-1.
4) Gellinek, C. *Der Hörsaal im Hochschulbau* (Berlin, 1934), p. 32. TC.
5) Graubner, G. *Theaterbau – Aufgabe und Planung* (München, 1968), p. 18. TC.

Chapter 16 162–171

1) According to the definition in MVStättV 2005 § 6.
2) DIN 18065 'Gebäudetreppen', item 6.3.2.
3) MVStättV § 8, clause 4.
4) MVStättV 2005 § 10 'Gänge und Stufengänge'.
5) MVStättV 2005 § 7.
6) DIN 18065 item 6.3.3.
7) MVStättV section 2 § 11.
8) HHP-West. *Brandschutzgutachten Stadion Köln.* (Bielefeld 04/2004) item 4.1.7.
9) Ibid. 'Abweichungsantrag 34'.
10) ARGEBAU, Fachkommission Bauaufsicht. *Begründung und Erläuterung der MVStättV 2005* (Berlin, 2005), § 10 'Stufengänge' (§§ 13, 14, 15 and 94 VStättVO 1978).
11) MVStättV 2005 § 10.
12) As presently stipulated in MVStättV § 11 section 1.13; DIN 18065 'Gebäudetreppen'. NABau (Berlin, 1997) item 8.0 'Toleranzen'. TC.
14) Item 5.3.2 'Sitzplätze auf Sitzstufen'.
15) Table drawn and translated from EN/DIN 13200-3 'Sitze und Produktmerkmale'.

16) http://www.ruhr-uni-bochum.de/rubens/rubens36/13.htm. TC

Chapter 18   178–195
1) Item 5.2. TC.
2) *Stadion 2006,* extended FIFA Technical Recommendations (Frankfurt, 2002) item 8.2, 'Werbebanden'. TC.
3) http://www.apa.de.

Chapter 19   196–207
1) DFB guideline: *Richtlinine zur Verbesserung der Sicherheit bei Bundesligaspielen* (2004) § 5.
2) EN/DIN 13200-3 'Abschrankungen/Anforderungen', 2005 (draft), item 5.3.2. TC.
3) MVStättV 2005 § 27.
4) DFB safety guidelines § 7, latest version of Feb 2004.
5) STVO – Allgemeine Verkehrsregeln § 22 'Ladung', clause 1–2.
6) According to information from the professional association for vehicle owners [*Berufsgenossenschaft für Fahrzeughaltung*] Hamburg.
7) MVStättV 2005 § 27, section 1.
8) *Nationales Konzept 'Sport und Sicherheit'; Richtlinien zur Verbesserung der Sicherheit bei Bundesligaspielen; Richtlinien zur Verbesserung der Sicherheit bei Regionalliga-Spielen des Deutschen Fußball Bundes.*
9) ARGEBAU, Fachkommission Bauaufsicht. *Begründung und Erläuterung zur MVStättV 2005.* (June 2005). TC.
10) FIFA safety guidelines (2004) § 5.
11) MVStättV § 6 Grundsätze der Rettungswegeführung.

Chapter 20   208–225
1) MVStättV 2005 § 29.
2) Translated citation from VStättVO NRW (2002). This passage has been dropped for MVStättV 2005 and is currently under revision.
3) 500 more persons than permitted by the previous regulation VStättVO.
4) MVStättV 2005 § 7 'Bemessung der Rettungswege'.

Chapter 21   226–235
1) DIN 18035 'Sportplätze', item 4.5.4.
2) Those checks should take place prior to the qualifying round and prior to the second half of the championship maTChes.
3) *Handbook for illumination of football grounds by artificial light* (Philips Co., 2002), item 2.6, p.11.
4) MVStättV 2005 § 14 section 1.
5) MVStättV 2005 § 32.

Chapter 22   236–243
1) DIN 18035 'Sportplätze', part 2–7.
2) Knauf, H.-P. *Kunststoffrasen – Das Spielfeld der Zukunft?* Materialprüfungsanstalt, Otto-Graf-Institut, Stuttgart.
3) DIN 18035, part 7 'Kunststoffrasenflächen'.
4) Interview with E. Lehmacher, Labor für Baustoffe und Bauweisen des Sportplatz- und Landschaftsbaus GmbH.
5) Translated excerpt from the essay 'Kunststoffrasen der 3. Generation' published by Polytan Co.
6) Figures in weigth percent as stated in the press conference 'GreenGoal', 31 March 2003.

Chapter 23   244–255
1) Führer, W. et al. *Der Entwurf von Tragwerken* (RWTH Aachen, 1995), p. 51.
2) Peil, U. 'Statik der Dachtragwerke von Stadien' *Stahlbau* (issue no. 3/2005), p. 169. TC.
3) Göppert, G. 'Gottlieb-Daimler-Stadion', *Stahlbau Spezial 2005*, p. 192. TC.

Chapter 24   256–337
1) According to MVStättV 2005, going depth is at least 26 cm and riser height max. 19 cm.

Chapter 25   338–355
1) MVStättV 2005.
2) from http://www.sport1.at/coremedia/generator/id=1807014.html. TC

Epilogue   357
1) Freely adapted from *Die Welt als Entwurf*, Otl Aicher on Hans Gugelot, 1991.
2) *Architektur + Sport 2006*, p. 155. TC.

Bibliography

B001 Alberti, Leon Battista. *Zehn Bücher über die Baukunst*. Book 1–10. German transl. by Max Theuer; Wiss. Buchgesellschaft, Darmstadt, 1975.
B002 Architekturmuseum TU München. *Architektur + Sport*. 'Vom antiken Stadion zur modernen Arena'. Edition Minerva, Wolfratshausen, 2006.
B003 Augustin, Eduard et al. *Fußball unser*. 'Was man nicht alles wissen muss'. Süddeutsche Zeitung, München, 2005.
B004 Bruchhaus, Gundolf. *Architektur-Zusammenhänge* (Festschrift Gottfried Böhm). 'Proportion und Harmonie'. Scaneg Verlag, München, 1990.
B005 Burk, Annelie und Reinhard. *Checkliste der aktuellen Medizin. Checkliste Augenheilkunde*. Georg Thieme Verlag, New York, 2005.
B006 Burris-Meyer, Harold. *Theatres & Auditoriums* Progressive Architecture Library, Reinhold, New York, 1949.
B007 DSB – Deutscher Sportbund. *Bad Blankenburger Sportstättentagung. Planung, Bau und Sanierung von Sportplätzen*. BISP Bonn, Frankfurt, 2004.
B008 ENGEL, Heino. *Tragsysteme – Structure Systems*. Verlag Gerd Hatje, Ostfildern, 1997.
B009 Führer, Ingendaaij, Stein. *Der Entwurf von Tragwerken*. 2. Auflage, 'Hilfen zur Gestaltung und Optimierung'. Aachen, 1995.
B010 Gellinek, Christian. *Der Hörsaal im Hochschulbau*. Dissertation an der Technischen Hochschule zu Berlin, Berlin, 1934.
B011 Gellinek, Phillipp-Otto. *Sichtverhältnisse in Zuschauerräumen von Theatern*. Dissertation an der Fakultät für Bauwesen an der TH Hannover, Hannover, 1956.
B012 Graubner, Gerhard. *Theaterbau. Aufgabe und Planung*. Callwey Verlag, München, 1968.
B013 Grehn, Franz. *Augenheilkunde*. 29. Auflage, neue AO, Spinger-Verlag, Heidelberg, 2005.
B014 Gussmann, Hans. *Theatergebäude – Technik des Theaterbaus*. II. Band, VEB Verlag Technik, Berlin, 1954.
B015 Hausmann, Axel. *Der Goldene Schnitt. Göttliche Proportionen und noble Zahlen*. Books on Demand, Aachen, 2002.
B016 Herzhoff, Thomas. *Untersuchung … stereoskoper Videoendoskopien … auf die operative Tätigkeit*. Aachen, 1999.
B017 Holgate, Alan. *The Art of Structural Engineering. The Work of Jörg Schlaich and his Team*. Edition Axel Menges, Stuttgart, 1997.
B018 IAKS/IOC. *Internationaler IAKS Fachkongress*. Vorträge für Planung, Bau, Modernisierung und Management, 2005.
B019 John, Geraint & Sheard, Rod. *Stadia – a Design and Development Guide*. 3. Auflage, Architectural Press, Bath, GB, 2000.

B020 Langen, Gabi & Deres, Thomas. *Müngersdorfer Stadion Köln*. Emons, Köln, 1998.
B021 Lachenmayr, Bernhard et al. *Auge – Brille – Refraktion*. Begleitschrift zum 'Schober-Kurs', Ferdinand Enke Verlag, Stuttgart, 1996.
B022 Lauenstein, Hajo. *Arithmetik und Geometrie in Raffaels Schule von Athen. Die geheimnisvolle Schlüsselrolle der Tafeln …* Peter Lang Europ. Verlag, Frankfurt a. M., 1998.
B023 MARG, Volkwin (Hrsg.). *Olympiastadion Berlin – Sanierung und Modernisierung 2000 – 2004*. Berlin, 2004.
B024 MARG, Volkwin (Hrsg.). *Stadien und Arenen*. von Gerkan, Marg und Partner, Hatje Cantz, Hamburg, 2005.
B025 MARG, Volkwin. *Konstruktion und Deutung*. Ausstellungskatalog, Aedes, Berlin, 2006.
B026 Marschik, Matthias et al. *Das Stadion. Geschichte, Architektur, Politik, Ökonomie*. Turia + Kant, Wien, 2005.
B027 Merkt, Andreas. *Fußballgott – Elf Einwürfe*. Kiepenheuer & Witsch, Köln, 2006.
B028 Müller, Werner & Vogel, Gunter. *dtv-Atlas zur Baukunst*. Band 1 und 2, 7. Auflage, 'Baugeschichte von Mesopotamien bis Byzanz'. dtv-Verlag, München, 1987.
B029 Naredi-Rainer, Paul von. *Architektur und Harmonie*. 2. Auflage, 'Zahl, Maß und Proportion in der abendl. Baukunst'. DuMont Verlag, Köln, 1982.
B030 NEUFERT, Peter. *Bauentwurfslehre*. 33. Auflage, Vieweg Verlag, Braunschweig, 1992.
B031 Pollio, Marcus Vitruvius. *Zehn Bücher über Architektur*. Viertes und Fünftes Buch (transl. by PRESTEL), Heitz & Mündel, Strassburg, 1912.
B032 Pollio, Marcus Vitruvius. *Baukunst*. Bücher I – V, Erster Band (transl. by August RODE, Leipzig 1796), Artemis-Verlag, Zürich/München, 1987.
B033 REIM, Martin. *Augenheilkunde*. 4. Auflage, Enke Reihe zur AO, Ferdinand Enke Verlag, Stuttgart, 1993.
B034 Sedlacek, Gerhard (Hrsg.). *Arenen im 21. Jahrhundert. Leistungsschau des Stadionbaus*. Ernst & Sohn, Berlin, 2005.
B035 SCOTTISH OFFICE. *Guide to Safety at Sports Grounds*. 4th ed., The Stationary Office, London, 1997.
B036 Sheard, Rod. *Sports architecture*. Spon Press, London, 2001.
B037 Sheard, Rod. *The Stadium – Architecture for New Global Culture*. Periplus, Singapore, 2005.
B038 Skrentny, Werner (Hrsg.). *Das große Buch der deutschen Fussball-Stadien*. Verlag Die Werkstatt, Göttingen, 2001.
B039 Stadionwelt. *Faszination Stadion 2006 – Die WM-Stadien*. 'Geschichte – Potraits – Ausblick'. Stadionwelt ©, Brühl, 2005.
B040 Stahl-Informations-Zentrum. *Arenen für die Fußballweltmeisterschaft 2006*. Dokumentation 590, Stahl-Informations-Zentrum, Düsseldorf, 2006.
B041 Stick, Gernot. *Stadien der Fußballweltmeisterschaft 2006*. Birkhäuser, Berlin, 2005.

B042 Verspohl, Franz-Joachim. *Stadionbauten von der Antike bis zur Gegenwart – Regie und Selbsterfahrung der Massen*. Anabas-Verlag, Gießen, 1976.
B043 Goethe, J. W. *The Italian Journey*. Transl. by W. H. Auden and Elizabeth Mayer, Penguin Books, 1962.
B044 Le Corbusier. *The Modulor*, first edition, Faber & Faber, London, 1954.
B045 Panofsky, Erwin. *Meaning in the Visual Arts*, New York, 1955.
B046 Vitruvius. *De architectura libri Decem*. Transl. C. Fensterbusch, Darmstadt, 1991.
B047 Vitruvius. *The Ten Books on Architecture*. Transl. by Morris Hicky Morgan, Dover Books, New York, 1960.

Z001 Göppert, Knut. 'Adaptive Tragwerke – Wandelbare Dachkonstruktionen für Sportbauten'. Bautechnik, 82. Jahrgang, März 2005, Heft 3, p. 157 – 161, Verlag Ernst & Sohn, Berlin, 2005.
Z002 Knauf, Hans-Peter. 'Kunstrasen – Das Spielfeld der Zukunft? Kunststoffrasenbeläge – Entwicklung, Aufbau und Materialien'. *sb – sportstättenbau und bäderanlagen*, 37. Jahrgang, Mai 2003, p. 54 – 59, sb 67 Verlagsgesellschaft mbH, Köln, 2003.
Z003 Lowe, Barry. 'Vom Sportfeld zur Sportstätte der Zukunft. Stadionplanung gestern und morgen'. *[Umrisse] – Zeitschrift für Baukultur*, 3. Jahrgang, p. 7 – 15, Verlagsgruppe Wiederspahn, Wiesbaden, 2004.
Z004 Marg, Volkwin. 'Stadionbauten – Regie und Selbsterfahrung der Massen; Beispiel Berlin'. *Bauingenieur*, Band 79, Mai 2004, p. 201–204, Springer VDI Verlag, München, 2004.
Z005 Peil, Udo. 'Statik der Dachtragwerke von Stadien'. *Stahlbau* 74. Jahrgang, März 2005, p. 159 – 177, Verlag Ernst & Sohn, Berlin, 2005.
Z006 Schmidt, Thomas. 'Olympiastadien des 20. Jahrhunderts. Typologie und bauhistorische Entwicklungslinien'. *sb – sportstättenbau und bäderanlagen*, 37. Jahrgang, Mai 2003, p. 14 – 21, sb 67 Verlagsgesellschaft mbH, Köln, 2003.
Z007 SB-Redaktion. 'Finanzierung und Betrieb von Sportanlagen'. *sb – sportstättenbau und bäderanlagen*, (Schriftenreihe), sb 67 Verlagsgesellschaft mbH, Köln, 2005
Z008 Stadionwelt. *Stadionwelt – Das Fan- und Stadionmagazin*. Magazin (Schriftenreihe 1–16), Stadionwelt©, Brühl, 2005.
Z009 Detail-Redaktion. 'Konzeptheft Stadien'. *DETAIL – Zeitschrift für Architektur und Baudetail*, Inst. f. int. Arch.-Dokum., München, 2005.
Z010 DBZ-Redaktion. 'Textiles Bauen'. *DBZ – Deutsche Bauzeitschrift*, Bertelsmann, Gütersloh, 2000.
Z011 Baumeister-Redaktion. '4XL – Planen und Praktizieren'. *Baumeister*, Callwey, München, 2005.

D001 MVStättV 'Muster-Versammlungsstättenverordnung, Muster-Verordnung über den Bau und Betrieb von Versammlungsstätten'. Juni 2005. ARGEBAU, Fachkommission Bauaufsicht, Beuth Verlag, Berlin.
D002 MVStättV 'Muster-Versammlungsstättenverordnung, Begründung und Erläuterung zur MVStättV'. Juni 2005. ARGEBAU, Fachkommission Bauaufsicht, Beuth Verlag, Berlin.
D003 VstättVO 'Versammlungsstättenverordnung, Verordnung über den Bau und Betrieb von Versammlungsstätten'. Sept. 2002 (gültige Fassung), Nordrhein-Westfalen.
D004 VStättVO 'Versammlungsstättenverordnung, Verordnung über den Bau und Betrieb von Versammlungsstätten'. Aug. 1974 (alte Fassung), Baden-Württemberg.
D005 18. BImSchV 'Sportanlagenlärmschutz-Verordnung, 18. Verordnung zur Durchführung des Bundes-Immissionsschutzgesetzes. Juli 1991, Bundesregierung.
D006 DIN EN 13200-1 'Zuschaueranlagen' – Teil 1: Kriterien für die räumliche Anordnung von Zuschauer-plätzen. [Engl. Version: 'Spectator Facilities' – Part 1: Layout criteria for spectator viewing area – Specification] May 2004, DIN Normenausschuss Bauwesen, Beuth Verlag, Berlin.
D007 DIN EN 13200-3 'Zuschaueranlagen – Teil 3 (Entwurf): Abschrankungen, Anforderungen'. Sept. 2003, DIN Normenausschuss Bauwesen, Beuth Verlag, Berlin.
D008 DIN EN 13200-4 'Zuschaueranlagen – Teil 4 (Entwurf): Sitze, Produktmerkmale'. Nov. 2004, DIN Normenausschuss Bauwesen, Beuth Verlag, Berlin.
D009 DIN EN 13200-5 'Zuschaueranlagen – Teil 5 (Entwurf): Ausfahrbare (ausziehbare) Tribünen'. Jan. 2005, DIN Normenausschuss Bauwesen, Beuth Verlag, Berlin.
D010 DIN EN 13200-6 'Zuschaueranlagen – Teil 6 (Entwurf): Demontierbare (provisorische) Tribünen'. Jan. 2005, DIN Normenausschuss Bauwesen, Beuth Verlag, Berlin.
D011 DIN 18030 'Barrierefreies Bauen (Entwurf): Planungsgrundlagen (Ersatz: DIN 18024 + 18025)'. Nov. 2004, DIN Normenausschuss Bauwesen, Beuth Verlag, Berlin.
D012 DIN 18035-1 'Sportplätze – Teil 1: Freianlagen für Spiele und Leichtathletik, Planung und Maße'. Feb. 2003, DIN Normenausschuss Bauwesen, Beuth Verlag, Berlin.
D013 DIN 18065 'Gebäudetreppen – Definition, Meßregeln, Hauptmaße'. Jan. 2000, DIN Normenausschuss Bauwesen, Beuth Verlag, Berlin.
D014 DIN 18203-1 'Toleranzen im Hochbau – Teil 1: Vorgefertigte Teile aus Beton, Stahlbeton und Spannbeton'. April 1997, DIN Normenausschuss Bauwesen, Beuth Verlag, Berlin.
D015 DIN 33402-2 'Ergonomie – Teil 2 (Entwurf): Körpermaße des Menschen – Werte'. Jan. 2005, DIN Normenausschuss Bauwesen, Beuth Verlag, Berlin.
D016 DIN 4844-1 'Sicherheitsfarben und Sicherheitszeichen – Teil 1: Gestaltungsgrundlagen für Sicherheitszeichen'. Mai 2004, DIN Normenausschuss Bauwesen, Beuth Verlag, Berlin.

D017 DIN EN 12193 'Sportstättenbeleuchtung – Licht und Beleuchtung'. Nov. 1999, DIN Normenausschuss Bauwesen, Beuth Verlag, Berlin.
D018 EAR 'Empfehlungen für Anlagen des ruhenden Verkehrs'. FGSV Straßen- und Verkehrswesen, Arbeitsgruppe Straßenentwurf, 2004.

V001 FIFA/UEFA Pflichtenheft 'Stadion 2006 – Profile und Anforderungen; Städte und Stadien zur FIFA Fußball-Weltmeisterschaft 2006'. Nov. 2002.
V002 FIFA/UEFA 'Technische Empfehlungen und Anforderungen für den Neubau oder die Modernisierung von Fußballstadien'. 3. Auflage, 1995.
V003 FIFA 'Safety Guidelines. Sicherheitsrichtlinien der FIFA'. Jan. 2004.
V004 UEFA 'Anforderungen an die Austragungsorte von EM-Endrunden u. von Endspielen der UEFA-Klubwettbewerbe'. 2005.
V005 UEFA Pflichtenheft 'EM 2008 – Anforderungen an die Austragungsorte'. Nov. 2001.
V006 UEFA 'Verbindliche Sicherheitsvorkehrungen'. 2004.
V007 UEFA 'Reglement der UEFA Championsleague 2005/2006'. 2005.
V008 DFB 'Richtlinien zur Verbesserung der Sicherheit bei Bundesligaspielen'. Jan. 2004.
V009 DFB 'Regeln 2005/2006'. 2005.
V010 DFL 'LO – Lizenzordnung'. Juli 2004.
V011 IAAF 'Track and Field Facilities – Manual'. Köln, 1995.
V012 IAKS/IOC 'Planungsgrundlagen Sportplätze/Stadien'. Nummer 33, 1993.

L001 Nixdorf, Stefan. *Sichtlinien und Sicherheit. Tribünenprofile moderner Sport- und Veranstaltungsstätten*. Dissertation, Faculty of Architecture, RWTH Aachen, 2006.
L002 Faculty of Architecture RWTH Aachen. *Theater und Konzerthäuser – Gebäudetypen Eins*. Lehrstuhl für Gebäudelehre und Grundlagen des Entwerfens, Aachen, 2005.

B001	B = Books
Z001	Z = Magazines/journals (articles)
D001	D = EN or DIN standards
V001	V = Guidelines of the sports associations
L001	L = Teaching aids
FIFA	Fédération Internationale de Football Association [World Football Association]
UEFA	Union of European Football Associations
DFB	Deutscher Fußball-Bund [German Football Federation]
DFL	Deutsche Fußball Liga GmbH [Operational business „Die Liga – Fußballverband e. V." (1st and 2nd Bundesliga)]
IAAF	International Association of Athletics Federations [World governing body for track and field]
IOC	International Olympic Committee (IOK) [Non-government organisation and support of the Olympic Games]
IAKS	International Association for Sports and Leisure Facilities
LOC	Local Organization Committee (FIFA)
RWTH	Rheinisch-Westfälische Technische Hochschule [RWTH Aachen University]
DIN	Deutsches Institut für Normung e. V. [German Institute for Standardization]

Sources

In the framework of scientific research, the author has gathered extensive image material for documentation and illustration purposes, including citations from other publications.

To the best of our knowledge, we, the publishers have sought the permission for reproduction of these images.
The relevant sources are listed below, indicating also references which appear in the bibliography.

003 Z006 p. 15
004 B041 p. 63
006 B041 p. 71
007 B032 p. 212
008 B015 p. 367
009 B001 p. 319
010 B032 p. 284
011 Roland Reiman
012 B041 p. 100
014 Z011 p. 16
015 B012 p. 22
016 B012 p. 22
017 B041 p. 273
018 B041 p. 285
019 B041 p. 41
020 B041 p. 31
021 B041 p. 31
022 B041 p. 164
023 Z006 p. 16
024 gmp Architects
026 B041 p. 243
027 gmp Architects
028 B020 p. 36 Sportmuseum Köln
029 B020 p. 103 Hamburger Aero-Lloyd
030 B023 p. 17 Archiv V. Kluge
031 http://wiki.phantis.com
033 PFEIFER, Memmingen, Majowiecki
034 gmp Architects
035 asp Architekten, Manfred Storck, Stuttgart
036 www.allianz-arena.de
037 www.stadion-koeln.de
038, 039 www.stadionwelt.de
041 www.allianz-arena.de
042 FSW Düsseldorf GmbH, Jörg Faltin
043 DIN 13200-1
044 gmp Architects Aachen
045 gmp Architects
047 www.stadionwelt.de
048 skidata AG, Runig Werner
052 gmp Architects Aachen
053 gmp Architects Aachen
054 picture AFP
055 1. FC-Köln, van der Kooi
067–069 www.stadionwelt.de
070 EHEIM Möbel, C. Eichler

071 B038 p. 140/141
072, 073 www.stadionwelt.de
076 Auslobung WBW Köln, KSS
080 B038 p. 98
082 www.stadionwelt.de
088 B038 p. 178
089 B038 p. 32
091 www.stadionwelt.de
092 asp Architekten, Manfred Storck, Stuttgart
093 gmp Architects, Fritz Busam
094 www.stadionwelt.de
096 RWTH Aachen, Weng
098 www.stadionwelt.de
100 Z006 p. 15
101, 102 Olympiastadion Berlin, Fritz Busam
103 gmp Architects Aachen
106 www.allianz-arena.de
107 gmp Architects
108 Niering & Seifert Co.
109 www.allianz-arena.de
110 gmp Architects, Heiner Leiska, Hamburg
111 gmp Architects Aachen
112 Auslobung WBW Köln, KSS
115 B038 p. 139
119, 120 www.stadionwelt.de
121 gmp Architects
122–124 gmp Architects, Heiner Leiska, Hamburg
125 gmp Architects
126 gmp Architects, Heiner Leiska, Hamburg
128, 129 Niering & Seifert Co.
130 www.arena-aufschalke.de
134 gmp Architects, Heiner Leiska, Hamburg
135 LOC 2006 Köln, Polte
136 www.arena-aufschalke.de
137, 138 Niering & Seifert Co.
139, 140 www.stadionwelt.de
143 B038 p. 28
144 B038 p. 112
145 www.stadionwelt.de
151 FIFA-Pflichtenheft 2006 FIFA
152 UEFA-Pflichtenheft 2008
154 LOC 006 Köln, Polte
155 LOC 006 Frankfurt
158 www.stadionwelt.de
159 gmp Architects Berlin
165, 166 gmp Architects Aachen
171 UEFA-Pflichtenheft 2008
181 www.stadionwelt.de
182 www.stadion-koeln.de
183 www.stadionwelt.de
185, 186 www.stadion-koeln.de
187 www.stadionwelt.de
190 www.stadion-koeln.de
191–193 gmp Architects Aachen
194 www.stadionwelt.de

197 FIFA-Website WM 2006
198 FIFA-OC 2006
200 EN 13200-1
202 B038 p. 27
202–221 B038
216, 217 www.stadionwelt.de
219 www.stadionwelt.de
220 Schwaar & Partner, Bern
224 B038 p. 32
240–244 RWTH Aachen, F. Stumpen
248 B015 p. 151
249 B015 p. 62
250 B015 pp. 45, 56
258 www.mannpharma.de
259 www.mannpharma.de
260 B033 p. 33
263 www.mannpharma.de
264 B033 p. 33
265 Martin Lea, Hyvarinen
266 www.ch-forschung.ch, W. E. Hill
267 FIFA Website World Cup 2006, Popperfoto
270 D016 figure 1
276 www.stadionwelt.de
277 Souto de Moura Architects
279, 280 gmp Architects Aachen
281 asp Architekten, Manfred Storck, Stuttgart
282 Zwarts & Jansma Architects, Amsterdam
283 B038 p. 162
284 Schultz + Partner Architekten
288 gmp Architects, J. Hempel
289 gmp Architects Aachen
297 EHEIM Möbel, C. Eichler
307 www.stadion-koeln.de
312 www.stadion-koeln.de
315 DIN 18065
316 DIN 13200-3
317 DIN 13200-3
320 www.stadion-koeln.de
321 DIN 18203-1
322 gmp Architects, Heiner Leiska, Hamburg
324 gmp Architects, F. Busam
347 www.apa.de
348 www.stadionwelt.de
349–354 www.apa.de
356 www.stadionwelt.de
357 B038 p. 101
358–363 www.stadionwelt.de
371 www.stadionwelt.de
372 SecuFence Co., Neumann
374 www.stadionwelt.de
379 www.stadionwelt.de
380–382 SecuFence Co., Neumann
383 www.stadionwelt.de
384 SecuFence Co., Neumann
387, 388 www.stadionwelt.de

389 www.allianz-arena.de
390 SecuFence Co., Neumann
391 www.stadionwelt.de
394 www.thestreaker.org.uk
395 www.bbv-net.de Agentur AP
397a B028 p. 240
397b B028 p. 56
400 www.stadion-koeln.de
401–404 gmp Architects Aachen
405, 406 www.arena-aufschalke.de
407 www.stadionwelt.de
409 B038 p. 77
410 www.arena-aufschalke.de
411, 412 www.stadionwelt.de
414, 415 www.sportsvenue-technology.com
417, 418 Schüßler-Plan Ingenieurgesellschaft mbH
419, 420 RWTH Aachen, Kai Grosche
421 www.stadionwelt.de, HOK s+v+e
423 Andrew Clem
425 www.sportsvenue-technology.com
426 EN 13200-5
428 Z009 p. 930
420 gmp Architects Berlin
435 gmp Architects Berlin
437 Schlaich Bergermann und Partner (SBP), Stuttgart
439 www.stadionwelt.de, Euroluftbild.de
440 www.max-boegl.de
441 B019 p. 62
442 gmp Architects Berlin
444 gmp Architects, Heiner Leiska
445–447 B036 'Sports Architecture' HOK s+v+e
448 Herzog & de Meuron
449 Schuster Architekten, Düsseldorf, O. Allstedt
451 gmp Architects Berlin
452 Wiel Arets Architects & Associates, Maastricht
453 Euroborg NV, Groningen
454a www.allianz-arena.de
458 www.stadionwelt.de
459 gmp Architects, Heiner Leiska, Hamburg
460 gmp Architects, Marcus Bredt, Berlin
461, 462 gmp Architects, Heiner Leiska, Hamburg
463 gmp Architects, F. Busam, Berlin
464, 465 gmp Architects
466 www.stadionwelt.de
467, 468 Philips AG
469 gmp Architects, Philips AG
470 www.stadionwelt.de
471 www.proletzigrund.ch, Bétrix & Consolascio
472 www.stadionwelt.de
473 FIFA-Pflichtenheft (1995)
474 Philips AG
475 DIN 12193
476 Schuster Architekten, Düsseldorf, O. Allstedt
477 Z002 p. 56
478–480 www.polytan.de

481a FIFA Organization
481b FIH Organization
481c IAAF Organization
481d RAL Gütezeichen
482 gmp Architects Aachen
483 FIFA Organization
484–486 Schulitz + Partner
487, 488 www.stadion-koeln.de
489–491 www.stadionwelt.de
492 www.stadionwelt.de, S. Katzenberg, Hamburg
494 Frei Otto
495 B008
498, 499 www.stadionwelt.de
500 www.penninehelis.co.uk
501 B036 Sports Architecture, HOK s+v+e
502 Z005, p. 17
503 www.worldstadiums.com
504 RWTH Aachen, Taha Anwar
505 www.tokyo-dome.co.jp Tokyo Dome Corp.
506 Peter Kulka Architektur Dresden GmbH
507, 508 Schulitz + Partner, spa Hannover
509, 510 www.stadionwelt.de
512, 513 gmp Architects, SBP, Stuttgart
514 www.stadionwelt.de
515 B038 p. 66
516 B038 p. 146
517 gmp Architects, Heiner Leiska, Hamburg
518 www.allianz-arena.de
522 B038 p. 36
524 B038 p. 25/26 Euroluftbild.de
527 B038 p. 30/31 Roland Solich, Neuss
532 B038 p. 53 BVB-Archiv, Gerd Kolbe
534 B038 p. 40/41 Euroluftbild.de
537 B038 p. 46/47 Roland Solich, Neuss
542 B038 p. 69 Inst. für Stadtgeschichte, Frankfurt a. M.
544 fvgffm 005
547 B038 p. 64 Roland Solich, Neuss
552 B038 p. 84 Stadt Gelsenkirchen
554 B038 p. 74/75 Euroluftbild.de
557 B038 p. 78/79 Roland Solich, Neuss
562 B038 p. 84 HSV Museum
564 B038 p. 88/89 Euroluftbild.de
567 B038 p. 94/95 Roland Solich, Neuss
572 B038 p. 116 Historisches Museum Hannover
574 B038 p. 104/105 Euroluftbild.de
577 B038, p. 110/111 Roland Solich, Neuss
582 B038, p. 133 Stadt Kaiserslautern
584 B038, p. 120/121 Euroluftbild.de
587 B038 p. 126/127 Roland Solich, Neuss
592 B020 p. 37 Deutsches Sportmuseum Köln
594 www-stadion-koeln.de KSS
597 gmp Architects, Heiner Leiska, Hamburg
602 B038 p. 165 Westend Publ. Rel.
604 B038 p. 152/153 Euroluftbild.de
607 B038 p. 158/159 Roland Solich, Neuss

612 www.bvb-nord.de
614 B038 p. 168/169 Euroluftbild.de
617 B038 p. 174/175 Roland Solich, Neuss
622 B038 p. 197 Stadt Nürnberg
624 B038 p. 184/185 Euroluftbild.de
627 B038 p. 190/191 Roland Solich, Neuss
632 B038 p. 213 Archiv der Stadt Stuttgart
634 B038 p. 200/201 Euroluftbild.de
637 B038 p. 206/207 Roland Solich, Neuss
648 www.stadionwelt.de
649 www.ansa.it
650 www.derstandard.at
651 www.derstandard.at
652 Schlaich Bergermann und Partner (SBP), Stuttgart
653 www.derstandard.at
654 www.fussballtempel.net
655 www.derstandard.at
656 www.lifeinkorea.com
657 www.derstandard.at
658 www.city.kashima.ibaraki.jp
659 www.derstandard.at
660 www.derstandard.at
661 www.ansa.it
662 Ken Terasaki SS Kyushu
663 www.sportsvenue-technology.com
664 www.dfb.de, AFP
665 2002.fifaworldcup.yahoo.com
666 www.startline.encyber.com, fukuroi
667 www.stadiumguide.com
668–680 www.stadionwelt.de,
681 www.proletzigrund.ch
682 www.stade.ch
683 Stadtpresse Klagenfurt/Eggenberger
684 Generalunternehmer Porr/Alpine
685 www.stadionwelt.de
686 B038, p. 235 Rakete GmbH
687 gmp Architects Berlin
688–692 B038 pp. 234–236 Rakete GmbH
693 gmp Architects Berlin
694, 695 B038 p. 237 Rakete GmbH

Acknowledgements

The publisher and the author would like to thank especially for the provision and kind permission for publication of images:

APA Firmengruppe, Neuwied
ASP Architekten, Stuttgart
Axel Hausmann
Bauaufsicht Stadt Köln, Cologne
Bundesinstitut für Sportwissenschaft, Bonn
Thomas Schmidt, Berlin
EHEIM Möbel GmbH, Möhringen
Engelbert Lehmacher Laboratory Construction Materials and Methods for Sports Fields and Landscape GmbH, Osnabrück
FIFA Committee Germany 2006
GCA Ingenieure AG, Unterhaching
gmp Architects and Volkwin Marg
Herzog & De Meuron Architects, Basle
HHP West Consultants, Fire Engineering, Bielefeld
HOK Sport Venue Event
International Association for Sports and Leisure Facilities IAKS, Cologne
International Olympic Committee IOC
Institut für Sportstättenberatung, Bad Münstereiffel
Manfred Storck, Stuttgart
Max Bögl GmbH & Co. KG, Sengenthal
MICHAELIS GmbH, Television and Event Equipment, Erkrath
Niering & Seifert Planning Office for Catering Technology GbR, Cologne
Peter Kulka
Philips GmbH, Lighting Sports & Area, Berlin
Polytan Sport Surface Systems GmbH, Burgheim
Roland Reimann, Darmstadt
Schuster Architekten, Düsseldorf
Schüßler Plan, Düsseldorf
SecuFence AG, Westhausen-Lippach
Skidata, Munich
Stadionwelt.de Thomas Krämer, Ingo Partecke, Pascal Reichardt
Steinhart + CoGmbH Traffic Consulting, Aachen
Wiel Arets Architects, Maastricht

Allianz Arena München Stadion GmbH
Borussia Dortmund GmbH & Co. KGaA
Commerzbank Arena Stadion Frankfurt Management GmbH
FC Schalke 04 Stadion Betriebsgesellschaft mbH, Gelsenkirchen
Fritz-Walter-Stadion Kaiserslautern GmbH
HSV Arena GmbH & Co, Hamburg
Kölner Sportstätten GmbH RheinEnergieStadion
Landeshauptstadt Stuttgart, Sportamt
Niedersachsenstadion Projekt- und Betriebsgesellschaft AWD-Arena, Hannover
Olympiastadion Berlin GmbH
Stadion Nürnberg Betriebs GmbH
ZSL Betreibergesellschaft mbH, Leipzig

Special warmest thanks from the author go to gmp Architects Hamburg, Berlin, Aachen for supporting this work, to Stadionwelt® for the provision of images, to the FIFA Committee Germany Fußball-Weltmeisterschaft 2006™ for supporting the scientific work and provision of planning documents of the twelve World Cup Stadia, and to all interviewees for comprehensive information and critical reply.

Architects of the Football World Cup stadia, Germany 2006

FIFA World Cup Stadium Berlin
gmp Architects, Berlin

FIFA World Cup Stadium Frankfurt
gmp Architects, Frankfurt

FIFA World Cup Stadium Hamburg
MOS-Architekten, Hamburg

FIFA World Cup Stadium Dortmund
Planungsgruppe Schröder/Schulte-Ladbeck/Strothmann, Dortmund

FIFA World Cup Stadium Gelsenkirchen
HGB Engineering, Rijswijk

FIFA World Cup Stadium Hannover
SPA Schulitz + Partner Architects, Braunschweig

FIFA World Cup Stadium Kaiserslautern
Volker Fiebiger GmbH Architekten und Ingenieure, Kaiserslautern

FIFA World Cup Stadium Leipzig
ARGE Zentralstadion Leipzig
Wirth + Wirth GmbH, Leipzig, Basel
Glöckner Architekten, Nürnberg
Körber Barton Fahle Planungs GmbH, Freiburg

FIFA World Cup Stadium Nuremberg
HPP Hentrich-Petschnigg & Partner KG, Düsseldorf

FIFA World Cup Stadium Cologne
gmp Architects, Aachen

FIFA World Cup Stadium Munich
Herzog & De Meuron Architects, Basel

FIFA World Cup Stadium Stuttgart
Arat, Siegel und Partner Architects, Stuttgart

List of advertisers

agn Niederberghaus & Partner GmbH, 49479 Ibbenbüren   349
Bayer MaterialScience AG, 51368 Leverkusen   159
Dyneon GmbH & Co. KG, 41453 Neuss   263
Hightex GmbH, 83253 Rimsting/Chiemsee   259
Hochtief Construction AG, 45128 Essen   177
Lindapter GmbH, 45141 Essen   350
Max Bögl Stahl- und Anlagenbau GmbH & Co. KG, 92301 Neumarkt   275
Odenwald Faserplattenwerk GmbH, 63916 Amorbach   55
Schaeffler KG, 33803 Steinhagen   283
Seatbox AG, 9034 Eggersriet/St. Gallen   354
Verseidag-Indutex GmbH, 47723 Krefeld   255
Voigt & Schweitzer Markenverbund-Holding GmbH & Co. KG, 45881 Gelsenkirchen   Loose insert

Dr.-Ing. Stefan Nixdorf Architect
Aachen
snixdorf@web.de

agn Architekten · Ingenieure · Generalplaner
www.agn.de

Cover image: Tip-up seat, Allianz Arena, Munich
(Seat design: HdM Architects, Basle;
EHEIM Möbel GmbH, Möhringen)

The contents of this book are intended to provide orientation
during the planning process and have no legislative effect.
In case of implementation, all measures and requirements are
to be verified according to the current technical rules and
regulations.

Bibliographic information published by Deutsche
Nationalbibliothek.
Deutsche Nationalbibliothek lists this publication
in the Deutsche Nationalbibliografie; detailed bibliographic
data available on the website <http://dnd.d-n.b.de>.

ISBN 978-3-433-01851-4

© 2008 Ernst & Sohn Verlag für Architektur und
technische Wissenschaften GmbH & Co. KG, Berlin

All rights reserved (including those of translation into other
languages). No part of this book may be reproduced in any
form – by photoprinting, microfilm, or any other means –
nor transmitted or translated into a machine language without
written permission from the publishers. Registered names,
trademarks, etc. used in this book, even when not specifically
marked as such, are not to be considered unprotected by law.

Conception: Stefan Nixdorf
Typodesign: Sophie Bleifuß
Translation: Ariane Baldus, Berlin
Typesetting: Daniela Leupelt, Uta-Beate Mutz, Michael Uszinski
Production: HillerMedien, Berlin
Printing: Medialis, Berlin

Printed in Germany